FOUNDATIONS OF FINSLER GEOMETRY AND SPECIAL FINSLER SPACES

MAKOTO MATSUMOTO

KAISEISHA

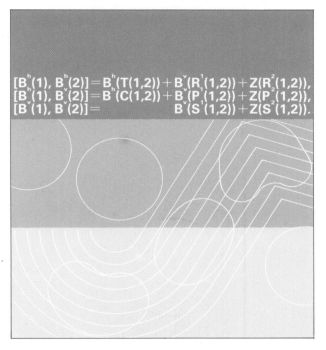

Makoto Matsumoto

FOUNDATIONS OF FINSLER GEOMETRY AND SPECIAL FINSLER SPACES

Saikawa 3-23-2, Otsushi, Shigaken
Printed in Japan

ISBN 4-106165-11-7

PREFACE

The theory of spaces with a generalized metric was initiated by P.Finsler in 1918 under the influence of geometrization of variation calculus and was developed independently by J.L.Synge, J.H.Taylor and, in particular, L.Berwald in the middle of 1920's as a generalization of Riemannian geometry. Next É.Cartan laid the foundations of Finsler geometry in 1933 by introducing a metrical connection from the viewpoint that a Finsler space is locally euclidean, and since then, important contributions to Finsler geometry have resulted one after another on the analogy of Riemannian geometry. In the late 1940's, however, another and essential viewpoint that a Finsler space is locally Minkowskian became a common recognition and O.Varga, H.Busemann, H.Rund and so on have vigorously made great progress of Finsler geometry. Their treatments of this subject were no longer Riemannian but really Finslerian.

The global theory of connections in fiber bundles enabled us in the 1960's to treat connections in Finsler geometry from more generalized and systematical standpoints. On the other hand, there have been the extensive studies and advancement of theoretical physics and engineerings, which have needed differential-geometric interpretation of structures appearing in these subjects and have stimulated the research of various generalizations of Riemannian geometry. We have, however, only few books referring to Finsler geometry and even the most substantial and coherent book written by H.Rund was published more than 20 years ago. These last 20 years are indeed filled with noteworthy results of Finsler geometry.

This book was planned to provide a systematical description of global theory of connections in Finsler geometry, to discuss various kinds of special Finsler spaces and to introduce some topics containing many important open problems. The readers are expected to know the basic facts on differentiable manifolds, fiber bundles and linear connections in principal bundles. Chapter I consists of an exposition of these basic facts in form and extent sufficient for the later use of terminology and notations.

Chapters II and III give an adaptable formulation of connections appearing in Finsler geometry. Roughly speaking, the distinctive feature of this formulation is to understand the essential role of non-linear connections. On this foundation is naturally developed the theory of torsions and curvatures which is independent of Finsler metrics, quite analogously to the case of linear connections. After discussing certain desirable conditions of connections, the concept of Finsler metric is first introduced with concrete examples which are interesting from various points of view. Then É.Cartan's standard connection is uniquely determined by a system of five axioms which are just analogous to those for the Riemannian connection in a Riemannian space. Chapter III ends by considering relations between well-known connections and presenting a new connection.

Chapter IV shows clearly the merits of our formulation of Finsler connection. That is, by means of the non-linear connection constituting a Finsler connection a special linear connection is induced in the tangent bundle from this Finsler connection, and in consequence the torsion and curvature tensors of the Finsler connection may be regarded as fragments, in a sense, of only one torsion tensor and only one curvature tensor of the above special linear connection. This chapter also contains geometrical examinations of the concept of isotropy in Finsler space and gives an almost complex structure in the tangent bundle.

The first four chapters are quite complete in themselves and may be considered as a standard theory of Finsler connections. Although attention in this book is confined to Finsler connections alone, the most concepts involved can be easily extended to more general vector bundles and metrics. For instance, the so-called Cartan geometry of hyperareal spaces may be successfully treated in an analogous way.

In Chapter V are various special Finsler spaces studied. This chapter contains new ideas and many interesting results announced at the Symposium on Finsler Geometry which was started in 1969 by encouragement from Professor Dr. A.Kawaguchi and the author and has been held every year up to now. The author wishes that more of spacial Finsler spaces should be found and studied, although certain classes of important special Finsler spaces were already proposed by L.Berwald and others. There are two reasons for this; one is the need for many typical models of Finsler spaces such as spheres, non-euclidean spaces, projective spaces and so on in Riemannian geometry, and the other is the wish to offer physicists, technologists and others useful special Finsler metrics for applications.

The theory of transformations has been generalized by several authors from the viewpoint of Finsler geometry and some interesting results have been shown recently. Chapter VI is devoted to a collective description of such generalized transformations based on the theory of Finsler bundles, but their full development is yet expected a matter of future.

The first half of Chapter VII deals with the study of parallel displacements and paths. This theory is systematically described on the basis given in Chapters II and III. The second half is concerned with geodesics and conjugate points, although there is not very much space left and almost all the proofs of differential-topological results must be omitted. The global theory of geodesics started in the 1960's, but it may be said that this theory lies still in the scope of Riemannian geometry.

The Bibliography at the end contains only the papers and books quoted in the text and those of which the author may suppose that they are of interest for the readers. The author has great interest in applications of Finsler geometry to other fields of science and the literatures of application are collected in a separate place. He is afraid that a fair amount of papers of application may escape his notice, and he is grateful to Professor Dr. Y.Takano who informed him a lot of papers of theoretical physics.

In the writing of this book the author has had invaluable assistance and criticism by Dr. M.Hashiguchi, Dr. S.Hōjō, Mr. S.Numata, Dr. T.Okada and Mr. K.Okubo. He is also happy to thank here the hearty support and collaboration of all the members of the Symposium on Finsler Geometry mentioned above. During these nealy twenty years about twenty-five persons have constantly joined the Symposium and a number of interesting papers of various fields of Finsler geometry have been published by almsot all the members.

The original manuscript of this book was written in 1975 to be published from VEB Deutscher Verlag der Wissenschaften by the cordial recommendation of Professor Dr. R.Sulanke in Berlin. The author is also grateful to Professor Dr. L.Tamássy in Debrecen and Professor Dr. R.Miron in Iaşi who made every possible effort to publish this revised and supplemented book.

Finally the author desires to express his appretiation of the courtesies extended by the Kaiseisha Press.

Kyoto, April 30, 1986. Makoto MATSUMOTO

CONTENTS

CHAPTER I

LINEAR CONNECTIONS

The purpose of the present chapter is to give an outline of the theory of connections in fiber bundles and to introduce various notations and terminology for the later use. It is, however, not intended to expose an exhaustive foundational treatment of the subject. It is sufficient for understanding the contents of this book to be familiar with the concepts of linear connections in the linear frame bundles, associated connections in the tangent bundles and some generalizations.

The concept of connections is treated in the last two sections with somewhat detailed calculations. Although it seems too troublesome to show such long calculations, a mediation between classical and modern treatments is necessary for the benefit of readers who have studied Finsler geometry only in the tensor calculus.

§ 1. Differentiable manifolds

A *topological space* S is a set which is equipped with a *topological structure* $\tau(S)$. The topological structure $\tau(S)$ is a collection of subsets of S which satisfies the following three conditions:

(1) S and the empty set ϕ belong to $\tau(S)$.

(2) The intersection $U_1 \cap U_2$ of any $U_1, U_2 \in \tau(S)$ belongs to $\tau(S)$.

(3) The union of any subcollection (finite or infinite) of $\tau(S)$ belongs to $\tau(S)$.

A subset belonging to $\tau(S)$ is called an *open set* of S. An open set containing a point $s \in S$ is called a *neighborhood* of s. The complement $S - U$ of an open set U is called a *closed set* of S.

If any two distinct points $s_1, s_2 \in S$ have respective neighborhoods U_1, U_2 satisfying $U_1 \cap U_2 = \phi$, then S is called a *Hausdorff space*. Every point of a Hausdorff space is closed.

A topological space S is called *connected*, if there do not exist two non-empty open sets U_1, U_2 satisfying $S = U_1 \cup U_2$ and $U_1 \cap U_2 = \phi$.

An *n-dimensional chart* (U, u) of a Hausdorff space M is by definition a pair of an open set U of M and a homeomorphism $u: U \to D$ of U with an open set D of $R^n = \{(x^1, \cdots, x^n) \mid x\text{'s} \in R\}$. To each point $x \in U$ there corresponds a point $u(x) \in D$ which is an ordered set $(x^1, \cdots, x^n)$ of n real numbers x's. We use the abbreviation

(x^i) to denote this ordered set. The set (x^i) is called the *coordinate system* of x with respect to the above chart (U, u). The open set U is called the *coordinate neighborhood* carrying the coordinate system (x^i).

Assume we have such two n-dimensional charts (U, u), $(\bar{U}, \bar{u})$ of a Hausdorff space M that $U \cap \bar{U}$ is non-empty. Then each point $x \in U \cap \bar{U}$ has two coordinate systems (x^i), $(\bar{x}^a)$ with respect to respective charts, and the homeomorphism $\bar{u} \circ u^{-1}: u(U \cap \bar{U}) \to \bar{u}(U \cap \bar{U})$ is written by n real-valued functions $\bar{x}^a = \bar{x}^a(x^1, \cdots, x^n)$ of n variables x's, called the *coordinate transformation*. If every function $\bar{x}^a$ is of differentiability class C^r, $r = 1, \cdots, \infty, \omega$(ω means to be analytic), these two charts are said to be *C^r-related*. If $U \cap \bar{U}$ is empty, the word "C^r-related" also is assumed to be applied to them.

DEFINITION 1.1. An *n-dimensional differentiable structure* $\delta(M)$ of class C^r, $r = 1, \cdots, \infty, \omega$, of a Hausdorff space M is a family $\{(U, u)\}$ of n-dimensional charts of M which satisfies the following three conditions:

(1) $\delta(M)$ covers M, i.e., each point of M is contained in the coordinate neighborhood U of at least one chart (U, u) belonging to $\delta(M)$.

(2) $\delta(M)$ is of class C^r, i.e., any two charts of $\delta(M)$ are C^r-related.

(3) $\delta(M)$ is maximal, i.e., any chart of M which is C^r-related to every chart of $\delta(M)$ belongs to $\delta(M)$.

A Hausdorff space M, equipped with an n-dimensional differentiable structure $\delta(M)$, is called an *n-dimensional differentiable manifold of class C^r*.

A chart of $\delta(M)$ is said to be *admissible* to the differentiable manifold M. It is noted that the maximal condition (3) is not necessary, because we can construct the maximal structure from a structure $\delta(M)$ satisfying (1) and (2) only, by adding to $\delta(M)$ all the charts which are C^r-related to every chart of $\delta(M)$.

Let $f: D \to R$ be a mapping of a domain D of a differentiable manifold M of class C^r into R, that is, f is a real-valued function defined on D. There exists an admissible chart $(U, u) \in \delta(M)$ such that $D \cap U$ is non-empty. Then we get a real-valued function $f^* = f \circ u^{-1}: u(D \cap U) \to R$ of n real variables x^i, where (x^i) is the coordinate system with respect to (U, u). This f^* is called a *coordinate expression* of f.

DEFINITION 1.2. If a coordinate expression f^* of a function $f: D \to R$ with respect to an admissible chart (U, u) is of class C^s, $s \leq r$, at a point $u(x)$, $x \in D$, the function f is called *differentiable of class C^s at x*.

The differentiability of f is well-defined, independently on the choice of an admissible chart with respect to which f^* is defined. If $f: D \to R$ is differentiable of class C^s at every point of D, then f is called *differentiable of class C^s on D*.

Throughout the following, the differentiability of manifolds and of functions on them is always assumed to be of class C^{∞}.

The set of all the functions which are defined in some neighborhood of a point $x \in M$ and differentiable at x is denoted by $\zeta(M)_x$.

EXAMPLE 1.1. Every coordinate x^i of a coordinate system (x^i) belongs to $\zeta(M)_x$ where the point x is to be contained in the coordinate neighborhood carrying (x^i).

We consider a mapping μ: $M \to N$ of an m-dimensional differentiable manifold M into another n-dimensional differentiable manifold N. Take admissible charts (U, u) and (V, v) of M and N respectively, in which $x \in U$ and $y = \mu(x) \in V$ have coordinate systems (x^i) and (y^a) respectively. Then we get n real-valued functions $y^a = y^a(x^i)$ of m variables x^i, which locally express the mapping $v \circ \mu \circ u^{-1}$.

DEFINITION 1.3. Corresponding to $g \in \zeta(N)_y$, $y = \mu(x)$, we get a real-valued function $f = g \circ \mu$ which is defined on a neighborhood of x. If for any $g \in \zeta(N)_y$ the function f belongs to $\zeta(M)_x$, the mapping μ is called *differentiable at x*.

Then a mapping μ is differentiable at x, if and only if the local expressions $y^a = y^a(x^i)$ are differentiable in the usual sense.

EXAMPLE 1.2. Let M×N be the topological product of two differentiable manifolds M, N of respective dimensions m, n. Then M×N has a natural differentiable structure with respect to which (U×V, u×v) is an admissible chart, where (U, u) and (V, v) are admissible charts of M and N respectively and $u \times v$: $(x, y) \in U \times V \mapsto (u(x), v(y)) \in R^m \times R^n = R^{m+n}$. Therefore M×N is of dimension m + n and called the *product manifold* of M and N. Then we naturally have two differentiable mappings

$$\rho_1: (x, y) \in M \times N \mapsto x \in M, \qquad \rho_2: (x, y) \in M \times N \mapsto y \in N,$$

which are called the *first and second canonical mappings* respectively.

EXAMPLE 1.3. From a differentiable mapping μ: $M \times N \to L$ of a product manifold M×N into a differentiable manifold L, we get two differentiable mappings

$$_x\mu: y \in N \mapsto \mu(x, y) \in L, \qquad \mu_y: x \in M \mapsto \mu(x, y) \in L,$$

where in $_x\mu$, for instance, x is to be a fixed point of M. These are called the *left and right-fixed mappings of* μ respectively.

Now we are concerned with an n-dimensional real vector space V. With respect to a base $B = \{e_a\}$ of V any vector $v \in V$ is uniquely written as $v = v^a e_a$. Consequently we get a mapping u: $v \in V \mapsto (v^a) \in R^n$. The pair (V, u) gives rise to a typical differentiable structure of V, and (v^a) is the coordinate system of v with respect to the chart (V, u). Such a coordinate system as above obtained from a base is called *linear*.

DEFINITION 1.4. All the linear mappings $V \to R$ of a vector space V into R constitute a real vector space, which is called the *dual vector space* of V and denoted by V_*.

A mapping $v_*: V \to R$ is linear, if the following two conditions hold:

(1) $v_*(u + v) = v_*(u) + v_*(v)$, (2) $v_*(r\cdot v) = r\cdot v_*(v)$,

for any u, $v \in V$ and any $r \in R$. The set V_* is a real vector space with the summation σ_* and the multiplication μ_*, defined by

(1) $(u_* + v_*)(v) = u_*(v) + v_*(v)$, (2) $(r\cdot u_*)(v) = r\cdot u_*(v)$,

for any $v \in V$.

Let $B = \{e_a\}$ be a base of an n-dimensional vector space V. Any linear mapping $v_* \in V_*$ is uniquely determined by n values $v_a = v_*(e_a)$. In fact, for any $v = v^a e_a$ we have $v_*(v) = v^a v_a$, $v = v^a e_a$. Therefore, if n linear mappings $e^a \in V_*$ are defined by

$$e^a(e_b) = \delta^a_b \quad \text{(Kronecker's deltas)},$$

then $B_* = \{e^a\}$ is a base of V_* and the above v_* is written as $v_* = v_a e^a$. Consequently the dimension of V_* is equal to that of V. B_* is called the *dual base* of V_*, corresponding to B.

The original vector space V can be identified with the dual space $(V_*)_*$ of V_*. Thus the value $v_*(v)$ may be written $v(v_*)$; so we write this value in the inner product form:

$$v_*(v) = v(v_*) = (v, v_*) = (v_*, v).$$

DEFINITION 1.5. Let V be an n-dimensional vector space and V_* be the dual vector space of V. All the multilinear mappings $\underbrace{V\times\cdots\times V}_{s}\times\underbrace{V_*\times\cdots\times V_*}_{r} \to R$ constitute a vector space of dimension n^{r+s}, called the *tensor space of (r, s)-type*, constructed from V and denoted by $\underbrace{V_*\otimes\cdots\otimes V_*}_{s}\otimes\underbrace{V\otimes\cdots\otimes V}_{r}$ or V^r_s simply. A vector of V^r_s is called a *tensor of (r, s)-type*. (The symbol $\otimes$ is called the *tensor product*.)

Here we are concerned with a tensor space $V^1_2 = V_*\otimes V_*\otimes V$ of (1, 2)-type for brevity. A tensor $w \in V^1_2$ is a multilinear mapping $V\times V\times V_* \to R$, i.e., it is linear on the first V, the second V and V_*. The vector space structure of V^1_2 is defined similarly to the case of V_*. Taking a base $B = \{e_a\}$ of V and the dual base $B_* = \{e^a\}$ of V_*, a base

$B_2^1 = \{e_c^{ab}\}$ of V_2^1 is defined by

$$e_c^{ab}(e_i,\ e_j,\ e^k) = \delta_i^a \delta_j^b \delta_c^k.$$

This base B_2^1 is said to be constructed from the base B of V. The tensors e_c^{ab} are sometimes written as $e^a \otimes e^b \otimes e_c$. The tensor product $u_* \otimes v_* \otimes w$ of u_*, $v_* \in V_*$ and $w \in V$ is by definition a tensor of (1, 2)-type given by $(u_a v_b w^c) e_c^{ab}$, where $u_* = u_a e^a$, $v_* = v_a e^a$ and $w = w^a e_a$.

Now we return to a differentiable manifold M of dimension n and attach to every point of M a real vector space of dimension n as follows:

DEFINITION 1.6. A *tangent vector* X (or a vector simply) at a point x of a differentiable manifold M is a mapping $\zeta(M)_x \to R$ which satisfies the following two conditions:

(1) X is a linear mapping, i.e., for any f, $g \in \zeta(M)_x$ and any $r \in R$ we have

$$X(f + g) = X(f) + X(g), \qquad X(r \cdot f) = r \cdot X(f).$$

(2) X is a derivation, i.e., for any f, $g \in \zeta(M)_x$ we have

$$X(f \cdot g) = X(f) \cdot g(x) + f(x) \cdot X(g).$$

The set of all the tangent vectors at x is called the *tangent (vector) space* of M at the point x and denoted by M_x.

The vector space structure of M_x is given by the summation σ_x and the multiplication μ_x defined by

(1) $(X + Y)(f) = X(f) + Y(f)$, $\qquad X,\ Y \in M_x$,

(2) $(r \cdot X)(f) = r \cdot X(f)$, $\qquad X \in M_x, \quad r \in R$,

for any $f \in \zeta(M)_x$. For a constant function c we have $X(c) = 0$ for any X.

Next, we refer to the admissible chart (U, u) carrying a coordinate system (x^i). For a point $x_o = (x_o^i) \in U$ a differentiable function f is written

$$f(x) = f(x_o) + f_i(x_o)(x^i - x_o^i) + f_{ij}(x)(x^i - x_o^i)(x^j - x_o^j),$$

where $f_i = \partial f / \partial x^i$ and f_{ij} are differentiable functions. We consider X(f) for $X \in M_{x_o}$. First $X(f(x_o)) = 0$. Secondly $X(f_{ij}(x)(x^i - x_o^i)(x^j - x_o^j)) = 0$ from $(x^i - x_o^i)_{x_o} = 0$. Consequently we obtain $X(f) = f_i(x_o) X(x^i)$, that is,

$$X(f) = \left(\frac{\partial f}{\partial x^i}\right)_{x_o} X^i, \qquad X^i = X(x^i)_{x_o}.$$

Therefore X is regarded as the differentiation operator $X^i (\partial / \partial x^i)_{x_o}$.

In fact, n differentiation operators $(\partial / \partial x^i)_x$ satisfy both of the conditions for a tangent vector and their independence is easily shown. Thus the collection $B_x =$

$\{(\partial/\partial x^i)_x\}$ is a base of the tangent vector space M_x and called the *natural base* in the coordinate system (x^i). The above X^i are called the *components* of X in this coordinate system.

DEFINITION 1.7. A *tangent vector field* X on a differentiable manifold M is an indication $x \in M \mapsto X_x \in M_x$. If the function $X(f)\colon x \in M \mapsto X_x(f) \in R$ is differentiable for any differentiable function f, the field X is called *differentiable*.

A tangent vector field X is differentiable, if and only if every component X^i is a differentiable function.

DEFINITION 1.8. For two differentiable tangent vector fields X, Y, the *Lie bracket* [X, Y] is a tangent vector given by

$$[X, Y](f) = X(Y(f)) - Y(X(f))$$

at x for any $f \in \zeta(M)_x$.

For $X = X^i(\partial/\partial x^i)$ and $Y = Y^i(\partial/\partial x^i)$ we have

$$[X, Y] = \left(X^r \frac{\partial Y^i}{\partial x^r} - Y^r \frac{\partial X^i}{\partial x^r}\right)\frac{\partial}{\partial x^i}\,. \tag{1.1}$$

The following two equations are important later on:

$$[fX, gY] = fg[X, Y] + \{f\cdot X(g)\}Y - \{g\cdot Y(f)\}X, \tag{1.2}$$

where the right-hand side is a linear combination of [X, Y], Y and X with coefficients fg, $f\cdot X(g)$ and $g\cdot Y(f)$ respectively.

$$[X, [Y, Z]] + [Y, [Z, X]] + [Z, [X, Y]] = 0, \tag{1.3}$$

which is called the *Jacobi identity*.

DEFINITION 1.9. From the tangent vector space M_x we construct the dual vector space $(M_x)_*$ and tensor space $(M_x)^r_s$ of (r, s)-type. An element of $(M_x)^r_s$ is called a *tensor of (r, s)-type at x*. In particular, a tensor of (0, 1)-type is called a *tangent covector*.

Let $B_x = \{(\partial/\partial x^i)_x\}$ be the natural base of M_x in a coordiante system (x^i). The dual base $(B_x)_*$ is denoted by $\{(dx^i)_x\}$. Therefore we have $(\partial/\partial x^j, dx^i) = \delta^i_j$.

An indication $T\colon x \in M \mapsto T_x \in (M_x)^r_s$ is called a *tensor field of (r, s)-type*. The differentiability of a tensor field is defined as follows: First a tangent covector field X_* is differentiable, if for any differentiable tangent vector field X the function $X_*(X)$ is differentiable. A differentiable X_* is called a *differential one-form*. Next, a tensor field T of (1, 1)-type, for instance, is differentiable, if for any differentiable tangent vector field X and any differential one-form X_* the function $T(X_*, X)$ is differentiable. Referring to a coordinate system (x^i), we have

$$T = T^i_j \frac{\partial}{\partial x^i} \otimes dx^j, \qquad T(X_*, X) = T^i_j X_i X^j$$

for $X_* = X_i dx^i$ and $X = X^i \partial/\partial x^i$. The tensor field T is differentiable, if and only if every component T^i_j is a differentiable function.

Finally, a tensor field of (0, 0)-type is a real-valued function, called a *scalar field*.

REMARK 1.1. We have already given the concept of differentiability of manifolds, functions, mappings and tensor fields. Throughout the following, all the entities we consider are assumed to be differentiable. Therefore the adjective "differentiable" will be omitted, when there is no confusion.

We are concerned with a mapping μ: M → N of an m-dimensional manifold M into an n-dimensional manifold N. For any $g \in \zeta(N)_y$, $y = \mu(x)$, we obtain $f = g \circ \mu \in \zeta(M)_x$.

DEFINITION 1.10. The *differential* μ' at $x \in M$ of a mapping μ: M → N is a linear mapping $M_x \to N_y$, $y = \mu(x)$, given by

$$(\mu' X)(g) = X(g \circ \mu), \qquad X \in M_x,$$

for any $g \in \zeta(N)_y$.

We recall a local expression $y^a = y^a(x^i)$ of the mapping μ. For $X = X^i \partial/\partial x^i$ we have

$$(\mu' X)(g) = X^i \frac{\partial}{\partial x^i}\{g(y(x))\} = X^i \frac{\partial g}{\partial y^a}\frac{\partial y^a}{\partial x^i},$$

which implies $\mu' X = X^i(\partial y^a/\partial x^i)\partial/\partial y^a$. Thus for the natural bases $B_x = \{\partial/\partial x^i\}$ and $B_y = \{\partial/\partial y^a\}$ we get

$$(1.4) \qquad \mu' : \frac{\partial}{\partial x^i} \mapsto \frac{\partial y^a}{\partial x^i}\frac{\partial}{\partial y^a}.$$

DEFINITION 1.11. A *differentiable curve* C on a manifold M is a differentiable mapping $(a, b) \subset R \to M$ of an open interval (a, b) of R into M. The image C((a, b)) is also called a differentiable curve on M.

From a tangent vector d/dt at a point $t \in (a, b)$ we get a tangent vector

$$C'\left(\frac{d}{dt}\right) = \frac{dx^i}{dt}\frac{\partial}{\partial x^i} \in M_{C(t)},$$

where $C(t) = (x^i(t))$. This is called the *tangent vector of C*. As a consequence, the tangent vector space M_x is regarded as the set of tangent vectors at x of all the curves through x.

REMARK 1.2. In the following, we sometimes treat of a curve defined on a closed interval [a, b] of R. In such a case, the curve is assumed to be defined on some open interval containing [a, b].

PROPOSITION 1.1. *If a mapping $\mu: M \to N$ is a constant mapping, namely, the image $\mu(M)$ is a single point, $\mu'(X) = 0$ holds for any tangent vector X at any point of M. The converse is true, provided M is connected.*

Proof. If μ is constant, the function $g \circ \mu$ in Definition 1.10 is a constant function, so that $(\mu' X)(g) = X(g \circ \mu) = 0$.

To show the converse, take an admissible chart (U, u) of M carrying a coordinate system (x^i). Let C be a curve on M such that $C(0) = x_o \in U$, and put $y_o = \mu(x_o)$. Next, let (V, v) be an admissible chart of N carrying a coordinate system (y^a) such that $y_o \in V$. Then we obtain a curve $\bar{C} = \mu \circ C$ on N such that $\bar{C}(0) = y_o$. From the assumption we have $\bar{C}'(d/dt) = \mu' \circ C'(d/dt) = 0$, i.e., $dy^a/dt = 0$ at every point of $\bar{C}$, which implies $y^a(t) = y^a_o$; so $\bar{C}$ is a single point y_o. Since any point $x \in U$ can be connected with x_o by a curve in U, the mapping μ is constant on U: $\mu(U) = y_o$. Therefore the set $M_o = \{x \in M \mid \mu(x) = y_o\}$ is a sum of open sets; so M_o is open. But a single point y_o is closed, so that M_o is closed. Thus M_o must coincide with M itself, because M is assumed to be connected.

DEFINITION 1.12. Let f be a real-valued function on a manifold M. The *differential* df of f is a differential one-form, given by $df(X) = X(f)$ at $x \in M$ for any $X \in M_x$.

For $X = X^i \partial/\partial x^i$ we have $df(X) = X^i \partial f/\partial x^i$, so that df is written as

$$df = \frac{\partial f}{\partial x^i} dx^i.$$

REMARK 1.3. We now notice that the notation dx^i is used to denote two concepts, one being an element $(dx^i)_x$ of the natural dual base $(B_x)_* = \{(dx^i)_x\}$ and the other the differential dx^i of the function x^i. But there is no confusion, because, with respect to this dual base, the differential dx^i of x^i is written $dx^i = (\partial x^i/\partial x^j)(dx^j)_x$, which is equal to $(dx^i)_x$ as an element of $(B_x)_*$.

More generally, we define the differential $d\phi$ of a vector-valued function $\phi: x \in M \mapsto \phi^a(x) e_a \in V$, where V is a vector space with a base $B = \{e_a\}$, such that $d\phi: M_x \mapsto V$ is the collection of differentials $d\phi^a$ of real-valued functions ϕ^a, namely, $d\phi =$

$d\phi^a e_a$. Thus for $X = X^i\partial/\partial x^i$ we have

$$d\phi(X) = d\phi^a(X)e_a = \left(X^i \frac{\partial\phi^a}{\partial x^i}\right)e_a \in V.$$

On the other hand, regarding ϕ as a mapping $M \to V$, we have the differential ϕ': $M_x \mapsto V_v$, $v = \phi(x)$, namely,

$$\phi'(X) = X^i \frac{\partial\phi^a}{\partial x^i}\left(\frac{\partial}{\partial v^a}\right)_v$$

where (v^a) is the linear coordinate system with respect to the above base B. Consequently, if we introduce a mapping

(1.5) $$I_v: V \to V_v, \qquad u^a e_a \mapsto u^a\left(\frac{\partial}{\partial v^a}\right)_v,$$

we get

(1.6) $$\phi' = I_v \circ d\phi.$$

§ 2. Lie transformation groups

A mapping μ: $M \to M$ of a manifold M into itself is called a *transformation* of M, if μ is onto and one-to-one differentiable mapping and the inverse μ^{-1} is also differentiable.

DEFINITION 2.1. A family $\{\mu_t\}$ of transformations of a manifold M, parametrized by $t \in R$, is called a *one-parameter transformation group* of M, if the following two conditions hold:

(1) $\mu_{t+s} = \mu_t \circ \mu_s$ for any $t, s \in R$.

(2) The mapping μ: $M\times R \to M$, $(x, t) \mapsto \mu_t(x)$, is differentiable.

Every transformation μ_t for a fixed t is a right-fixed mapping of the above μ. It follows immediately from (1) that $\{\mu_t\}$ is abelian, μ_o = identity and $\mu_{-t} = (\mu_t)^{-1}$.

EXAMPLE 2.1. We are concerned with the summation σ: $(v, u) \in V\times V \mapsto v + u \in V$. A right-fixed mapping σ_u is evidently a transformation of V. The family $\{\sigma_{tu}\}$, $t \in R$, is a one-parameter transformation group of V.

On the other hand, from the multiplication μ: $(r, v) \in R\times V \mapsto rv \in V$ we get another transformation $_r\mu$: $v \mapsto rv$, $r \neq 0, 1$, having a single fixed point $0 \in V$. The family $\{_{tr}\mu\}$, $t \in R$, is not a one-parameter transformation group.

We consider a one-parameter transformation group $\{\mu_t\}$ of a manifold M and the mapping μ: $(x, t) \mapsto \mu_t(x)$. A left-fixed mapping $_x\mu$: $R \to M$, $t \mapsto \mu_t(x)$, is a curve on

M through the point x because of $\mu_o(x) = x$. This curve or the set $\{{}_x\mu(t) \mid t \in R\}$ is called the *locus* of x. We get the tangent vector ${}_x\mu'(d/dt)_o$ at x of the locus. The tangent vector field X on M, defined by $x \in M \mapsto X_x = {}_x\mu'(d/dt)_o$, is said to *be induced by* $\{\mu_t\}$. It is observed that for $f \in \zeta(M)_x$

$$ {}_x\mu'\left(\frac{d}{dt}\right)_o(f) = \left(\frac{d}{dt}\right)_o(f\circ{}_x\mu) = \left(\frac{d}{dt}\right)_o(f\circ\mu_t(x)). $$

Thus we may write the induced vector field X as

$$ (2.1) \qquad X_x(f) = d_0t\{f\circ\mu_t(x)\}, $$

or, for easier understanding,

$$ (2.1') \qquad X_x(f) = \lim_{t\to 0}\frac{1}{t}\{f\circ\mu_t(x) - f(x)\}. $$

<u>EXAMPLE 2.2.</u> We return to the one-parameter transformation group $\{\sigma_{tu}\}$, given in Example 2.1. From (2.1) the induced vector field X is such that

$$ X_v(f) = d_0t\{f(v + tu)\} = \left(\frac{\partial f}{\partial v^a}\right)_v u^a, $$

where $u = u^a e_a$ and (v^a) is the linear coordinate system with respect to the base B $= \{e_a\}$. We denote this X by $\vec{u}$, namely,

$$ \vec{u}_v = u^a\left(\frac{\partial}{\partial v^a}\right)_v , $$

and call it the *tangent vector field parallel to* u. From (1.5) it is now seen that

$$ I_v(u) = \vec{u}_v. $$

This is regarded as another definition of the mapping I_v, so that I_v is well-defined, independently on the choice of base.

We shall show the inverse correspondence of "one-parameter transformation group → induced vector field". From (2.1) it follows that the locus ${}_x\mu(t) = (x^i(t))$ of x $= (x^i(0))$ satisfies $X^i\partial f/\partial x^i = (\partial f/\partial x^i)(dx^i/dt)$, i.e.,

$$ (2.2) \qquad \frac{dx^i}{dt} = X^i, $$

where the right-hand side are components of the induced vector field X. Conversely, if we regard (2.2) as a system of differential equations with a given tangent vector field $X = X^i\partial/\partial x^i$, the solution $\mu_t(x) = (x^i(t))$ for an initial condition $(x^i(0)) = x_o$ gives the locus of x_o and we get the one-parameter transformation group $\{\mu_t\}$ by which X is induced. This $\{\mu_t\}$ is said to *be generated by* X. It should be remarked that, according to the theory of differential equations, the solution $\{\mu_t\}$ is local in general, that is, it is given on a suitable small neighborhood of t = 0.

PROPOSITION 2.1. *Let X be the tangent vector field induced by a one-parameter transformation group* $\{\mu_t\}$. *The tangent vector field* $\nu'X$, *which is given by the differential* ν' *of a transformation* ν, *generates the one-parameter transformation group* $\{\nu\circ\mu_t\circ\nu^{-1}\}$.

Proof. Putting $\bar{x} = \nu(x)$, for any $f \in \zeta(M)_x$ the equation (2.1) yields

$$(\nu'X)_{\bar{x}}(f) = X_x(f\circ\nu) = d_0t\{f\circ\nu\circ\mu_t(x)\} = d_0t\{f\circ(\nu\circ\mu_t\circ\nu^{-1}(\bar{x}))\},$$

which shows the result.

COROLLARY 2.1. *The tangent vector field X, induced by* $\{\mu_t\}$, *is invariant by a transformation* ν, *i.e.,* $\nu'X = X$, *if and only if* ν *commutes with* μ_t *for any* $t \in R$.

EXAMPLE 2.3. In the notations of Example 2.2, we are concerned with the tangent vector field $\vec{u}$ parallel to a vector $u \in V$ which generates $\{\sigma_{tu}\}$. Because the transformation ${}_v\sigma\colon w \in V \mapsto v + w \in V$ commutes with any σ_{tu}, we have ${}_v\sigma'\vec{u} = \vec{u}$, namely, $\vec{u}$ is invariant by any ${}_v\sigma$.

On the other hand, in the notations of Example 2.1, we have ${}_r\mu\circ\sigma_{tu}\circ{}_r\mu^{-1} = \sigma_{t(ru)}$, so that Proposition 2.1 yields ${}_r\mu'(\vec{u}) = \overrightarrow{ru}$.

PROPOSITION 2.2. *Let X, Y be tangent vector fields and let* $\{\mu_t\}$ *be the one-parameter transformation group generated by X. The Lie bracket* $[X, Y]$ *is written*

$$[X, Y] = -\,d_0t(\mu_t'Y) = \lim_{t\to 0}\frac{1}{t}\{Y - \mu_t'Y\}.$$

Proof. It is known that for a small $t \in R$ any $f \in \zeta(M)_x$ is written

$$f(\mu_t(\bar{x})) = f(\bar{x}) + t\cdot g(t, \bar{x}), \qquad \mu_t(\bar{x}) = x,$$

where $g(t, \bar{x}) \in \zeta(M)_{\bar{x}}$ for every t and $X_x(f) = g(0, x)$. Then

$$(\mu_t'Y_{\bar{x}})(f) = Y_{\bar{x}}(f\circ\mu_t) = Y(f)_{\bar{x}} + t\cdot Y_{\bar{x}}(g(t, \bar{x})).$$

Therefore we have

$$-\,\{d_0t(\mu_t'Y_{\bar{x}})\}(f) = \lim_{t\to 0}\frac{1}{t}\{Y(f)_x - (\mu_t'Y_{\bar{x}})(f)\}$$

$$= \lim_{t\to 0}\frac{1}{t}\{Y(f)_x - Y(f)_{\bar{x}}\} - \lim_{t\to 0} Y_{\bar{x}}(g(t, \bar{x})),$$

which is equal to $X(Y(f))_x - Y(X(f))_x$.

From Proposition 2.1 it follows that if Y generates a one-parameter transformation group $\{\nu_s\}$, we have

$$(\mu_t'Y_{\bar{x}})(f) = \lim_{s\to 0}\frac{1}{s}\{f\circ\mu_t\circ\nu_s\circ\mu_t^{-1}(x) - f(x)\}.$$

On the other hand, (2.1') yields

$$Y_x(f) = \lim_{s\to 0} \frac{1}{s} \{f\circ\nu_s(x) - f(x)\}.$$

Therefore Proposition 2.2 gives

PROPOSITION 2.3. *Let X, Y be tangent vector fields which generate one-parameter transformation groups* $\{\mu_t\}$, $\{\nu_s\}$ *respectively. Then we have*

$$[X, Y](f) = \lim_{s,t\to 0} \frac{1}{st} \{f\circ\nu_s - f\circ\mu_t\circ\nu_s\circ\mu_t^{-1}\}, \qquad f \in \zeta(M)_x.$$

As a corollary of Propositions 2.2 and 2.3, we get

COROLLARY 2.2. *In the notations of Proposition 2.3, the Lie bracket [X, Y] vanishes, if and only if one of the following two conditions holds:*

(1) *Y is invariant by* μ_t *for any* $t \in R$.

(2) μ_t *and* ν_s *commute with each other for any* $t, s \in R$.

The concepts of Lie group and Lie algebra are important for our modern formulation of differential geometry. Since these theories are well-known, we here show an outline of the general theories and are mainly concerned with the general linear groups for the later use.

We consider an algebraic group G. That is, G is a set equipped with a group operation τ: $(g, h) \in G\times G \mapsto gh \in G$. There exist the so-called unit element $e \in G$ such that $eg = ge = g$ for any $g \in G$ and the inverse g^{-1} such that $g^{-1}g = gg^{-1} = e$.

DEFINITION 2.2. If a group G is a differentiable manifold and the mapping $(g, h) \in G\times G \mapsto gh^{-1} \in G$ is differentiable, then G is called a *Lie group*.

The group operation τ is differentiable. From it two mappings are obtained:

$${}_g\tau: h \mapsto gh, \qquad \tau_g: h \mapsto hg,$$

which are called the *left* and *right translation* by $g \in G$ respectively. From these translations we get

$$I_g = {}_g\tau\circ\tau_{g^{-1}} = \tau_{g^{-1}}\circ {}_g\tau: h \mapsto ghg^{-1},$$

which is called the *inner automorphism* by $g \in G$.

We are specially concerned with a Lie group GL(n, R), denoted by G(n) simply. This G(n) is the set of all the non-singular n×n matrices, and is evidently a group with the usual matrix product

$$\tau: G(n)\times G(n) \to G(n), \qquad (g = (g^a_b),\ h = (h^a_b)) \mapsto gh = (g^a_c h^c_b),$$

where all the indices run from 1 to n. Further $G(n)$ is regarded as a subspace of R^m, $m = n^2$; a point $g = (g_b^a)$ has a coordinate system

$$(g_1^1, g_2^1, \cdots, g_n^1, g_1^2, \cdots, g_n^2, \cdots, g_n^n)$$

in R^m. Since $g = (g_b^a)$ is provided by the inequality $\det(g_b^a) \neq 0$, $G(n)$ is an open subspace; so it is equipped with the so-called induced differentiable structure from R^m. Consequently $G(n)$ itself is a differentiable manifold of dimension n^2. Thus $G(n)$ is a Lie group, called the *general linear group of order* n. The above coordinate system $(g_1^1, \cdots, g_n^n)$ is called *canonical*.

We shall find the differential of a left translation $_h\tau$, $h \in G(n)$:

$$_h\tau \colon g = (g_b^a) \mapsto hg = \bar{g} = (\bar{g}_b^a), \qquad \bar{g}_b^a = h_c^a g_b^c,$$

$$_h\tau' \colon \left(\frac{\partial}{\partial g_b^a}\right)_g \mapsto \frac{\partial \bar{g}_d^c}{\partial g_b^a}\left(\frac{\partial}{\partial g_d^c}\right)_{\bar{g}} = \frac{\partial(h_e^c g_d^e)}{\partial g_b^a}\left(\frac{\partial}{\partial g_d^c}\right)_{\bar{g}} = h_e^c \delta_a^e \delta_d^b \left(\frac{\partial}{\partial g_d^c}\right)_{\bar{g}} .$$

Consequently we get

(2.3)
$$_h\tau' \colon \quad \frac{\partial}{\partial g_b^a} \mapsto h_a^c \frac{\partial}{\partial g_b^c} .$$

This may be rather easier to understand, if we write it as

(2.3')
$$_h\tau' \colon \quad X_b^a \frac{\partial}{\partial g_b^a} \mapsto h_a^c X_b^a \frac{\partial}{\partial g_b^c} ,$$

because the matrix product $hX = (h_a^c X_b^a)$ takes place as components. As to a right translation, we similarly get

(2.4)
$$\tau_h' \colon \quad X_b^a \frac{\partial}{\partial g_b^a} \mapsto X_a^c h_b^a \frac{\partial}{\partial g_b^c} .$$

Combining the above two formulas, we get

(2.5)
$$I_h' \colon \quad X_b^a \frac{\partial}{\partial g_b^a} \mapsto h_c^a X_d^c (h^{-1})_b^d \frac{\partial}{\partial g_b^a} ,$$

where $(h^{-1})_b^d$ are elements of the inverse matrix h^{-1} of h.

We return to a general Lie group G. Let G^1 be the set of all the tangent vector fields on G which are left-invariant, namely, invariant by any left translation. Therefore $_g\tau' A_h = A_{gh}$ is the characteristic equation for $A \in G^1$, where g, h are any points of G. If we consider the one-parameter transformation group $\{a_t\}$ generated by A, Corollary 2.1 yields

(2.6)
$$_g\tau \circ a_t = a_t \circ {}_g\tau$$

for any $g \in G$ and any $t \in R$.

We introduce the mapping

$$I_G: A \in G^l \mapsto A_e \in G_e,$$

where G_e is the tangent vector space of G at the unit e and A_e is the value of A at e. The inverse mapping I_G^{-1} of I_G is obviously given by

$$I_G^{-1}: X_e \in G_e \mapsto X \in G^l, \qquad X_g = {}_g\tau' X_e,$$

where X_g is the value of the field X at an arbitrary point $g \in G$. Consequently I_G is an isomorphism of G^l with G_e and so G^l may be identified with G_e by I_G.

REMARK 2.1. Similarly to the case of G^l, we can deal with the set G^r consisting of all the right-invariant tangent vector fields on G. Since G^r also is identified with G_e, we are concerned with G^l only in the following.

DEFINITION 2.3. The set of all the left-invariant tangent vector fields on a Lie group G is called the *Lie algebra* of G, denoted by G'.

Since the unit e of G is fixed by any inner automorphism I_g, the differential I'_g is a linear isomorphism of G_e. Thus we get the so-called *adjoint representation*

$$ad(g) = I_G^{-1} \circ I'_g \circ I_G : G' \to G'$$

of $g \in G$.

The equation ${}_g\tau' A = A$, $A \in G'$, is obvious from the definition of G'. On the other hand, we consider $\tau'_g A$ for $A \in G'$.

$$(ad(g^{-1})A)_e = I_G\{ad(g^{-1})A\} = I'_{g^{-1}}(I_G A) = I'_{g^{-1}}(A_e),$$

$$(ad(g^{-1})A)_h = {}_h\tau'(I'_{g^{-1}}A_e) = {}_h\tau' \circ {}_{g^{-1}}\tau' \circ \tau'_g(A_e) = \tau'_g({}_{hg^{-1}}\tau' A_e) = \tau'_g(A_{hg^{-1}}).$$

Therefore we get

$$(2.7) \qquad ad(g^{-1})A = \tau'_g A.$$

Next we consider the one-parameter transformation group $\{a_t\}$ generated by $A \in G'$. From (2.6) we have $a_t \circ {}_g\tau(e) = {}_g\tau \circ a_t(e)$, namely, $a_t(g) = g a_t(e)$. Thus

PROPOSITION 2.4. *$a_t = \tau_{a_t(e)}$, namely, every transformation a_t of the one-parameter transformation group $\{a_t\}$ generated by $A \in G'$ is nothing but the right translation by $a_t(e)$.*

We have to show the algebraic property of the Lie algebra G'. From Propositions 2.2 and 2.4 and (2.7), for A, $B \in G'$ we have

$$[A, B]_e = - d_0 t(a'_t B) = - d_0 t\{\tau'_{a_t(e)} B\} = - d_0 t\{ad(a_t(e)^{-1})B\},$$

namely,

$$(2.8) \qquad [A, B] = \lim_{t \to 0} \frac{1}{t}\{B - ad(a_t(e)^{-1})B\}.$$

Both B and $ad(a_t(e)^{-1})B$ belong to G', so that [A, B] belongs to G'. Consequently G' is an algebra with the Lie bracket operator.

Now we are specially concerned with the Lie algebra G'(n) of the general linear group G(n). In the canonical coordinate system (g^a_b) in G(n) we obtain the natural base $B_g = \{\partial/\partial g^a_b\}$. Since I_G is the isomorphism of G'(n) with $G(n)_e$, any $A \in G'(n)$ is identified with a tangent vector

$$(2.9) \qquad A_e = A^a_b\, E^b_a,$$

where the notations $E^b_a = (\partial/\partial g^a_b)_e$ are used throughout the following. The matrix (A^a_b) may be non-singular or singular. Consequently this algebra G'(n) is regarded as the set of all the real n×n matrices. From (2.3') we have

$$(2.9') \qquad A_g = {}_g\tau' A_e = g^a_c A^c_b \left(\frac{\partial}{\partial g^a_b}\right)_g .$$

From (2.7), (2.9') and (2.4) we see

$$(ad(g^{-1})A)_h = \tau'_g(A_{hg^{-1}}) = \{h^a_c (g^{-1})^c_d A^d_e\} g^e_b \left(\frac{\partial}{\partial g^a_b}\right)_h ,$$

which implies

$$(2.10) \qquad (ad(g^{-1})A)_e = \{(g^{-1})^a_c A^c_d g^d_b\} E^b_a .$$

Therefore (2.8) gives

$$[A, B]_e = - d_0 t\{(\alpha^{-1})^a_c B^c_d \alpha^d_b\} E^b_a , \qquad \alpha^a_b = (a_t(e))^a_b .$$

Since $(d\alpha^a_b/dt)_0 = A^a_b$ from (2.2) and $a_t(e)^{-1} = a_{-t}(e)$, the above is equal to $(A^a_c B^c_d \delta^d_b - \delta^a_c B^c_d A^d_b) E^b_a$, which implies

$$(2.11) \qquad [A, B] = AB - BA,$$

where the right-hand side is constructed by multiplications and subtraction of matrices $A = (A^a_b)$ and $B = (B^a_b)$.

The concept of Lie transformation group plays essential roles in the theory of fiber bundles and connections, as it will be shown later on.

<u>DEFINITION 2.4</u>. A pair (M, G) of a manifold M and a Lie group G is called a *Lie transformation group*, if the following three conditions hold:

(1) Every $g \in G$ is a transformation of $M : x \in M \mapsto xg \in M$.

(2) The mapping $\xi: M\times G \to M$, $(x, g) \mapsto xg$, is differentiable.

(3) $(xg)h = x(gh)$ holds for any $x \in M$ and any $g, h \in G$.

We sometimes say that G is a Lie transformation group of M with an action ξ. In almost all the cases, we shall restrict our consideration to Lie transformation groups which satisfy further conditions as follows:

(4) G is effective, i.e., if $xg = x$ for any x, then $g = e$.

(5) G is without fixed point, i.e., if $xg = x$ for a point x, then $g = e$.

The right-fixed mapping $\xi_g: x \mapsto xg$ of the action ξ is nothing but the transformation g and called the *right translation* of M by g. The subset $x/G = \{xg \mid g \in G\}$ of M for a fixed point $x \in M$ is called the *locus* of x by G.

We consider a left-fixed mapping ${}_x\xi: g \in G \mapsto xg \in M$ of the action ξ, called a *fundamental mapping*. The image of G by ${}_x\xi$ is the locus x/G; so we get the mapping ${}_x\xi: G \to x/G$. If G is without fixed point, this mapping ${}_x\xi$ is one-to-one, so that the inverse ${}_x\xi^{-1}: x/G \to G$ is defined.

If any two points of a locus are identified with each other, we get the set of all the equivalence classes M/G, called the *quotient space*. The mapping $\pi: x \in M \mapsto x/G \in M/G$ is called the *projection*.

REMARK 2.2. We must remark on the action of G on M. In the above we write a Lie transformation group as (M, G); in this case G is said to *act on M on the right*. But we also need such a case that it is convenient to write (G, M) and $x \in M \mapsto gx \in M$; in such a case G is said to *act on M on the left*. Although it seems that "on the right or on the left" is not an essential problem, we have to pay attention to the condition (3): In the case of "on the left", (3) should be written as $h(gx) = (hg)x$, where hg, not gh, takes place. If G is not abelian, hg is different from gh, so that the problem must not be neglected.

Now we are concerned with the general linear group $G(n)$ of order n and an n-dimensional vector space V^n with a fixed base $B = \{e_a\}$. The pair $(G(n), V^n)$ is a Lie transformation group with the action

$$\xi: (g, v) \in G(n)\times V^n \mapsto gv \in V^n, \quad g = (g^a_b),\ v = v^a e_a,\ gv = (g^a_b v^b)e_a. \tag{2.12}$$

It is observed that $h(gv) = (h^a_c g^c_b v^b)e_a$, so that $h(gv) = (hg)v$ holds and the action of $G(n)$ on V^n is on the left. If $gv = v$ for any $v \in V$, then $g = e$ is necessary, so that $G(n)$ is effective. It is obvious that $G(n)$ is not without fixed point; $0 \in V$ is a fixed point.

Next we consider the dual space V^n_* of V^n, and define the Lie transformation group $(G(n), V^n_*)$ by

$$\xi_*: (g, v_*) \in G(n)\times V^n_* \mapsto gv_* \in V^n_*, \tag{2.13}$$

$$g = (g^a_b), \quad v_* = v_a e^a, \quad gv_* = v_b (g^{-1})^b_a e^a,$$

where $B_* = \{e^a\}$ is the dual base. For $v = v^a e_a \in V^n$ we have $(gv_*)(v) = v_b(g^{-1})^b_a v^a$, that is, $(gv_*)(v) = v_*(g^{-1}v)$, or more clearly

$$(gv_*, v) = (v_*, g^{-1}v). \tag{2.13'}$$

This is regared as another definition of ξ_*. We observe

$$(h(gv_*), v) = (gv_*, h^{-1}v) = (v_*, g^{-1}(h^{-1}v)) = (v_*, (g^{-1}h^{-1})v)$$

$$= (v_*, (hg)^{-1}v) = ((hg)v_*, v),$$

which shows the "on the left" property of $(G(n), V^n_*)$.

In general, the action ξ^r_s: $w \mapsto gw$ of $g \in G(n)$ on $(V^n)^r_s$ is defined by

$$(gw)(v_1, \cdots, v_s, v^1_*, \cdots, v^r_*)$$

(2.14)

$$= w(g^{-1}v_1, \cdots, g^{-1}v_s, g^{-1}v^1_*, \cdots, g^{-1}v^r_*),$$

for any v's $\in V^n$ and any v_*'s $\in V^n_*$. In particular, we are concerned with $(V^n)^1_1$ for brevity:

$$(gw)(v, v_*) = w(g^{-1}v, g^{-1}v_*) = (w^a_b e^b_a)((g^{-1})^c_d v^d e_c, v_e g^e_f e^f)$$

$$= \{g^e_a w^a_b (g^{-1})^b_d\} v^d v_e,$$

which shows

(2.14') $$gw = \{g^a_c w^c_d (g^{-1})^d_b\} e^b_a.$$

REMARK 2.3. The actions ξ^r_s depend, of course, on the choice of the base $B = \{e_a\}$ of V^n. Throughout the following, we denote by V^n an n-dimensional vector space *with a fixed base* $B = \{e_a\}$ in case where one of actions ξ^r_s takes place in our consideration.

We shall consider the behavior of parallel vector fields $\vec{u}$ by the action ξ. Putting $g = (g^a_b)$, from (2.12) we get

(2.15) $${}_g\xi': \frac{\partial}{\partial v^a} \mapsto g^b_a \frac{\partial}{\partial v^b},$$

so that from Example 2.2 we have ${}_g\xi'\vec{u} = {}_g\xi'(u^a\partial/\partial v^a) = g^b_a u^a \partial/\partial v^b$. Consequently

(2.16) $${}_g\xi'\vec{u} = \overrightarrow{gu}.$$

We have to remark on the Lie transformation group $(G(n), G'(n))$ with the adjoint representation ad(g). From (2.10), if we identify $G'(n)$ with G_e, we have ad(g): $A^a_b E^b_a \mapsto g^a_c A^c_d (g^{-1})^d_b E^b_a$, $g = (g^a_b)$. We now introduce another identification $I: G'(n) \to (V^n)^1_1$, defined by $I(E^a_b) = e^a_b$. Therefore (2.14') yields $I(\mathrm{ad}(g)A) = gI(A)$, namely

(2.17) $$I \circ \mathrm{ad}(g) = gI.$$

We return to a general Lie transformation group (M, G) and consider a fundamental mapping ${}_x\xi$: $g \in G \mapsto xg \in M$.

DEFINITION 2.5. Let (M, G) be a Lie transformation group with an action ξ. The *fundamental vector field* M(A), corresponding to $A \in G'$, is a tangent vector field on M, the value $M(A)_x$ at $x \in M$ being given by $M(A)_x = {}_x\xi' A_e$.

Consider the one-parameter transformation group $\{a_t\}$ generated by $A \in G'$. From (2.1) it follows that for $f \in \zeta(M)_x$

$$({}_x\xi' A_e)(f) = A_e(f \circ {}_x\xi) = d_0 t\{f \circ {}_x\xi \circ a_t(e)\} = d_0 t\{f \circ \xi_{a_t(e)}(x)\},$$

which shows

PROPOSITION 2.5. *A fundamental vector field M(A) generates the one-parameter group of right translations* $\{\xi_{a_t(e)}\}$ *of M, where* $\{a_t\}$ *is the one-parameter transformation group generated by A.*

If G is without fixed point, we have the inverse ${}_x\xi^{-1}$ of ${}_x\xi$ for any point $x \in M$. Therefore

PROPOSITION 2.6. *If G acts on M without fixed point, any fundamental vector field M(A),* $A \neq 0$*, has no zero point.*

Next we consider the behavior of M(A) by the action of G. First, for $g, h \in G$ and $x \in M$

$$\xi_g \circ {}_x\xi(h) = (xh)g = xg(g^{-1}hg) = {}_{xg}\xi \circ I_{g^{-1}}(h),$$

which implies $\xi_g \circ {}_x\xi = {}_{xg}\xi \circ I_{g^{-1}}$. Then

$$\xi'_g M(A)_x = \xi'_g \circ {}_x\xi' A_e = {}_{xg}\xi' \circ I'_{g^{-1}} A_e = {}_{xg}\xi'(\mathrm{ad}(g^{-1})A)_e.$$

Thus Definition 2.4 gives

$$\xi'_g M(A) = M(\mathrm{ad}(g^{-1})A). \tag{2.18}$$

REMARK 2.4. If we are concerned with G acting on M on the left, M(A) is defined by $M(A)_x = \xi'_x A_e$. Therefore, from ${}_g\xi \circ \xi_x = \xi_{gx} \circ I_g$, (2.18) should be modified as

$${}_g\xi' M(A) = M(\mathrm{ad}(g)A). \tag{2.18'}$$

PROPOSITION 2.7. *The Lie bracket in a Lie algebra G' causes the one in the set of all the fundamental vector fields, namely,* $M([A, B]) = [M(A), M(B)]$.

Proof. First we observe that for g, h, k $\in$ G and x $\in$ M

$$ {}_x\xi \circ \tau_g \circ {}_h\tau(k) = x(hkg) = \xi_g \circ {}_{xh}\xi(k), $$

which implies ${}_x\xi \circ \tau_g \circ {}_h\tau = \tau_g \circ {}_{xh}\xi$. From Definition 2.5 and Proposition 2.2 we have

$$ M([A, B])_x = {}_x\xi'[A, B]_e = {}_x\xi'\{\lim_{t\to 0} \frac{1}{t} (B_e - a_t' B_h)\}, $$

where $\{a_t\}$ is the one-parameter transformation group generated by A and $h = a_{-t}(e)$. From Proposition 2.4 the above is equal to

$$ \lim_{t\to 0} \frac{1}{t}\{ {}_x\xi' B_e - {}_x\xi' \circ \tau_{a_t(e)} \circ {}_h\tau' B_e\} = \lim_{t\to 0} \frac{1}{t} \{M(B)_x - \xi'_{a_t(e)} \circ {}_{xh}\xi' B_e\} $$

$$ = \lim_{t\to 0} \frac{1}{t} \{M(B)_x - \xi'_{a_t(e)} M(B)_{xh}\}, $$

which is equal to $[M(A), M(B)]_x$ from Propositions 2.2 and 2.5.

In particular, we are concerned with fundamental vector fields on $(V^n)^r_s$ which are introduced from the Lie transformation group $(G(n), (V^n)^r_s)$. First we consider the fundamental vector field $V^n(A)$ on V^n, corresponding to $A = (A^a_b) \in G'(n)$. From (2.12) we get $\xi_v: (g^a_b) \mapsto (g^a_b v^b)$, $v = v^a e_a$, so that

$$ (2.19) \qquad \xi'_v: \frac{\partial}{\partial g^a_b} \mapsto v^b \frac{\partial}{\partial v^a}. $$

Therefore we have

$$ (2.20) \qquad V^n(A)_v = A^a_b v^b \frac{\partial}{\partial v^a}. $$

It is observed that if (A^a_b) is a singular matrix, we can take a non-zero $v = (v^a)$ satisfying $A^a_b v^b = 0$, so that $V^n(A)_v = 0$ (cf. Proposition 2.6). Of course, $V^n(A)_0 = 0$.

Further we are concerned with (2.18'): From (2.15), (2.20) and (2.10) we have

$$ {}_g\xi' V^n(A)_v = g^b_a A^a_c v^c \frac{\partial}{\partial v^b} = \{g^b_a A^a_c (g^{-1})^c_d\} g^d_e v^e \frac{\partial}{\partial v^b} = V^n(\mathrm{ad}(g)A)_{gv}. $$

In general, it is easy to show that a fundamental vector field $(V^n)^1_2(A)$, for instance, with respect to the Lie transformation group $((V^n)^1_2, G(n))$ is written

$$ (2.21) \qquad (V^n)^1_2(A)_w = (A^a_d w^d_{bc} - A^d_b w^a_{dc} - A^d_c w^a_{bd}) \frac{\partial}{\partial w^a_{bc}}, $$

where $w = w^a_{bc} e^{bc}_a \in (V^n)^1_2$.

<u>DEFINITION 2.6</u>. The *action* $\tau_A : w \in (V^n)^r_s \mapsto A\cdot w \in (V^n)^r_s$ of $A \in G'(n)$ on $(V^n)^r_s$ is given by $A\cdot w = I_w^{-1}\{(V^n)^r_s(A)\}$, where $I_w: (V^n)^r_s \mapsto ((V^n)^r_s)_w$ is defined by (1.5).

Thus (2.21) yields the explicit form of $A\cdot w$:

$$ (2.22) \qquad A\cdot w = (A^a_d w^d_{bc} - A^d_b w^a_{dc} - A^d_c w^a_{bd})\, e^{bc}_a. $$

From the definition of τ_A and Proposition 2.5 it follows that for a function f on $(V^n)^r_s$

$$(I_w(A\cdot w))(f) = (V^n)^r_s(A)_w(f) = d_0 t\{f\circ_{a_t(e)}\xi(w)\}.$$

On the other hand, Example 2.2 yields

$$(I_w(A\cdot w))(f) = (\overrightarrow{A\cdot w})_w(f) = d_0 t\{f(w + tA\cdot w)\}.$$

Consequently we get a geometrical meaning of A·w:

(2.22') $\qquad A\cdot w = d_0 t\{a_t(e)w\}.$

Finally, Propositions 2.2 and 2.5 and (2.16) lead to

$$[(V^n)^r_s(A),\ \vec{w}] = -\ d_0 t\{\overrightarrow{{}_{a_t(e)}\xi' w}\} = -\ d_0 t\{\overrightarrow{a_t(e)w}\}.$$

Thus (2.22') gives

(2.23) $\qquad [(V^n)^r_s(A),\ \vec{w}] = -\ \overrightarrow{A\cdot w}.$

§ 3. Fiber bundles

To give a global foundation of the theory of connections, we need the concept of fiber bundles. A connection is defined as a distribution satisfying some conditions in a principal bundle.

DEFINITION 3.1. A collection P(M) = (P, M, π, G) is called a *(differentiable) principal bundle* over M, if the following four conditions hold:

(1) G is a Lie transformation group of a manifold P with an action β: $(p, g) \in P\times G \mapsto pg \in P$ (on the right), and G is assumed to be effective.

(2) M is a manifold and identified with the quotient space P/G. The projection π: P $\to$ P/G is differentiable.

(3) With every point $x \in M$ is associated a pair (U, η) of a neighborhood U of x and a mapping η: $\pi^{-1}(U) \to G$, where $\pi^{-1}(U) = \{p \in P \mid \pi(p) \in U\}$.

(4) The mapping η satisfies an equation $\eta(pg) = \eta(p)g$ for any $p \in \pi^{-1}(U)$ and any $g \in G$, and the mapping $\phi = (\pi, \eta)$: $p \in \pi^{-1}(U) \mapsto (\pi(p), \eta(p)) \in U\times G$ is a diffeomorphism.

The manifold M is called the *base space*, P is the *total space* or *bundle space* and G is the *structure group*. π is called the *projection* of P(M) and η is the *transit* over U. The diffeomorphism ϕ is called the *local diffeomorphism* over U, by means of which every subset $\pi^{-1}(U)$ of P is identified with the product manifold U×G. Owing to this property, the total space P is called locally product. The subset $\pi^{-1}(x) = \{p \in P \mid \pi(p) = x\}$ for a fixed point $x \in M$ is called the *fiber* over x. This is nothing but the locus p/G of $p \in P$ contained in $\pi^{-1}(x)$.

From the action β of G on P we get a fundamental mapping ${}_p\beta$: $g \in G \mapsto pg \in P$ and a right-translation β_g: $p \in P \mapsto pg \in P$.

We are concerned with (U, η) in (3). Take $x \in U$ and put $p_o = \phi^{-1}(x, e)$, where e is the unit of G. Thus $\eta(p_o) = e$ and the restriction $\eta|_x$ of η to $\pi^{-1}(x)$ satisfies $\eta|_x(pg) = \eta(p_o)g = g$. Consequently, if G is assumed to be without fixed point, this $\eta|_x$ is onto and one-to-one, and the inverse $\eta|_x^{-1}$: $g \in G \mapsto p_o g \in \pi^{-1}(x)$ is differentiable, because it is the restriction of the transformation g to $\pi^{-1}(x)$. Therefore $\eta|_x$ is a diffeomorphism $\pi^{-1}(x) \to G$, by means of which every fiber $\pi^{-1}(x)$ is identified with G. It is remarked that this identification of $\pi^{-1}(x)$ with G depends on the choice of (U, η).

A *cross section* is by definition a mapping γ: $M \to P$ satisfying $\pi \circ \gamma$ = identity. It is remarked that there is not necessarily a cross section in any principal bundle, but there exists a so-called local cross section in any principal bundle. In fact, owing to the diffeomorphism ϕ: $\pi^{-1}(U) \to U \times G$, a mapping γ: $x \in U \mapsto \phi(x, g)$ for arbitrary fixed $g \in G$ is a cross section defined on U only.

REMARK 3.1. Throughout the following, owing to the fact stated in Proposition 2.6, the structure group G of a principal bundle P(M) is assumed to be without fixed point.

DEFINITION 3.2. In a principal bundle P(M) a tangent vector $X \in P_p$ of the total space P is called *vertical,* if $\pi' X = 0$ holds. The set of all the vertical tangent vectors at a point p is called the *vertical subspace* of P_p, denoted by P_p^v.

It is clear from the linearity of the differential π' that all the vertical tangent vectors at a point really constitute a vector space.

We now consider a fundamental vector field P(A) on P which is given by $P(A)_p = {}_p\beta'(A_e)$ from Definition 2.5. Since $\pi \circ {}_p\beta(g) = \pi(pg) = \pi(p)$ for any $g \in G$, the mapping $\pi \circ {}_p\beta$ is constant, so that $\pi' P(A) = 0$ from Proposition 1.1.

PROPOSITION 3.1. *Every fundamental vector* $P(A)_p$ *is vertical. Conversely, any vertical vector* $X \in P_p^v$ *is written in the form* $P(A)_p$ *for a certain* $A \in G'$.

Proof. We show the converse. For a point $p \in P$, let (U, η) be a pair in (3) of Definition 3.1, where $\pi(p) \in U$, and put $g = \eta(p) \in G$. Then we get a tangent vector $A_e = {}_{g^{-1}}\tau' \circ \eta' X \in G_e$, where ${}_g\tau$ is the left translation by g. From this A_e an element $A = I_G^{-1}(A_e)$ of G' is obtained. $X = P(A)_p$ is our conclusion. In fact, from $\eta \circ {}_p\beta \circ {}_{g^{-1}}\tau$ = identity we have

$$\eta' P(A)_p = \eta' \circ {}_p\beta' A_e = \eta' \circ {}_p\beta' \circ {}_{g^{-1}}\tau' \circ \eta' X = \eta' X.$$

Introducing two canonical mappings μ_1: $U \times G \to U$ and μ_2: $U \times G \to G$, we have $\mu_1 \circ \phi = \pi$,

$\mu_2 \circ \phi = \eta$. Therefore the above shows $\mu_2' \circ \phi'(P(A)_p - X) = 0$. On the other hand, we see $\mu_1' \circ \phi'(P(A)_p - X) = \pi'(P(A)_p - X) = 0$. Consequently $\phi'(P(A)_p - X) = 0$; so $P(A)_p - X = 0$, because ϕ is a diffeomorphism.

COROLLARY 3.1. *In a principal bundle P(M) having the structure group G without fixed point, the vertical subspace* P_p^v *has the base* $\{P(A_1)_p, \cdots, P(A_m)_p\}$, *corresponding to a base* $\{A_1, \cdots, A_m\}$ *of G'.*

It follows from Proposition 2.5 that every P(A) is induced from the one-parameter group of right translations $\{\beta_{a_t(e)}\}$, and the locus of p by all the right translations is nothing but the fiber through p. Therefore Proposition 3.1 leads to

COROLLARY 3.2. *The vertical subspace* P_p^v *is the tangent vector space of the fiber* $\pi^{-1}(x)$ *through p.*

DEFINITION 3.3. An indication $\delta: x \in M \mapsto \delta(x)$ (a subspace of M_x) on a manifold M is called a *distribution* in M. If every $\delta(x)$ has a common dimension r, then δ is called r-dimensional. Further, if we have locally r differentiable tangent vector fields which constitute a base of $\delta(x)$ at every point x, then δ is called differentiable.

As a consequence of Corollary 3.1 the distribution $P^v: p \in P \mapsto P_p^v$ is of dimension dimG and differentiable.

DEFINITION 3.4. The distribution $P^v: p \in P \mapsto P_p^v$ in the total space P of a principal bundle P(M) is called the *vertical distribution.*

We have a well-known construction theorem of a principal bundle over a given manifold. To construct a principal bundle, the concept of transition function is used effectively.

First, consider two transits η_α, η_β over neighborhoods U_α, U_β respectively. For a point $p \in \pi^{-1}(U_\alpha \cap U_\beta)$ we get two elements $\eta_\alpha(p)$, $\eta_\beta(p)$ of the structure group G. Introduce a mapping

$$\eta_{\beta\alpha} : x \in U_\alpha \cap U_\beta \mapsto \eta_\beta(p)\{\eta_\alpha(p)\}^{-1} \in G,$$

where p is an arbitrary point of the fiber $\pi^{-1}(x)$. This $\eta_{\beta\alpha}$ is well-defined, independently on the choice of p. These mappings $\eta_{\beta\alpha}$ are called the *transition functions* of P(M) under consideration. The relation

$$(3.1) \qquad \eta_{\gamma\beta}(x)\eta_{\beta\alpha}(x) = \eta_{\gamma\alpha}(x), \qquad x \in U_\alpha \cap U_\beta \cap U_\gamma$$

is obvious.

PROPOSITION 3.2. *If there exist an open covering* $\{U_\alpha\}$ *of a manifold M and a family of mappings* $\eta_{\beta\alpha}$: $U_\alpha \cap U_\beta \to G$, *where G is a Lie group and* $\eta_{\beta\alpha}$ *satisfy (3.1), then we can construct a principal bundle P(M) over M which has the structure group G and the transition functions* $\eta_{\beta\alpha}$. (The proof is omitted.)

We are now concerned with an n-dimensional manifold M. A (*linear*) *frame* z at a point x of M is by definition a base $z = \{z_a\}$, $a = 1, \cdots, n$, of the tangent vector space M_x, and the point x is called the *origin* of z. With reference to a coordinate system (x^i) on M, every tangent vector z_a at $x = (x^i)$ is written $z_a = z^i_a(\partial/\partial x^i)_x$ by a non-singular matrix $(z^i_a) \in G(n)$.

Let L be the set of all the linear frames at all the points of M. Then a mapping π_L: $L \to M$ is defined such that the image $\pi_L(z)$ is the origin x of z. Let $\delta(M) = \{(U, u)\}$ be the differentiable structure of M. Then we get a differentiable structure $\delta(L) = \{(\bar{U}, \bar{u})\}$ of L such that $\bar{U} = \pi_L^{-1}(U)$ and $\bar{u}$: $\bar{U} \to R^n \times G(n) = R^m$, $m = n + n^2$, defined by $\bar{u}(z) = (x^i, z^i_a)$ where (x^i) is the coordinate system of $\pi(z) = x$ with respect to the chart (U, u) and (z^i_a) is such as above shown. Thus L is an $(n + n^2)$-dimensional manifold and π_L: $(x^i, z^i_a) \mapsto (x^i)$ is clearly differentiable. This (x^i, z^i_a) is called the *induced coordinate system* from (x^i).

DEFINITION 3.5. Let L be the set of all the linear frames at all the points of a manifold M and π_L: $z \in L \mapsto x \in M$ where x is the origin of z. The collection $L(M) = (L, M, \pi_L, G(n))$ is called the *linear frame bundle* over M.

We have to show the structure of L(M) as a principal bundle. The general linear group G(n) is a Lie transformation group of L on the right with the action

(3.2) $\quad \underline{\beta}$: $(z, g) \in L \times G(n) \mapsto zg \in L, \qquad z = (z_a),\ g = (g^a_b), \qquad zg = (z_b g^b_a).$

The relation $(zg)h = z(gh)$, $g, h \in G(n)$, is obvious. Further $zg = z$ clearly implies $g = e$ because of $(z^i_a) \in G(n)$. Consequently G(n) acts on L without fixed point.

Then a right translation $\underline{\beta}_g$ is obtained. In terms of an induced coordinate system we have

(3.3) $\quad \underline{\beta}_g$: $z = (x^i, z^i_a) \mapsto zg = (x^i, z^i_b g^b_a).$

Further a fundamental mapping ${}_z\underline{\beta}$ is written

(3.4) $\quad {}_z\underline{\beta}$: $g = (g^a_b) \mapsto zg = (x^i, z^i_b g^b_a).$

The fiber $\pi_L^{-1}(x)$, namely, the set of all the linear frames with the origin x is a locus of a point $z = (x^i, z^i_a) \in \pi_L^{-1}(x)$ by the action $\underline{\beta}$. In fact, for a point $\bar{z} = (x^i, \bar{z}^i_a) \in \pi_L^{-1}(x)$ we have $\bar{z} = zg$ where $g = ((z^{-1})^a_i \bar{z}^i_b) \in G(n)$. Consequently M is regarded as the quotient space L/G(n).

Finally we take an admissible chart $(U, u) \in \delta(M)$ carrying a coordinate system (x^i), and a mapping

$$\eta : \pi_L^{-1}(U) \to G(n), \qquad z = (x^i, z_a^i) \mapsto (z_a^i).$$

The relation $\eta(zg) = \eta(z)g$ is obvious. The mapping

$$\phi = (\pi_L, \eta) : (x^i, z_a^i) \mapsto ((x^i), (z_a^i))$$

is clearly a diffeomorphism $\pi_L^{-1}(U) \to U\times G(n)$. Consequently we obtain the principal bundle $L(M)$.

We are concerned with the transition functions of $L(M)$ defined on $U\cap\bar{U}$, where U, $\bar{U}$ are coordinate neighborhoods. We have

$$\eta : \pi_L^{-1}(U) \to G(n), \qquad z = (x^i, z_a^i) \mapsto (z_a^i),$$

$$\bar{\eta} : \pi_L^{-1}(\bar{U}) \to G(n), \qquad z = (\bar{x}^i, \bar{z}_a^i) \mapsto (\bar{z}_a^i),$$

$$z_a = z_a^i \frac{\partial}{\partial x^i} = \bar{z}_a^j \frac{\partial}{\partial \bar{x}^j} = \bar{z}_a^j \frac{\partial x^i}{\partial \bar{x}^j}\frac{\partial}{\partial x^i},$$

which implies $z_a^i = (\partial x^i/\partial \bar{x}^j)\bar{z}_a^j$. Thus the transition function is given by

$$\bar{\eta}(z)\{\eta(z)\}^{-1} = (\bar{z}_a^i(z^{-1})_j^a) = \frac{\partial \bar{x}^i}{\partial x^j} \in G(n).$$

Consequently the transition function is nothing but the Jacobian matrix of the coordinate transformation $(x^i) \to (\bar{x}^i)$.

We consider the vertical subspace L_z^v in terms of an induced coordinate system (x^i, z_a^i). Any tangent vector $X \in L_z$ is written

$$X = X^i\left(\frac{\partial}{\partial x^i}\right)_z + X_a^i\left(\frac{\partial}{\partial z_a^i}\right)_z.$$

The projection π_L is given by $(x^i, z_a^i) \mapsto (x^i)$, so that the differential π_L' is

(3.5) $$\pi_L' : \left(\frac{\partial}{\partial x^i}\right)_z \mapsto \left(\frac{\partial}{\partial x^i}\right)_x, \qquad \left(\frac{\partial}{\partial z_a^i}\right)_z \mapsto 0.$$

Thus $\pi_L'(X) = X^i(\partial/\partial x^i)_x$ and X is vertical if and only if X is written in the form $X = X_a^i(\partial/\partial z_a^i)_z$.

Next we consider a fundamental vector field $\underline{Z}(A)$ on L, $A = (A_b^a) \in G'(n)$. The differential ${}_z\underline{\beta}'$ is given by

(3.4') $${}_z\underline{\beta}' : \frac{\partial}{\partial g_b^a} \mapsto z_a^i \frac{\partial}{\partial z_b^i}.$$

Thus the fundamental vector $\underline{Z}(A)_z = {}_z\underline{\beta}'(A_e)$ is written

(3.6) $$\underline{Z}(A)_z = z_a^i A_b^a \frac{\partial}{\partial z_b^i}, \qquad A = (A_b^a), \qquad z = (x^i, z_a^i).$$

For a given vertical vector $X = X_a^i\partial/\partial z_a^i$ we have $X = \underline{Z}(A)_z$ where $A_b^a = (z^{-1})_i^a X_b^i$, as it is shown in Proposition 3.1.

Next, from (3.3) the differential $\underline{\beta}'_g$ is written

$$\text{(3.3')} \qquad \underline{\beta}'_g : \frac{\partial}{\partial x^i} \mapsto \frac{\partial}{\partial x^i}, \qquad \frac{\partial}{\partial z^i_a} \mapsto g^a_b \frac{\partial}{\partial z^i_b}.$$

Then it is observed from (3.6) that

$$\underline{\beta}'_g Z(A)_z = z^i_a A^a_b g^b_c \frac{\partial}{\partial z^i_c} = (z^i_a g^a_e)\{(g^{-1})^e_d A^d_b g^b_c\} \frac{\partial}{\partial z^i_c},$$

so that (2.10) leads to

$$\text{(3.7)} \qquad \underline{\beta}'_g Z(A) = \underline{Z}(\mathrm{ad}(g^{-1})A),$$

although this is known from (2.18).

The total space L of L(M) has an important differential one-form. We first introduce a mapping

$$\text{(3.8)} \qquad \alpha : \pi_L^{-1}(x)\times V^n \to M_x, \qquad (z, v) \mapsto zv = z^i_a v^a \left(\frac{\partial}{\partial x^i}\right)_x,$$

where $x = (x^i)$, $z = (x^i, z^i_a)$ and $v = v^a e_a$ (cf. (2.12)). Since the left-fixed mapping ${}_z\alpha$ of α is given by

$$\text{(3.9)} \qquad {}_z\alpha : v = v^a e_a \in V^n \mapsto zv = z^i_a v^a \frac{\partial}{\partial x^i} \in M_x,$$

we have the inverse ${}_z\alpha^{-1}$ of ${}_z\alpha$ such that

$$\text{(3.10)} \qquad {}_z\alpha^{-1} : X = X^i \frac{\partial}{\partial x^i} \in M_x \mapsto z^{-1}X = \{(z^{-1})^a_i X^i\} e_a \in V^n.$$

DEFINITION 3.6. The *basic form* θ of an n-dimensional manifold M is a V^n-valued differential one-form on the total space L of the linear frame bundle L(M), given by $\theta_z = {}_z\alpha^{-1} \circ \pi'_L$ at $z \in L$.

We consider θ in terms of an induced coordinate system. Put

$$\theta = \theta^a e_a \in V^n, \qquad \theta^a = \theta^a_i dx^i + \theta^{ab}_i dz^i_b,$$

where θ^a are n real-valued differential one-forms on L. Then we see

$$X = X^i \frac{\partial}{\partial x^i} + X^i_a \frac{\partial}{\partial z^i_a} \in L_z, \qquad \pi'_L X = X^i \frac{\partial}{\partial x^i},$$

$${}_z\alpha^{-1} \circ \pi'_L X = \{(z^{-1})^a_i X^i\} e_a,$$

which gives $\theta^a_i X^i + \theta^{ab}_i X^i_b = (z^{-1})^a_i X^i$ for any X^i and X^i_a. Thus we get $\theta^a_i = (z^{-1})^a_i$ and $\theta^{ab}_i = 0$, so that

$$\text{(3.11)} \qquad \theta_z = \{(z^{-1})^a_i dx^i\} e_a.$$

We shall show the formula

(3.12) $$\theta\circ\beta'_{\underline{g}} = g^{-1}\theta,$$

where g^{-1} is the action of $g^{-1} \in G(n)$ on $\theta \in V^n$ given by (2.12). This means ${}_{zg}\alpha^{-1}\circ \pi'_L\circ\beta'_{\underline{g}} = {}_{g^{-1}}\xi\circ{}_z\alpha^{-1}\circ\pi'_L$. Since $\pi_L\circ\beta_{\underline{g}} = \pi_L$ is obvious, (3.12) is proved by ${}_{zg}\alpha^{-1} = {}_{g^{-1}}\xi\,{}_z\alpha^{-1}$, which is a consequence of

(3.13) $${}_{zg}\alpha = {}_z\alpha\circ{}_g\xi.$$

We have to be concerned with a strange concept "Finslerian tensor field" in future. The components of a Finslerian tensor field depend not only on a position x, but also on a direction $\dot{x}$; so it is not a tensor field in a strict sense. For a rigorous formulation of this concept, we here reconsider the concept of tensor field, based on the linear frame bundle.

We deal with a tensor field T of (r, s)-type on an n-dimensional manifold M. In the following we put r = s = 1 for brevity. Let L(M) be the linear frame bundle over M. From the tensor field T we define a $(V^n)^1_1$-valued function $\tilde{T}$ on the total space L by

(3.14) $$\tilde{T}_z(v_*,\ v) = T_x(zv_*,\ zv), \qquad x = \pi_L(z),$$

at $z \in L$ for any $v_* \in V^n_*$ and any $v \in V^n$, where $zv = \alpha(z, v)$ is defined by (3.8) and zv_* is similarly defined by

(3.8') $$zv_* = v_a(z^{-1})^a_i dx^i, \qquad z = (x^i,\ z^i_a), \qquad v_* = v_a e^a,$$

similarly in form to (2.13).

We write $\tilde{T}$ in terms of an induced coordinate system $(x^i,\ z^i_a)$ and components $(T^i_{\ j})$ of T with respect to the coordinate system (x^i):

$$T_x(zv_*,\ zv) = T^i_{\ j}\frac{\partial}{\partial x^i}\otimes dx^j\left(v_a(z^{-1})^a_h dx^h,\ z^k_b v^b\frac{\partial}{\partial x^k}\right) = \{(z^{-1})^a_i T^i_{\ j} z^j_b\}v_a v^b,$$

which gives

(3.14') $$\tilde{T}_z = T^a_{\ b}e^b_a, \qquad T^a_{\ b} = (z^{-1})^a_i T^i_{\ j} z^j_b.$$

We now examine the behavior of the function $\tilde{T}$ under a right translation $\beta_{\underline{g}}$. Since $(zg)v = z(gv)$ and $(zg)v_* = z(gv_*)$ are easily shown, we see

$$\tilde{T}_{zg}(v_*,\ v) = T_x((zg)v_*,\ (zg)v) = T_x(z(gv_*),\ z(gv)) = \tilde{T}_z(gv_*,\ gv).$$

Thus (2.14) leads to $\tilde{T}_{zg} = g^{-1}\tilde{T}_z$, i.e., $\tilde{T}\circ\beta_{\underline{g}} = g^{-1}\tilde{T}$.

<u>DEFINITION 3.7</u>. A *tensor function* $\tilde{T}$ *of (r, s)-type* in a manifold M of dimension n is a $(V^n)^r_s$-valued function on the total space L of the linear frame bundle L(M) which satisfies $\tilde{T}\circ\beta_{\underline{g}} = g^{-1}\tilde{T}$ for any $g \in G(n)$. The tensor function $\tilde{T}$, introduced by (3.14) from a tensor field T, is called the *associated tensor function with* T.

PROPOSITION 3.3. *There is a one-to-one correspondence between the set of all the tensor fields of (r, s)-type on a manifold M and the set of all the tensor functions of (r, s)-type in M; the correspondence is given by (3.14) and (3.15).*

Proof. The inverse correspondence "$\tilde{T} \to T$" is given by

$$T_x(X_*, X) = \tilde{T}_z(z^{-1}X_*, z^{-1}X), \qquad z \in \pi_L^{-1}(x), \tag{3.15}$$

for any $X_* \in (M_x)_*$ and any $X \in M_x$, where z is an arbitrary point of the fiber $\pi_L^{-1}(x)$, $z^{-1}X = {}_z\alpha^{-1}(X)$ (cf. (3.10)) and $z^{-1}X_*$ is similarly defined by

$$z^{-1}X_* = (X_i z^i_a)e^a \in (V^n)_*, \qquad z = (x^i, z^i_a), \qquad X_* = X_i dx^i. \tag{3.10'}$$

T_x as above is well-defined, independently on the choice of z, because

$$\tilde{T}_{zg}((zg)^{-1}X_*, (zg)^{-1}X) = g^{-1}\tilde{T}_z(g^{-1}(z^{-1}X_*), g^{-1}(z^{-1}X)),$$

which is equal to $\tilde{T}_z(z^{-1}X_*, z^{-1}X)$ from (2.14).

It is clear that (3.15) is the inverse correspondence of (3.14).

In particular, for a tangent vector field $X = X^i\partial/\partial x^i$, its associated tensor function $\tilde{X}$ is given by $\tilde{X}_z = (z^{-1})^a_i X^i e_a = z^{-1}X$.

REMARK 3.3. Proposition 3.3 shows that a tensor field may be identified with a tensor function. We can find various merits of the concept "tensor function". For instance, it is well-known that the geometrical differentiation, so-called covariant differentiation, of a tensor field is not a usual partial differentiation, but it is impressive that the former differentiation is simply a usual one of the associated tensor function by some vector field, called the basic vector field, as it will be shown in § 5. Further the concept of Lie derivative plays important roles in the theory of transformations, and the Lie derivative of T is simply a usual derivative of $\tilde{T}$ (cf. § 32).

We return to a general principal bundle P(M). Its structure group G acts on the total space P on the right, $\beta_g : p \mapsto pg$.

Now assume that this G also is a Lie transformation group of another manifold F on the left:

$$\gamma : (g, f) \in G\times F \mapsto gf \in F.$$

From γ we define

$$\bar{\gamma} : ((p, f), g) \in (P\times F)\times G \mapsto (pg, g^{-1}f) \in P\times F,$$

by means of which G becomes a Lie transformation group of the product manifold P×F. We denote by B the quotient space (P×F)/G; a point of B is an equivalence class b = {(p, f)} with respect to $\bar{\gamma}$. From {(pg, f)} = {(p, gf)} it follows that if we denote

$\{(p, f)\}$ by pf, we get $(pg)f = p(gf)$.

Next, introduce a mapping

$$\pi_B : pf \in B \mapsto \pi(p) \in M.$$

This is well-defined, independently on the choice of a point belonging to $\{(p, f)\}$, because $\pi_B((pg)(g^{-1}f)) = \pi(pg) = \pi(p)$. Referring to the first canonical mapping σ_1: $P\times F \to P$ and the canonical mapping α: $(p, f) \in P\times F \mapsto pf \in (P\times F)/G = B$, the above shows $\pi_B \circ \alpha = \pi \circ \sigma_1$. It is remarked that we again use the notation α(cf. (3.8)), but there is no confusion, as it will be shown later on.

DEFINITION 3.8. The collection $B(M) = (B, M, \pi_B, F, G)$ as above defined from $P(M)$ and F is called the *associated bundle with P(M) having the standard fiber F.*

In the above consideration the *total space* B of B(M) is only the quotient space $(P\times F)/G$, but it is shown that B is a differentiable manifold and all the mappings we consider are differentiable. (The proof is omitted.) The mapping π_B is called the *projection* of B(M), and the subspace $\pi_B^{-1}(x)$ of B for a fixed point $x \in M$ is the *fiber* over x. Every fiber is diffeomorphic with the standard fiber F.

Similarly to the case of P(M), a tangent vector $X \in B_b$ is called *vertical* if $\pi_B' X = 0$ holds, and the subspace of B_b consisting of all the vertical vectors is called the *vertical subspace* of B_b, denoted by B_b^v. Then, similarly to Corollary 3.2, we have $B_b^v = (\pi_B^{-1}(x))_b$. A left-fixed mapping $_p\alpha$: $F \to B$ of α is called an *admissible mapping* and a right-fixed mapping α_f: $P \to B$ is an *associated mapping.*

Let $L(M) = (L, M, \pi_L, G(n))$ be the linear frame bundle over an n-dimensional manifold M, and let $(V^n)^r_s$ be the tensor space of (r, s)-type constructed from the n-dimensional vector space V^n. The general linear group G(n) also is a Lie transformation group of $(V^n)^r_s$ with the action ξ^r_s given by (2.14).

DEFINITION 3.9. The associated bundle $T^r_s(M) = (T^r_s, M, \pi_T, (V^n)^r_s, G(n))$ with the linear frame bundle L(M), having the standerd fiber $(V^n)^r_s$, is called the *tensor bundle of (r, s)-type over the manifold M.*

To consider the bundle structure of $T^r_s(M)$ in detail, we shall restrict our consideration to the most important tensor bundle $T^1_0(M)$, called the *tangent (vector) bundle* and denoted by T(M) simply. We shall follow the procedure to construct the associated bundle. First, the action ξ of G(n) on V^n is extended to

$$\bar{\xi} : ((z, v), g) \in (L\times V^n)\times G(n) \mapsto (zg, g^{-1}v) \in L\times V^n.$$

The total space T of T(M) is the quotient space $(L\times V^n)/G(n)$ by this action $\bar{\xi}$. Then, introducing the first canonical mapping σ_1: $L\times V^n \to L$ and the canonical mapping α: $(z. v) \in L\times V^n \mapsto zv \in T$, we have $\pi_T \circ \alpha = \pi_L \circ \sigma_1$, where π_T: $zv \in T \mapsto \pi_L(z) \in M$ is the projection of T(M).

Recall the mapping α given by (3.8). This α clearly satisfies $(zg)(g^{-1}v) = zv$, so that the equivalence class $\{(z, v)\}$ is nothing but this tangent vector $y = zv$. Consequently we recognize that the total space T is the set of all the tangent vectors at all the points of M and the canonical mapping α is

(3.8") $$\alpha : (z, v) \in L\times V \mapsto y = zv \in T,$$
$$z = (x^i, z^i_a), \quad v = v^a e_a, \quad y = (x^i, y^i = z^i_a v^a).$$

Since $y = zv$ is a tangent vector at the origin x of z and $\pi_L(z) = x$, we get

(3.16) $$\pi_T : y \in T \mapsto x \in M, \qquad y \in M_x.$$

The differentiable structure $\delta(T)$ is naturally induced from $\delta(M)$ as follows: Let (U, u) be an admissible chart of M carrying a coordinate system (x^i). Then the chart (U^*, u^*) of T is defined by $U^* = \pi_T^{-1}(U)$ and $u^*(y) = (x^i, y^i)$ for $y = y^i\partial/\partial x^i \in U^*$, and $\delta(T)$ is the collection of such charts (U^*, u^*). The coordinate system (x^i, y^i) as thus defined is called the *induced coordinate system from* (x^i).

From the canonical mapping α we get an admissible mapping

(3.9') $${}_z\alpha : v = (v^a) \in V^n \mapsto y = zv = (x^i, y^i = z^i_a v^a) \in T.$$

The differential ${}_z\alpha'$ is given by

(3.9") $${}_z\alpha' : \frac{\partial}{\partial v^a} \mapsto z^i_a \frac{\partial}{\partial y^i}.$$

On the other hand, we get an associated mapping

(3.17) $$\alpha_v : z \in L \mapsto y = zv \in T, \qquad z = (x^i, z^i_a), \quad y = (x^i, z^i_a v^a).$$

The differential α'_v is given by

(3.17') $$\alpha'_v : \frac{\partial}{\partial x^i} \mapsto \frac{\partial}{\partial x^i}, \qquad \frac{\partial}{\partial z^i_a} \mapsto v^a \frac{\partial}{\partial y^i}.$$

Finally we shall introduce an isomorphism of the tangent vector space M_x with the vertical subspace T^v_y of the tangent vector space T_y where $y = zv$ is a point of the fiber $\pi_T^{-1}(x)$. For a tangent vector $X \in M_x$ the inverse mapping ${}_z\alpha^{-1} : \pi_T^{-1}(x) \to V^n$ of an admissible mapping ${}_z\alpha$ yields a vector ${}_z\alpha^{-1}(X) = z^{-1}X = u$ of the standard fiber V^n. Then we get a tangent vector $\vec{u}$ on V^n parallel to u. Therefore we get ${}_z\alpha'\vec{u}_v$ which is a vertical vector, because $\pi_T \circ {}_z\alpha$ is a constant mapping.

<u>DEFINITION 3.10</u>. The vertical vector ${}_z\alpha'(\overrightarrow{{}_z\alpha^{-1}X})_v$ at a point $y = zv \in T$ is called the *vertical lift* of $X \in M_x$ and denoted by $1^v_y(X)$.

The vertical lift $1^v_y(X)$ is well-defined, independently on the choice of $z \in \pi_L^{-1}(x)$, as it is easily verified from (3.13) and (2.16). Further the mapping $1^v_y : M_x \longrightarrow T^v_y$ is obviously an isomorphism.

We write the vertical lift $1^V_y(X)$ in terms of components. Putting $X = X^i\partial/\partial x^i$, from (3.10) and Example 2.2 we have

$$ {}_z\alpha^{-1}(X) = (z^{-1})^a_i X^i e_a, \qquad \overrightarrow{({}_z\alpha^{-1}X)}_v = (z^{-1})^a_i X^i \frac{\partial}{\partial v^a}, $$

Consequently (3.9") gives

$$ (3.18) \qquad 1^V_y(X) = X^i \frac{\partial}{\partial y^i}, \qquad X = X^i \frac{\partial}{\partial x^i}. $$

DEFINITION 3.11. The *intrinsic vertical vector field* 1^V is a vertical vector field on the total space T of the tangent bundle T(M) such that the value $(1^V)_y$ at $y \in T$ is equal to $1^V_y(y)$.

From (3.18) we immediately have $1^V = y^i\partial/\partial y^i$ at $y = (x^i, y^i)$.

§ 4. Linear connections

The concept of connections was introduced by generalizing Levi-Civita's parallelism in a Riemannian space. Although Cartan's method to define this concept was local, it has made much contribution to the progress of differential geometries. This concept was laid a rigorous and elegant foundation by Ehresmann, based on the theory of fiber bundles. Here is presented the concept of connections in a modern way, and the linear connections are considered in detail from the standpoint of later applications. Coordinate systems and components of tensors are not necessary to our theory, but these tools are very useful and show how to relate our theory to classical one.

We have the vertical distribution $P^V: p \in P \mapsto P^V_p$ in the total space P of a principal bundle P(M) = (P, M, π, G). Since P^V_p is spanned by fundamental vectors, we see $\dim P^V_p = \dim G' = \dim G$.

DEFINITION 4.1. A distribution $\Gamma: p \in P \mapsto \Gamma_p$ in the total space P of a principal bundle P(M) is called a *connection* in P or on the base space M, if the following two conditions hold:

(Γ 1) $\qquad P_p = \Gamma_p \oplus P^V_p \qquad$ (direct sum).

(Γ 2) $\qquad \beta'_g\Gamma = \Gamma \qquad$ for any $g \in G$.

The condition (Γ 2) says that Γ is invariant by any right translation β_g. It follows from (Γ 1) that $\dim \Gamma_p = \dim M$ and any tangent vector $X \in P_p$ is uniquely written

$$X = hX + vX, \qquad hX \in \Gamma_p, \quad vX \in P_p^v,$$

from which we have

(4.1) $\qquad \pi' \circ h = \pi'.$

If $vX = 0$, namely, $X \in \Gamma_p$, then X is called *horizontal* and Γ_p is the *horizontal subspace* of P_p. From (Γ 2) it is easy to show the commutation formulas

(4.2) $\qquad h \circ \beta'_g = \beta'_g \circ h, \qquad v \circ \beta'_g = \beta'_g \circ v.$

If hX or vX is differentiable for any differentiable vector field X on P, then Γ is called differentiable. Throughout the following, we shall be concerned with differentiable connections only.

DEFINITION 4.2. A G'-valued differential one-form ω on the total space P of a principal bundle P(M) with a structure group G is called a *connection form* , if the following two conditions hold :

(ω 1) $\qquad \omega(P(A)) = A \quad$ for any $A \in G'$.

(ω 2) $\qquad \omega \circ \beta'_g = ad(g^{-1})\omega \quad$ for any $g \in G$.

Assume that a connection Γ be given. Then we define a G'-valued one-form ω by

(ω 1) $\qquad \omega(P(A)) = A \quad$ for any $A \in G'$,

(ω 3) $\qquad \omega(X) = 0 \quad$ for any $X \in \Gamma$.

It follows from (Γ 1) and Proposition 3.1 that ω is well-defined by these two conditions. This ω is really a connection form, because from (2.18), (ω 1) and (Γ 2)

$$\omega \circ \beta'_g(P(A)) = \omega(P(ad(g^{-1})A)) = ad(g^{-1})A = ad(g^{-1})\omega(P(A)),$$

$$\omega \circ \beta'_g(X) = 0 = ad(g^{-1})\omega(X), \qquad X \in \Gamma_p,$$

which implies (ω 2) from (ω 1) and (ω 3). This ω is called the *connection form of the given connection* Γ.

Conversely a connection form ω induces a connection Γ, whose connection form is the given ω. Because, in the viewpoint of (ω 3), this Γ should be defined by

(Γ 3) $\qquad \Gamma_p = \{X \in P_p \mid \omega(X) = 0\}.$

This is really a connection. In fact, for any $X \in P_p$ it is observed from (ω 1) that we have $\omega(X - P(\omega(X))_p) = 0$, which implies a decomposition $X = Y + P(\omega(X))_p$, $Y \in \Gamma_p$, $P(\omega(X))_p \in P_p^v$. Further, if $X \in \Gamma_p \cap P_p^v$, then $\omega(X) = 0$ from $X \in \Gamma_p$, and from $X \in P_p^v$ there exists an $A \in G'$ such that $X = P(A)$. Thus $\omega(X) = A$; so $A = 0$ and $X = 0$. Consequently (Γ 1) is verified. Next, for $X \in \Gamma_p$, from (ω 2) we see $\omega \circ \beta'_g(X) = ad(g^{-1})\omega(X) = 0$, so that $\beta'_g(\Gamma_p) \subset \Gamma_{pg}$. Similarly we get $\beta'_{g^{-1}}(\Gamma_{pg}) \subset \Gamma_p$, namely, $\Gamma_{pg} \subset \beta'_g(\Gamma_p)$. Therefore ($\Gamma$ 2) is verified.

As a consequence of the one-to-one correspondence between the set of all the connections and the set of all the connection forms, the theory of connections will be developed by considering Γ or ω. We shall mainly consider Γ in the following.

DEFINITION 4.3. The *lift* $1_p(X)$ of a tangent vector $X \in M_x$ at a point $x = \pi(p)$ of the base space M is a horizontal vector at $p \in P$ satisfying $\pi'(1_p X) = X$.

First we must show the existence and uniqueness of $1_p(X)$. Recall a pair (U, η) and the diffeomorphism $\phi = (\pi, \eta)$ in Definition 3.1. We take any tangent vector $Y \in G_g$, $g = \eta(p)$, and construct $\bar{X} = (\phi^{-1})'(X, Y)$. Then $\phi'(\bar{X}) = (X, Y)$, $\pi'(\bar{X}) = X$ and the horizontal component $h\bar{X}$ is obviously equal to $1_p(X)$ from (4.1). If we have $\bar{X}_1$, $\bar{X}_2 \in \Gamma_p$ such that $\pi'(\bar{X}_1) = \pi'(\bar{X}_2) = X$, the horizontal vector $\bar{X}_1 - \bar{X}_2$ satisfies $\pi'(\bar{X}_1 - \bar{X}_2) = 0$, so that $\bar{X}_1 - \bar{X}_2$ is vertical. Thus the condition (Γ 1) shows $\bar{X}_1 = \bar{X}_2$.

The mapping $1_p: M_x \to \Gamma_p$ is obviously an isomorphism. From the condition (Γ 2) we get

(4.3) $$\beta'_g \circ 1_p = 1_{pg}$$

for any $p \in P$ and any $g \in G$.

DEFINITION 4.4. A curve $\bar{C}$: [0, 1] $\to$ P is called a *lift of a given curve* C: [0, 1] $\to$ M, if it satisfies

(1) $\bar{C}$ covers C, i.e., $\pi \circ \bar{C} = C$,

(2) $\bar{C}$ is horizontal, i.e., every tangent vector $\bar{C}'(d/dt)$ of $\bar{C}$ is horizontal.

It is obvious that the tangent vector $\bar{C}'(d/dt)$ of the lift of C is a lift of the tangent vector $C'(d/dt)$ of C.

PROPOSITION 4.1. *For a given curve C on the base space M, there exists a unique lift $\bar{C}$ issuing from a given point $p_o \in \pi^{-1}(C(0))$. The lift of C issuing from $p_o g$, $g \in G$, is given by $\beta_g \circ \bar{C}$.*

Proof. The first half will be shown, owing to the theory of ordinary differential equations of first order, if we refer to a coordinate system. The second half is evident from (4.3).

DEFINITION 4.5. Let C: $t \in [0, 1] \mapsto x(t) \in M$ be a curve of the base space M and p_o be a point of the fiber $\pi^{-1}(x(0))$. Every point p(t) of the lift $\bar{C}$: $t \in [0, 1] \mapsto p(t) \in P$ of C issuing from $p_o = p(0)$ is said to be obtained from p_o by *parallel displacement along C*.

Now a connection Γ in the linear frame bundle L(M) over a manifold M is called a *linear connection* on M. We consider a linear connection $\underline{\Gamma}$ in terms of an induced coordinate system.

The connection form $\underline{\omega}$ of $\underline{\Gamma}$ is written in the form $\underline{\omega} = \underline{\omega}^a_b E^b_a$ (cf. (2.9)), where the Lie algebra G'(n) is identified with $G(n)_e$. Every real-valued differential one-form $\underline{\omega}^a_b$ is written in terms of an induced coordinate system (x^i, z^i_a) as

$$\underline{\omega}^a_b = \underline{\omega}^a_{bi}(z)dx^i + \underline{\omega}^{ac}_{bi}(z)dz^i_c$$

at $z \in L$, where coefficients $\underline{\omega}^a_{bi}$, $\underline{\omega}^{ac}_{bi}$ are real-valued functions on L. We are first concerned with the condition (ω 1); for $A = (A^a_b)$ we get the fundamental vector $\underline{Z}(A)_z = z^i_a A^a_b \partial/\partial z^i_b$ from (3.6) and $\underline{\omega}^a_b(\underline{Z}(A)_z) = \underline{\omega}^{ac}_{bi} z^i_d A^d_c$, which must be equal to A^a_b for any A. Thus we get $\underline{\omega}^{ac}_{bi}(z) = (z^{-1})^a_i \delta^c_b$.

To consider the condition (ω 2), we are concerned with a mapping μ: M → N with a local expression $y^\alpha = y^\alpha(x^i)$ and a differential one-form $u = u_\alpha dy^\alpha$ on N. Then we obtain a differential one-form $u \circ \mu'$ on M such that $(u\circ\mu')(X) = u(\mu' X)$ for $X \in M_x$. Putting $X = X^i \partial/\partial x^i$, we get

$$(u\circ\mu')(X) = \left(u_\alpha dy^\alpha,\ X^i \frac{\partial y^\beta}{\partial x^i}\frac{\partial}{\partial y^\beta}\right) = u_\alpha \frac{\partial y^\alpha}{\partial x^i} X^i,$$

which shows $u\circ\mu' = u_\alpha(\partial y^\alpha/\partial x^i)dx^i$. Consequently we have

$$dy^\alpha \mapsto \frac{\partial y^\alpha}{\partial x^i} dx^i. \tag{4.4}$$

This is fundamental to construct $u\circ\mu'$ from $u = u_\alpha dy^\alpha$.

Applying this law to a right translation $\underline{\beta}_g$: $(x^i, z^i_a) \mapsto (x^i, z^i_b g^b_a)$, $g = (g^a_b)$, we get "$dx^i \mapsto dx^i$ and $dz^i_a \mapsto g^b_a dz^i_b$". Therefore

$$\underline{\omega}^a_b(zg)\circ\underline{\beta}'_g = \underline{\omega}^a_{bi}(zg)dx^i + \underline{\omega}^{ac}_{bi}(zg)g^d_c dz^i_d.$$

On the other hand, (2.10) leads to

$$\mathrm{ad}(g^{-1})\underline{\omega}^a_b(z) = (g^{-1})^a_c\{\ \underline{\omega}^c_{di}(z)dx^i + \underline{\omega}^{ce}_{di}(z)dz^i_e\ \}g^d_b.$$

Consequently (ω 2) means

$$\underline{\omega}^a_{bi}(zg) = (g^{-1})^a_c \underline{\omega}^c_{di}(z) g^d_b, \qquad \underline{\omega}^{ac}_{bi}(zg)g^d_c = (g^{-1})^a_c \underline{\omega}^{cd}_{ei}(z) g^e_b.$$

The latter holds from $\underline{\omega}^{ac}_{bi}(z) = (z^{-1})^a_i \delta^c_b$ as it has been shown. The former shows that the quantities

$$\underline{\Gamma}_k{}^j{}_i(z) = (z^{-1})^a_k \underline{\omega}^b_{ai}(z) z^j_b \tag{4.5}$$

satisfy $\underline{\Gamma}_k{}^j{}_i(zg) = \underline{\Gamma}_k{}^j{}_i(z)$, namely, $\underline{\Gamma}_k{}^j{}_i$ are constant on every fiber and regarded as functions on every coordinate neighborhood of the base space M.

Consequently $\underline{\omega}^a_b$ are written

$$(4.6) \qquad \underline{\omega}^a_b(z) = (z^{-1})^a_i\{ dz^i_b + z^j_b\underline{\Gamma}_j{}^i{}_k(x)dx^k \}, \qquad x = \pi_L(z).$$

These local functions $\underline{\Gamma}_j{}^i{}_k(x)$ are called the *connection coefficients* or *connection parameters* of the linear connection $\underline{\Gamma}$.

From (Γ 3) we get a general form of a horizontal vector:

$$(4.7) \qquad X = X^i\left(\frac{\partial}{\partial x^i} - z^j_a\underline{\Gamma}_j{}^k{}_i\frac{\partial}{\partial z^k_a}\right) \in \underline{\Gamma}_z.$$

This is also regarded as the lift $1_z(X)$ of $X = X^i\partial/\partial x^i$ to a point $z = (x^i, z^i_a)$.

DEFINITION 4.6. Let L(M) be the linear frame bundle over an n-dimensional manifold M with a linear connection $\underline{\Gamma}$. The *basic vector field* $\underline{B}(v)$, corresponding to $v \in V^n$, is a horizontal vector field on L, given by $\underline{B}(v)_z = 1_z(zv)$.

Here it is noted that V^n is the standard fiber of the tangent vector bundle T(M) and $zv = {}_z\alpha(v)$ is the image of v by the admissible mapping ${}_z\alpha$: $V^n \to T$.

We consider $\underline{B}(v)$ in terms of an induced coordinate system. Putting $z = (x^i, z^i_a)$ and $v = v^ae_a$, we have $zv = z^i_av^a\partial/\partial x^i$ from (3.9'), so that (4.7) yields

$$(4.8) \qquad \underline{B}(v)_z = z^i_av^a\left(\frac{\partial}{\partial x^i} - z^j_b\underline{\Gamma}_j{}^k{}_i\frac{\partial}{\partial z^k_b}\right).$$

For any $X \in \underline{\Gamma}_z$ of the form (4.7), we have a unique $v = (v^a)$ such that $X^i = z^i_av^a$. Therefore

PROPOSITION 4.2. *The horizontal subspace* $\underline{\Gamma}_z$ *with respect to a linear connection* $\underline{\Gamma}$ *is spanned by n basic vectors* $\underline{B}(e_a)_z$, *where* $B = \{e_a\}$ *is the fixed base of* V^n.

We now show two important equations

$$(4.9) \qquad \underline{\beta}'_g\underline{B}(v) = \underline{B}(g^{-1}v),$$

$$(4.10) \qquad \theta\ \underline{B}(v) = v.$$

From (4.3) it follows that

$$\underline{\beta}'_g\underline{B}(v)_z = \underline{\beta}'_g\circ 1_z(zv) = 1_{zg}((zg)(g^{-1}v)) = \underline{B}(g^{-1}v)_{zg},$$

which implies (4.9). Next, from Definition 3.6 we have

$$\theta\underline{B}(v)_z = {}_z\alpha^{-1}\circ\pi'_L\circ 1_z(zv) = {}_z\alpha^{-1}(zv) = v,$$

so that (4.10) is verified.

From the property of the connection form $\underline{\omega}$ and (4.10) it follows that the pair $(\underline{Z}(A), \underline{B}(v))$ is regarded as the dual of the pair $(\underline{\omega}, \theta)$ in the following sense:

$$(4.11) \qquad \begin{array}{ll} \underline{\omega}(\underline{Z}(A)) = A, & \underline{\omega}(\underline{B}(v)) = 0, \\ \theta(\underline{Z}(A)) = 0, & \theta(\underline{B}(v)) = v. \end{array}$$

PROPOSITION 4.3. *A horizontal vector* $X \in \underline{\Gamma}_z$ *with respect to a linear connection* $\underline{\Gamma}$ *is equal to the basic vector* $\underline{B}(v)_z$, *corresponding to* $v = \theta_z(X)$.

Proof. This $\underline{B}(v)_z$ is horizontal and

$$\pi'_{\underline{L}}\underline{B}(v)_z = zv = z({}_z\alpha^{-1}\circ\pi'_L X) = \pi'_L X,$$

which shows $X = \underline{B}(v)_z$ from the uniqueness of a lift.

We are concerned with the parallel displacement with respect to a linear connection $\underline{\Gamma}$. A curve $\bar{C}$: $t \in [0, 1] \mapsto z(t) = (x^i(t), z^i_a(t))$ of the total space L is horizontal if and only if the tangent vector $(dx^i/dt)\partial/\partial x^i + (dz^i_a/dt)\partial/\partial z^i_a$ is horizontal. Comparing this with (4.7), we get

$$\frac{dz^i_a}{dt} + z^j_a \underline{\Gamma}_j{}^i{}_k(x)\frac{dx^k}{dt} = 0. \tag{4.12}$$

This is the system of differential equations satisfied by a lift of a curve C: $t \in [0, 1] \mapsto x(t) = (x^i(t))$ of the base space M. Therefore the linear frame z(t) satisfying (4.12) is said to be obtained from a linear frame z(0) by parallel displacement along C.

DEFINITION 4.7. A *path* with respect to a linear connection $\underline{\Gamma}$ is the projection of an integral curve of a basic vector field $\underline{B}(v)$ onto the base space M, corresponding to a fixed $v \in V$.

Let $\bar{C}$: $t \in [0, 1] \mapsto z(t) = (x^i(t), z^i_a(t))$ be an integral curve of $\underline{B}(v)$ given by (4.8). Then we have

$$\frac{dx^i}{dt} = z^i_a v^a, \qquad \frac{dz^k_b}{dt} = -z^i_a v^a z^j_b \underline{\Gamma}_j{}^k{}_i.$$

Since v^a are constant, we get $d^2x^i/dt^2 = -z^j_c v^c z^k_a \underline{\Gamma}_j{}^i{}_k v^a$. Consequently the differential equations of a path are given by

$$\frac{d^2x^i}{dt^2} + \underline{\Gamma}_j{}^i{}_k(x)\frac{dx^j}{dt}\frac{dx^k}{dt} = 0. \tag{4.13}$$

We consider two linear connections $\underline{\Gamma}$ and $\underline{\Gamma}'$ on a same manifold M. Let $\underline{B}(v)$ and $\underline{B}'(v)$ be respective basic vector fields, corresponding to a same v. Because $\pi'_{\underline{L}}\underline{B}(v)_z = \pi'_{\underline{L}}\underline{B}'(v)_z = zv$, the difference $\underline{B}'(v) - \underline{B}(v)$ is vertical, so that we have a G(n)'-valued function D*(v), corresponding to v, such that

$$\underline{B}'(v) = \underline{B}(v) + \underline{Z}(D^*(v)). \tag{4.14}$$

From (4.9) and (3.7) we observe

$$\underline{\beta}'_g\underline{B}'(v)_z = \underline{B}'(g^{-1}v)_{zg} = \underline{B}(g^{-1}v)_{zg} + \underline{Z}(\mathrm{ad}(g^{-1})D^*(v)_z)_{zg}.$$

This and (4.14) implies $D^*(g^{-1}v)_{zg} = ad(g^{-1})D^*(v)_z$. Recalling the identification I of G'(n) with $(V^n)^1_1$, we get $\tilde{D}(v) = ID^*(v) \in (V^n)^1_1$ and (2.17) shows $\tilde{D}(g^{-1}v)_{zg} = g^{-1}\tilde{D}(v)_z$. If we define a $(V^n)^1_2$-valued function $\tilde{D}$ on L such that $\tilde{D}(u_*, u, v) = \tilde{D}(v)(u_*, u)$ for any $u_* \in V^n_*$ and any $u, v \in V^n$, we observe

$$\tilde{D}_{zg}(u_*, u, v) = \tilde{D}(v)_{zg}(u_*, u) = (g^{-1}\tilde{D}(gv)_z)(u_*, u)$$

$$= \tilde{D}_z(gu_*, gu, gv) = g^{-1}\tilde{D}_z(u_*, u, v),$$

which implies $\tilde{D}_{zg} = g^{-1}\tilde{D}_z$, namely, $\tilde{D}$ is a tensor function. Consequently we get the tensor field D of (1, 2)-type with which $\tilde{D}$ is associated.

DEFINITION 4.8. The tensor field D of (1, 2)-type, defined from D* of (4.14), is called the *difference tensor of* $\underline{\Gamma}'$ *from* $\underline{\Gamma}$.

PROPOSITON 4.4. *Assume that a linear connection* $\underline{\Gamma}$ *on a manifold M be fixed, and let* $\underline{B}(v)$ *be a basic vector field with respect to* $\underline{\Gamma}$. *Then there exists a one-to-one correspondence between the set* $\{\underline{\Gamma}'\}$ *of all the linear connections on M and the set {D} of all the tensor fields of (1, 2)-type of M; the correspondence is given by (4.14).*

The proof is easily shown by the tensor function property of D; the set $\{\underline{B}'(v) \mid v \in V\}$, defined by (4.14), gives rise to a linear connection $\underline{\Gamma}'$. From (4.14) and (4.8) the components $D_j{}^i{}_k$ of the difference tensor D are written

$$D_j{}^i{}_k = \underline{\Gamma}_j{}^i{}_k - \underline{\Gamma}'_j{}^i{}_k, \tag{4.15}$$

where the usage of indices of $D_j{}^i{}_k$ are

$$\tilde{D}(v)_z = \{D_b{}^a{}_c v^c\}e^b_a, \qquad D_j{}^i{}_k = z^i_a D_b{}^a{}_c (z^{-1})^b_j (z^{-1})^c_k, \qquad \text{(cf. (3.14'))}.$$

Let P(M) = (P, M, π, G) be a general principal bundle with a connection Γ, and let B(M) = (B, M, π_B, F, G) be an associated fiber bundle with P(M). By an associated mapping α_f: $p \in P \mapsto pf \in B$ we get a subspace

$$\Gamma^*_b = \alpha'_f(\Gamma_p), \qquad b = pf, \tag{4.16}$$

of the tangent space B_b. If the point b = pf is regarded as $(pg)(g^{-1}f)$, from the condition (Γ 2) of Definition 4.1, we have

$$\alpha'_{g^{-1}f}(\Gamma_{pg}) = \alpha'_{g^{-1}f}\circ\beta'_g(\Gamma_p) = \alpha'_f(\Gamma_p),$$

so that Γ^*_b is well-defined for a point $b \in B$.

DEFINITION 4.9. The distribution Γ^*: $b \in B \mapsto \Gamma^*_b$ in the total space B of the associated bundle B(M) as introduced by (4.16) is called the *associated connection with* Γ. A tangent vector $X \in \Gamma^*_b$ is called *horizontal*.

The direct sum property

$$B_b = \Gamma^*_b \oplus B^v_b \tag{4.17}$$

is essential, where B^v_b is the vertical subspace. To show this, we consider the lift 1_p with respect to Γ. For $X \in B_b$ we put $Y = \alpha'_f \circ 1_p \circ \pi'_B X$ which belongs to Γ^*_b from (4.16). Then

$$\pi'_B(X - Y) = \pi'_B X - \pi'_B \circ \alpha'_f \circ 1_p \circ \pi'_B X = \pi'_B X - \pi' \circ 1_p \circ \pi'_B X,$$

which is equal to zero from $\pi' \circ 1_p$ = identity. Thus $X = Y + Z$, $Z \in B^v_b$, which shows $B_b = \Gamma^*_b + B^v_b$. Further, if $X \in \Gamma^*_b \cap B^v_b$, from $X \in \Gamma^*_b$ we have $\bar{X} \in \Gamma_p$ such that $X = \alpha'_f \bar{X}$. Then $\pi'_B X = \pi'_B \circ \alpha'_f \bar{X} = \pi' \bar{X}$, which is equal to zero from $X \in B^v_b$, so that $\bar{X} \in P^v_p$ and hence $\bar{X} = 0$. Thus (4.17) is proved.

DEFINITION 4.10. Let $\bar{C}: t \in [0, 1] \mapsto p(t) \in P$ be a lift of a curve $C: t \in [0, 1] \mapsto x(t) \in M$. The curve $C^* = \alpha_f \circ \bar{C}: t \in [0, 1] \mapsto p(t)f \in B$ is called a *lift of C into B*. A point $b(t) = p(t)f$ of the lift C^* is said to be obtained from $b(0)$ by *parallel displacement along C*.

The lift C^* is horizontal with respect to the associated connection Γ^* because of $\bar{C}'(d/dt) \in \Gamma$ and $C^{*\prime}(d/dt) = \alpha'_f \circ \bar{C}'(d/dt) \in \Gamma^*$.

We are now concerned with the parallel displacement of a tangent vector along a curve C in M with respect to a linear connection $\underline{\Gamma}$. Let $\bar{C}(t) = (x^i(t), z^i_a(t))$ be a lift of $C(t) = (x^i(t))$ into L, which is given by (4.12). Then a lift C^* of C into the total space T of the tangent vector bundle T(M) is given by $\alpha_v \circ \bar{C}(t) = (x^i(t), y^i(t) = z^i_a(t)v^a)$, $v = (v^a)$. Thus (4.12) leads to

$$\frac{dy^i}{dt} = \frac{dz^i_a}{dt} v^a = - z^j_a v^a \underline{\Gamma}_j{}^i{}_k \frac{dx^k}{dt},$$

which implies

$$\frac{dy^i}{dt} + y^j \underline{\Gamma}_j{}^i{}_k(x) \frac{dx^k}{dt} = 0. \tag{4.18}$$

This is the system of differential equations for the parallel displacement of a tangent vector y(0) along $C(t) = (x^i(t))$.

We consider the associated connection $\underline{\Gamma}^*$. From (3.17') and (4.8) we have

$$\alpha'_v \underline{B}(u)_z = z^i_a u^a \left(\frac{\partial}{\partial x^i} - z^j_b \underline{\Gamma}_j{}^k{}_i v^b \frac{\partial}{\partial y^k} \right), \qquad u = (u^a).$$

Since the induced coordinate system of $y = zv$ is $(x^i, y^i = z^i_a v^a)$, the above is written

$$(4.19) \qquad \alpha'_{\underline{v}}\underline{B}(u)_z = z^i_a u^a\left(\frac{\partial}{\partial x^i} - y^j \underline{\Gamma}^k_{j\ i}\frac{\partial}{\partial y^k}\right).$$

It is clear that $\underline{\Gamma}^*_y$, y = zv, is spanned by such tangent vectors.

§ 5. Structure equations

Throughout the present section, we restrict our consideration to linear connections in the linear frame bundle L(M).

DEFINITION 5.1. Let K be a tensor field of (r, s)-type on a manifold M with a linear connection $\underline{\Gamma}$. The *covariant derivative* $\underline{\nabla}K$ of K with respect to $\underline{\Gamma}$ is a tensor field of (r, s + 1)-type on M, with which the tensor function $\underline{\nabla}\tilde{K}$ is associated, where $\underline{\nabla}\tilde{K}(v) = \underline{B}(v)\tilde{K}$ and $\tilde{K}$ is the associated tensor function with K.

To write the above equation explicitly, we deal with K of (1, 1)-type, for instance. The notations $\underline{\nabla}\tilde{K}(v) \in (V^n)^1_1$ and $\underline{\nabla}\tilde{K} \in (V^n)^1_2$ are used as

$$(5.1) \qquad \underline{\nabla}\tilde{K}(v)(u_*, u) = \underline{B}(v)\{\tilde{K}(u_*, u)\} = \underline{\nabla}\tilde{K}(u_*, u, v),$$

where $u, v \in V^n$ and $u_* \in V^n_*$. It is noted that v of $\underline{\nabla}\tilde{K}(v)$ is put at the last place in $\underline{\nabla}\tilde{K}(\cdots)$.

First of all, we have to show the tensor function property of $\underline{\nabla}\tilde{K}$, namely, $\underline{\nabla}\tilde{K}\circ\underline{\beta}_g = g^{-1}\underline{\nabla}\tilde{K}$ for any right translation $\underline{\beta}_g$. From (4.9), Definition 3.7 and (2.14) we observe

$$\begin{aligned}\underline{\nabla}\tilde{K}_{zg}\circ\underline{\beta}_g(u_*, u, v) &= \underline{B}(v)_{zg}\{\tilde{K}_{zg}(u_*, u)\} = \{\underline{\beta}'_g\underline{B}(gv)_z\}\{\tilde{K}_{zg}(u_*, v)\}\\ &= \underline{B}(gv)_z\{\tilde{K}_{zg}(u_*, u)\circ\underline{\beta}_g\} = \underline{B}(gv)_z\{g^{-1}\tilde{K}_z(u_*, u)\}\\ &= \underline{B}(gv)_z\{\tilde{K}_z(gu_*, gu)\} = \underline{\nabla}\tilde{K}_z(gu_*, gu, gv),\end{aligned}$$

which is equal to $g^{-1}\underline{\nabla}\tilde{K}_z(u_*, u, v)$, so that our assertion is verified.

In particular, we treat of the covariant derivative $\underline{\nabla}K$ of a tensor field $K = K^i_{\ j}\partial/\partial x^i \otimes dx^j$ of (1, 1)-type in terms of coordinate system. From (3.14') the associated tensor function $\tilde{K}$ with K is given by $\tilde{K} = \{(z^{-1})^a_i K^i_{\ j} z^j_b\}e^b_a$ at a point $z = (x^i, z^i_a)$. Putting $u_* = (u_a)$, $u = (u^a)$, (4.8) gives

$$\begin{aligned}\underline{B}(v)_z\tilde{K}(u_*, u) &= z^k_c v^c\left(\frac{\partial}{\partial x^k} - z^h_d \underline{\Gamma}^m_{h\ k}\frac{\partial}{\partial z^m_d}\right)\{(z^{-1})^a_i K^i_{\ j} z^j_b u_a u^b\}\\ &= z^k_c v^c\left[(z^{-1})^a_i\frac{\partial K^i_{\ j}}{\partial x^k}z^j_b - z^h_d\underline{\Gamma}^m_{h\ k}\{-(z^{-1})^a_m(z^{-1})^d_i K^i_{\ j} z^j_b + (z^{-1})^a_i K^i_{\ j}\delta^j_m\delta^d_b\}\right]u_a u^b\\ &= z^k_c v^c\left[(z^{-1})^a_i\frac{\partial K^i_{\ j}}{\partial x^k}z^j_b + K^h_{\ j}\underline{\Gamma}^m_{h\ k}(z^{-1})^a_m z^j_b - K^i_{\ m}\underline{\Gamma}^m_{h\ k}(z^{-1})^a_i z^h_b\right]u_a u^b\end{aligned}$$

$$= (z^k_c v^c)\{(z^{-1})^a_i u_a\}(z^j_b u^b)\left(\frac{\partial K^i_j}{\partial x^k} + K^h_j \Gamma^i_{h\,k} - K^i_h \Gamma^h_{j\,k}\right),$$

which shows

$$\nabla K = K^i_{j;k} \frac{\partial}{\partial x^i} \otimes dx^j \otimes dx^k,$$

(5.2)

$$K^i_{j;k} = \frac{\partial K^i_j}{\partial x^k} + K^h_j \Gamma^i_{h\,k} - K^i_h \Gamma^h_{j\,k}.$$

From the viewpoint of Corollary 3.1 and Proposition 4.2, the equation

$$\underline{Z}(A)\tilde{K} = - A\cdot\tilde{K} \tag{5.3}$$

is important, where $\tilde{K}$ is a tensor function. The proof of (5.3) is given by Proposition 2.5, Definition 3.7 and (2.22'):

$$\underline{Z}(A)\tilde{K} = d_0 t\{\tilde{K}\circ \underline{\beta}_{a_t(e)}\} = d_0 t\{a_t(e)^{-1}\tilde{K}\} = - A\cdot\tilde{K}.$$

We are concerned with Lie brackets of fundamental vector fields $\underline{Z}(A)$ and basic vector fields $\underline{B}(v)$ with respect to a linear connection $\underline{\Gamma}$. The first Lie bracket we consider is given by

$$[\underline{Z}(A),\ \underline{Z}(B)] = \underline{Z}([A,\ B]) \tag{5.4}$$

for two fixed elements $A,\ B \in G'(n)$. This is shown by Proposition 2.7 and has no relation to $\underline{\Gamma}$.

Secondly we consider $[\underline{Z}(A),\ \underline{B}(v)]$. From Propositions 2.2 and 2.5 and (4.9) we get

$$[\underline{Z}(A),\ \underline{B}(v)] = - d_0 t\{\underline{\beta}'_{a_t(e)}\underline{B}(v)\} = - d_0 t\{\underline{B}(a_t(e)^{-1}v)\},$$

so that (2.22') yields

$$[\underline{Z}(A),\ \underline{B}(v)] = \underline{B}(A\cdot v). \tag{5.5}$$

Finally we put

$$[\underline{B}(u),\ \underline{B}(v)] = \underline{Z}(\underline{R}^*(u,\ v)) + \underline{B}(\tilde{\underline{T}}(u,\ v)), \tag{5.6}$$

that is, the vertical component of $[\underline{B}(u),\ \underline{B}(v)]$ is the fundamental vector, corresponding to $\underline{R}^*(u,\ v) \in G'(n)$, and the horizontal component is the basic vector, corresponding to $\tilde{\underline{T}}(u,\ v) \in V^n$. Thus we obtain two functions $\underline{R}^*\colon L \to G'(n)\otimes(V^n)^0_2$ and $\tilde{\underline{T}}\colon L \to (V^n)^1_2$ such that $\underline{R}^*(u,\ v) \in G'(n)$ and $\tilde{\underline{T}}(u,\ v) \in V^n$ for any $u,\ v \in V^n$. From $\underline{R}^*$ we obtain the function $\tilde{\underline{R}}\colon L \to (V^n)^1_3$ by the identification $I\colon G'(n) \to (V^n)^1_1$. The equation (5.6) is called the *structure equation* of the linear connection $\underline{\Gamma}$.

The present purpose is to show that $\tilde{\underline{R}}$ and $\tilde{\underline{T}}$ are tensor functions. To do so, it is observed from (4.9) that

$$\underline{\beta}'_g[\underline{B}(u),\ \underline{B}(v)]_z = [\underline{\beta}'_g\underline{B}(u),\ \underline{\beta}'_g\underline{B}(v)]_{zg} = [\underline{B}(g^{-1}u),\ \underline{B}(g^{-1}v)]_{zg}$$

$$= \underline{Z}(\underline{R}^*_{zg}(g^{-1}u,\ g^{-1}v)) + \underline{B}(\tilde{\underline{T}}_{zg}(g^{-1}u,\ g^{-1}v)).$$

On the other hand, from (3.7) and (4.9) the above is equal to

$$\underline{\beta}'_g\underline{Z}(\underline{R}^*_z(u,\ v)) + \underline{\beta}'_g\underline{B}(\tilde{\underline{T}}_z(u,\ v)) = \underline{Z}(\mathrm{ad}(g^{-1})\underline{R}^*_z(u,\ v)) + \underline{B}(g^{-1}\tilde{\underline{T}}_z(u,\ v)).$$

Therefore we get

(5.7) $$\underline{R}^*_{zg}(g^{-1}u,\ g^{-1}v) = \mathrm{ad}(g^{-1})\underline{R}^*_z(u,\ v),$$

(5.8) $$\tilde{\underline{T}}_{zg}(g^{-1}u,\ g^{-1}v) = g^{-1}\tilde{\underline{T}}_z(u,\ v).$$

Now, (5.8) together with (2.13') and (2.14) show

$$\tilde{\underline{T}}_{zg}(u,\ u_*,\ v) = (\tilde{\underline{T}}_{zg}(u,\ v),\ u_*) = (g^{-1}\tilde{\underline{T}}_z(gu,\ gv),\ u_*) = (\tilde{\underline{T}}_z(gu,\ gv),\ gu_*)$$

$$= \tilde{\underline{T}}_z(gu,\ gu_*,\ gv) = (g^{-1}\tilde{\underline{T}}_z)(u,\ u_*,\ v),$$

which shows the tensor function property $\tilde{\underline{T}}\circ\underline{\beta}_g = g^{-1}\tilde{\underline{T}}$ of $\tilde{\underline{T}}$. The tensor field $\underline{T}$ of (1, 2)-type as thus obtained, with which $\tilde{\underline{T}}$ is associated, is called the *torsion tensor field* of the linear connection $\underline{\Gamma}$.

As to (5.7), from (2.17) it is written as $\tilde{\underline{R}}_{zg}(g^{-1}u,\ g^{-1}v) = g^{-1}\tilde{\underline{R}}_z(u,\ v)$ in terms of $\tilde{\underline{R}}$, so that the procedure similar to the case of $\tilde{\underline{T}}$ leads to the tensor field $\underline{R}$ of (1, 3)-type, with which $\tilde{\underline{R}}$ is associated. This $\underline{R}$ is called the *curvature tensor field* of $\underline{\Gamma}$.

We now apply the equations (5.5) and (5.6) to a tensor function $\tilde{K}$. First

(5.9) $$\underline{\nabla}\tilde{K}_{zg}(v) = g^{-1}\underline{\nabla}\tilde{K}_z(gv)$$

is easily show by Definition 5.1. Applying (5.5) to $\tilde{K}$, we get

$$[\underline{Z}(A),\ \underline{B}(v)]_z(\tilde{K}) = \underline{B}(A\cdot v)_z(\tilde{K}).$$

From (5.3) and Proposition 2.5 the left-hand side is written

$$\underline{Z}(A)_z(\underline{\nabla}\tilde{K}(v)) + \underline{B}(v)_z(A\cdot\tilde{K}) = d_0t\{\underline{\nabla}\tilde{K}_{za_t(e)}\} + A\cdot\underline{B}(v)_z(\tilde{K}).$$

From (5.9) and (2.22') this is equal to

$$d_0t\{a_t(e)^{-1}\underline{\nabla}\tilde{K}_z(a_t(e)v)\} + A\cdot\underline{\nabla}\tilde{K}_z(v) = \{-\ A\cdot\underline{\nabla}\tilde{K}_z(v) + \underline{\nabla}\tilde{K}_z(A\cdot v)\} + A\cdot\underline{\nabla}\tilde{K}_z(v),$$

which is equal to the right-hand side. Therefore the above equation is trivial.

Next, applying the structure equation (5.6) to $\tilde{K}$, we get

$$[\underline{B}(u),\ \underline{B}(v)](\tilde{K}) = \underline{Z}(\underline{R}^*(u,\ v))(\tilde{K}) + \underline{B}(\tilde{\underline{T}}(u,\ v))(\tilde{K}).$$

The left-hand side is equal to

$$\underline{B}(u)(\underline{\nabla}\tilde{K}(v)) - \underline{B}(v)(\underline{\nabla}\tilde{K}(u)) = \underline{\nabla}(\underline{\nabla}\tilde{K}(v))u - \underline{\nabla}(\underline{\nabla}\tilde{K}(u))v.$$

From (5.3) the right-hand side is equal to

$$-\ \underline{R}^*(u,\ v)\cdot\tilde{K} + \underline{\nabla}\tilde{K}(\tilde{\underline{T}}(u,\ v)).$$

Therefore, if we put $\underline{\nabla}\underline{\nabla}\tilde{K}(v, u) = \underline{\nabla}(\underline{\nabla}\tilde{K}(v))u$, we get

(5.10) $$\underline{\nabla}\underline{\nabla}\tilde{K}(u, v) - \underline{\nabla}\underline{\nabla}\tilde{K}(v, u) = \underline{R}^*(u, v)\cdot\tilde{K} - \underline{\nabla}\tilde{K}(\underline{\tilde{T}}(u, v)).$$

This is called the *Ricci identity* of the linear connection $\underline{\Gamma}$.

We consider the Ricci identity in terms of components. For this purpose we are concerned with a tensor field $K = K^i{}_j \partial/\partial x^i \otimes dx^j$.

$$\tilde{K}_z = K^a{}_b e^b_a \in (V^n)^1_1, \qquad K^a{}_b = (z^{-1})^a_i K^i{}_j z^j_b,$$

$$\underline{\nabla}\tilde{K}_z = K^a{}_{b;c} e^{bc}_a \in (V^n)^1_2, \qquad K^a{}_{b;c} = (z^{-1})^a_i K^i{}_{j;k} z^j_b z^k_c,$$

$$\underline{\nabla}\underline{\nabla}\tilde{K}_z(u, v) = (K^a{}_{b;c;d} u^c v^d) e^b_a \in (V^n)^1_1, \qquad K^a{}_{b;c;d} = (z^{-1})^a_i K^i{}_{j;k;h} z^j_b z^k_c z^h_d.$$

As to $\underline{R}^*$, we see

$$\underline{\tilde{R}}_z = \underline{R}_a{}^b{}_{cd} e^a_b \otimes e^{cd} \in (V^n)^1_3, \qquad \underline{R}_a{}^b{}_{cd} = (z^{-1})^a_i \underline{R}_h{}^i{}_{jk} z^h_b z^j_c z^k_d,$$

$$\underline{R}^*_z = \underline{R}_a{}^b{}_{cd} E^a_b \otimes e^{cd} \in G'(n)\otimes(V^n)^0_2,$$

$$\underline{R}^*_z(u, v) = \underline{R}_a{}^b E^a_b, \qquad \underline{R}_a{}^b = \underline{R}_a{}^b{}_{cd} u^c v^d,$$

$$\underline{R}^*_z(u, v)\cdot\tilde{K} = (\underline{R}_e{}^a K^e{}_b - \underline{R}_b{}^e K^a{}_e) e^b_a \in (V^n)^1_1.$$

As to $\underline{\tilde{T}}$, we see

$$\underline{\tilde{T}}_z = \underline{T}_b{}^a{}_c e^{bc}_a \in (V^n)^1_2, \qquad \underline{T}_b{}^a{}_c = (z^{-1})^a_i \underline{T}_j{}^i{}_k z^j_b z^k_c,$$

$$\underline{\tilde{T}}_z(u, v) = (\underline{T}_b{}^a{}_c u^b v^c) e_a \in V^n,$$

$$\underline{\nabla}\tilde{K}_z(\underline{\tilde{T}}(u, v)) = (K^a{}_{b;e} \underline{T}_c{}^e{}_d u^c v^d) e^b_a \in (V^n)^1_1.$$

Since u, v are arbitrary, we conclude that the Ricci identity is written in the form

(5.10') $$K^h{}_{i;j;k} - K^h{}_{i;k;j} = K^m{}_i \underline{R}_m{}^h{}_{jk} - K^h{}_m \underline{R}_i{}^m{}_{jk} - K^h{}_{i;m} \underline{T}_j{}^m{}_k.$$

We consider the components $\underline{R}_i{}^h{}_{jk}$ and $\underline{T}_j{}^i{}_k$ with respect to a coordinate system (x^i). If we apply (5.10') to a tangent vector field $X = X^i\partial/\partial x^i$, we get

$$X^i{}_{;j;k} - X^i{}_{;k;j} = X^h \underline{R}_h{}^i{}_{jk} - X^i{}_{;h} \underline{T}_j{}^h{}_k.$$

Direct calculation of the left-hand side of this equation leads to

(5.11) $$\underline{R}_h{}^i{}_{jk} = \mathfrak{A}_{(jk)}\left\{\frac{\partial \underline{\Gamma}_h{}^i{}_j}{\partial x^k} + \underline{\Gamma}_h{}^m{}_j \underline{\Gamma}_m{}^i{}_k\right\},$$

(5.12) $$\underline{T}_j{}^i{}_k = \mathfrak{A}_{(jk)}\{\underline{\Gamma}_j{}^i{}_k\}.$$

REMARK 5.1. In these equations we use the notation $\mathfrak{A}_{(jk)}$ to denote the interchange of indices j, k and subtraction. For instance,

$$\mathfrak{A}_{(jk)}\{A_i{}^r{}_j B_{rk}\} = A_i{}^r{}_j B_{rk} - A_i{}^r{}_k B_{rj} .$$

Throughout the following, we use this notation and also $\mathfrak{A}_{(12)}$.

We shall write more general equations than (5.4), (5.5) and (5.6), where $u, v \in V^n$ and $A, B \in G'(n)$ are assumed, of course, to be fixed. Now suppose that $u, v: L \to V^n$ and $A, B: L \to G'(n)$ are functions on L. Then, from (1.2) we easily get

$$(5.13) \qquad [\underline{Z}(A), \underline{Z}(B)] = \underline{Z}([A, B]) + \underline{Z}(\underline{Z}(A)B) - \underline{Z}(\underline{Z}(B)A),$$

$$(5.14) \qquad [\underline{Z}(A), \underline{B}(v)] = \underline{B}(A\cdot v) + \underline{B}(\underline{Z}(A)v) - \underline{Z}(\underline{B}(v)A),$$

$$(5.15) \qquad [\underline{B}(u), \underline{B}(v)] = \underline{Z}(\underline{R}^*(u, v)) + \underline{B}(\underline{\tilde{T}}(u, v)) + \underline{B}(\underline{B}(u)v) - \underline{B}(\underline{B}(v)u).$$

Now we apply the Jacobi identity (1.3) to fundamental vector fields and basic vector fields. We first show that two Jacobi identities

$$(5.16) \qquad [\underline{B}(v), [\underline{Z}(A_1), \underline{Z}(A_2)]] + \mathfrak{A}_{(12)}\{[\underline{Z}(A_1), [\underline{Z}(A_2), \underline{B}(v)]]\} = 0,$$

$$(5.17) \qquad [\underline{Z}(A), [\underline{B}(v_1), \underline{B}(v_2)]] + \mathfrak{A}_{(12)}\{[\underline{B}(v_1), [\underline{B}(v_2), \underline{Z}(A)]]\} = 0$$

reduce to trivial equations. In fact, from (5.4) and (5.5) the left-hand side of (5.16) is written

$$[\underline{B}(v), \underline{Z}([A_1, A_2])] + \mathfrak{A}_{(12)}\{[\underline{Z}(A_1), \underline{B}(A_2\cdot v)]\}$$

$$= - \underline{B}([A_1, A_2]\cdot v) + \mathfrak{A}_{(12)}\{\underline{B}(A_1\cdot(A_2\cdot v))\},$$

which is equal to zero from (2.11) and (2.22).

As to (5.17), from (5.6) and (5.5) the left-hand side is written

$$[\underline{Z}(A), \underline{Z}(\underline{R}^*(v_1, v_2)) + \underline{B}(\underline{\tilde{T}}(v_1, v_2))] - \mathfrak{A}_{(12)}\{[\underline{B}(v_1), \underline{B}(A\cdot v_2)]\}.$$

From (5.13), (5.14) and (5.6) this is rewritten

$$\{\underline{Z}([A, \underline{R}^*(v_1, v_2)]) + \underline{Z}(\underline{Z}(A)\underline{R}^*(v_1, v_2))\} + \{\underline{B}(A\cdot\underline{\tilde{T}}(v_1, v_2))$$
$$+ \underline{B}(\underline{Z}(A)\underline{\tilde{T}}(v_1, v_2))\} - \mathfrak{A}_{(12)}\{\underline{Z}(\underline{R}^*(v_1, A\cdot v_2)) + \underline{B}(\underline{\tilde{T}}(v_1, A\cdot v_2))\}.$$

This is of the form $\underline{Z}(\underline{A}) + \underline{B}(\underline{v})$ with

$$\underline{A} = [A, \underline{R}^*(v_1, v_2)] + \underline{Z}(A)\underline{R}^*(v_1, v_2) - \mathfrak{A}_{(12)}\{\underline{R}^*(v_1, A\cdot v_2)\},$$

$$\underline{v} = A\cdot\underline{\tilde{T}}(v_1, v_2) + \underline{Z}(A)\underline{\tilde{T}}(v_1, v_2) - \mathfrak{A}_{(12)}\{\underline{\tilde{T}}(v_1, A\cdot v_2)\}.$$

As for $\underline{A}$, putting $g = a_t(e)$, Proposition 2.5 and (5.7) lead to

$$\underline{Z}(A)_z\underline{R}^*(v_1, v_2) = d_0 t\{\underline{R}^*_{zg}(v_1, v_2)\} = d_0 t\{ad(g^{-1})\underline{R}^*_z(gv_1, gv_2)\}.$$

From (2.8) and (2.22') this is rewritten

$$- [A, \underline{R}^*_z(v_1, v_2)] + \underline{R}^*_z(A\cdot v_1, v_2) + \underline{R}^*_z(v_1, A\cdot v_2).$$

Consequently we have $\underline{A} = 0$.

Next, we consider $\underline{v}$. From Proposition 2.5, (5.8) and (2.22') we have

$$\underline{Z}(A)_{z}\underline{\tilde{T}}(v_1, v_2) = d_0 t\{\underline{\tilde{T}}_{zg}(v_1, v_2)\} = d_0 t\{g^{-1}\underline{\tilde{T}}_z(gv_1, gv_2)\}$$

$$= - A\cdot\underline{\tilde{T}}_z(v_1, v_2) + \underline{\tilde{T}}_z(A\cdot v_1, v_2) + \underline{\tilde{T}}_z(v_1, A\cdot v_2),$$

which shows $\underline{v} = 0$.

As a consequence, the Jacobi identity we must consider is only one:

$$(5.18) \qquad \mathfrak{S}_{(123)}\{[\underline{B}(v_1), [\underline{B}(v_2), \underline{B}(v_3)]]\} = 0.$$

REMARK 5.2. In (5.18) we use the notation $\mathfrak{S}_{(123)}$ to denote the cyclic permutation of indices 1, 2, 3 and summation. Throughout the following, we use this notation and also $\mathfrak{S}_{(ijk)}$ such that

$$\mathfrak{S}_{(ijk)}\{A^r{}_i B_{rhjk}\} = A^r{}_i B_{rhjk} + A^r{}_j B_{rhki} + A^r{}_k B_{rhij}.$$

From the structure equation (5.6) we first see

$$[\underline{B}(v_1), [\underline{B}(v_2), \underline{B}(v_3)]] = [\underline{B}(v_1), \underline{Z}(\underline{R}^*(v_2, v_3)) + \underline{B}(\underline{\tilde{T}}(v_2, v_3))],$$

and (5.14) and (5.15) show

$$= -\{\underline{B}(\underline{R}^*(v_2, v_3)\cdot v_1) + \underline{Z}(\underline{B}(v_1)\underline{R}^*(v_2, v_3))\} + \{\underline{Z}(\underline{R}^*(v_1, \underline{\tilde{T}}(v_2, v_3)))$$

$$+ \underline{B}(\underline{\tilde{T}}(v_1, \underline{\tilde{T}}(v_2, v_3))) + \underline{B}(\underline{B}(v_1)\underline{\tilde{T}}(v_2, v_3))\}.$$

Here it is observed that, referring to the identification I: $G'(n) \to (V^n)^1_1$,

$$\underline{R}^*(v_2, v_3)\cdot v_1 = \underline{\tilde{R}}(v_1, v_2, v_3) \in V^n,$$

$$\underline{B}(v_1)\underline{R}^*(v_2, v_3) = I^{-1}\underline{B}(v_1)\underline{\tilde{R}}(v_2, v_3) = I^{-1}\underline{\underline{\nabla\tilde{R}}}(v_2, v_3, v_1) \in G'(n),$$

$$\underline{R}^*(v_1, \underline{\tilde{T}}(v_2, v_3)) = I^{-1}\underline{\tilde{R}}(v_1, \underline{\tilde{T}}(v_2, v_3)) \in G'(n),$$

$$\underline{B}(v_1)\underline{\tilde{T}}(v_2, v_3) = \underline{\underline{\nabla\tilde{T}}}(v_2, v_3, v_1) \in V^n.$$

Therefore (5.18) yields

$$(5.19) \qquad \mathfrak{S}_{(123)}\{\underline{\underline{\nabla\tilde{R}}}(v_1, v_2, v_3) + \underline{\tilde{R}}(v_1, \underline{\tilde{T}}(v_2, v_3))\} = 0,$$

$$(5.20) \qquad \mathfrak{S}_{(123)}\{\underline{\underline{\nabla\tilde{T}}}(v_1, v_2, v_3) + \underline{\tilde{T}}(v_1, \underline{\tilde{T}}(v_2, v_3)) - \underline{\tilde{R}}(v_1, v_2, v_3)\} = 0.$$

These are called the *Bianchi identities with respect to the curvature tensor and the torsion tensor* respectively.

To write these Bianchi identities in the components, we observe, for instance,

$$\underline{\underline{\nabla\tilde{R}}}(v_1, v_2, v_3) = (\underline{R}_d{}^e{}_{ab;c}v_1^a v_2^b v_3^c)E_e^d \in G'(n),$$

$$\tilde{\underline{T}}(v_1, \tilde{\underline{T}}(v_2, v_3)) = \{\underline{T}_a{}^d{}_e v_1^a (\underline{T}_b{}^e{}_c v_2^b v_3^c)\} e_d \in V^n.$$

Thus we get

$$(5.19') \qquad \mathfrak{S}_{(ijk)}\{ \underline{R}_h{}^m{}_{ij;k} + \underline{R}_h{}^m{}_{ir}\underline{T}_j{}^r{}_k \} = 0,$$

$$(5.20') \qquad \mathfrak{S}_{(ijk)}\{ \underline{T}_i{}^m{}_{j;k} + \underline{T}_i{}^m{}_r \underline{T}_j{}^r{}_k - \underline{R}_i{}^m{}_{jk} \} = 0.$$

At the end of the present chapter, we are concerned with the Riemannian connections. An n-dimensional *Riemannian space* M is by definition a manifold equipped with a *Riemannian metric* $ds^2 = g_{ij}(x)dx^i dx^j$, where $g_{ij}(x)$ are components of a symmetric tensor field of (0, 2)-type. Then the inner product (X, Y) of two tangent vectors $X = X^i\partial/\partial x^i$, $Y = Y^i\partial/\partial x^i$ at a point $x \in M$ is defined by $(X, Y) = g_{ij}(x)X^iY^j$. The so-called *fundamental tensor field* $g = (g_{ij})$ is assumed to satisfy the regularity condition: $\det(g_{ij}) \neq 0$. In almost all the geometrical theories the positive definiteness condition is imposed: $g_{ij}X^iX^j > 0$ for any $X \neq 0$.

PROPOSITON 5.1. *There exists a unique linear connection γ on a Riemannian space M with the fundamental tensor field $g = (g_{ij})$ which satisfies the following two conditions:*

(1) γ is metrical, i.e, $\underline{\nabla} g = 0$.

(2) γ is without torsion, i.e., $\underline{T} = 0$.

This connection γ is called the *Riemannian connection* on M. If we denote by $\gamma_j{}^i{}_k$ the connection coefficients of γ, the above two conditions are respectively written

$$(5.21) \qquad \frac{\partial g_{ij}}{\partial x^k} - g_{hj}\gamma_i{}^h{}_k - g_{ih}\gamma_j{}^h{}_k = 0,$$

$$(5.22) \qquad \underline{T}_j{}^i{}_k = \gamma_j{}^i{}_k - \gamma_k{}^i{}_j = 0.$$

Here we introduce an useful process, called the *Christoffel process with respect to i, j, k,* by means of which from quantities A_{hijk}, for instance, we make

$$A_{hijk} + A_{hjki} - A_{hkij}.$$

Now, applying the Christoffel process with respect to i, j, k to the left-hand side of (5.21) and paying attention to (5.22), we immediately obtain

$$(5.23) \qquad \gamma_j{}^i{}_k = \frac{1}{2} g^{hi}\left(\frac{\partial g_{hj}}{\partial x^k} + \frac{\partial g_{hk}}{\partial x^j} - \frac{\partial g_{jk}}{\partial x^h}\right),$$

where (g^{hi}) is the inverse matrix of (g_{ij}). The connection coefficients $\gamma_j{}^i{}_k$ as thus determined are called the *Christoffel symbols*.

We consider the curvature tensor $\underline{R}$ of the Riemannian connection γ. If we introduce the covariant components $\underline{R}_{hijk} = g_{im}\underline{R}_h{}^m{}_{jk}$ of $\underline{R}$, the conditions (1), (2) and (5.10') yield

$$g_{hi;j;k} - g_{hi;k;j} = 0 = - g_{mi}\underline{R}_h{}^m{}_{jk} - g_{hm}\underline{R}_i{}^m{}_{jk},$$

which shows $\underline{R}_{hijk} + \underline{R}_{ihjk} = 0$. This and (5.20') lead to

$$(5.24) \qquad \underline{R}_{hijk} = - \underline{R}_{ihjk} = - \underline{R}_{hikj}, \qquad \mathfrak{S}_{(ijk)}\{\underline{R}_{hijk}\} = 0.$$

Further (5.19') and (1) give

$$(5.25) \qquad \mathfrak{S}_{(ijk)}\{\underline{R}_{hmij;k}\} = 0.$$

It is well-known that the set (5.24) of identities is the complete set of identities satisfied by $\underline{R}_{hijk}$, and (5.24) yields very useful identity

$$(5.26) \qquad \underline{R}_{hijk} = \underline{R}_{jkhi}.$$

CHAPTER II

FINSLER CONNECTIONS

In this chapter we shall give an elegant and comprehensive foundation of connections on Finsler spaces which have been defined by several authors from their own standpoints. Differential-geometric concepts and quantities are mainly introduced in the principal bundle over the tangent bundle induced from the linear frame bundle. The torsion and curvature tensors of line-element spaces are obtained as a natural generalization of those of a linear connection.

Although our theory can be satisfactorily developed without reference to any coordinate systems and components of tensors, these classical tools are used in the following to show close relations between our theory and classical one.

§ 6. Finsler bundles

We are first concerned with the concept of induced bundle. Let $P(M) = (P, M, \pi, G)$ be a principal bundle and let $\tau: N \to M$ be a mapping of a manifold N into the base space M of P(M). A subspace Q of the product manifold $N\times P$ is defined by $Q = \{(n, p) \in N\times P \mid \tau(n) = \pi(p)\}$. Then we may obtain another principal bundle (Q, N, π_1, G) over the manifold N with the same structure group G, which is called the *induced bundle from P(M) by* τ and denoted by $\tau^{-1}P(M)$.

DEFINITION 6.1. Let $L(M) = (L, M, \pi_L, G(n))$ be the linear frame bundle over an n-dimensional manifold M and $\pi_T: T \to M$ be the projection of the tangent bundle $T(M) = (T, M, \pi_T, V^n, G(n))$ over M. Then the induced bundle $\pi_T^{-1}L(M) = (F, T, \pi_1, G(n))$ is called the *Finsler bundle* of M and denoted by F(M).

We exactly show the bundle structure of the Finsler bundle F(M), instead of showing that of a general induced bundle. The total space F of F(M) is given by

(6.1) $$F = \{(y, z) \in T\times L \mid \pi_T(y) = \pi_L(z)\},$$

that is, a point $u = (y, z) \in F$ is a pair of a tangent vector $y \in M_x$ at a point x of M and a linear frame z with the origin x. The action β of G(n) on the total space F is essentially equal to the action $\underline{\beta}$ of G(n) on L, namely, $\beta: (u = (y, z), g) \in F\times G(n) \mapsto ug = (y, zg) \in F$, where $zg = \underline{\beta}(z, g)$. A right translation β_g of F, $g \in G(n)$, is given by

(6.2) $$\beta_g: u = (y, z) \in F \mapsto ug = (y, zg) \in F.$$

On the other hand, a fundamental mapping ${}_u\beta$, $u \in F$, is written

$$(6.3)\qquad {}_u\beta\colon g \in G(n) \mapsto ug \in F.$$

From Definition 2.5 a fundamental vector field $Z(A)$ of F, corresponding to $A \in G'(n)$, is given by

$$(6.4)\qquad Z(A)_u = {}_u\beta'(A),$$

and from (2.18) we have

$$(6.5)\qquad \beta_g' Z(A) = Z(ad(g^{-1})A),$$

similar to (3.7).

Next, the projection $\pi_1\colon F \to T$ of the Finsler bundle $F(M)$ is simply given by

$$(6.6)\qquad \pi_1\colon u = (y, z) \in F \mapsto y \in T,$$

and another simple mapping

$$(6.7)\qquad \pi_2\colon u = (y, z) \in F \mapsto z \in L$$

is called the *induced mapping*. As for these mappings, we easily have

$$(6.8)\qquad \begin{array}{llll} (1) & \pi_1 \circ \beta_g = \pi_1, & (2) & \pi_2 \circ \beta_g = \underline{\beta}_g \circ \pi_2, \\ (3) & \pi_1 \circ {}_u\beta = \text{const.}, & (3) & \pi_2 \circ {}_u\beta = {}_{z}\underline{\beta}, \qquad z = \pi_2(u). \end{array}$$

From (6.4) and (4) of (6.8) we get

$$(6.9)\qquad \pi_2' Z(A) = \underline{Z}(A),$$

where the right-hand side with bar is a fundamental vector field on L.

Let (x^i) be a coordinate system on M, and let (x^i, y^i) and (x^i, z^i_a) be respective coordinate systems on T and L, induced from (x^i). Then a coordinate system (x^i, y^i, z^i_a) of a point $u = (y, z) \in F$ is obtained, where $y = (x^i, y^i)$ and $z = (x^i, z^i_a)$. This coordinate system on F is also called the *induced coordinate system* from (x^i).

In terms of the induced coordinate system (x^i, y^i, z^i_a), a right translation β_g is written

$$(6.2')\qquad \beta_g\colon u = (x^i, y^i, z^i_a) \mapsto ug = (x^i, y^i, z^i_b g^b_a), \qquad g = (g^a_b),$$

and its differential β_g' is given by

$$(6.2'')\qquad \beta_g'\colon \frac{\partial}{\partial x^i} \mapsto \frac{\partial}{\partial x^i}, \quad \frac{\partial}{\partial y^i} \mapsto \frac{\partial}{\partial y^i}, \quad \frac{\partial}{\partial z^i_a} \mapsto g^a_b \frac{\partial}{\partial z^i_b}.$$

As to a fundamental mapping ${}_u\beta$, we obtain

$$(6.3')\qquad {}_u\beta\colon g = (g^a_b) \mapsto ug = (x^i, y^i, z^i_b g^b_a), \qquad u = (x^i, y^i, z^i_a),$$

and its differential ${}_u\beta'$ is given by

$$(6.3'')\qquad {}_u\beta'\colon \frac{\partial}{\partial g^a_b} \mapsto z^i_a \frac{\partial}{\partial z^i_b}.$$

Therefore (6.4) and (2.9) immediately give

$$(6.4') \qquad Z(A)_u = z^i_a A^a_b \frac{\partial}{\partial z^i_b} .$$

From the basic form θ on L, given by Definition 3.6, is induced a V^n-valued differential one-form

$$(6.10) \qquad \theta^h = \theta \circ \pi'_2$$

on F, which is called the *h-basic form*. An important equation

$$(6.11) \qquad \theta^h \circ \beta'_g = g^{-1}\theta^h$$

is derived from (3.12) and (2) of (6.8).

REMARK 6.1. There are many papers which develop Finsler geometry and its generalizations based on fiber bundles. It seems that the Finsler bundle F(M) was perhaps first considered by Auslander [1] in 1955. Akbar-Zadeh has published various papers on Finsler geometry since 1954 and gave a somewhat elegant formulation of Finsler geometry in his paper [7] based on the Finsler bundle. In 1960 Matsumoto [1] explicitly introduced the concept of Finsler bundle to formulate Finsler geometry, although his formulation was not complete yet. He arrived at a firm formulation of Finsler geometry in 1963 [2] by the help of Okada [1] and developed it systematically in his monograph [12].

In Chapter I we saw the theory of linear connections which is developed in the linear frame bundle L(M). A point of L is thought of as a pair (x, z) of a point x of a manifold M and a linear frame z at x. Thus it may be said that the ordinary differential geometry of M is successfully treated by adding the space of linear frames to M. On the other hand, according to Cartan's viewpoint [4], Finsler geometry of M is to study the space of line-elements (x, y). If the homogeneity in y is neglected, the space of line-elements is nothing but the total space T of the tangent bundle T(M). Therefore it will be natural to think that, by adding a linear frame z, the consideration of spaces of elements (x, y, z) will be an effective way toward studying Finsler geometry. In the present section the space of (x, y, z) has been exactly formulated as the induced bundle $\pi_T^{-1}L(M) = F(M)$.

We have many papers which treat Finsler geometry and its generalizations in such a modern way as above. For example, Dazord [7], Grifone [3, 5], Ichijyo [1], Kashiwabara [1], Kurita [2], Ōtsuki [2], Soós [6], Tashiro [1] and Vilms [3, 4].

DEFINITION 6.2. A *Finsler tensor field K of (r, s)-type* on an n-dimensional manifold M is a $(V^n)^r_s$-valued function on the total space F of the Finsler bundle F(M) which satisfies a condition $K \circ \beta_g = g^{-1}K$ for any $g \in G(n)$.

Therefore a Finsler tensor field is considered as an quantity similar to a tensor function from the standpoint of Definition 3.7. We have to show that Definition 6.2 is a rigorous formulation of a strange concept of tensor field appearing in the geometry of line-element space. Let K be a Finsler tensor field of (1, 1)-type, for instance. With respect to the base $B^1_1 = \{e^a_b\}$ of $(V^n)^1_1$ (cf. Remark 2.3), K is written in the form $K = K^a_b e^b_a$, where K^a_b are real-valued functions on F. The condition in Definition 6.2 is written as

$$K^a_b(ug) = (g^{-1})^a_c K^c_d(u) g^d_b, \qquad g = (g^a_b),$$

which is equivalent to the fact that the quantities K^i_j defined by

$$(6.12) \qquad K^i_j(u) = z^i_a K^a_b(u)(z^{-1})^b_j, \qquad u = (x^i, y^i, z^i_a),$$

satisfy $K^i_j(ug) = K^i_j(u)$ (cf. (3.14')), so that K^i_j are regarded as functions of variables x^i and y^i only. These K^i_j are nothing but the components of K in the classical sense. Similarly to (5.3), an important equation

$$(6.13) \qquad Z(A)K = -\,A\cdot K$$

is easily shown for a Finsler tensor field K.

REMARK 6.2. If we introduce a natural generalization $G'(n)\otimes(V^n)^r_s \to (V^n)^1_1\otimes(V^n)^r_s = (V^n)^{r+1}_{s+1}$ of the identification $I: G'(n) \to (V^n)^1_1$, we can be concerned with the concept of Finsler tensor field of (r+1, s+1)-type which is a $G'(n)\otimes(V^n)^r_s$-valued function on F. Cf. the curvature $\underline{R}^*$ in (5.6).

EXAMPLE 6.1. Let $\tilde{K}: L \to (V^n)^r_s$ be the associated tensor function with a tensor field K on the base space. Then the function $\bar{K} = \tilde{K}\circ\pi_2$ on the Finsler bundle F satisfies $\bar{K}\circ\beta_g = g^{-1}\bar{K}$ from (2) of (6.8), so that $\bar{K}$ is a Finsler tensor field, called the *induced tensor field from K*. In the classical sense $\bar{K}$ is the same with K.

DEFINITION 6.3. A Finsler tensor field ε of (1, 0)-type, given by $\varepsilon: u = (y, z) \in F \mapsto z^{-1}y \in V^n$, is called the *supporting element*.

Here $z^{-1}y = {}_z\alpha^{-1}(y)$ (cf. (3.10)), the image of $y \in T$ by the inverse of admissible mapping ${}_z\alpha: V^n \to T$. The tensor property of ε is easily shown by (3.13). Since $\varepsilon: (x^i, y^i, z^i_a) \mapsto (z^{-1})^a_i y^i e_a$, it follows from (6.12) that y^i are nothing but the components of ε. This is the reason for calling ε "supporting element" which is due to Cartan [4].

The supporting element ε will play various important roles in Finsler geometry. Here we shall show one of them.

PROPOSITION 6.1. *A mapping*

$$I = (\pi_2, \varepsilon): u = (y, z) \in F \mapsto (z, z^{-1}y) \in L\times V^n$$

gives an isomorphism of F with the product manifold $L\times V^n$.

This is obvious, because the inverse I^{-1} of I is defined by

$$(6.14)\qquad I^{-1}:\ (z,\ v)\in L\times V^n \mapsto (zv,\ z)\in F.$$

In terms of an induced coordinate system $(x^i,\ y^i,\ z^i_a)$ the differential ε' of ε is given by

$$(6.15)\qquad \varepsilon':\ \begin{array}{l} \dfrac{\partial}{\partial x^i} \mapsto 0, \qquad \dfrac{\partial}{\partial y^i} \mapsto (z^{-1})^a_i \dfrac{\partial}{\partial v^a}, \\ \dfrac{\partial}{\partial z^i_a} \mapsto -(z^{-1})^a_j y^j (z^{-1})^b_i \dfrac{\partial}{\partial v^b}, \end{array}$$

where (v^a) is the linear coordinate system of V^n induced from the fixed base B. Then the differential I' of I is written

$$(6.16)\qquad I':\ \begin{array}{l} \dfrac{\partial}{\partial x^i} \mapsto \left(\dfrac{\partial}{\partial x^i},\ 0\right), \qquad \dfrac{\partial}{\partial y^i} \mapsto \left(0,\ (z^{-1})^a_i \dfrac{\partial}{\partial v^a}\right), \\ \dfrac{\partial}{\partial z^i_a} \mapsto \left(\dfrac{\partial}{\partial z^i_a},\ -(z^{-1})^a_j y^j (z^{-1})^b_i \dfrac{\partial}{\partial v^b}\right), \end{array}$$

and the differential $(I^{-1})'$ of the inverse I^{-1} is

$$(6.14')\qquad (I^{-1})':\ \begin{array}{l} \left(\dfrac{\partial}{\partial x^i},\ 0\right) \mapsto \dfrac{\partial}{\partial x^i}, \qquad \left(\dfrac{\partial}{\partial z^i_a},\ 0\right) \mapsto v^a \dfrac{\partial}{\partial y^i} + \dfrac{\partial}{\partial z^i_a}, \\ \left(0,\ \dfrac{\partial}{\partial v^a}\right) \mapsto z^i_a \dfrac{\partial}{\partial y^i}. \end{array}$$

The isomorphism $I:\ F \to L\times V^n$ gives rise to an alternative interpretation of mappings and quantities relating to F. Those of $L\times V^n$, corresponding to them by I, are denoted by putting the symbol $\times$:

$$(6.17)\qquad \begin{array}{ll} (1) & \overset{\times}{\pi}_1 = \pi_1\circ I^{-1}:\ (z,\ v)\in L\times V^n \mapsto zv\in T, \\ (2) & \overset{\times}{\pi}_2 = \pi_2\circ I^{-1}:\ (z,\ v)\in L\times V^n \mapsto z\in L, \\ (3) & \overset{\times}{\beta}_g = I\circ\beta_g\circ I^{-1}:\ (z,\ v)\in L\times V^n \mapsto (zg,\ g^{-1}v)\in L\times V^n, \\ (4) & {}_{(z,v)}\overset{\times}{\beta} = I\circ{}_u\beta:\ g\in G(n) \mapsto (zg,\ g^{-1}v)\in L\times V^n, \qquad u = I^{-1}(z,\ v), \\ (5) & \overset{\times}{\varepsilon} = \varepsilon\circ I^{-1}:\ (z,\ v)\in L\times V^n \mapsto v\in V^n. \end{array}$$

The mapping $\overset{\times}{\pi}_1$ is nothing but the canonical mapping α (cf. (3.8)) and $\overset{\times}{\beta}_g$ is equal to the action of G(n) on $L\times V^n$. Further, mappings $\overset{\times}{\pi}_2$ and $\overset{\times}{\varepsilon}$ are solely canonical mappings of the product manifold $L\times V^n$ into its components.

From (1) of (6.17) we have $\pi_1 = \alpha\circ I$, so that $\pi'_1 = \alpha'\circ(\pi_2,\ \varepsilon)'$. Therefore at $u = I^{-1}(z,\ v)$ we get

(6.18) $$\pi_1' = \alpha_v'\circ\pi_2' + {}_z\alpha'\circ\varepsilon',$$

which shows the difference of differentials of two mappings π_1 and $\alpha_v\circ\pi_2$: $F \to T$. On the other hand, it is obvious that $\pi_T\circ\pi_1 = \pi_L\circ\pi_2$: $F \to M$.

PROPOSITION 6.2. *A tangent vector $X \in F_u$ is uniquely determined by each one of (X_2, X_3) and (X_1, X_2) as follows:*

(1) $X_2 = \pi_2'X$ and $X_3 = \varepsilon'X$ such that $X = (I^{-1})'(X_2, X_3)$.

(2) $X_1 = \pi_1'X$ and X_2 such that $X = (I^{-1})'(X_2, ({}_z\alpha^{-1})'(X_1 - \alpha_v'X_2))$ at $u = I^{-1}(z, v)$.

This is clear from Proposition 6.1 and (6.18). As a corollary of Proposition 6.2 we obtain

PROPOSITION 6.3. *A tangent vector $X \in F_u$ vanishes, if and only if (1) $\pi_2'X = 0$ and $\varepsilon'X = 0$, or (2) $\pi_1'X = 0$ and $\pi_2'X = 0$.*

We apply Proposition 6.2 to a fundamental vector $Z(A)$. Then $X_1 = \pi_1'Z(A) = 0$ is clear and $X_2 = \pi_2'Z(A) = \underline{Z}(A)$ from (6.9). It follows from (6.13) that $Z(A)\varepsilon = - A\cdot\varepsilon$, so that (1.6) gives $\varepsilon'Z(A) = I_v(- A\cdot\varepsilon)$ at $u = I^{-1}(z, v)$. Therefore from Example 2.2 and (1) of Proposition 6.2 we get

(6.19) $$Z(A) = (I^{-1})'(\underline{Z}(A), - \overrightarrow{A\cdot\varepsilon}).$$

REMARK 6.3. The notion of supporting element ε as above defined appears in Matsumoto [1], in which it is called the characteristic vector, but the complete recognition of its importance is first seen in Matsumoto [12].

§ 7. Distributions in the Finsler bundles

The Finsler bundle $F(M) = (F, T, \pi_1, G(n))$ is a principal bundle over the tangent bundle $T(M) = (T, M, \pi_T, V^n, G(n))$. Therefore we have first the *vertical distribution*

$$F^v: u \in F \mapsto F_u^v = \{X \in F_u \mid \pi_1'X = 0\}$$

on F, where F_u^v is the vertical subspace of F_u. There are other important subspaces of F_u which are naturally obtained from the structure of induced bundle. We show those subspaces in the following.

DEFINITION 7.1. The *quasi-vertical subspace* F_u^q of the tangent space F_u is given by

$$F^q_u = \{X \in F_u \mid \pi'_T \circ \pi'_1 X = 0\}.$$

The distribution F^q: $u \in F \mapsto F^q_u$ is called the *quasi-vertical distribution* on F.

A *quasi-vertical vector* $X \in F^q_u$ is characterized by another equation $\pi'_L \circ \pi'_2 X = 0$. It is obvious that F^v_u is contained in F^q_u.

DEFINITION 7.2. The *induced-vertical subspace* F^i_u of the tangent space F_u is given by

$$F^i_u = \{X \in F_u \mid \pi'_2 X = 0\}.$$

The distribution F^i: $u \in F \mapsto F^i_u$ is called the *induced-vertical distribution* on F.

It is obvious from $\pi_T \circ \pi_1 = \pi_L \circ \pi_2$ that F^i_u is contained in F^q_u. The vertical subspace F^v_u is spanned by fundamental vector fields from Corollary 3.1. We shall introduce important vector fields on F, by which the induced-vertical subspace F^i_u is spanned.

DEFINITION 7.3. The *induced-fundamental vector field* Y(v), corresponding to $v \in V^n$, is given by the properties

$$(1) \quad \pi'_2 Y(v) = 0, \qquad (2) \quad \varepsilon' Y(v) = \vec{v}.$$

At a point $u \in F$ the condition (2) means $\varepsilon' Y(v)_u = I_{\varepsilon(u)}(v)$ from Example 2.2. It follows from Proposition 6.2 that Y(v) is written as $Y(v) = (I^{-1})'(0, \vec{v})$. Because an induced-vertical vector $X \in F^i_u$ has a form $X = (I^{-1})'(0, \vec{X_3})$ from Proposition 6.2 and for the fixed base $B = \{e_a\}$ of V^n we obtain a base $\{\vec{e}_a\}$ of a tangent space of V^n at every point, it is clear that $\{Y(e_a)\}$ is a base of F^i_u.

PROPOSITION 7.1. *An induced-fundamental vector field Y(v) satisfies*

$$(1) \quad \pi'_1 Y(v)_u = {}_z\alpha'(\vec{v}) = l^v_y(zv), \qquad u = (y, z),$$

$$(2) \quad Y(v)\varepsilon = v, \qquad (3) \quad \beta'_g Y(v) = Y(g^{-1}v).$$

Proof. (1) follows from (6.18) and Definition 3.10. (2) is a consequence of (1.6) and (2) of Definition 7.3. (3) is shown from (2) of (6.8), the tensor property of ε and (2.16) as follows:

$$\pi'_2(\beta'_g Y(v)) = \beta'_g \circ \pi'_2 Y(v) = 0,$$

$$\varepsilon'(\beta'_g Y(v)) = (g^{-1}\varepsilon)' Y(v) = {}_{g^{-1}}\xi' \vec{v} = \overrightarrow{g^{-1}v},$$

so that Definition 7.3 leads to (3).

From Proposition 2.5 a fundamental vector field Z(A) generates the one-parameter group $\{\beta_{a_t(e)}\}$ of right translations. On the other hand, the one-parameter transformation groups which is generated by an induced-fundamental vector field is given as follows:

PROPOSITION 7.2. *An induced-fundamental vector field Y(v) on F generates a one-parameter transformation group $\{S_{tv}\}$, where the transformation S_v of F, corresponding to $v \in V^n$, is defined by $(y, z) \mapsto (y + zv, z)$.*

Proof. Putting the symbol × on S_v as in (6.17), a natural transformation of L×V arises as follows:

$$S_v^{\times} = I \circ S_v \circ I^{-1} : (z, w) \in L\times V \mapsto (z, w + v) \in L\times V. \tag{7.1}$$

Therefore Example 2.2 and Definition 7.3 complete the proof.

From Corollary 2.1 we get

$$S_w' Y(v) = Y(v) \tag{7.2}$$

for any $w \in V^n$, and from Corollary 2.2

$$[Y(v_1), Y(v_2)] = 0 \tag{7.3}$$

for any $v_1, v_2 \in V^n$.

Consider a Lie bracket [Z(A), Y(v)], applying Proposition 6.2. From (6.9) and Definition 7.3 we have

$$\pi_2'[Z(A), Y(v)] = [\underline{Z}(A), 0] = 0,$$

and from (6.19), Definitions 7.3 and 2.6 and further (2.23)

$$\varepsilon'[Z(A), Y(v)] = [-\overrightarrow{A\cdot\varepsilon}, \vec{v}] = -[V^n(A), \vec{v}] = \overrightarrow{A\cdot v}.$$

Therefore Proposition 6.2 and Definition 7.3 yield

$$[Z(A), Y(v)] = Y(A\cdot v), \tag{7.4}$$

which is quite similar in form to (5.5).

PROPOSITION 7.3. *The quasi-vertical subspace F_u^q is the direct sum of the induced-vertical subspace F_u^i and vertical subspace F_u^v: $F_u^q = F_u^i \oplus F_u^v$.*

Proof. It is obvious that both F_u^i and F_u^v are contained in F_u^q and that $F_u^i \cap F_u^v$ is only zero because of (2) of Proposition 6.3. Next, for any $X \in F_u^q$, $u = (y, z)$, we have a vertical vector $X_1 = \pi_1' X \in T_y^v$; so there exists $v \in V^n$ such that $X_1 = {}_z\alpha'\vec{v}$, and (1) of Proposition 7.1 shows that $Z = X - Y(v)$ is vertical. Thus X is written as a sum of the vertical vector Z and the induced-vertical vector Y(v).

In terms of an induced coordinate system (x^i, y^i, z^i_a) we obtain an expression of $Y(v)$:

$$(7.5) \qquad Y(v)_u = z^i_a v^a \frac{\partial}{\partial y^i}, \qquad v = (v^a), \quad u = (x^i, y^i, z^i_a),$$

which easily follows from $Y(v) = (I^{-1})'(0, v^a\partial/\partial v^a)$ and (6.14'). Therefore the induced-vertical subspace is locally spanned by $\partial/\partial y^i$, while the vertical subspace is done by $\partial/\partial z^i_a$.

REMARK 7.1. The concept of induced-fundamental vector field appears in Matsumoto and Okada [1]. Before then, in Matsumoto [4] the *intrinsic induced-fundamental vector field* $Y(\varepsilon) = y^i(\partial/\partial y^i)_u$ plays a role in considering the C_2-condition (cf. § 13).

We are concerned with a connection Γ in the Finsler bundle $F(M)$ and its connection form ω. The connection Γ is defined as a distribution $u \in F \mapsto \Gamma_u \subset F_u$ which satisfies the two conditions

$$(7.6) \qquad (1) \quad F_u = \Gamma_u \oplus F^v_u, \qquad (2) \quad \beta'_g \Gamma_u = \Gamma_{ug}.$$

The connection form ω of Γ is given by

$$(7.7) \qquad (1) \quad \omega Z(A) = A, \qquad (2) \quad \omega\Gamma = 0,$$

and satisfies

$$(7.8) \qquad \omega\circ\beta'_g = ad(g^{-1})\omega.$$

In terms of an induced coordinate system (x^i, y^i, z^i_a) the connection form ω is written

$$\omega = \omega^a_b E^b_a \in G'(n), \qquad \omega^a_b = \omega^a_{bi}dx^i + \omega^a_{b(i)}dy^i + \omega^{ac}_{bi}dz^i_c.$$

In a similar way by means of which (4.6) is derived, from (1) of (7.7) and (7.8) it is seen that $\omega^{ac}_{bi} = (z^{-1})^a_i\delta^c_b$ and that quantities

$$\Gamma_j{}^i{}_k(u) = z^i_a\omega^a_{bk}(u)(z^{-1})^b_j, \qquad C_j{}^i{}_k(u) = z^i_a\omega^a_{b(k)}(u)(z^{-1})^b_j$$

are really functions of x^i, y^i alone. Therefore we conclude

$$(7.7') \qquad \omega^a_b = (z^{-1})^a_i\{dz^i_b + z^j_b(\Gamma_j{}^i{}_k dx^k + C_j{}^i{}_k dy^k)\}.$$

Then the equation (2) of (7.7) yields an expression of a horizontal vector $X \in \Gamma_u$ as follows:

$$(7.9) \qquad X = X^i\left(\frac{\partial}{\partial x^i} - \Gamma_k{}^j{}_i z^k_b\frac{\partial}{\partial z^j_b}\right) + X^{(i)}\left(\frac{\partial}{\partial y^i} - C_k{}^j{}_i z^k_b\frac{\partial}{\partial z^j_b}\right),$$

where X^i and $X^{(i)}$ are arbitrary numbers.

REMARK 7.2. It will be shown in § 17 that the quantities $\Gamma_j{}^i{}_k$ and $C_j{}^i{}_k$ in (7.7') coincide with those introduced by Cartan [4], if we restrict our consideration to the connection due to Cartan. Akbar-Zadeh [3, 7], Matsumoto [1] and almost all the authors who have studied Finsler geometry based on fiber bundles consider a Finsler connection only as a connection Γ in F(M) with some regularity in a sense. In particular, Akbar-Zadeh [7] treats even the holonomy group of a Finsler space as the one of this Γ. It seems, however, that such a formulation of connections appearing in the classical theories of Finsler spaces are a little rough. The regularity is only a pretext for the sake of convenience of coming computations. In such a formulation the only way to show the tensor property of $C_j{}^i{}_k$ in (7.7') will be the classical one, i.e., to perform a coordinate transformation. See § 8.

§ 8. Various connections relating to the Finsler bundles

To define the concept of Finsler connection, we shall here introduce two distributions Γ^h, Γ^v in the Finsler bundle F(M), a distribution N in the tangent bundle T(M) and a family of distributions Γ_V in the linear frame bundle L(M).

DEFINITION 8.1. A *horizontal connection* Γ^h in the Finsler bundle F(M) is a distribution $u \in F \mapsto \Gamma^h_u \subset F_u$ which satisfies

$$(1) \quad F_u = \Gamma^h_u \oplus F^q_u, \qquad (2) \quad \beta'_g \Gamma^h_u = \Gamma^h_{ug}.$$

It is noted that F^q in (1) is the quasi-vertical distribution. From Proposition 7.3 we obtain a decomposition of F_u as follows:

$$(8.1) \qquad F_u = \Gamma^h_u \oplus F^i_u \oplus F^v_u.$$

If the manifold M is of dimension n, then Γ^h_u, F^i_u and F^v_u are of dimension n, n and n^2 respectively, while a connection Γ in F(M) is of 2n dimensions.

DEFINITION 8.2. A *non-linear connection* N in the tangent bundle T(M) is a distribution $y \in T \mapsto N_y \subset T_y$ such that $T_y = N_y \oplus T^v_y$.

From (4.17) it is observed that the concept of non-linear connection is a generalization of the connection associated with a connection in a principal bundle.

REMARK 8.1. This modern formulation of a non-linear connection appears in Barthel [11], where he further imposes a homogeneity condition. Cf. (5) of Definition 12.1 and Proposition 12.4.

Applying the differential π'_1 of the projection π_1: F $\to$ T to (8.1), we have

$$T_y = \pi_1'(\Gamma_u^h) \oplus \pi_1'(F_u^i), \qquad y = \pi_1(u),$$

and Proposition 7.1 gives $\pi_1'(F_u^i) = T_y^v$, so that we have

$$T_y = \pi_1'(\Gamma_u^h) \oplus T_y^v.$$

Further, from (2) of Definition 8.1 it follows that $\pi_1'(\Gamma_{ug}^h) = \pi_1'(\Gamma_u^h)$, that is, $\pi_1'(\Gamma_u^h)$ is uniquely determined at each point $y \in T$. Consequently we have

PROPOSITION 8.1. *The projection $\pi_1'(\Gamma^h)$ into T of a horizontal connection Γ^h in $F(M)$ is a non-linear connection in $T(M)$.*

We denote the lifts with respect to Γ^h and $N = \pi_1'(\Gamma^h)$ by 1_u^h and 1_y respectively. It is seen that $1_u^h \circ 1_y(X) \in \Gamma_u^h$ for any $X \in M_x$, $y = \pi_1(u)$, $x = \pi_T(y)$.

The h-basic form θ^h on F is introduced in § 6. It is defined without relation to any connection. We now introduce another V^n-valued differential one-form on F:

DEFINITION 8.3. With respect to a non-linear connection N, the *v-basic form* θ^v on $F(M)$ is a V^n-valued differential one-form such that $\theta_u^v = I^{-1} \circ ({}_z\alpha^{-1})' \circ v_N \circ \pi_1'$ at a point $u = (y, z) \in F$, where we denote by $v_N(X)$ for $X \in T_y$ the vertical component of X with respect to N.

From Definition 3.10 it follows that θ^v is also written in another form

$$\theta_u^v = z^{-1} \circ (1_y^v)^{-1} \circ v_N \circ \pi_1'. \tag{8.2}$$

Similarly to (6.11) an equation

$$\theta^v \circ \beta_g' = g^{-1}\theta^v \tag{8.3}$$

is easily shown from (3.13) and (2.16). In particular, if N is the projection of a horizontal connection Γ^h, then $\theta^v(X) = 0$ for $X \in \Gamma_u^h$.

In terms of an induced coordinate system (x^i, y^i, z_a^i) we put

$$\theta^v = (\theta^v)^a e_a \in V^a, \qquad (\theta^v)^a = \theta_i^a dx^i + \theta_{(i)}^a dy^i + \theta_i^{ab} dz_b^i.$$

Given a vertical vector $X = X_a^i \partial/\partial z_a^i$, we obtain $(\theta^v)^a(X) = \theta_i^{ab} X_b^i = 0$, so that $\theta_i^{ab} = 0$. Next, given an induced-vertical vector $X = X^i \partial/\partial y^i$, we obtain $(\theta^v)^a(X) = \theta_{(i)}^a X^i = (z^{-1})_i^a X^i$, so that $\theta_{(i)}^a = (z^{-1})_i^a$. In the similar way by means of which (4.6) is derived, we obtain $\theta_i^a(ug) = (g^{-1})_b^a \theta_i^b(u)$ from (8.3), so that the quantities $N_i^j(u) = z_a^j \theta_i^a(u)$ are functions of x^i and y^i alone. Consequently $(\theta^v)^a$ is written

$$(\theta^v)^a = (z^{-1})_i^a (dy^i + N_j^i dx^j). \tag{8.4}$$

From (8.4) we get an expression of a horizontal vector $X \in N_y$ with respect to the non-linear connection N:

$$X = X^i \left(\frac{\partial}{\partial x^i} - N^j{}_i \frac{\partial}{\partial y^j} \right). \tag{8.5}$$

The functions $N^j{}_i$ of x^k and y^k are called the *coefficients of the non-linear connection N*.

DEFINITION 8.4. A *vertical connection* Γ^v in the Finsler bundle F(M) is a distribution $u \in F \mapsto \Gamma^v_u \subset F^q_u$ which satisfies

(1) $F^q_u = \Gamma^v_u \oplus F^v_u$, (2) $\beta'_g(\Gamma^v_u) = \Gamma^v_{ug}$.

It should be remarked that Γ^v_u is contained in the quasi-vertical subspace F^q_u. We consider a subspace F(x) of the Finsler bundle such that $F(x) = \{u \in F \mid \pi_T \circ \pi_1(u) = x \in M\}$ for a fixed point x. The subspace F(x) has the structure of a principal bundle over the fiber $\pi_T^{-1}(x)$ with the structure group G(n). Its right translation is, of course, equal to β_g of F(M). Because of $F^q_u = \{X \in F_u \mid \pi'_1(X) \in T^v_y,\ y = \pi_1(u)\}$, the tangent space $F(x)_u$ of F(x) at u is nothing but F^q_u. Consequently

PROPOSITION 8.2. *A vertical connection Γ^v is such that the restriction $\Gamma^v|F(x)$ to every F(x) is an ordinary connection in the principal bundle F(x).*

The principal bundle F(x) is further regarded as the linear frame bundle $L(\pi_T^{-1}(x))$ over the fiber $\pi_T^{-1}(x)$ by identifying its point u = (y, z) with a linear frame $1^v_y(z) = \{1^v_y(z_a)\}$ with the origin y, where 1^v_y is the vertical lift (Definition 3.10) and $z = \{z_a\}$ (cf. Definition 19.1). Therefore $\Gamma^v|F(x)$ is regarded as a linear connection on the fiber $\pi_T^{-1}(x)$.

PROPOSITION 8.3. *The induced-vertical distribution F^i is a kind of a vertical connection in the Finsler bundle F(M).* (This F^i is called the *flat vertical connection*.)

The proof follows immediately from Proposition 7.3 and (3) of Proposition 7.1. The linear connection F^i in F(x) in the above sense is really flat, namely, without torsion and curvature. To show this, we consider a basic vector field $B^v(v)$, corresponding to $v \in V^n$, of a general vertical connection Γ^v. From Definition 4.6 we immediately obtain

$$B^v(v)_u = 1^v_u \circ {}_z\alpha'(\vec{v}), \qquad u = (y, z), \tag{8.6}$$

where 1^v_u denotes the lift with respect to Γ^v. It is obvious that $\{B^v(e_a)_u\}$ is a base of Γ^v_u. $B^v(v)_u$ is characterized by $B^v(v)_u \in \Gamma^v_u$ and $\pi'_1(B^v(v)_u) = {}_z\alpha'(\vec{v})$ (cf. Definition 3.10), and satisfies

$$(8.7) \qquad \beta'_g(B^v(v)_u) = B^v(g^{-1}v)_{ug}.$$

In fact, from (2) of Definition 8.4, (2.16) and (3.13) we have

$$\beta'_g \circ 1^v_u \circ {}_z\alpha'(\vec{v}) = 1^v_{ug} \circ {}_z\alpha'({}_g\xi'(\overrightarrow{g^{-1}v})) = 1^v_{ug} \circ {}_{zg}\alpha'(\overrightarrow{g^{-1}v}),$$

which is equal to $B^v(g^{-1}v)_{ug}$.

It follows from (1) of Proposition 7.1 that an induced-fundamental vector field $Y(v)$ is nothing but a basic vector field of the vertical connection F^i. Therefore, by comparing (7.3) with a general structure equation (5.6) it is concluded that F^i is flat.

Let C be the difference tensor field of F^i from a general vertical connection Γ^v. Then (4.14) yields

$$(8.8) \qquad Y(v) = B^v(v) + Z(C(v)).$$

Similarly to the case of (4.14), it is seen that the $(V^n)^1_2$-valued function C on F is a Finsler tensor field of (1, 2)-type and called the *Cartan tensor field* of Γ^v. From (6.4') and (7.5) we get

$$(8.6') \qquad B^v(v)_u = z^i_a v^a\left(\frac{\partial}{\partial y^i} - C_j{}^k{}_i z^j_b \frac{\partial}{\partial z^k_b}\right),$$

where $C_j{}^k{}_i$ are components of the Cartan tensor field C such that $C(v)_u = (C_b{}^a{}_c v^c)E^b_a \in G'(n)$, $C_b{}^a{}_c = (z^{-1})^a_i C_j{}^i{}_k z^j_b z^k_c$ (cf. (6.12)).

The Cartan tensor field C is regarded as $(C_j{}^i{}_k)$, appeared already in (7.7'). To show this, we first prove

<u>PROPOSITION 8.4</u>. *Assume that a pair* (Γ^h, Γ^v) *of horizontal and vertical connections be given in* $F(M)$. *Then the distribution* $\Gamma\colon u \in F \mapsto \Gamma_u = \Gamma^h_u \oplus \Gamma^v_u$ *is a connection in* $F(M)$.

Proof. It follows from Definitions 8.1 and 8.4 that

$$(8.9) \qquad F_u = \Gamma^h_u \oplus \Gamma^v_u \oplus F^v_u,$$

and that $\beta'_g(\Gamma_u) = \Gamma_{ug}$. Thus the proof is complete.

Next, if we denote by 1_u the lift with respect to the sum $\Gamma = \Gamma^h \oplus \Gamma^v$, then (8.6) gives

$$B^v(v)_u = 1_u\left(z^i_a v^a \frac{\partial}{\partial y^i}\right).$$

Then, from (7.9) it follows that with respect to this connection Γ we have

$$= z^i_a v^a \left(\frac{\partial}{\partial y^i} - C_k{}^j{}_i z^k_b \frac{\partial}{\partial z^j_b} \right),$$

which shows that $(C_j{}^i{}_k)$ in (8.6') coincides with $(C_j{}^i{}_k)$ in (7.7').

It follows from (8.8), (6.13) and (2) of Proposition 7.1 that

(8.10) $$B^V(v)\varepsilon = v + C(\varepsilon, v),$$

where the notation $C(u, v)$ for $u, v \in V^n$ is such that $C(u, v) = C(v)\cdot u = (C_c{}^a{}_b v^b)u^c$, $u = (u^a)$, $v = (v^a)$ (cf. (2.22)).

REMARK 8.1. If we are concerned with the connection due to Cartan, $C(\varepsilon, v)$ in (8.10) vanishes (cf. Definition 13.3).

Although the connection-like property of Γ^h is first clearly written in this book, Γ^h already appears in Okada [1] as a characteristic distribution. He introduces the notion of Finsler connection as a pair (Γ^h, Γ^v). The connection-like property of the induced fundamental distribution F^i is noticed by Matsumoto [12], in which the Cartan tensor field C is introduced by (8.8). The connection-like property of F^i is, however, equivalent to the well-known fact that the partial differentiation of components of a tensor field by y^i gives rise to a new tensor field, as it has been noticed since the early period of Finsler geometry. Cf. Berwald [2], Synge [1], Taylor [1] and (9.21).

We are in a position to introduce a family of distribution Γ_V in the linear frame bundle $L(M)$. Because of its complicated character, we shall make some preparations for it.

First, it follows from (8.8) that

(8.11) $$\pi_2' B^V(v) = - \underline{Z}(C(v)).$$

Hence $\pi_2'(\Gamma^v)$ is vertical in the linear frame bundle $L(M)$.

On the other hand, putting

(8.12) $$\Gamma_{(v)z} = \pi_2'(\Gamma^h_u), \qquad u = I^{-1}(z, v),$$

a distribution $\Gamma_{(v)}: z \in L \mapsto \Gamma_{(v)z}$ is introduced, corresponding to a fixed $v \in V^n$. Consequently we get a family $\Gamma_V = \{\Gamma_{(v)} \mid v \in V^n\}$ of distributions, which is parametrized by all the vectors of V^n. From (2) of (6.8) and Definition 8.1 it follows that

$$\underline{\beta}_g'(\Gamma_{(v)z}) = \pi_2' \circ \beta_g'(\Gamma^h_u) = \pi_2'(\Gamma^h_{ug}).$$

Because $I(ug) = I \circ \beta_g \circ I^{-1}(z, v) = \beta^\times_g(z, v) = (zg, g^{-1}v)$ from (3) of (6.17), the above equation and (8.12) yield

(8.13) $$\underline{\beta}_g'(\Gamma_{(v)z}) = \Gamma_{(g^{-1}v)zg}.$$

Therefore the single $\Gamma_{(v)}$ for a fixed v is not invariant by a right translation in general. As a consequence of the above consideration, we are led to the following definition:

DEFINITION 8.5. A *V-connection* Γ_V in the linear frame bundle L(M) is a family $\{\Gamma_{(v)} \mid v \in V^n\}$ of distributions which is parametrized by $v \in V^n$ and satisfies

$$(1) \quad L_z = \Gamma_{(v)z} \oplus L_z^V, \qquad (2) \quad \underline{\beta}'_g(\Gamma_{(v)z}) = \Gamma_{(g^{-1}v)zg}.$$

In fact, $\Gamma_{(v)}$, obtained by (8.12), satisfies the condition (1) also. To show this, we consider the non-linear connection $N = \pi_1'(\Gamma^h)$. Then, corresponding to $X \in L_z$, we obtain $Y = \pi_2' \circ 1_u^h \circ 1_y \circ \pi_L'(X)$, $y = zv$, $u = (zv, z)$. Then $Y \in \Gamma_{(v)z}$ from (8.12) and X - Y is clearly vertical, so that we have $L_z = \Gamma_{(v)z} + L_z^V$. Further, if $X \in \Gamma_{(v)z} \cap L_z^V$, there exists $Z \in \Gamma_u^h$ such that $X = \pi_2'(Z)$. Because of $X \in L_z^V$, we have $\pi_L' \circ \pi_2'(Z) = 0$, so that $Z \in F_u^q$. Therefore (1) of Definition 8.1 gives Z = 0, hence X = 0. Consequently

PROPOSITION 8.5. *From a horizontal connection Γ^h we obtain a V-connection which is defined by (8.12).*

Further we show that there is a close relation between a V-connection and a non-linear connection as follows:

PROPOSITION 8.6. *A distribution N which is obtained from a V-connection $\Gamma_V = \{\Gamma_{(v)} \mid v \in V^n\}$ by $N_y = \alpha_v'(\Gamma_{(v)z})$, $y = zv$, is a non-linear connection in the tangent bundle T(M).* (This N is called the *associated non-linear connection with* Γ_V.)

It must be first proved that the above N_y is well-defined. In fact, taking a point $(zg, g^{-1}v) \in L \times V^n$, $g \in G(n)$, for a given point $y = zv$, from (8.13) we see

$$\alpha'_{g^{-1}v}(\Gamma_{(g^{-1}v)zg}) = \alpha'_{g^{-1}v} \circ \underline{\beta}'_g(\Gamma_{(v)z}) = \alpha'_v(\Gamma_{(v)z}),$$

which is the asserted fact.

The definition of the above N is quite similar to (4.16). Corresponding to $X \in T_y$, we get $1_{(v)z} \circ \pi_T'(X) \in \Gamma_{(v)z}$, where $1_{(v)z}: M_x \to \Gamma_{(v)z}$, $x = \pi_L(z)$, is the lift to $z \in L$ with respect to $\Gamma_{(v)}$. Then $Y = \alpha_v' \circ 1_{(v)z} \circ \pi_T'(X) \in N_y$ and X - Y is clearly vertical. If $X \in N_y \cap T_y^V$, there exists $Z \in \Gamma_{(v)z}$ such that $X = \alpha_v'(Z)$. Because of $X \in T_y^V$, we have $\pi_T' \circ \alpha_v'(Z) = 0 = \pi_L'(Z)$, hence Z = 0. This shows that N is a non-linear connection.

REMARK 8.2. The concept of V-connection is introduced by Sulanke [1] and formulated in the above form by Matsumoto [5]. Although the condition (2) of Definition

8.5 may be rather difficult to understand, a similar idea is very effective to get a generalization of transformation to Finsler spaces (cf. Definition 33.1).

Similarly to the case of ordinary connection, the connection forms $\omega_{(v)}$ of a V-connection Γ_V are introduced by

(8.14) (1) $\omega_{(v)}(\underline{Z}(A)) = A$, (2) $\omega_{(v)}(\Gamma_{(v)}) = 0$,

but the equation, showing the "adjoint type" property in Definition 4.2, should be modified as

(8.15) $$\omega_{(v)} \circ \underline{\beta}'_g = ad(g^{-1})\omega_{(gv)},$$

which is obvious from (2) of Definition 8.5.

In a similar way to the case of (4.6), the connection form $\omega_{(v)} = \omega_{(v)b}{}^a E_a^b \in G'(n)$ is written

(8.16) $$\omega_{(v)b}^{\ a} = (z^{-1})^a_i \{dz^i_b + F_j{}^i{}_k(zv) z^j_b dx^k\},$$

where $F_j{}^i{}_k$ are functions of x^i and $y^i = z^i_a v^a$, $v = (v^a)$.

The basic vector field $B_{(v)}(w)$ with respect to $\Gamma_{(v)}$, corresponding to $w \in V^n$, is similarly defined by

(8.17) $$B_{(v)}(w)_z = 1_{(v)z}(zw),$$

and (2) of (8.14) leads to

(8.17') $$B_{(v)}(w)_z = z^i_a w^a \left(\frac{\partial}{\partial x^i} - F_j{}^k{}_i z^j_b \frac{\partial}{\partial z^k_b} \right).$$

The functions $F_j{}^k{}_i$ are called the *coefficients of the V-connection* Γ_V. From (8.17') and (3.17') we obtain

$$\alpha'_V B_{(v)}(w)_z = z^i_a w^a \left(\frac{\partial}{\partial x^i} - F_j{}^k{}_i z^j_b v^b \frac{\partial}{\partial y^k} \right).$$

Because $(x^i, y^i = z^i_b v^b)$ is the induced coordinate system of $y = zv$, comparing this with (8.5) it is seen that

(8.18) $$N^k{}_i = y^j F_j{}^k{}_i(y)$$

are nothing but the coefficients of the non-linear connection N associated with Γ_V (cf. (9.22)).

REMARK 8.3. By means of Proposition 8.6 a V-connection is regarded as a non-linear connection *with three indices* $(F_j{}^i{}_k)$, while a general non-linear connection N is *with two indices* $(N^i{}_j)$. If a non-linear connection N is assumed to be homogeneous, it is always associated with a V-connection. See Theorem 15.3 and Matsumoto [10].

In case of the connection due to Cartan, coefficients $F_j{}^i{}_k$ of the V-connection are denoted by the well-known notations $\Gamma^*{}_j{}^i{}_k$ which will be shown in § 17.

§ 9. Finsler connections

We are ready for stating the definition of Finsler connection which is an inclusive generalization of connections appeared in classical theories of Finsler spaces. Three but equivalent definitions are presented. These have respective merits according as what we consider.

DEFINITION 9.1. A *Finsler connection* $F\Gamma$ on a manifold M is a pair (Γ, N) of a connection Γ in the Finsler bundle F(M) and a non-linear connection N in the tangent bundle T(M).

REMARK 9.1. At this stage of our theory, no relation is imposed between Γ and N of $F\Gamma$. Thus the concept of Finsler connection as above is of great generality.

If we intend to introduce a Finsler connection as an application of the theory of connections in fiber bundles, it may be natural to understand it as a connection in some fiber bundle over the tangent bundle T(M). Almost all the authors really put on this point of view. They did not notice the importance on non-linear connections in Finsler geometry. In the classical treatments due to Berwald [2] and Cartan [4], the well-known quantities $G^i{}_j(x, y)$ are nothing but coefficients of a non-linear connection. A.Kawaguchi [3] is the first who emphasized the importance of non-linear connections in Finsler geometry.

The respective lifts with respect to Γ and N of a Finsler connection $F\Gamma = (\Gamma, N)$ are denoted by

$$l_u: T_y \to \Gamma_u, \quad y = \pi_1(u), \qquad l_y: M_x \to N_y, \quad x = \pi_T(y).$$

According to (8.5), a lift $l_y(X)$ of $X = X^i \partial/\partial x^i \in M_x$ is given by

$$(9.1) \qquad l_y(X) = X^i\left(\frac{\partial}{\partial x^i} - N^j{}_i \frac{\partial}{\partial y^j}\right).$$

It follows from (7.9) that the further lift $l_u \circ l_y(X)$ is written

$$(9.2) \qquad l_u \circ l_y(X) = X^i\left(\frac{\partial}{\partial x^i} - N^j{}_i \frac{\partial}{\partial y^j} - F_k{}^j{}_i z^k_a \frac{\partial}{\partial z^j_a}\right),$$

where we put

$$(9.3) \qquad F_k{}^j{}_i = \Gamma_k{}^j{}_i - C_k{}^j{}_r N^r{}_i.$$

On the other hand, a lift $l_u(X^v)$ of a vertical vector $X^v = X^{(i)}\partial/\partial y^i \in T^v_y$ is written

$$(9.4)\qquad 1_u(X^v) = X^{(i)}\left(\frac{\partial}{\partial y^i} - C_k{}^j{}_i z_a^k \frac{\partial}{\partial z_a^j}\right).$$

According to (9.2), it seems that $F_k{}^j{}_i$ are more essential than $\Gamma_k{}^j{}_i$ in the theory of Finsler connections.

THEOREM 9.1. *There exists a one-to-one correspondence between the set of all the Finsler connections $F\Gamma = (\Gamma, N)$ on a manifold M and the set of all the pairs (Γ^h, Γ^v) of horizontal and vertical connections in the Finsler bundle F(M).*

Proof. Assume that a Finsler connection $F\Gamma = (\Gamma, N)$ be given on M. From $F\Gamma$ a pair (Γ^h, Γ^v) of distributions in F is defined by

$$(9.5)\qquad (1)\quad \Gamma_u^h = 1_u(N_y), \qquad (2)\quad \Gamma_u^v = 1_u(T_y^V), \qquad u = (y, z),$$

where 1_u is the lift with respect to Γ. If we consider a tangent vector $X \in F_u$, a unique decomposition of $\pi_1'(X)$ is obtained such that

$$\pi_1'(X) = X_1 + X_2, \qquad X_1 \in N_y, \qquad X_2 \in T_y^V, \qquad y = \pi_1(u),$$

with respect to N. The tangent vector $X - 1_u(X_1) - 1_u(X_2)$ is obviously vertical and $1_u(X_1) \in \Gamma_u^h$, $1_u(X_2) \in \Gamma_u^v$ by (9.5), so that we get $F_u = \Gamma_u^h + \Gamma_u^v + F_u^V$. From $N_y \cap T_y^V = 0$ we have $\Gamma_u^h \cap \Gamma_u^v = 0$, and $(\Gamma_u^h \oplus \Gamma_u^v) \cap F_u^V = 0$ is obvious from $\Gamma_u \cap F_u^V = 0$. Therefore we obtain

$$(9.6)\qquad F_u = \Gamma_u^h \oplus \Gamma_u^v \oplus F_u^V, \qquad \Gamma_u = \Gamma_u^h \oplus \Gamma_u^v.$$

Further, from $\pi_T' \circ \pi_1'(\Gamma_u^v) = 0$ it follows that $\Gamma_u^v \subset F_u^q$, which implies $\Gamma_u^v \oplus F_u^V \subset F_u^q$. Consequently $\Gamma_u^v \oplus F_u^V = F_u^q$ is shown by considering their dimensions; so (1) of Definitions 8.1 and 8.4 are verified. Further (2) of them are obvious from the general property (4.3) of the lift 1_u. Therefore the pair (Γ^h, Γ^v) of horizontal and vertical connections are obtained by (9.5).

Conversely, if such a pair (Γ^h, Γ^v) be given, from Proposition 8.4 we have the sum $\Gamma = \Gamma^h \oplus \Gamma^v$, which is a connection in F(M), and the projection

$$(9.7)\qquad N_y = \pi_1'(\Gamma_u^h), \qquad u = (y, z),$$

becomes a non-linear connection N from Proposition 8.1. Thus we obtain the pair (Γ, N), a Finsler connection. It will be clear that the above correspondence is one-to-one.

As a consequence of Theorem 9.1, we have the *second definition of a Finsler connection* as follows:

DEFINITION 9.2. A *Finsler connection* $F\Gamma$ on a manifold M is a pair (Γ^h, Γ^v) of a horizontal connection Γ^h and a vertical connection Γ^v in the Finsler bundle F(M).

According to the second definition $F\Gamma = (\Gamma^h, \Gamma^v)$, we already have the notions of a basic vector field $B^v(v)$ and the Cartan tensor field C with respect to the *vertical part* Γ^v, which are called a *v-basic vector field* and the *Cartan tensor field* of the Finsler connection $F\Gamma = (\Gamma^h, \Gamma^v)$ respectively. The vertical part Γ^v is spanned by v-basic vector fields. On the other hand, the *horizontal part* Γ^h of $F\Gamma$ is spanned by following tangent vector fields:

DEFINITION 9.3. The *h-basic vector field* $B^h(v)$ of a Finsler connection $F\Gamma = (\Gamma, N)$, corresponding to $v \in V^n$, is a tangent vector field of F given by $B^h(v)_u = 1_u \circ 1_y(zv)$, $u = (y, z)$.

The lift $1_y(zv)$ is horizontal with respect to N and from (9.5) it is seen that $1_u \circ 1_y(zv) \in \Gamma^h_u$. The equation similar to (8.7), namely,

$$\beta'_g(B^h(v)) = B^h(g^{-1}v) \tag{9.8}$$

is shown as follows: For $u = (y, z)$

$$\beta'_g B^h(v)_u = \beta'_g \circ 1_u \circ 1_y(zv) = 1_{ug} \circ 1_y((zg)(g^{-1}v)) = B^h(g^{-1}v)_{ug},$$

because of $ug = (y, zg)$.

We have the connection form ω of Γ, the v-basic form θ^v with respect to N of a Finsler connection $F\Gamma$ and the h-basic form θ^h which is defined in no relation to any connection. h and v-basic vector fields $B^h(v)$, $B^v(v)$ and a fundamental vector field Z(A) are thought of as the dual of those one-forms in the following sense:

$$\begin{aligned} &\omega(Z(A)) = A, && \omega(B^h(v)) = 0, && \omega(B^v(v)) = 0, \\ &\theta^h(Z(A)) = 0, && \theta^h(B^h(v)) = v, && \theta^h(B^v(v)) = 0, \\ &\theta^v(Z(A)) = 0, && \theta^v(B^h(v)) = 0, && \theta^v(B^v(v)) = v. \end{aligned} \tag{9.9}$$

These equations are easily verified from their definitions. Cf. (4.11).

From (9.2) an expression of $B^h(v)$ in terms of an induced coordinate system (x^i, y^i, z^i_a) is given by

$$B^h(v)_u = z^i_a v^a \left(\frac{\partial}{\partial x^i} - N^j_i \frac{\partial}{\partial y^j} - F_k{}^j{}_i z^k_b \frac{\partial}{\partial z^j_b} \right). \tag{9.10}$$

REMARK 9.2. The first definition of Finsler connection explicitly appears in Matsumoto [10]. On the other hand, the second one appears in Okada [1, 2], called a pair-connection. From the second definition and Proposition 8.3 it is immediately noticed that a pair (Γ^h, F^i) of an arbitrary horizontal connection Γ^h and the flat vertical connection F^i is a simple Finsler connection in a sense. In this connection an induced-fundamental vector field $Y(v)$ plays the role of a v-basic vector field $B^v(v)$; so the Cartan tensor field C vanishes.

As it is shown above, a Finsler connection $F\Gamma$ is regarded as a pair (Γ, N) as well as a pair (Γ^h, Γ^v). Thus, throughout the following, we shall write $F\Gamma = (\Gamma, N) = (\Gamma^h, \Gamma^v)$.

By Proposition 8.5, from Γ^h of $F\Gamma = (\Gamma^h, \Gamma^v)$ the induced mapping π_2 gives

$$(9.11)\qquad \Gamma_{(v)z} = \pi_2'(\Gamma^h_u), \qquad u = I^{-1}(z, v),$$

which constitutes a V-connection $\Gamma_V = \{\Gamma_{(v)} \mid v \in V^n\}$.

THEOREM 9.2. *There exists a one-to-one correspondence between the set of all the Finsler connections on a manifold M and the set of all the triads (Γ_V, N, Γ^v) of a V-connection Γ_V in the linear frame bundle L(M), a non-linear connection N in the tangent bundle T(M) and a vertical connection Γ^v in the Finsler bundle F(M).*

Proof. Assume a triad (Γ_V, N, Γ^v) be given. Then we put

$$(9.12)\qquad \Gamma^h_u = \{X \in F_u \mid \pi_1'(X) \in N_y,\ \pi_2'(X) \in \Gamma_{(v)z},\ u = (zv, z),\ y = zv\}.$$

The distribution Γ^h as thus defined is really a horizontal connection. In fact, from $X \in F_u$ two tangent vectors

$$X_1 = h_N \circ \pi_1'(X), \qquad X_2 = h_{(v)} \circ \pi_2'(X)$$

are obtained, where h_N (resp. $h_{(v)}$) denotes the horizontal component of a tangent vector of T (resp. L) with respect to N (resp. $\Gamma_{(v)}$). From Proposition 6.2 we get a unique tangent vector $Y \in F_u$ such that $\pi_1'(Y) = X_1$, $\pi_2'(Y) = X_2$, and (9.12) shows $Y \in \Gamma^h_u$. Then $X - Y \in F^q_u$ is easily verified; so $F_u = \Gamma^h_u + F^q_u$ is shown. If $X \in \Gamma^h_u \cap F^q_u$, then $\pi_2'(X) \in \Gamma_{(v)z}$ from (9.12), and $\pi_2'(X) \in L^v_z$ from $X \in F^q_u$. Thus $\pi_2'(X) = 0$ and similarly $\pi_1'(X) = 0$. Hence Proposition 6.3 shows $X = 0$ and we get the direct sum $F_u = \Gamma^h_u \oplus F^q_u$. Further, taking $X \in \Gamma^h_u$, the tangent vector $\beta_g'(X)$ is such that

$$\pi_1'(\beta_g' X) = \pi_1'(X) \in N_y,$$

$$\pi_2'(\beta_g' X) = \underline{\beta}_g' \circ \pi_2'(X) \in \underline{\beta}_g'(\Gamma_{(v)z}) = \Gamma_{(g^{-1}v)zg},$$

which implies $\beta_g'(X) \in \Gamma^h_w$, $w = I^{-1}(zg, g^{-1}v) = ug$, from (3) of (6.17). Consequently (1), (2) of Definition 8.1 are proved and thus the pair (Γ^h, Γ^v) is a Finsler connection.

Conversely, from (9.12) we obtain (9.11) and (1) of (9.5), so that the above correspondence is one-to-one.

As a consequence of Theorem 9.2, we obtain the *third definition of a Finsler connection* as follows:

DEFINITION 9.4. A *Finsler connection* $F\Gamma$ on a manifold M is a triad (Γ_V, N, Γ^v) of a V-connection Γ_V in the linear frame bundle L(M), a non-linear connection N in the tangent bundle T(M) and a vertical connection Γ^v in the Finsler bundle F(M).

According to the third definition, we first have coefficients $F_j{}^i{}_k$ of Γ_V in (8.17'), secondly coefficients $N^i{}_j$ of N in (8.5) and thirdly coefficients $C_j{}^i{}_k$ of Γ^v in (8.6'). These quantities $(F_j{}^i{}_k, N^i{}_j, C_j{}^i{}_k)$ are called the *coefficients of the Finsler connection* $F\Gamma = (\Gamma_V, N, \Gamma^v)$.

If a Finsler connection $F\Gamma$ is considered as a triad (Γ_V, N, Γ^v), we have the connection forms $\omega_{(v)}$ of Γ_V, the v-basic form θ^v with respect to N and the Cartan tensor field C of Γ^v. The following shows how to construct the connection Γ of $F\Gamma = (\Gamma, N)$ from these quantities:

PROPOSITION 9.1. *The connection form ω of the connection Γ of a Finsler connection $F\Gamma = (\Gamma, N) = (\Gamma_V, N, \Gamma^v)$ is written*

$$\omega = \omega_{(v)} \circ \pi'_2 + C(\theta^v)$$

at a point $u = (zv, z)$, where $\omega_{(v)}$ is the connection form of $\Gamma_{(v)} \in \Gamma_V$, C is the Cartan tensor field of Γ^v and θ^v is the v-basic form with respect to N.

Proof. It follows from (9.12) and Definition 9.3 that

$$\pi'_2 B^h(w)_u \in \Gamma_{(v)z}, \qquad \pi'_L \circ \pi'_2 B^h(w)_u = \pi'_T \circ \pi'_1 B^h(w)_u = zw,$$

which implies

$$\pi'_2 B^h(w)_u = B_{(v)}(w)_z, \qquad u = (zv, z). \tag{9.13}$$

Putting $\gamma = \omega_{(v)} \circ \pi'_2 + C(\theta^v)$ at $u = (zv, z)$, we obtain $\gamma(Z(A)) = A$ from (8.14), $\gamma(B^h(w)) = 0$ from (9.13) and $\gamma(B^v(w)) = 0$ from (8.11) and (8.14). Consequently $\gamma = \omega$ is shown from (7.7).

REMARK 9.3. The third definition of a Finsler connection first appears in Matsumoto [10]. We enumerate distributions appearing in above three definitions of a Finsler connection:

$\Gamma \cdots$ in F(M), $\quad N \cdots$ in T(M), $\quad \Gamma^h \cdots$ in F(M),

$\Gamma^v \cdots$ in F(M), $\quad \Gamma_V \cdots$ in L(M).

Thus the first definition refers to F(M), T(M), the second to F(M) only and the third to L(M), T(M), F(M) all. As a consequence it seems that the second is the best, while the third is rather intricate, but every definition has respective merits, as it will be gradually seen in the following.

According to Propositions 8.3 and 8.6, we obtain a simple Finsler connection (Γ_V, N, F^i), where N is the non-linear connection associated with Γ_V; the coefficients ($F_j{}^i{}_k$, $N^i{}_j$, $C_j{}^i{}_k$) of this connection are such that $N^i{}_j = y^k F_k{}^i{}_j$ from (8.18) and $C_j{}^i{}_k = 0$. If $F_j{}^i{}_k$ are equal to Cartan's $\Gamma^*{}_j{}^i{}_k$, this simple connection is nothing but the connection due to Rund [11] (cf. § 18).

The concept of Finsler connection in the present section is defined from very general standpoint; so it will be applicable to the theory of manifolds with more general metric than a Finsler metric. See, for instance, Kern [3].

DEFINITION 9.5. The *h and v-covariant derivatives* $\nabla^h K$, $\nabla^v K$ of a Finsler tensor field K with respect to a Finsler connection $F\Gamma = (\Gamma^h, \Gamma^v)$ are respectively given by

(1) $\nabla^h K(v) = B^h(v)(K)$, (2) $\nabla^v K(v) = B^v(v)(K)$.

The *0-covariant derivative* $\nabla^0 K$ is given by

(3) $\nabla^0 K(v) = Y(v)(K)$.

These definitions are quite similar to Definition 5.1 with respect to a linear connection.

From (8.10) the v-covariant derivative of the supporting element ε is of an interesting form

$$\nabla^v \varepsilon(v) = v + C(\varepsilon, v). \tag{9.14}$$

Moreover the equation

$$\nabla^0 \varepsilon(v) = v \tag{9.15}$$

is obvious from (2) of Proposition 7.1.

On the other hand, putting

$$\nabla^h \varepsilon(v) = D(v), \tag{9.16}$$

the tensor field D of (1, 1)-type is obtained and called the *deflection tensor field* of the Finsler connection. From Proposition 6.2, (9.13) and (9.16) we see

$$B^h(w)_u = (I^{-1})'(B_{(v)}(w)_z, \overrightarrow{D(w)}_v), \qquad u = (zv, z), \tag{9.17}$$

for any $w \in V^n$.

We shall find components of $\nabla^h K$ of a tensor field K of (1, 1)-type, for instance. From (9.10) and (6.12) we have

$$\nabla^h K(v) = z^i_a v^a \left(\frac{\partial}{\partial x^i} - N^j{}_i \frac{\partial}{\partial y^j} - F_k{}^j{}_i z^k_b \frac{\partial}{\partial z^j_b} \right) \{ (z^{-1})^c_h K^h{}_m z^m_d \} e^d_c$$

$$= z^i_a v^a \Big((z^{-1})^c_h (\partial_i K^h{}_m) z^m_d - N^j{}_i (z^{-1})^c_h (\dot{\partial}_j K^h{}_m) z^m_d$$

$$- F_k{}^j{}_i z^k_b \{ - (z^{-1})^c_j (z^{-1})^b_h K^h{}_m z^m_d + (z^{-1})^c_h K^h{}_m \delta^m_j \delta^b_d \} \Big) e^d_c$$

$$= (z^i_a v^a)(z^{-1})^c_h z^m_d \Big(\partial_i K^h{}_m - N^j{}_i (\dot{\partial}_j K^h{}_m) + K^k{}_m F_k{}^h{}_i - K^h{}_k F_m{}^k{}_i \Big) e^d_c,$$

where and throughout the following we put $\partial_i \phi = \partial\phi/\partial x^i$ and $\dot{\partial}_i \phi = \partial\phi/\partial y^i$ for a function $\phi(x, y)$. Therefore, putting

(9.18) $$K^h{}_{m|i} = \partial_i K^h{}_m - N^j{}_i \dot{\partial}_j K^h{}_m + K^k{}_m F_k{}^h{}_i - K^h{}_k F_m{}^k{}_i,$$

these are components of $\nabla^h K$ and we have

$$\nabla^h K(v) = \{(z^{-1})^c_h K^h{}_{m|i} z^m_d (z^i_a v^a)\} e^d_c.$$

In (9.18) the differentiation operators

(9.19) $$\delta_i = \partial_i - N^j{}_i \dot{\partial}_j$$

appear. It is noteworthy that δ_i are first defined when a non-linear connection $N = (N^j{}_i)$ is given, and that for a scalar field S we have $S_{|i} = \delta_i S$.

Similarly, from (8.6') we get the components $K^h{}_m|_i$ of $\nabla^v K$:

(9.20) $$K^h{}_m|_i = \dot{\partial}_i K^h{}_m + K^j{}_m C_j{}^h{}_i - K^h{}_j C_m{}^j{}_i.$$

Finally the components $K^h{}_{m \cdot i}$ of $\nabla^0 K$ are solely given by

(9.21) $$K^h{}_{m \cdot i} = \dot{\partial}_i K^h{}_m.$$

Here we write another expression of (9.14), (9.15) and (9.16) in terms of components:

(9.14') $$y^i|_j = \delta^i_j + y^k C_k{}^i{}_j,$$

(9.15') $$y^i{}_{\cdot j} = \delta^i_j,$$

(9.16') $$y^i{}_{|j} = D^i{}_j,$$

where $D^i{}_j$ are components of the deflection tensor D. From (9.16) and (9.18) $D^i{}_j$ are given by

(9.22) $$D^i{}_j = y^r F_r{}^i{}_j - N^i{}_j.$$

<u>REMARK 9.4</u>. As the above differentiation operators δ_i are defined with respect to a non-linear connection, the importance of non-linear connection in Finsler geometry is really clear.

The deflection tensor D first appears in Okada [1], but its importance may be first noticed by Matsumoto [6]. The name "deflection" is given by Matsumoto [10]. This tensor field plays some important roles in determining a Finsler connection from a given Finsler metric. See Hashiguchi [2] and Theorem 17.2.

EXAMPLE 9.1. We consider a linear connection $\underline{\Gamma}$ in the linear frame bundle L(M). A distribution Γ in the Finsler bundle F(M) is introduced by

$$\Gamma_u = \{X \in F_u \mid \pi_2'(X) \in \underline{\Gamma}_z,\ u = (y, z)\}.$$

This is a connection in F(M). In fact, the differential one-form $\omega = \underline{\omega}\circ\pi_2'$ which is induced from the connection form $\underline{\omega}$ of $\underline{\Gamma}$ satisfies

$$\omega Z(A) = \underline{\omega Z}(A) = A, \qquad \omega\circ\beta_g' = \underline{\omega}\circ\underline{\beta}_g'\circ\pi_2' = \mathrm{ad}(g^{-1})\omega,$$

which shows that ω is a connection form. If $X \in \Gamma_u$, then we have $\omega X = \underline{\omega}\circ\pi_2' X = 0$. Conversely, if $\omega X = 0$, then $\underline{\omega}\circ\pi_2' X = 0$, so that $\pi_2' X \in \underline{\Gamma}_z$ and $X \in \Gamma_u$. Consequently it is concluded that Γ is a connection with the connection form ω. Then, by pairing this Γ with the associated connection N with $\underline{\Gamma}$, a Finsler connection $F\Gamma = (\Gamma, N)$ is obtained and called the *linear Finsler connection induced from* $\underline{\Gamma}$.

If we denote the linear Finsler connection $F\Gamma$ by the pair (Γ^h, Γ^v), then Γ^v coincides with the flat vertical connection F^i. In fact, $\pi_2' B^v(v) \in \underline{\Gamma}_z$ in this case, so that (8.11) yields C = 0.

If we denote this $F\Gamma$ by the triad (Γ_V, N, Γ^v), then we have $\Gamma_V = \underline{\Gamma}$, $\Gamma^v = F^i$ and N is associated with $\underline{\Gamma}$.

As for the h-part Γ^h of the linear Finsler connection, it is seen that

$$B^h(v) = (I^{-1})'(\underline{B}(v), 0), \tag{9.23}$$

where $\underline{B}(v)$ is a basic vector field with respect to $\underline{\Gamma}$. In fact, from (2) of (6.17) we observe

$$\omega B^h(v) = \underline{\omega}\circ\pi_2'\circ(I^{-1})'(\underline{B}(v), 0) = \underline{\omega B}(v) = 0,$$

which implies that $B^h(v)$ of (9.23) is contained in Γ. Further it follows from (1) of (6.17) and $\overset{\times}{\pi}_1 = \alpha$ that at $u = (y, z) = I^{-1}(z, v)$

$$\pi_1' B^h(w)_u = (\overset{\times}{\pi}_1)'(\underline{B}(w), 0) = \alpha_{\underline{v}}'\underline{B}(w)_z = \alpha_v'\circ\underline{l}_z(zw) = \underline{l}_y(zw),$$

where $\underline{l}_z$ (resp. $\underline{l}_y$) is the lift with respect to $\underline{\Gamma}$ (resp. N). Consequently Definition 9.3 gives the fact that we desire to show.

Therefore Γ^h is written as $(I^{-1})'(\underline{\Gamma}, 0)$. From (9.17) we have D = 0. It is easily shown that the h-covariant derivative $\nabla^h\bar{K}$ of an induced tensor field $\bar{K} = \tilde{K}\circ\pi_2$, defined in Example 6.1, is given by $\underline{\nabla}\tilde{K}(v)\circ\pi_2$, where $\underline{\nabla}$ is the covariant differentiation with respect to $\underline{\Gamma}$. Therefore, roughly speaking, the linear Finsler connection is the pair of the linear connection $\underline{\Gamma}$ and the flat vertical connection F^i in the viewpoint of covariant differentiations.

§ 10. Torsions and curvatures of Finsler connections

We shall define the torsion and curvature tensor fields of a Finsler connection $F\Gamma = (\Gamma^h, \Gamma^v)$. It has been seen that the h-part Γ^h (resp. v-part Γ^v) is spanned by h (resp. v)-basic vector fields. Thus, generalizing the concept of structure equation (5.6) of a linear connection to a Finsler connection, those essential tensor fields will be naturally obtained as follows.

Similarly to (5.4), we first get a trivial equation

$$[Z(1), Z(2)] = Z([1, 2]), \tag{10.1}$$

where $A_\alpha \in G'(n)$, $\alpha = 1, 2$, are denoted by their indices α only.

Secondly, similarly to (5.5), we obtain

$$\begin{aligned}[Z(A), B^h(v)] &= B^h(A\cdot v),\\ [Z(A), B^v(v)] &= B^v(A\cdot v),\end{aligned} \tag{10.2}$$

which are easily verified by Proposition 2.2, (8.7), (9.8) and (2.22').

Here we write generalizations of (10.1) and (10.2), owing to (1.2), assuming that A_α, A and v are functions on F:

$$[Z(1), Z(2)] = Z([1, 2]) + \mathfrak{A}_{(12)}\{Z(Z(1)A_2)\}, \tag{10.1'}$$

$$\begin{aligned}[Z(A), B^h(v)] &= B^h(A\cdot v) + B^h(Z(A)v) - Z(B^h(v)A),\\ [Z(A), B^v(v)] &= B^v(A\cdot v) + B^v(Z(A)v) - Z(B^v(v)A).\end{aligned} \tag{10.2'}$$

Now we are concerned with three structure equations similar to (5.6). First we put

$$[B^h(1), B^h(2)] = B^h(T(1, 2)) + B^v(R^1(1, 2)) + Z(R^2(1, 2)), \tag{10.3}$$

where $v_\alpha \in V^n$, $\alpha = 1, 2$, are denoted by the indices α only. By the usual process of applying the differential β'_g of a right translation β_g to (10.3), the tensor property of T, R^1 and R^2 in (10.3) is easily shown.

Secondly, we put

$$[B^h(1), B^v(2)] = B^h(C(1, 2)) + B^v(P^1(1, 2)) + Z(P^2(1, 2)). \tag{10.4}$$

The h-part of the above is really equal to the h-basic vector field, corresponding to the value $C(1, 2) \in V^n$ of the Cartan tensor field C. To prove this, taking a function f on the base space M, we observe

$$(\pi'_T \circ \pi'_1[B^h(1), B^v(2)])(f) = [B^h(1), B^v(2)](f \circ \pi_T \circ \pi_1)$$

$$= B^h(1)\left(\{\pi'_T \circ \pi_1 B^v(2)\}f\right) - B^v(2)\left(\{\pi'_T \circ \pi'_1 B^h(1)\}f\right) = -B^v(2)\phi,$$

where ϕ is a function on F defined by $\phi(u) = (zv_1)f = z^i_a v^a_1(\partial_i f)$, $u = (y, z) = (x^i,$

y^i, z^i_a). From (8.8) the above is written $- Y(2)\phi + Z(C(2))\phi$. The first term vanishes because of Proposition 7.2 or (7.5), and the second term is equal to $(zC(1, 2))f$ from (6.4') and (3.9). Thus the proof is completed.

Thirdly, by (8.8) the Lie bracket $[B^v(1), B^v(2)]$ is written in the form

$$[B^v(1), B^v(2)] = [Y(1) - Z(C(1)), Y(2) - Z(C(2))].$$

From (7.3), (7.4) and (10.1') the right-hand side is rewritten

$$\mathfrak{A}_{(12)}\{Y(C(1, 2)) - Z(Y(1)C(2)) + Z(Z(C(1))C(2))\} + Z([C(1), C(2)]).$$

According to Definition 9.5 and (6.13), we have

$$Y(1)C(2) = \nabla^0 C(2, 1), \qquad Z(C(1))C(2) = - C(1)\cdot C(2);$$

in the latter it is noted that $C(1) \in G'(n)$ and $C(2) \in (V^n)^1_1$ (cf. Definition 2.6). Thus we have

$$\begin{aligned}[B^v(1), B^v(2)] &= \mathfrak{A}_{(12)}\{Y(C(1, 2)) - Z(\nabla^0 C(2, 1)) - Z(C(1)\cdot C(2))\} \\ &\qquad + Z([C(1), C(2)]) \\ &= \mathfrak{A}_{(12)}\{B^v(C(1, 2)) + Z(C(C(1, 2))) - Z(\nabla^0 C(2, 1)) \\ &\qquad - Z(C(1)\cdot C(2))\} + Z([C(1), C(2)]),\end{aligned}$$

which is summarizing up in the form

$$(10.5) \qquad [B^v(1), B^v(2)] = B^v(S^1(1, 2)) + Z(S^2(1, 2)),$$

where we put

$$(10.6) \qquad S^1(1, 2) = \mathfrak{A}_{(12)}\{C(1, 2)\},$$

and

$$S^2(1, 2) = \mathfrak{A}_{(12)}\{C(C(1, 2)) - C(1)\cdot C(2) - \nabla^0 C(2, 1)\} + [C(1), C(2)].$$

We rewrite $S^2(1, 2)$ as follows: From (2.22) we have

$$C(1)\cdot C(2) = C(C(2), 1) - C(C(1), 2) - C(C(2, 1)),$$

and further

$$[C(1), C(2)] = C(C(2), 1) - C(C(1), 2).$$

Therefore $S^2(1, 2)$ is finally written as

$$(10.7) \qquad S^2(1, 2) = \mathfrak{A}_{(12)}\{\nabla^0 C(1, 2) + C(C(1), 2)\}.$$

The equations (10.3), (10.4) and (10.5) are called the *structure equations of the Finsler connection* $F\Gamma$.

Consequently the eight Finsler tensor fields T, R^1, R^2, C, P^1, P^2, S^1 and S^2 are obtained and called as follows:

Torsion tensors : $T\ (T_{j\ k}^{\ i})$ ··· *(h)h-torsion*, $R^1\ (R^i_{\ jk})$ ··· *(v)h-torsion*,

$C\ (C_{j\ k}^{\ i})$ ··· *(h)hv-torsion*, $P^1\ (P^i_{\ jk})$ ··· *(v)hv-torsion*,

$S^1\ (S^i_{\ jk})$ ··· *(v)v-torsion*,

Curvature tensors: $R^2\ (R_{h\ jk}^{\ i})$ ··· *h-curvature*,

$P^2\ (P_{h\ jk}^{\ i})$ ··· *hv-curvature*, $S^2\ (S_{h\ jk}^{\ i})$ ··· *v-curvature*,

where respective components are shown in the parentheses: The usage of their indices are such that, for instance,

$$T(v_1,\ v_2) = \{(z^{-1})^a_i T_{j\ k}^{\ i}(z^j_b v^b_1)(z^k_c v^c_2)\}e_a,$$

$$P^1(v_1,\ v_2) = \{(z^{-1})^a_i P^i_{\ jk}(z^j_b v^b_1)(z^k_c v^c_2)\}e_a,$$

$$R^2(v_1,\ v_2) = \{(z^{-1})^a_i R_{h\ jk}^{\ i} z^h_b (z^j_c v^c_1)(z^k_d v^d_2)\}E^b_a.$$

Respective components $T_{j\ k}^{\ i}$, $R^i_{\ jk}$, $R_{h\ jk}^{\ i}$, $S^i_{\ jk}$, $S_{h\ jk}^{\ i}$ of T, R^1, R^2 of the *h-type* and S^1, S^2 of the *v-type* are skew-symmetric in the indices j, k, but $C_{j\ k}^{\ i}$, $P^i_{\ jk}$, $P_{h\ jk}^{\ i}$ of C, P^1 and P^2 of the *hv-type* are not so in general.

REMARK 10.1. It is noteworthy that the Cartan tensor field C is one of the torsion tensors, as it has been indicated by Cartan [4].

If we are concerned with the connection due to Cartan, the (h)h- and (v)v-torsion tensors T, S^1 vanish (cf. Definition 17.2) and the (v)h- and (v)hv-torsion tensors R^1, P^1 are derived directly from the h- and hv-curvature tensors R^2, P^2 respectively (cf. Theorem 13.3).

Because of Proposition 8.2 the h-part $B^h(\cdot)$ does not take place in the right-hand side of (10.5).

For the later use we write generalizations of the structure equations (10.3), (10.4) and (10.5), assuming that v_α, $\alpha = 1, 2$, are V^n-valued functions on F:

(10.3') $$[B^h(1),\ B^h(2)] = B^h(T(1,\ 2)) + B^v(R^1(1,\ 2)) + Z(R^2(1,\ 2)) + \mathfrak{A}_{(12)}\{B^h(B^h(1)v_2)\},$$

(10.4') $$[B^h(1),\ B^v(2)] = B^h(C(1,\ 2)) + B^v(P^1(1,\ 2)) + Z(P^2(1,\ 2)) + B^v(B^h(1)v_2) - B^h(B^v(2)v_1),$$

(10.5') $$[B^v(1),\ B^v(2)] = B^v(S^1(1,\ 2)) + Z(S^2(1,\ 2)) + \mathfrak{A}_{(12)}\{B^v(B^v(1)v_2)\}.$$

Next, by generalizing (5.10), we shall derive the Ricci identities which are commutation formulas of covariant differentiations. Taking a Finsler tensor field K, the structure equation (10.3) first yields

$$\mathfrak{A}_{(12)}\{\nabla^h\nabla^h K(2, 1)\} = \nabla^h K(T(1, 2)) + \nabla^v K(R^1(1, 2)) - R^2(1, 2)\cdot K. \tag{10.8}$$

Secondly, (10.4) similarly gives

$$\nabla^h\nabla^v K(2, 1) - \nabla^v\nabla^h K(1, 2) = \nabla^h K(C(1, 2)) + \nabla^v K(P^1(1, 2)) - P^2(1, 2)\cdot K. \tag{10.9}$$

Thirdly, (10.5) leads to

$$\mathfrak{A}_{(12)}\{\nabla^v\nabla^v K(2, 1)\} = \nabla^v K(S^1(1, 2)) - S^2(1, 2)\cdot K. \tag{10.10}$$

These identities (10.8), (10.9) and (10.10) are called the *Ricci identities of the Finsler connection.*

We shall write these commutation formulas of covariant differentiations in terms of components for the later use. Take a Finsler tensor field K of (1, 1)-type, for instance, and deal with the identity (10.9), for instance. Then, according to the usage of indices, we have

$$\nabla^h\nabla^v K(2, 1) = \{(z^{-1})^a_h K^h{}_{i\big|k|j} z^i_b (z^k_d v^d_2)(z^j_c v^c_1)\} e^b_a,$$

$$\nabla^v K(P^1(1, 2)) = \{(z^{-1})^a_h K^h{}_{i\big|r} z^i_b (P^r{}_{jk}(z^j_c v^c_1)(z^k_d v^d_2))\} e^b_a,$$

$$P^2(1, 2)\cdot K = \{(z^{-1})^a_h (P_r{}^h{}_{jk} K^r{}_i - P_i{}^r{}_{jk} K^h{}_r) z^i_b (z^j_c v^c_1)(z^k_d v^d_2)\} e^b_a,$$

and so on. Thus (10.9) is written

$$K^h{}_{i|j\big|k} - K^h{}_{i\big|k|j} = K^r{}_i P_r{}^h{}_{jk} - K^h{}_r P_i{}^r{}_{jk} - K^h{}_{i|r} C_j{}^r{}_k - K^h{}_{i\big|r} P^r{}_{jk}. \tag{10.9'}$$

Similarly, (10.8) and (10.10) are respectively written

$$\mathfrak{A}_{(jk)}\{K^h{}_{i|j|k}\} = K^r{}_i R_r{}^h{}_{jk} - K^h{}_r R_i{}^r{}_{jk} - K^h{}_{i|r} T_j{}^r{}_k - K^h{}_{i\big|r} R^r{}_{jk}, \tag{10.8'}$$

$$\mathfrak{A}_{(jk)}\{K^h{}_{i\big|j\big|k}\} = K^r{}_i S_r{}^h{}_{jk} - K^h{}_r S_i{}^r{}_{jk} - K^h{}_{i\big|r} S^r{}_{jk}. \tag{10.10'}$$

We shall find the components of torsion and curvature tensors in terms of coefficients $(F_j{}^i{}_k, N^i{}_j, C_j{}^i{}_k)$ of the Finsler connection. First of all, (10.6) and (10.7) immediately yield respective components $S^i{}_{jk}$, $S_h{}^i{}_{jk}$ of S^1, S^2 as follows:

$$S^i{}_{jk} = \mathfrak{A}_{(jk)}\{C_j{}^i{}_k\}, \tag{10.6'}$$

$$S_h{}^i{}_{jk} = \mathfrak{A}_{(jk)}\{\dot\partial_k C_h{}^i{}_j + C_h{}^r{}_j C_r{}^i{}_k\}. \tag{10.7'}$$

These are quite similar in form to (5.12) and (5.11) respectively in case of a linear connection; it is clearly conjectured from Proposition 8.2.

To find the components of other torsion and curvature tensors, we prepare simple commutation formulas of differentiation operators ∂_i, $\dot{\partial}_i$, δ_i:

$$\partial_j\delta_i - \delta_i\partial_j = -(\partial_j N^k{}_i)\dot{\partial}_k,$$

(10.11)
$$\dot{\partial}_j\delta_i - \delta_i\dot{\partial}_j = -(\dot{\partial}_j N^k{}_i)\dot{\partial}_k,$$

$$\delta_j\delta_i - \delta_i\delta_j = -(\delta_j N^k{}_i - \delta_i N^k{}_j)\dot{\partial}_k.$$

Then, applying (10.8') to a Finsler scalar field S, we have

$$\mathfrak{A}_{(jk)}\{S_{|j|k}\} = -S_{|r}T_j{}^r{}_k - S|_r R^r{}_{jk}.$$

By applying (10.11) the left-hand side is written

$$-S_{|r}(F_j{}^r{}_k - F_k{}^r{}_j) - S|_r(\delta_k N^r{}_j - \delta_j N^r{}_k).$$

Therefore we get

(10.12)
$$T_j{}^i{}_k = \mathfrak{A}_{(jk)}\{F_j{}^i{}_k\},$$

(10.13)
$$R^i{}_{jk} = \mathfrak{A}_{(jk)}\{\delta_k N^i{}_j\}.$$

From (10.12) it is observed that the (h)h-torsion tensor T is regarded as the torsion tensor of the V-connection Γ_V and from (10.13) that the (v)h-torsion tensor R^1 may be regarded as the curvature tensor of the non-linear connection N.

REMARK 10.2. It follows from (8.5) that a non-linear connection N is locally spanned by n tangent vectors

$$X_{i)} = \frac{\partial}{\partial x^i} - N^j{}_i(x, y)\frac{\partial}{\partial y^j}.$$

If N is integrable, namely, there exists a subspace T_N: $y^i = y^i(x)$ of T such that $X_{i)}$, $i = 1,\cdots, n$, are tangent vectors of T_N, then a tangent vector of T_N is given by

$$\frac{\partial x^j}{\partial x^i}\frac{\partial}{\partial x^j} + \frac{\partial y^j}{\partial x^i}\frac{\partial}{\partial y^j} = \frac{\partial}{\partial x^i} + \frac{\partial y^j}{\partial x^i}\frac{\partial}{\partial y^j},$$

so that we get

$$\frac{\partial y^j}{\partial x^i} = -N^j{}_i(x, y).$$

Conversely, if the above system of differential equations are completely integrable, we get a subspace $y^i = y^i(x; x_o, y_o)$ through an initial point (x^i_o, y^i_o) which are solutions of the system, and this subspace T_N is an integral manifold of N. The integrability condition of the above system is given by

$$\frac{\partial}{\partial x^k}\left(\frac{\partial y^j}{\partial x^i}\right) - \frac{\partial}{\partial x^i}\left(\frac{\partial y^j}{\partial x^k}\right) = - R^j{}_{ik} = 0.$$

Therefore R^1 is considered as the curvature tensor of the non-linear connection N.

Next, similarly we have

$$S_{|j|k} - S_{|k|j} = - S_{|r}C_j{}^r{}_k - S|_r(\dot{\partial}_k N^r{}_j - F_k{}^r{}_j),$$

so that (10.9') yields

$$P^i{}_{jk} = \dot{\partial}_k N^i{}_j - F_k{}^i{}_j. \tag{10.14}$$

Comparing (10.14) with (9.22), it will be conjectured that there may be some relation between the deflection tensor D and the (v)hv-torsion tensor P^1 (cf. Theorem 12.2).

To find components of the curvature tensors R^2 and P^2, we deal with a Finsler tensor field X of (1, 0)-type and compute

$$X^i{}_{|j|k} - X^i{}_{|k|j}, \qquad X^i{}_{|j}|_k - X^i|_{k|j}$$

directly. Comparing these results with the Ricci identities applied to X and referring to (10.12), (10.13) and (10.14), we finally obtain

$$R_h{}^i{}_{jk} = \mathfrak{A}_{(jk)}\{\delta_k F_h{}^i{}_j + F_h{}^r{}_j F_r{}^i{}_k\} + C_h{}^i{}_r R^r{}_{jk}, \tag{10.15}$$

$$P_h{}^i{}_{jk} = \dot{\partial}_k F_h{}^i{}_j - \delta_j C_h{}^i{}_k + F_h{}^r{}_j C_r{}^i{}_k - C_h{}^r{}_k F_r{}^i{}_j + C_h{}^i{}_r \dot{\partial}_k N^r{}_j. \tag{10.16}$$

In (10.15) we find a tensor K having the components

$$K_h{}^i{}_{jk} = \mathfrak{A}_{(jk)}\{\delta_k F_h{}^i{}_j + F_h{}^r{}_j F_r{}^i{}_k\}, \tag{10.17}$$

which is similar in form to the right-hand side of (5.11). Then we get

$$R_h{}^i{}_{jk} = K_h{}^i{}_{jk} + C_h{}^i{}_r R^r{}_{jk}, \tag{10.18'}$$

which may be written as

$$R^2(1,\ 2) = K(1,\ 2) + C(R^1(1,\ 2)). \tag{10.18}$$

On the other hand, the right-hand side of (10.16) has somewhat complicated form. If $\delta_j C_h{}^i{}_k$ is changed for the h-covariant derivative $C_h{}^i{}_{k|j}$, then we obtain

$$P_h{}^i{}_{jk} = F_h{}^i{}_{jk} - C_h{}^i{}_{k|j} + C_h{}^i{}_r P^r{}_{jk}, \tag{10.19'}$$

where we put

(10.20) $$F_h{}^i{}_{jk} = \dot\partial_k F_h{}^i{}_j.$$

Then (10.19') may be written in the form

(10.19) $$P^2(1, 2) = F(1, 2) - \nabla^h C(2, 1) + C(P^1(1, 2)).$$

From this equation it is observed that F, the components being given by (10.20), is a Finsler tensor field of (1, 3)-type.

We shall be concerned with two tensor fields K and F as above introduced. Changing $B^v(\cdot)$ in (10.3) and (10.4) for an induced-fundamental vector field $Y(\cdot)$ and paying attention to the important relation (8.8), we obtain

(10.21) $$[B^h(1), B^h(2)] = B^h(T(1, 2)) + Y(R^1(1, 2)) + Z(K(1, 2)),$$

(10.22) $$[B^h(1), Y(2)] = Y(P^1(1, 2)) + Z(F(1, 2)).$$

Thus K and F appear in the right-hand sides. It should be emphasized that (10.21), (10.22) and (7.3) are nothing but the structure equations of a simple Finsler connection (Γ^h, F^i) with the flat vertical connection F^i. Therefore K and F are the h- and hv-curvature tensor fields of this Finsler connection respectively. Further, it is also seen that three torsion tensors T, R^1 and P^1 of a Finsler connection (Γ^h, Γ^v) are determined by its h-part Γ^h alone, independently on the v-part Γ^v, because these tensors again appear in the structure equations (10.21) and (10.22) of the Finsler connection (Γ^h, F^i) without any modification.

<u>EXAMPLE 10.1</u>. We consider the linear Finsler connection $(\underline{\Gamma}, N, F^i)$ introduced in Example 9.1, where N is associated with the linear connection $\underline{\Gamma}$. It is clear that $S^1 = 0$, $S^2 = 0$. Paying attention to (9.17), Definition 7.3 and (6.19), from (10.21) we obtain

$$I'([B^h(1), B^h(2)]) = (B_{(v)}(T(1, 2)), \overrightarrow{D(T(1, 2))}) + (0, \overrightarrow{R^1(1, 2)}) + (\underline{Z}(K(1, 2)), -\overrightarrow{K(1, 2)\cdot\varepsilon}).$$

From (9.23) the above is rewritten

$$([\underline{B}(1), \underline{B}(2)], 0) = (\underline{B}(T(1, 2)), 0) + (0, \overrightarrow{R^1(1, 2)}) + (\underline{Z}(K(1, 2)), -\overrightarrow{K(1, 2)\cdot\varepsilon}).$$

Comparing this with (5.6), we get $T(1, 2) = \underline{\tilde T}(1, 2)$, $K(1, 2) = \underline{\tilde R}(1, 2)$ and $R^1(1, 2) = K(1, 2)\cdot\varepsilon$. Consequently we have

$$T = \underline{\tilde T}\circ\pi_2, \qquad K = \underline{\tilde R}\circ\pi_2, \qquad R^1 = (\underline{\tilde R}\circ\pi_2)\cdot\varepsilon.$$

In other words, the (h)h-torsion tensor T and h-curvature tensor $R^2(= K)$ are induced tensors from the torsion tensor and curvature tensor of $\underline{\Gamma}$ respectively. Similarly we get $P^1 = 0$, $P^2 (= F) = 0$ easily.

§ 11. Bianchi identities of Finsler connections

In the present section the Bianchi identities (5.19) and (5.20) are generalized to the case of a Finsler connection. For this purpose we consider Jacobi identities with respect to fundamental, h-basic and v-basic vector fields of the Finsler bundle F(M). The six Jacobi identities containing fundamental vector field

$$\mathfrak{S}_{(123)}\{[Z(1), [Z(2), Z(3)]]\} = 0,$$

$$[B^h(v), [Z(1), Z(2)]] + \mathfrak{A}_{(12)}\{[Z(1), [Z(2), B^h(v)]]\} = 0,$$

$$[B^v(v), [Z(1), Z(2)]] + \mathfrak{A}_{(12)}\{[Z(1), [Z(2), B^v(v)]]\} = 0,$$

$$[Z(A), [B^h(1), B^h(2)]] + \mathfrak{A}_{(12)}\{[B^h(1), [B^h(2), Z(A)]]\} = 0,$$

$$[Z(A), [B^h(1), B^v(2)]] + [B^h(1), [B^v(2), Z(A)]] + [B^v(2), [Z(A), B^h(1)]] = 0,$$

$$[Z(A), [B^v(1), B^v(2)]] + \mathfrak{A}_{(12)}\{[B^v(1), [B^v(2), Z(A)]]\} = 0$$

are all trivial. In fact, we are concerned with the fifth identity, for instance. From (10.4) and (10.2) the left-hand side is first written

$$[Z(A), B^h(C(1, 2)) + B^v(P^1(1, 2)) + Z(P^2(1, 2))]$$
$$- [B^h(1), B^v(A(2))] + [B^v(2), B^h(A(1))],$$

where $A(2) = A\cdot v_2$, for instance. From (10.1'), (10.2') and (10.4') the above is re-written

$$B^h(A\cdot(C(1, 2)) - (A\cdot C)(1, 2)) + B^v(A\cdot(P^1(1, 2)) - (A\cdot P^1)(1, 2))$$
$$+ Z([A, P^2(1, 2)] - (A\cdot P^2)(1, 2)) - B^h(C(1, A(2))) - B^v(P^1(1, A(2)))$$
$$- Z(P^2(1, A(2))) - B^h(C(A(1), 2)) - B^v(P^1(A(1), 2)) - Z(P^2(A(1), 2)).$$

The h-part of the above is the h-basic vector, corresponding to

$$A\cdot(C(1, 2)) - (A\cdot C)(1, 2) - C(1, A(2)) - C(A(1), 2),$$

which is reduced to zero because $(A\cdot C)(1, 2) = A\cdot(C(1, 2)) - C(A(1), 2) - C(1, A(2))$ from (2.22). Similarly it is seen that the v-part and vertical part are equal to zero also.

Consequently it is important to consider the following four Jacobi identities which are satisfied by h- and v-basic vector fields only.

The first identity is

$$\mathfrak{S}_{(123)}\{[B^h(1),\ [B^h(2),\ B^h(3)]]\} = 0.$$

From (10.3) it is first written as

$$\mathfrak{S}_{(123)}\{[B^h(1),\ B^h(T(2,\ 3)) + B^v(R^1(2,\ 3)) + Z(R^2(2,\ 3))]\} = 0,$$

and moreover from (10.3'), (10.4') and (10.2')

$$\begin{aligned}\mathfrak{S}_{(123)}\{&B^h(T(1,\ T(2,\ 3))) + B^v(R^1(1,\ T(2,\ 3))) + Z(R^2(1,\ T(2,\ 3)))\\ &+ B^h(\nabla^h T(2,\ 3,\ 1)) + B^h(C(1,\ R^1(2,\ 3))) + B^v(P^1(1,\ R^1(2,\ 3)))\\ &+ Z(P^2(1,\ R^1(2,\ 3))) + B^v(\nabla^h R^1(2,\ 3,\ 1)) - B^h(R^2(1,\ 2,\ 3))\\ &+ Z(\nabla^h R^2(2,\ 3,\ 1))\} = 0,\end{aligned}$$

where we use the notation $R^2(1,\ 2,\ 3) = R^2(2,\ 3)\cdot v_1$. In the following the similar notations $P^2(1,\ 2,\ 3)$ and $S^2(1,\ 2,\ 3)$ are used. Equating the h-, v-parts and vertical part of the above to zero, we obtain the *first group of Bianchi identities* as follows:

(11.1) $$\mathfrak{S}_{(123)}\{T(1,\ T(2,\ 3)) + \nabla^h T(1,\ 2,\ 3) + C(1,\ R^1(2,\ 3)) - R^2(1,\ 2,\ 3)\} = 0,$$

(11.2) $$\mathfrak{S}_{(123)}\{R^1(1,\ T(2,\ 3)) + P^1(1,\ R^1(2,\ 3)) + \nabla^h R^1(1,\ 2,\ 3)\} = 0,$$

(11.3) $$\mathfrak{S}_{(123)}\{R^2(1,\ T(2,\ 3)) + P^2(1,\ R^1(2,\ 3)) + \nabla^h R^2(1,\ 2,\ 3)\} = 0.$$

It will be useful to write these identities in terms of components. These are respectively written as follows:

(11.1') $$\mathfrak{S}_{(ijk)}\{T_i{}^h{}_r T_j{}^r{}_k + T_i{}^h{}_{j|k} + C_i{}^h{}_r R^r{}_{jk} - R_i{}^h{}_{jk}\} = 0,$$

(11.2') $$\mathfrak{S}_{(ijk)}\{R^h{}_{ir} T_j{}^r{}_k + P^h{}_{ir} R^r{}_{jk} + R^h{}_{ij|k}\} = 0,$$

(11.3') $$\mathfrak{S}_{(ijk)}\{R_m{}^h{}_{ir} T_j{}^r{}_k + P_m{}^h{}_{ir} R^r{}_{jk} + R_m{}^h{}_{ij|k}\} = 0.$$

The second identity is

$$[B^v(3),\ [B^h(1),\ B^h(2)]] + \mathfrak{A}_{(12)}\{[B^h(1),\ [B^h(2),\ B^v(3)]]\} = 0.$$

In the similar way to the first identity, the above gives the *second group of Bianchi identities* as follows:

(11.4) $$\nabla^v T(1,\ 2,\ 3) - C(T(1,\ 2),\ 3) + \mathfrak{A}_{(12)}\{T(1,\ C(2,\ 3)) + \nabla^h C(2,\ 3,\ 1)$$

$$+ C(1, P^1(2, 3)) - P^2(1, 2, 3)\} = 0,$$

(11.5) $$\nabla^v R^1(1, 2, 3) - P^1(T(1, 2), 3) + S^1(3, R^1(1, 2)) - R^2(3, 1, 2)$$
$$+ \mathfrak{A}_{(12)}\{R^1(1, C(2, 3)) + P^1(1, P^1(2, 3)) + \nabla^h P^1(2, 3, 1)\} = 0,$$

(11.6) $$\nabla^v R^2(1, 2, 3) - P^2(T(1, 2), 3) + S^2(3, R^1(1, 2))$$
$$+ \mathfrak{A}_{(12)}\{R^2(1, C(2, 3)) + P^2(1, P^1(2, 3)) + \nabla^h P^2(2, 3, 1)\} = 0.$$

It is noteworthy that (11.5) gives the h-curvature tensor field R^2 by torsion tensor fields and their covariant derivatives only.

The expressions of these identities in terms of components are respectively as follows:

(11.4') $$T_i{}^h{}_{j|k} - C_r{}^h{}_k T_i{}^r{}_j + \mathfrak{A}_{(ij)}\{T_i{}^h{}_r C_j{}^r{}_k + C_j{}^h{}_{k|i} + C_i{}^h{}_r P^r{}_{jk} - P_i{}^h{}_{jk}\} = 0,$$

(11.5') $$R^h{}_{ij|k} - P^h{}_{rk} T_i{}^r{}_j + S^h{}_{kr} R^r{}_{ij} - R_k{}^h{}_{ij}$$
$$+ \mathfrak{A}_{(ij)}\{R^h{}_{ir} C_j{}^r{}_k + P^h{}_{ir} P^r{}_{jk} + P^h{}_{jk|i}\} = 0,$$

(11.6') $$R_m{}^h{}_{ij|k} - P_m{}^h{}_{rk} T_i{}^r{}_j + S_m{}^h{}_{kr} R^r{}_{ij}$$
$$+ \mathfrak{A}_{(ij)}\{R_m{}^h{}_{ir} C_j{}^r{}_k + P_m{}^h{}_{ir} P^r{}_{jk} + P_m{}^h{}_{jk|i}\} = 0.$$

The third identity is

$$[B^h(3), [B^v(1), B^v(2)]] + \mathfrak{A}_{(12)} [B^v(1), [B^v(2), B^h(3)]] = 0.$$

Similarly we obtain the *third group of Bianchi identities* as follows:

(11.7) $$C(3, S^1(1, 2)) - S^2(3, 1, 2)$$
$$+ \mathfrak{A}_{(12)}\{C(C(3, 2), 1) - \nabla^v C(3, 2, 1)\} = 0,$$

(11.8) $$\nabla^h S^1(1, 2, 3) + P^1(3, S^1(1, 2)) + \mathfrak{A}_{(12)}\{P^1(C(3, 2), 1)$$
$$- S^1(1, P^1(3, 2)) - \nabla^v P^1(3, 2, 1) + P^2(1, 3, 2)\} = 0,$$

(11.9) $$\nabla^h S^2(1, 2, 3) + P^2(3, S^1(1, 2))$$
$$+ \mathfrak{A}_{(12)}\{P^2(C(3, 2), 1) - S^2(1, P^1(3, 2)) - \nabla^v P^2(3, 2, 1)\} = 0.$$

It is easy to show that (11.7) is another form of (10.7).

In terms of components the latter two identities (11.8) and (11.9) are written

(11.8')
$$S^h{}_{ij|k} + P^h{}_{kr}S^r{}_{ij}$$
$$+ \mathfrak{A}_{(ij)}\{P^h{}_{ri}C_k{}^r{}_j - S^h{}_{ir}P^r{}_{kj} - P^h{}_{kj|i} + P_i{}^h{}_{kj}\} = 0,$$

(11.9')
$$S_m{}^h{}_{ij|k} + P_m{}^h{}_{kr}S^r{}_{ij}$$
$$+ \mathfrak{A}_{(ij)}\{P_m{}^h{}_{ri}C_k{}^r{}_j - S_m{}^h{}_{ir}P^r{}_{kj} - P_m{}^h{}_{kj|i}\} = 0.$$

Finally we are concerned with the Jacobi identity

$$\mathfrak{S}_{(123)}\ [B^v(1),\ [B^v(2),\ B^v(3)]] = 0.$$

By a simpler computation than the above three cases, we get the *fourth group of Bianchi identities* as follows:

(11.10)
$$\mathfrak{S}_{(123)}\{S^1(1,\ S^1(2,\ 3)) + \nabla^v S^1(1,\ 2,\ 3) - S^2(1,\ 2,\ 3)\} = 0,$$

(11.11)
$$\mathfrak{S}_{(123)}\{S^2(1,\ S^1(2,\ 3)) + \nabla^v S^2(1,\ 2,\ 3)\} = 0.$$

These are quite similar in form to (5.20) and (5.19) respectively, as it will be evident from Proposition 8.2.

In terms of components these are written

(11.10')
$$\mathfrak{S}_{(ijk)}\{S^h{}_{ir}S^r{}_{jk} + S^h{}_{ij|k} - S_i{}^h{}_{jk}\} = 0,$$

(11.11')
$$\mathfrak{S}_{(ijk)}\{S_m{}^h{}_{ir}S^r{}_{jk} + S_m{}^h{}_{ij|k}\} = 0.$$

Among the above eleven Bianchi identities, those of which contain covariant derivatives of torsion tensors, that is, (11.1), (11.2), (11.4), (11.5), (11.7), (11.8) and (11.10) are called the *Bianchi identities with respect to torsions*, and remaining four identities which contain covariant derivatives of curvature tensors are the *Bianchi identities with respect to curvatures*.

REMARK 11.1. As above shown, we have many Bianchi identities of a general Finsler connection. In the next chapter, we shall be concerned with some desirable conditions which are satisfied by Finsler connections, and further introduce so-called Finsler metrics. Then the several Bianchi identities will solely reduce to results of the others (cf. §§ 17, 18).

From the procedure of our theory it is clear that the above eleven Bianchi identities constitute the complete set of such identities. Cartan's monograph [4] gives

(11.1) and (11.3) only. Even Rund's book [18] treats these two identities only. See C.I.Ispas [4, 5], Mishra and Sinha [1], Rastogi [1], and U.P.Singh and Prasad [2].

CHAPTER III

TYPICAL FINSLER CONNECTIONS

We have two famous special Finsler connections; one is the so-called euclidean connection due to Cartan [4] and the other due to Berwald [2]. Although Finsler geometry is originated by Finsler [1] undoubtedly in the viewpoint of geometrization of variation calculus, it is necessary for further progress of the geometry to introduce the concept of connection (parallelism).

Berwald constructed a connection in 1926 from a standpoint of so-called geometry of paths. On the other hand, Cartan produced a connection in 1933 along a line of his general concept of euclidean connection. Since 1950, Rund was making effort to introduce a connection in a Finsler space from a new standpoint, based on Minkowski geometry, but he finally arrived at the same horizontal connection with Cartan's.

To determine a Finsler connection from a given Finsler metric, some desirable conditions for Finsler connections are studied in this chapter, and we conclusively introduce the typical connections due to Berwald, Cartan and Rund. Those conditions first acquire precise meaning in the geometry of general Finsler connections as developed in the last chapter. In particular, Cartan's euclidean connection is determined by certain elegant conditions which are perfectly similar to those for the Riemannian connections in Riemannian spaces.

§ 12. Homogeneity of Finsler connections

We are concerned with an n-dimensional manifold M and a real vector space V^n of n dimensions with a fixed base $B = \{e_a\}$ (cf. Remark 2.3). Consider the multiplication μ: $(a, v) \in R \times V^n \mapsto av \in V^n$. By the notation μ^+ we now denote the restriction of μ to the set R^+ of all the positive numbers:

$$\mu^+ : (a, v) \in R^+ \times V^n \mapsto av \in V^n.$$

REMARK 12.1. In the following, we are concerned with transformations of the tangent bundle T(M) and of the Finsler bundle F(M) which are induced by μ^+. In this case we must restrict our consideration to non-zero tangent vectors y. If the base space M is identified with a closed subspace of T which consists of all the zero tangent vectors, the space T - M and the subbundle of F over T - M are concerned, instead of T and F themselves respectively. We shall, however, use the same letters T and F to denote such spaces.

From the mapping μ^+ we first obtain

(12.1) $$h\colon (a, y) \in R^{+}\times T \mapsto ay \in T.$$

In terms of μ^{+} the point ay is more exactly defined by $ay = {}_{z}\alpha\circ\mu^{+}(a, {}_{z}\alpha^{-1}(y))$ for any $z \in \pi_{L}^{-1}\circ\pi_{T}(y)$. A left-fixed mapping ${}_{a}h$ of h is called the *homogeneous transformation* of T by $a \in R^{+}$. It satisfies

(12.2) $$\pi_{T}\circ {}_{a}h = \pi_{T}, \qquad {}_{a}h'\circ 1_{y}^{V} = 1_{ay}^{V}\circ {}_{a}h.$$

Next, from the above h we obtain

(12.3) $$H\colon (a, u) \in R^{+}\times F \mapsto au \in F,$$

where $au = (ay, z)$ for $u = (y, z)$. A left-fixed mapping ${}_{a}H$ of H is called the *homogeneous transformation* of F by $a \in R^{+}$. Then

(12.4) $$\pi_{1}\circ {}_{a}H = {}_{a}h\circ\pi_{1}, \qquad \pi_{2}\circ {}_{a}H = \pi_{2}$$

hold evidently and are regarded as another definition of ${}_{a}H$.

PROPOSITION 12.1. *A homogeneous transformation* ${}_{a}H$ *of F satisfies*

(1) $\beta_{g}\circ {}_{a}H = {}_{a}H\circ\beta_{g}$, *(2)* ${}_{a}H\circ {}_{u}\beta = {}_{au}\beta$,

(3) $I\circ {}_{a}H = (1, {}_{a}\mu^{+})\circ I$, *(4)* $\varepsilon\circ {}_{a}H = {}_{a}\mu^{+}\circ\varepsilon$,

where $(1, {}_{a}\mu^{+})\colon (z, v) \in L\times V^{n} \mapsto (z, av) \in L\times V^{n}$.

These equations are easily proved from the definitions of β_{g}, ${}_{u}\beta$, I and ε.

A right-fixed mapping H_{u} of H satisfies

(12.5) $$\varepsilon\circ H_{u} = \mu^{+}_{\varepsilon(u)},$$

where μ^{+}_{v}, $v \in V^{n}$, is a right-fixed mapping of μ^{+}.

DEFINITION 12.1. In this definition $a \in R^{+}$ is any positive number.

(1) A vector-valued function f on F is called *(r)p-homogeneous* if $f\circ {}_{a}H = a^{r}f$ holds.

(2) A tangent vector field X of F is called *(r)p-homogeneous* if ${}_{a}H'X = a^{r}X$ holds.

(3) A vector-valued one-form α of F is called *(r)p-homogeneous* if $\alpha\circ {}_{a}H' = a^{r}\alpha$ holds.

(4) A distribution D on F is called *p-homogeneous* if ${}_{a}H'D = D$ holds.

(5) A distribution $\underline{D}$ on T is called *p-homogeneous* if ${}_{a}h'\underline{D} = \underline{D}$ holds.

REMARK 12.2. It is obvious that in terms of an induced coordinate system of F a homogeneous transformation ${}_{a}H$ is written $(x^{i}, y^{i}, z^{i}_{b}) \mapsto (x^{i}, ay^{i}, z^{i}_{b})$.

In Definition 12.1, the letter *p* of *p-homogeneous* is an abbreviation of *positively*. We must recognize the difference between "homogeneous" and "p-homogeneous". For instance, a function $f_2 = (y^1)^2 + \cdots (y^n)^2$ on F is (2)-homogeneous, while $f_1 = \sqrt{f_2}$ is (1)p-homogeneous, but not (1)-homogeneous.

We are concerned with the homogeneity of those entities which are defined without any connection.

PROPOSITION 12.2.

(1) A fundamental vector field Z(A) on F is (0)p-homogeneous.

(2) An induced-fundamental vector field Y(v) on F is (1)p-homogeneous.

(3) The supporting element ε is (1)p-homogeneous.

(4) The h-basic form θ^h on F is (0)p-homogeneous.

(5) The induced-vertical distribution F^i on F is p-homogeneous.

Proof. From (2) of Proposition 12.1 it follows that

$$ {}_aH'Z(A)_u = {}_aH' \circ \beta'_u A_e = \beta'_{au} A_e = Z(A)_{au}. $$

Next, from (12.4) we obtain

$$ \pi'_2 \circ {}_aH'Y(v)_u = \pi'_2 Y(v)_u = 0, $$

and from (4) of Proposition 12.1 and Example 2.3

$$ \varepsilon' \circ {}_aH'Y(v)_u = ({}_a\mu^+)' \circ \varepsilon' Y(v)_u = ({}_a\mu^+)'(\vec{v}) = \overrightarrow{av}. $$

Thus Definition 7.3 yields ${}_aH'Y(v) = Y(av) = aY(v)$, which shows (2).

Next, (3) is evident from (4) of Proposition 12.1. Further (4) is proved from (12.4) and (12.2) as follows:

$$ \theta^h_{au} \circ {}_aH' = {}_z\alpha^{-1} \circ \pi'_T \circ \pi'_1 \circ {}_aH' = {}_z\alpha^{-1} \circ \pi'_T \circ {}_ah' \circ \pi'_1 = {}_z\alpha^{-1} \circ \pi'_T \circ \pi'_1 = \theta^h_u. $$

Finally (5) follows immediately from (2) and the fact that F^i is spanned by induced-fundamental vectors.

PROPOSITION 12.3.

(1) Let X be an (r)p-homogeneous tangent vector field and let f be an (s)p-homogeneous function. Then X(f) is an (s - r)p-homogeneous function.

(2) If K is an (r)p-homogeneous Finsler tensor field, the 0-covariant derivative $\nabla^0 K$ of K is (r - 1)p-homogeneous.

Proof. (1) The assumptions are written as ${}_aH'X_u = a^r X_{au}$ and $f \circ {}_aH = a^s f$, so that we have

$$ X(f)_{au} = (a^{-r}\, {}_aH'X_u)(f) = a^{-r} X_u (f \circ {}_aH) = a^{s-r} X(f)_u. $$

(2) It has been shown that $Y(v)$ is (1)p-homogeneous, so that the proof follows from (1) and $\nabla^0 K(v) = Y(v)K$.

From (7.5) the intrinsic-fundamental vector field $Y(\varepsilon)$ is written as $y^i \partial/\partial y^i$, so that the well-known Euler's theorem on homogeneous functions shows that a Finsler tensor field K is (r)p-homogeneous if and only if K satisfies $(\dot{\partial}_i K)y^i = rK$, namely,

$$\nabla^0 K(\varepsilon) = rK. \tag{12.6}$$

REMARK 12.3. By the action (12.1), R^+ is regarded as a Lie transformation group of T, so that we get the quotient space T/R^+, which is the so-called sphere bundle over M. The concept "line-element space" due to Cartan [4] is nothing but the sphere bundle. See Auslander [1] and Kashiwabara [1]. In the sphere bundle, however, an induced coordinate system (x^i, y^i) on T is no longer regarded as a coordinate system, because only the ratio $y^1 : \cdots : y^n$ is essential. This brings a complicated situation into the tensor calculus in Finsler geometry. Consequently it will be better to consider the tangent bundle itself together with homogeneous transformations and to introduce a Finsler connection which is invariant by any homogeneous transformation, as in the following.

DEFINITION 12.2. A Finsler connection $F\Gamma = (\Gamma^h, \Gamma^v)$ is called *p-homogeneous* if both the h- and v-parts Γ^h, Γ^v are p-homogeneous distributions.

Then we have several other expressions of the homogeneity of Finsler connection as follows:

PROPOSITION 12.4. *A Finsler connection is p-homogeneous if and only if one of following three conditions holds:*

(1) Γ and N are p-homogeneous.

(2) Any h- and v-basic vector fields $B^h(v)$, $B^v(v)$ are (0) and (1)p-homogeneous respectively.

(3) The connection form ω and the v-basic form θ^v are (0) and (1)p-homogeneous respectively.

Proof. It is first remarked that the lift l_y, l_u with respect to a p-homogeneous Finsler connection $F\Gamma = (\Gamma, N)$ satisfy

$$_a h' \circ l_y = l_{ay}, \qquad _a H' \circ l_u = l_{au} \circ {}_a h', \tag{12.7}$$

which are easily verified. Then (1) is obvious from (9.5), (9.6) and (9.7). The proof of (2) is as follows: From (12.7) we have

$$ {}_aH'B^h(v)_u = {}_aH'\circ 1_u \circ 1_y(zv) = 1_{au}\circ 1_{ay}(zv) = B^h(v)_{au}, $$

and from (12.2)

$$ {}_aH'B^v(v)_u = {}_aH'\circ 1_u \circ 1^v_y(zv) = 1_{au}\circ 1^v_{ay}(z(av)) = aB^v(v)_{au}. $$

The duality (9.9) leads us to the proof of (3) at once. The sufficiency of these conditions will be clear.

PROPOSITION 12.5. *If K is an (r)p-homogeneous Finsler tensor field, the covariant derivatives $\nabla^h K$ and $\nabla^v K$ with respect to a p-homogeneous Finsler connection are (r) and (r - 1)p-homogeneous respectively.*

This is only a direct consequence of (1) of Proposition 12.3 and (2) of Proposition 12.4.

THEOREM 12.1. *If a Finsler connection $F\Gamma$ is p-homogeneous, the deflection tensor and the torsion and curvature tensors of $F\Gamma$ are p-homogeneous of the following degree:*

$D \cdots (1)$,	$T \cdots (0)$,	$C \cdots (-1)$,
$R^1 \cdots (1)$,	$P^1 \cdots (0)$,	$S^1 \cdots (-1)$,
$R^2 \cdots (0)$,	$P^2 \cdots (-1)$,	$S^2 \cdots (-2)$.

Proof. Proposition 12.5 and (3) of Proposition 12.2 show that D is (1)p-homogeneous. Next, from the first structure equation (10.3) we have

$$ {}_aH'[B^h(1), B^h(2)]_u = [B^h(1), B^h(2)]_{au} $$

$$ = B^h(T_{au}(1, 2))_{au} + B^v(R^1_{au}(1, 2))_{au} + Z(R^2_{au}(1, 2))_{au}. $$

On the other hand, we have

$$ = {}_aH'B^h(T_u(1, 2)) + {}_aH'B^v(R^1_u(1, 2)) + {}_aH'Z(R^2_u(1, 2)) $$

$$ = B^h(T_u(1, 2))_{au} + aB^v(R^1_u(1, 2))_{au} + Z(R^2_u(1, 2))_{au}. $$

Therefore $T_{au} = T_u$, $R^1_{au} = aR^1_u$ and $R^2_{au} = R^2_u$. Applying the similar way to the other structure equations (10.4) and (10.5), we are led to the complete proof.

We consider the homogeneity of a Finsler connection $F\Gamma$ in terms of coordinate system. Because of ${}_aH(x^i, y^i, z^i_b) = (x^i, ay^i, z^i_b)$, we get

$$ {}_aH': \quad \frac{\partial}{\partial x^i} \mapsto \frac{\partial}{\partial x^i}, \quad \frac{\partial}{\partial y^i} \mapsto a\frac{\partial}{\partial y^i}, \quad \frac{\partial}{\partial z^i_b} \mapsto \frac{\partial}{\partial z^i_b}. $$

Hence (8.6') gives

$$ {}_aH'B^v(v)_u = z^i_b v^b \left(a \frac{\partial}{\partial y^i} - C_j{}^k{}_i(y) z^j_c \frac{\partial}{\partial z^k_c} \right). $$

This is equal to $aB^v(v)_{au}$, so that $C_j{}^k{}_i(y) = aC_j{}^k{}_i(ay)$ and $C_j{}^k{}_i$ are (-1)p-homogeneous functions in y^i. Similarly from (9.10) $N^j{}_i$ and $F_j{}^i{}_k$ are (1) and (0)p-homogeneous functions respectively.

Conversely, the above p-homogeneity of coefficients $(F_j{}^i{}_k, N^i{}_j, C_j{}^i{}_k)$ of $F\Gamma$ yields the homogeneity of $F\Gamma$, which will be clear from (2) of Proposition 12.4.

THEOREM 12.2. *If a Finsler connection is p-homogeneous, we have*

(1) $P^1(v, \varepsilon) = - D(v)$,

(2) $P^2(v, \varepsilon) = - \nabla^h(C(\varepsilon))v \ \{= - \nabla^h C(\varepsilon, v) - C(D(v))\}$, *or* $F(v, \varepsilon) = 0$

for any $v \in V^n$.

Proof. If we refer to those components, the proof is easily obtained. Another proof is as follows: Let K be an (r)p-homogeneous Finsler tensor field. Then it is seen that

$$ (\nabla^0\nabla^h K)(v, \varepsilon) - (\nabla^h\nabla^0 K)(\varepsilon, v) = \nabla^0(\nabla^h K(v))\varepsilon - \{\nabla^h(\nabla^0 K(\varepsilon))v - \nabla^0 K(D(v))\}. $$

Then (12.6) and Proposition 12.5 lead to

$$ = r\nabla^h K(v) - \{r\nabla^h K(v) - \nabla^0 K(D(v))\}. $$

On the other hand, the equation (10.22) yields

$$ = - \nabla^0 K(P^1(v, \varepsilon)) + F(v, \varepsilon)\cdot K. $$

Therefore we obtain $D(v) = - P^1(v, \varepsilon)$ and $F(v, \varepsilon) = 0$ from arbitrariness of K.

REMARK 12.4. In the connection due to Cartan, Theorem 12.2 gives the well-known identities $P^1(v, \varepsilon) = 0$ $(P^i{}_{jk}y^k = 0)$ and $P^2(v, \varepsilon) = 0$ $(P_h{}^i{}_{jk}y^k = 0)$. Therefore we notice that these distinguished properties of Cartan's connection are results from the p-homogeneity only, independent on his axioms (cf. § 17). As to this situation, we recall a paragraph on the introduction of Finsler's thesis [1]:

Eine Behandlung der Geometrie unter möglichst allgemeinen Voraussetzungen hat zugleich den Vorteil, erkennen zu lassen, inwieweit die einzelnen Sätze derselben von speziellen Annahmen über die Massbestimmung oder die Zahl der Dimensionen unabhängig sind. (See Appendix.)

§ 13. The characteristic conditions for Finsler connections

In this section we shall introduce three charactetistic conditions, which are desirable to Finsler connections in a sense and satisfied by the well-known connections.

The meanings of these conditions are geometrically given as certain relations of connections which constitute a Finsler connection.

DEFINITION 13.1. If with respect to a Finsler connection $F\Gamma$ the intrinsic induced-fundamental vector field $Y(\varepsilon)$ is v-horizontal ($Y(\varepsilon) \in \Gamma^V$) at each point, $F\Gamma$ is said to *satisfy the* C_2*-condition.* If a p-homogeneous Finsler connection $F\Gamma$ satisfies the C_2-condition, $F\Gamma$ is called *strictly p-homogeneous.*

First we show that the C_2-condition is in close relation to the homogeneity. Let $\vec{a}$ be the tangent vector $(d/dt)_a$ at a point $a \in R^+$. Since the mapping $\mu_v^+ : a \in R^+ \mapsto av \in V^n$ is such that $av = (av^b)e_b$ for $v = v^b e_b$, the differential $(\mu_v^+)'$ is given by $\vec{a} \mapsto v^b(\partial/\partial v^b)_{av}$. In the notation of Example 2.2, this is written

$$(13.1) \qquad (\mu_v^+)'(\vec{a}) = \vec{v}.$$

Consider a tangent vector $H_u'(\vec{a}) \in F_{au}$. Because $\pi_2 \circ H_u$ is a constant mapping, from Definition 7.3 we have

$$\pi_2'\{H_u'(\vec{a}) - a^{-1}Y(\varepsilon)_{au}\} = - a^{-1}\pi_2'Y(\varepsilon)_{au} = 0.$$

From (12.5), Definition 7.3 and (13.1) it is observed that

$$\varepsilon'\{H_u'(\vec{a}) - a^{-1}Y(\varepsilon)_{au}\} = (\mu_{\varepsilon(u)}^+)'(\vec{a}) - a^{-1}\overrightarrow{\varepsilon(au)} = 0.$$

Consequently Proposition 6.3 yields

$$(13.2) \qquad H_u'(\vec{a}) = a^{-1}Y(\varepsilon).$$

As a consequence, we have another meaning of the C_2-condition as follows:

PROPOSITION 13.1. *The C_2-condition of a Finsler connection is equivalent to the fact that the tangent vector field $H_u'(\vec{a})$ is v-horizontal.*

Secondly, it follows from (8.8) that

$$Y(\varepsilon) = B^V(\varepsilon) + Z(C(\varepsilon)) \in \Gamma^V \oplus F^V,$$

so that the following is evident:

PROPOSITION 13.2. *The C_2-condition of a Finsler connection is such that the Cartan tensor field C satisfies $C(\varepsilon) = 0$.*

From Theorem 12.2 we immediately see

THEOREM 13.1. *If a Finsler connection is strictly p-homogeneous, the hv-curvature tensor field P^2 satisfies $P^2(v, \varepsilon) = 0$ for any $v \in V^n$.*

REMARK 13.1. It follows from Proposition 13.2 that in terms of components of the Cartan tensor C the C_2-condition is written as $C_j{}^i{}_k y^k = 0$. This condition is assumed by Cartan [5] in virtue of which the absolute differential

$$DX^i = dX^i + X^r(\Gamma_r{}^i{}_j dx^j + C_r{}^i{}_j d\dot{x}^j)$$

of a tangent vector field X^i should be invariant when we multiply $\dot{x}^j$ by a constant common factor. In Matsumoto [1] this condition is stated in terms of the parallelism. See Theorem 38.2. The equation $P^2(v, \varepsilon) = 0$ for any v is written $P_h{}^i{}_{jk} y^k = 0$ in terms of components $P_h{}^i{}_{jk}$ of P^2. See § 17.

Next we consider a mapping $\pi_v = \alpha_v \circ \pi_2$: $(y, z) \in F \mapsto zv \in T$. Then (6.18) is written

$$\pi_1' = \pi_v' + {}_z\alpha' \circ \varepsilon', \qquad u = (zv, z). \tag{13.3}$$

It is noted that $\pi_1(u) = \pi_v(u)$ for the point $u = (zv, z)$ and that $\pi_1'\Gamma^h = N$ from (9.7) and $\pi_1'\Gamma^v = T^v$ from (9.5). We examine $\pi_v'\Gamma^h$ and $\pi_v'\Gamma^v$ at $u = (zv, z)$ in the following. From (9.16) and (13.3) we have

$$\pi_1'B^h(w) = \pi_v'B^h(w) + {}_z\alpha'\overrightarrow{D(w)}, \tag{13.4}$$

and similarly from (9.14)

$$\pi_1'B^v(w) = \pi_v'B^v(w) + {}_z\alpha'(\overrightarrow{w + C(\varepsilon, w)}). \tag{13.5}$$

Now we are concerned with (13.4). From (9.7) it follows that the non-linear connection N is spanned by $\pi_1'B^h(w)$. On the other hand, it follows from (8.12) that each distribution of Γ_V is spanned by $\pi_2'B^h(w)$, so that Proposition 8.6 shows that the associated non-linear connection *N with Γ_V is spanned by $\pi_v'B^h(w)$. Because ${}_z\alpha'\overrightarrow{D(w)}$ is vertical, from (13.4) we obtain the following proposition:

PROPOSITION 13.3. *A necessary and sufficient condition for a Finsler connection $F\Gamma = (\Gamma_V, N, \Gamma^v)$ such that N is associated with Γ_V is that the deflection tensor field D vanishes.*

DEFINITION 13.2. If the non-linear connection N of a Finsler connection $F\Gamma = (\Gamma_V, N, \Gamma^v)$ is associated with the V-connection Γ_V of $F\Gamma$, that is, the deflection tensor field D vanishes, $F\Gamma$ is said to *satisfy the D-condition.*

It follows from (9.16) that D = 0 means $\nabla^h\varepsilon = 0$ ($y^i{}_{|j} = 0$). Theorems 12.2 and 13.1 lead us to an interesting theorem as follows:

THEOREM 13.2. *If a strictly p-homogeneous Finsler connection satisfies the D-condition, the (v)hv-torsion tensor P^1 and the hv-curvature tensor P^2 satisfy*

$$P^1(v, \varepsilon) = 0, \qquad P^2(v, \varepsilon) = 0 \qquad \text{for any } v \in V^n.$$

Let $*F\Gamma = (\Gamma, *N)$ be the Finsler connection obtained from a general Finsler connection $F\Gamma = (\Gamma, N) = (\Gamma_V, N, \Gamma^v)$ by changing N for the associated non-linear connection $*N$ with Γ_V. Since $F\Gamma$ and $*F\Gamma$ have the common Γ, they have the common lift l_u. Then, applying l_u to (13.4), it follows from (8.6) that

(13.6) $$B^h(v) = *B^h(v) + B^v(D(v)),$$

where $*B^h(v)$ is an h-basic vector field with respect to $*F\Gamma$. This shows a role of the deflection tensor D.

REMARK 13.2. In terms of connection coefficients the D-condition is written as $N^i{}_j = y^k F_k{}^i{}_j$ from (9.22). Cf. (8.18). All the well-known Finsler connections satisfy the D-condition and, as a consequence, it seems to the author that the importance of non-linear connection in Finsler geometry has been left behind by almost all the authors, because $N^i{}_j$ are solely given by $F_k{}^i{}_j$ as above. Although the D-condition may be desirable to a Finsler connection, it will be seen in Theorem 17.2 that this condition is never necessary to determine a Finsler connection from a Finsler metric.

We shall turn to the consideration of (13.5). Its left-hand side is equal to ${}_z\alpha'\vec{w}$ from (8.6), so that it is reduced to

(13.5') $$\pi'_v B^v(w) = -\ {}_z\alpha'\overrightarrow{C(\varepsilon, w)}.$$

As a consequence we have the following proposition:

PROPOSITION 13.4. *A necessary and sufficient condition for a Finsler connection $F\Gamma = (\Gamma^h, \Gamma^v)$ to satisfy $\pi'_v\Gamma^v = 0$ at any point $u = (zv, z)$ is that the Cartan tensor field C satisfies $C(\varepsilon, w) = 0$ for any $w \in V^n$.*

DEFINITION 13.3. If a Finsler connection $F\Gamma = (\Gamma^h, \Gamma^v)$ satisfies $\pi'_v\Gamma^v = 0$ at any point $u = (y, z)$, then $F\Gamma$ is said to *satisfy the C_1-condition.*

If a Finsler connection satisfies the D- and C_1-conditions, the torsion and curvature tensors behave conspicuously in their relations: In general, applying the Ricci identity (10.8) to the supporting element ε, we have

(13.7) $$R^2(\varepsilon, 1, 2) = R^1(1, 2) + C(\varepsilon, R^1(1, 2)) + D(T(1, 2)) + \mathfrak{A}_{(12)}\{\nabla^h D(1, 2)\}.$$

Similarly, from (10.9) and (10.10) we get

$$P^2(\varepsilon, 1, 2) = P^1(1, 2) + C(\varepsilon, P^1(1, 2)) - \nabla^h(C(\varepsilon, 2))v_1 + D(C(1, 2)) + \nabla^v D(1, 2), \tag{13.8}$$

$$S^2(\varepsilon, 1, 2) = S^1(1, 2) + C(\varepsilon, S^1(1, 2)) + \mathfrak{A}_{(12)}\{\nabla^v(C(\varepsilon, 1))v_2\}. \tag{13.9}$$

As a consequence, we conclude the following remarkable theorem:

THEOREM 13.3. *If a Finsler connection satisfies the D and C_1-conditions, we have*

(1) $R^1(1, 2) = R^2(\varepsilon, 1, 2)$, $\quad P^1(1, 2) = P^2(\varepsilon, 1, 2)$,

$S^1(1, 2) = S^2(\varepsilon, 1, 2)$,

(2) $\nabla^h R^1(1, 2, 3) = \nabla^h R^2(\varepsilon, 1, 2, 3)$, $\quad \nabla^h P^1(1, 2, 3) = \nabla^h P^2(\varepsilon, 1, 2, 3)$,

$\nabla^h S^1(1, 2, 3) = \nabla^h S^2(\varepsilon, 1, 2, 3)$,

(3) $\nabla^v R^1(1, 2, 3) = \nabla^v R^2(\varepsilon, 1, 2, 3) + R^2(3, 1, 2)$,

$\nabla^v P^1(1, 2, 3) = \nabla^v P^2(\varepsilon, 1, 2, 3) + P^2(3, 1, 2)$,

$\nabla^v S^1(1, 2, 3) = \nabla^v S^2(\varepsilon, 1, 2, 3) + S^2(3, 1, 2)$.

It is observed from Theorem 13.3 that the Bianchi identities (11.2), (11.5), (11.8) and (11.10) with respect to torsion tensors are merely results of (11.3), (11.6), (11.9) and (11.11) with respect to curvature tensors respectively.

REMARK 13.3. In Matsumoto [1] both the D and C_1-conditions are stated in terms of the parallelism as well as the C_2-condition. See Theorems 38.3 and 38.4. As to Cartan's theory the C_1-condition is a consequence of his axioms.

(1) of Theorem 13.3 is noteworthy, because the (v)h-, (v)hv- and (v)v-torsion tensors are simply given from the curvature tensors; in terms of components those are written in the form

$$R^i{}_{jk} = y^h R_h{}^i{}_{jk}, \qquad P^i{}_{jk} = y^h P_h{}^i{}_{jk}, \qquad S^i{}_{jk} = y^h S_h{}^i{}_{jk}.$$

Further the (h)h and (v)v-torsion tensors T, S^1 of Cartan's connection are assumed to vanish (cf. Definition 17.2). As a consequence, it may be said that the (h)hv-torsion tensor C is recognized as the only one essential torsion tensor.

§ 14. Difference of two Finsler connections

The concept of difference tensor of two linear connections (cf. § 4) is more important in the Finsler case, because several authors have introduced connections from their own standpoints. Let $F\Gamma$ and $*F\Gamma$ be two Finsler connections on a same manifold M. In the following, entities of $*F\Gamma$ are marked with asterisk.

We are first concerned with h-basic vector fields $B^h(v)$ and $*B^h(v)$, corresponding to a same $v \in V^n$. It follows from Definition 9.3 that $\pi_T' \circ \pi_1'(*B^h(v) - B^h(v)) = 0$, hence $*B^h(v) - B^h(v) \in F^q$. Therefore from (1) of Definition 8.4 it is seen that, with respect to $F\Gamma$, $*B^h(v)$ is written in the form

$$(14.1) \qquad *B^h(v) = B^h(v) + B^v(D^h(v)) + Z(A^h(v)),$$

where $D^h(v) \in V^n$ and $A^h(v) \in G'(n)$ are obtained, corresponding to $v \in V^n$ and each point of F. It is easy to show that D^h and A^h are Finsler tensor fields of (1, 1)-type and (1, 2)-type respectively.

Next, it follows from (8.6) that $\pi_1'(*B^v(v) - B^v(v)) = 0$, so that $*B^v(v) - B^v(v) \in F^v$. Therefore, with respect to $F\Gamma$, the v-basic vector field $*B^v(v)$ is written in the form

$$(14.2) \qquad *B^v(v) = B^v(v) + Z(A^v(v)),$$

where A^v is a Finsler tensor field of (1, 2)-type, similarly to the case of (14.1).

DEFINITION 14.1. The Finsler tensor fields D^h, A^h and A^v appearing in (14.1) and (14.2) are called the *difference tensor fields of $*F\Gamma$ from $F\Gamma$.*

It will be obvious that there exists a one-to-one correspondence between the set of all the Finsler connections on a manifold M and the set of all the triads (D^h, A^h, A^v) of Finsler tensor fields of (1,1)-, (1,2)- and (1,2)-type respectively such that D^h, A^h and A^v are the difference tensor fields of a Finsler connection $*F\Gamma$ from a fixed Finsler connection $F\Gamma$. Moreover, as to the homogeneity of Finsler connection, the following will be easily verified:

PROPOSITION 14.1. *Let $F\Gamma$ be a p-homogeneous Finsler connection. Then a Finsler connection $*F\Gamma$ is p-homogeneous, if and only if the difference tensor fields D^h, A^h and A^v of $*F\Gamma$ from $F\Gamma$ are p-homogeneous of respective degree 1, 0, -1.*

From (14.1) and (14.2) we get covariant derivatives $*\nabla^h K$ and $*\nabla^v K$ of a Finsler tensor field K with respect to $*F\Gamma$:

$$(14.3) \qquad *\nabla^h K(v) = \nabla^h K(v) + \nabla^v K(D^h(v)) - A^h(v)\cdot K,$$

$$(14.4) \qquad *\nabla^v K(v) = \nabla^v K(v) - A^v(v)\cdot K.$$

Taking a Finsler tensor field X of (1, 0)-type, the above relations are written

$$\partial_j X^i - {*N}^r{}_j \dot{\partial}_r X^i + X^r {*F}_r{}^i{}_j$$

$$= \partial_j X^i - N^r{}_j \dot{\partial}_r X^i + X^r F_r{}^i{}_j + (\dot{\partial}_r X^i + X^s C_s{}^i{}_r)(D^h)^r{}_j - (A^h)_r{}^i{}_j X^r,$$

$$\dot{\partial}_j X^i + X^r {*C}_r{}^i{}_j = \dot{\partial}_j X^i + X^r C_r{}^i{}_j - (A^v)_r{}^i{}_j X^r,$$

where $(D^h)^r{}_j$, $(A^h)_r{}^i{}_j$ and $(A^v)_r{}^i{}_j$ are components of the difference tensors D^h, A^h and A^v respectively such that for $v = (v^a)$

$$D^h(v) = \{(z^{-1})^a_r (D^h)^r{}_j (z^j_b v^b)\} e_a \in V^n,$$

$$A^h(v) = \{(z^{-1})^a_i (A^h)_r{}^i{}_j z^r_b (z^j_c v^c)\} E^b_a \in G'(n),$$

and the similar equation for A^v. Thus, by arbitrariness of X^i we obtain

$$*F_k{}^i{}_j = F_k{}^i{}_j + C_k{}^i{}_r (D^h)^r{}_j - (A^h)_k{}^i{}_j,$$

(14.5) $$*N^i{}_j = N^i{}_j - (D^h)^i{}_j,$$

$$*C_k{}^i{}_j = C_k{}^i{}_j - (A^v)_k{}^i{}_j.$$

The last equation of (14.5) is also derived as follows: Substituting from (8.8), the equation (14.2) is written

$$- Z(*C(v)) = - Z(C(v)) + Z(A^v(v)),$$

which implies

(14.6) $$*C(v) = C(v) - A^v(v).$$

Applying (14.1) to the supporting element ε, we get the relation between deflection tensors:

(14.7) $$*D(v) = D(v) + D^h(v) + C(\varepsilon, D^h(v)) - A^h(\varepsilon, v).$$

On the other hand, applying (14.2) to ε, we solely get an equation which is a consequence of (14.6).

If we construct the structure equations of $*F\Gamma$ by means of (14.1) and (14.2), disgusting and complicated calculations lead us to the following relations between the torsion and curvature tensors of $F\Gamma$ and $*F\Gamma$: Putting $D^h_\alpha = D^h(\alpha)$, $A^h_\alpha = A^h(\alpha)$ and $A^v_\alpha = A^v(\alpha)$, $\alpha = 1, 2$, where α denote the indices of $v_1, v_2 \in V^n$, we have

(14.8) $$*T(1, 2) = T(1, 2) - \mathfrak{A}_{(12)}\{A^h(1, 2) - C(1, D^h_2)\},$$

(14.9) $$*R^1(1, 2) = R^1(1, 2) - D^h(T(1, 2)) + S^1(D^h_1, D^h_2) - \mathfrak{A}_{(12)}\{\nabla^h D^h(1, 2)$$

$$+ \nabla^v D^h(1, D^h_2) + D^h(C(1, D^h_2)) - P^1(1, D^h_2)\},$$

$$(14.10) \qquad *P^1(1, 2) = P^1(1, 2) + A^h(2, 1) + S^1(D^h_1, 2) - D^h(C(1, 2)) - \nabla^v D^h(1, 2),$$

$$(14.11) \qquad *S^1(1, 2) = S^1(1, 2) - \mathfrak{A}_{(12)}\{A^v(1, 2)\},$$

$$(14.12) \qquad *R^2(1, 2) = R^2(1, 2) - A^h(T(1, 2)) + S^2(D^h_1, D^h_2) - A^v(R^1(1, 2))$$
$$+ A^v(D^h(T(1, 2))) - A^v(S^1(D^h_1, D^h_2)) - \mathfrak{A}_{(12)}\{\nabla^h A^h(1, 2)$$
$$- A^h(A^h_1, 2) + \nabla^v A^h(1, D^h_2) + A^h(C(1, D^h_2)) - P^2(1, D^h_2)$$
$$+ A^v(P^1(1, D^h_2)) - A^v(D^h(C(1, D^h_2))) - A^v(\nabla^v D^h(1, D^h_2))$$
$$- A^v(\nabla^h D^h(1, 2))\},$$

$$(14.13) \qquad *P^2(1, 2) = P^2(1, 2) - \nabla^v A^h(1, 2) - A^h(C(1, 2)) + S^2(D^h_1, 2) - A^h(A^v_2, 1)$$
$$+ A^v(A^h_1, 2) + \nabla^h A^v(2, 1) + \nabla^v A^v(2, D^h_1) - A^v(S^1(D^h_1, 2))$$
$$- A^v(P^1(1, 2)) + A^v(D^h(C(1, 2))) + A^v(\nabla^v D^h(1, 2)),$$

$$(14.14) \qquad *S^2(1, 2) = S^2(1, 2) - A^v(S^1(1, 2)) - \mathfrak{A}_{(12)}\{\nabla^v A^v(1, 2) - A^v(A^v_1, 2)\}.$$

<u>REMARK 14.1</u>. It will be a hard work to find the components of $*T_j{}^i{}_k$, $\cdots$, $*S_h{}^i{}_{jk}$ through direct computation from (14.5) by substituting into (10.15) and so on. Hashiguchi [5] finds these equations (14.8), $\cdots$, (14.14) in the above way to consider conformal changes of Finsler metrics. In case of conformal change the tensor $C_j{}^i{}_k = g^{ih}(\dot\partial_h g_{jk})/2$ is invariant. Thus, if we are concerned with Cartan's connection, we have $T = S^1 = A^v = 0$, so that these equations become considerably simple, and in consequence the difference of torsion and curvature tensors can be written down explicitly. It is, however, an open problem to find so-called conformal torsion and curvature tensors, similar to the Weyl curvature tensor in Riemannian geometry. As a consequence, we do not know yet the condition for a Finsler space to be conformal to a Minkowski space, for instance. See Hashiguchi and Ichijyo [2], Izumi [1], Tamássy and Matsumoto [1].

Now we are concerned with a general Finsler connection $F\Gamma = (\Gamma, N) = (\Gamma^h, \Gamma^v) = (\Gamma_V, N, \Gamma^v)$. Among the five distributions Γ, N, Γ^h, Γ^v, Γ_V constituting $F\Gamma$, only two Γ, Γ^v are connections in the strict sense, while Γ^h, N, Γ_V are not so. It will be interesting problems to develope theories of these connection-like distributions, in particular, the non-linear connection N. In the following we shall show an introduction to these theories as applications of the concept of difference tensors of two Finsler connections.

In the remainder of the present section, we only deal with the theory of horizontal connections Γ^h, because it is of the simplest behavior among Γ^h, N and Γ_V. Our starting point is as follows:

PROPOSITION 14.2. *Two Finsler connections* $F\Gamma$ *and* $*F\Gamma$ *have a common horizontal connection* Γ^h, *if and only if the difference tensors* D^h *and* A^h *vanish identically.*

This is evident from (14.1). In this case, we get $*B^h(v) = B^h(v)$. Eliminating A^v from (14.2) and (14.6), we obtain

$$*B^v(v) + Z(*C(v)) = B^v(v) + Z(C(v)),$$

which is nothing but an induced-fundamental vector field Y(v) from (8.8). However that may be, two tangent vector fields

$$\text{(14.15)} \qquad 'B^h(v) = B^h(v), \qquad 'B^v(v) = B^v(v) + Z(C(v))$$

may be said to be common to all the Finsler connections which have a same horizontal connection Γ^h.

It is observed that there exists such a Finsler connection $'F\Gamma$ that h-and v-basic vector fields are given by (14.15) respectively; $'F\Gamma$ is constructed from $F\Gamma$ by taking the difference tensors $(D^h, A^h, A^v) = (0, 0, C)$.

Accordingly, if the set of all the Finsler connections having a same horizontal connection Γ^h is denoted by $F\Gamma(\Gamma^h)$, the above Finsler connection $'F\Gamma$ is thought of as a representative of the set $F\Gamma(\Gamma^h)$ in such a sense that $'F\Gamma$ has the h- and v-basic vector fields which are commonly defined with respect to all the elements of $F\Gamma(\Gamma^h)$ and so constructed by this Γ^h only. Therefore it may be said that the theory of a horizontal connection Γ^h is regarded as the theory of this Finsler connection $'F\Gamma$.

DEFINITION 14.2. The Finsler connection $'F\Gamma = (\Gamma^h, F^i)$ with (0, 0, C) as the difference tensors from any $F\Gamma$ of the set $F\Gamma(\Gamma^h)$ is called the *Finsler* Γ^h*-connection* and denoted by $F\Gamma^h$. The process to construct $F\Gamma^h$ from a Finsler connection belonging to $F\Gamma(\Gamma^h)$ is called the *C-process*.

The deflection tensor $'D$, torsion and curvature tensors $'T, \cdots, 'S^2$ of the Finsler Γ^h-connection $F\Gamma^h$ are given in terms of tensors of $F\Gamma \in F\Gamma(\Gamma^h)$ as follows:

$$\begin{cases} 'D = D, \quad 'C = 0, \quad 'T = T, \quad 'R^1 = R^1, \\ 'P^1 = P^1, \quad 'S^1 = 0, \quad 'S^2 = 0, \end{cases}$$

$$(14.16)\quad \begin{cases} 'R^2(1,\ 2) = R^2(1,\ 2) - C(R^1(1,\ 2)) = K(1,\ 2) \\ 'P^2(1,\ 2) = P^2(1,\ 2) + \nabla^h C(2,\ 1) - C(P^1(1,\ 2)) = F(1,\ 2). \end{cases}$$

These are obtained from (14.8), $\cdots$, (14.14) by putting $(D^h,\ A^h,\ A^v) = (0,\ 0,\ C)$.

THEOREM 14.1. *The Finsler Γ^h-connection $F\Gamma^h$ is a unique element of the set $F\Gamma(\Gamma^h)$ such that the Cartan tensor field C vanishes.*

§ 15. Theory of non-linear connections

In the first half of the present section, we shall develop a theory of non-linear connections N. Let $F\Gamma(N)$ be the set of all the Finsler connections on a manifold M which have a common non-linear connection N.

PROPOSITION 15.1. *Two Finsler connections $F\Gamma$ and $*F\Gamma$ have a same non-linear connection N, if and only if the difference tensor field D^h vanishes.*

This is clear from (14.1) and (9.7). Putting $D^h = 0$ in (14.10), we get

$$A^h(2,\ 1) = *P^1(1,\ 2) - P^1(1,\ 2).$$

Making use of the Finsler tensor field δ of (1, 1)-type, so-called Kronecker delta, such that $\delta(v) = v$ for any $v \in V^n$, the above may be written $A^h(v) = *P^1(v,\ \delta) - P^1(v,\ \delta)$. Then (14.1) is rewritten as

$$*B^h(v) - Z(*P^1(v,\ \delta)) = B^h(v) - Z(P^1(v,\ \delta)),$$

so that the tangent vector fields

$$(15.1)\qquad N(v) = B^h(v) - Z(P^1(v,\ \delta)), \qquad Y(v) = B^v(v) + Z(C(v))$$

are common to all the Finsler connections belonging to $F\Gamma(N)$, where $Y(v)$ is an induced-fundamental vector field. There exists such a Finsler connection FN that h- and v-basic vector fields are given by (15.1) respectively, because FN is constructed from any Finsler connection $F\Gamma \in F\Gamma(N)$ by taking the difference tensors $(D^h(v),\ A^h(v),\ A^v(v)) = (0,\ -P^1(v,\ \delta),\ C(v))$.

DEFINITION 15.1. The Finsler connection $FN \in F\Gamma(N)$ such that h-and v-basic vector fields are given by (15.1) respectively is called the *Finsler N-connection.*

Therefore it may be said that FN is a representative of the set $F\Gamma(N)$ and the theory of a non-linear connection N is regarded as the theory of the Finsler N-connection FN.

DEFINITION 15.2. The process to construct a Finsler connection $*F\Gamma$ from a

Finsler connection $F\Gamma$ by taking the difference tensors $(D^h(v), A^h(v), A^v(v)) = (0, - P^1(v, \delta), 0)$ is called the *P^1-process*.

Putting $(D^h(v), A^h(v), A^v(v)) = (0, - P^1(v, \delta), 0)$ in (14.1), (14. 2), (14.6), ⋯, (14.14), we obtain $*B^h(v), \cdots, *S^2$ of the Finsler connection $*F\Gamma$ constructed from $F\Gamma$ by the P^1-process as follows:

$$
\begin{aligned}
&*B^h(v) = B^h(v) - Z(P^1(v, \delta)), \qquad *B^v(v) = B^v(v), \\
&*C = C, \quad *R^1 = R^1, \quad *P^1 = 0, \quad *S^1 = S^1, \\
&*S^2 = S^2, \qquad *D(v) = D(v) + P^1(v, \varepsilon), \\
&*T(1, 2) = T(1, 2) - \mathfrak{A}_{(12)}\{P^1(1, 2)\}, \\
&*R^2(1, 2) = R^2(1, 2) + P^1(T(1, 2), \delta) \\
&\qquad + \mathfrak{A}_{(12)}\{\nabla^h P^1(1, \delta, 2) + P^1(2, P^1(1, \delta))\}, \\
&*P^2(1, 2) = P^2(1, 2) + \nabla^v P^1(1, \delta, 2) + P^1(C(1, 2), \delta).
\end{aligned} \tag{15.2}
$$

PROPOSITION 15.2. *The Finsler N-connection FN is obtained from a Finsler connection $F\Gamma \in F\Gamma(N)$ by the C- and P^1-processes in this order or the P^1- and C-processes in this order.*

Proof. We denote by $'F\Gamma$ and $*F\Gamma$ the Finsler connections obtained from $F\Gamma$ by the C- and P^1-processes respectively. Then, from (14.15) and (15.2) it follows that

$$
\begin{aligned}
&'B^h(v) = B^h(v), \qquad 'B^v(v) = Y(v). \\
&*B^h(v) = B^h(v) - Z(P^1(v, \delta)), \qquad *B^v(v) = B^v(v).
\end{aligned}
$$

Next we denote by $*'F\Gamma$ the Finsler connection obtained from $'F\Gamma$ by the P^1-process. Then (15.2) and (14.16) give

$$
\begin{aligned}
&*'B^h(v) = 'B^h(v) - Z('P^1(v, \delta)) = B^h(v) - Z(P^1(v, \delta)), \\
&*'B^v(v) = 'B^v(v) = Y(v).
\end{aligned}
$$

On the other hand, we denote by $'*F\Gamma$ the Finsler connection obtained from $*F\Gamma$ by the C-process. Then

$$
\begin{aligned}
&'*B^h(v) = *B^h(v) = B^h(v) - Z(P^1(v, \delta)), \\
&'*B^v(v) = Y(v).
\end{aligned}
$$

Therefore (15.1) is obtained by either process.

THEOREM 15.1. *The Finsler N-connection FN is a unique element of the set* $F\Gamma(N)$ *such that the Cartan tensor field C and (v)hv-torsion tensor field* P^1 *vanish.*

Proof. From (14.16) and (15.2) we obtain $'C = 0$, $*'C = 'C = 0$, and further $'P^1 = P^1$, $*'P^1 = 0$ in the above notations.

The converse is shown as follows: Let $*F\Gamma$ and $F\Gamma$ be two Finsler connections which belong to $F\Gamma(N)$ and such that $*C = *P^1 = 0$. Then $D^h = 0$ from Proposition 15.1 and (14.6) gives $A^v(v) = C(v)$. Therefore, from (14.10) we get $A^h(v) = - P^1(v, \delta)$. Consequently $*F\Gamma$ is the Finsler N-connection constructed from $F\Gamma$.

REMARK 15.1. From (10.14) it follows that $P^1 = 0$ is written $\dot{\partial}_k N^i_j = F_k{}^i{}_j$. Thus it may be said that the V-connection of a Finsler N-connection FN is obtained from the non-linear connection N of FN.

The torsion and curvature tensors of FN, that is, those of the non-linear connection N, will be obtained from (14.6), (14.8), $\cdots$, (14.14) by putting $(D^h(v), A^h(v), A^v(v)) = (0, - P^1(v, \delta), C(v))$, or from (14.16) and (15.2) by means of Proposition 15.2, or from (15.1) by constructing the structure equations of FN. See Matsumoto [19]. Roughly speaking, those are obtained by putting $F_k{}^i{}_j = \dot{\partial}_k N^i_j$ and $C_j{}^i{}_k = 0$ in the general case.

It will be seen in § 18 that Berwald's connection is the Finsler N-connection with $N^i_j = G^i_j$.

Next, let $F\Gamma(\Gamma_V)$ be the set of all the Finsler connections on a manifold M which have a common V-connection Γ_V. From (14.1), (8.11) and (6.9) we get

$$\pi_2' {*B^h}(v) = \pi_2' B^h(v) - \underline{Z}(C(D^h(v)) - A^h(v)),$$

so that (9.11) leads to

PROPOSITION 15.3. *Two Finsler connection* $F\Gamma$ *and* $*F\Gamma$ *have a same V-connection* Γ_V, *if and only if* A^h *of the difference tensors is given by* $A^h(v) = C(D^h(v))$ *for any* $v \in V^n$.

In this case (14.7) yields $D^h = *D - D$, so that we have $A^h(v) = C(*D(v)) - C(D(v))$. Therefore (14.1) is rewritten

$$(15.3) \qquad *B^h(v) - B^v(*D(v)) - Z(C(*D(v))) = B^h(v) - B^v(D(v)) - Z(C(D(v))).$$

On the other hand, from (14.2) and (14.6) we have

$$*B^v(*D(v)) + Z(*C(*D(v))) = B^v(*D(v)) + Z(C(*D(v))).$$

Thus (15.3) may be written in the form

$$*B^h(v) - *B^v(*D(v)) - Z(*C(*D(v))) = B^h(v) - B^v(D(v)) - Z(C(D(v))).$$

Consequently the tangent vector fields

$$(15.4) \qquad F(v) = B^h(v) - B^v(D(v)) - Z(C(D(v))), \qquad Y(v) = B^v(v) + Z(C(v))$$

are common to all the Finsler connections belonging to the set $F\Gamma(\Gamma_V)$. There exists such a Finsler connection $F\Gamma_V$ that h- and v-basic vector fields are equal to the above $F(v)$ and $Y(v)$ respectively, because $F\Gamma_V$ is constructed from any Finsler connection $F\Gamma \in F\Gamma(\Gamma_V)$ by taking the difference tensors $(D^h(v), A^h(v), A^v(v)) = (-D(v), -C(D(v)), C(v))$. Thus we are led to

DEFINITION 15.3. The Finsler connection $F\Gamma_V \in F\Gamma(\Gamma_V)$ such that h- and v-basic vector fields are given by (15.4) respectively is called the *Finsler Γ_V-connection.*

Therefore it may be said that $F\Gamma_V$ is a representative of the set $F\Gamma(\Gamma_V)$ and the theory of Γ_V is regarded as the theory of the Finsler Γ_V-connection $F\Gamma_V$.

THEOREM 15.2. *The Finsler Γ_V-connection $F\Gamma_V$ is a unique element of the set $F\Gamma(\Gamma_V)$ such that the Cartan tensor field C and the deflection tensor field D vanish.*

Proof. Putting $(D^h(v), A^h(v), A^v(v)) = (-D(v), -C(D(v)), C(v))$ in (14.6) and (14.7), we obtain $*C = *D = 0$. Conversely, from (14.6) and $*C = 0$ it follows that $A^v(v) = C(v)$. Further, from (14.7) and $*D = 0$ we get

$$0 = D(v) + D^h(v) + C(\varepsilon, D^h(v)) - A^h(\varepsilon, v).$$

Thus Proposition 15.3 leads to $D^h = -D$, so that $A^h(v) = -C(D(v))$.

REMARK 15.2. The Finsler Γ_V-connection satisfies the D, C_1 and C_2-conditions. In particular, its non-linear connection N is associated with its V-connection. It will be seen in § 18 that Rund's connection on a Finsler space is the Finsler Γ_V-connection, where Γ_V is the V-connection of Cartan's connection.

THEOREM 15.3. *If a non-linear connection N is p-homogeneous, then N is associated with a V-connection.*

This is regarded as the converse of Proposition 8.6 on the assumption of homogeneity. In other words, the concept of V-connection is a kind of non-linear connection.

The proof is as follows: First, pairing a given N with any p-homogeneous connection Γ in the Finsler bundle F(M), a p-homogeneous Finsler connection $F\Gamma = (\Gamma, N)$ is obtained. Next, from this $F\Gamma$ we construct the Finsler N-connection $FN = (\Gamma_V, N, F^i)$.

From (15.1) it is seen that the difference tensors of FN from FΓ are such that $(D^h(v), A^h(v), A^v(v)) = (0, - P^1(v, \delta), C(v))$. Since this FΓ is p-homogeneous, it follows from Theorem 12.1 that P^1 and C of FΓ are p-homogeneous of degree 0 and - 1 respectively, so that FN is also p-homogeneous from Proposition 14.1. It follows from Theorem 15.1 that the (v)hv-torsion tensor P^1 of FN vanishes, so that the deflection tensor D of FN vanishes from Theorem 12.2. Consequently Proposition 13.3 completes the proof.

In terms of connection coefficients the above is obvious from Euler's theorem: $y^k \dot{\partial}_k N^i_j = N^i_j$. That is, connection coefficients $F_k{}^i{}_j$ of Γ_V with which N is associated are equal to $\dot{\partial}_k N^i_j$.

REMARK 15.3. The torsion and curvature tensors of a Finsler Γ_V-connection, namely, those of a V-connection, will be obtained by constructing the structure equations from (15.4).

The theory of non-linear connections has been studied by many authors. In the viewpoint of Finsler geometry, the author is interested in the papers: Barthel [11], A.Kawaguchi [3] and Rund [16]. As above shown, a non-linear connection N (N^i_j with two indices) can be treated as a Finsler N-connection with coefficients $(F_j{}^i{}_k = \dot{\partial}_j N^i_k, N^i_j, 0)$, and, on the other hand, a V-connection $\Gamma_V(F_j{}^i{}_k$ with three indices) can be studied as a Finsler Γ_V-connection with coefficients $(F_j{}^i{}_k, N^i_j = y^k F_k{}^i{}_j, 0)$. The method to consider a non-linear connection as such a Finsler connection and to apply a general theory of Finsler connections may be simpler than the usual tensor calculus. In fact, constructing the structure equations from (15.1) and (15.4), we easily obtain torsion and curvature tensors, and further the Jacobi identities give all the Bianchi identities. See Matsumoto [19].

§ 16. Finsler metrics

Let S be a Finsler tensor field of (0, 0)-type, namely, a Finsler scalar field. Since S satisfies $S \circ \beta_g = g^{-1} S$, that is, $S(ug) = S(u)$ for any point $u \in F$ and any $g \in G(n)$, the real-valued function $\underline{S}$ on the tangent bundle T(M) is uniquely determined such that $S = \underline{S} \circ \pi_1$. The function $\underline{S} = \underline{S}(x, y)$, $x = (x^i)$, $y = (y^i)$, is said to *be induced from* S and is denoted sometimes by the letter S itself.

From a Finsler scalar field S we obtain a Finsler tensor field $T = \nabla^0 \nabla^0 (S^2/2)$ which is of (0, 2)-type and symmetric. If the components T_{ij} of T constitute a non-singular matrix (T_{ij}), then S is called *regular*. The regularity is such that the tensor T gives rise to an isomorphism $v \in V^n \mapsto T(v) \in V^n_*$, where $(T(v), w) = T(v, w)$ for any $w \in V^n$.

DEFINITION 16.1. A manifold M is called a *Finsler space with a Finsler metric* L, if we have a (1)p-homogeneous and regular Finsler scalar field L, by means of which the *length* s of a curve C: $t \in [0, 1] \mapsto x(t) \in M$ is defined by the integral

(16.1) $$s = \int_0^1 \underline{L}(x(t), y(t))dt, \qquad y(t) = d(x(t))/dt.$$

The scalar L or the induced function $\underline{L}(x, y)$ is called the *fundamental function* and the Finsler tensor field $g = \nabla^0\nabla^0(\underline{L}^2/2)$ is the *fundamental tensor*. Then g gives an isomorphism $v \in V^n \mapsto g(v) \in V^n_*$ as above mentioned. We denote the components of g and the inverse isomorphism g^{-1} of g by g_{ij} and g^{ij} respectively. It follows from Proposition 12.3 that g and g^{-1} are (0)p-homogeneous.

PROPOSITION 16.1. *If K is an (r)p-homogeneous Finsler tensor field, we have*

$$(\underbrace{\nabla^0\cdots\nabla^0}_{p} K)(\underbrace{\varepsilon,\cdots,\varepsilon}_{q}, v_{q+1},\cdots, v_p)$$

$$= (r - p + q)(r - p + q - 1)\cdots(r - p + 1)(\underbrace{\nabla^0\cdots\nabla^0}_{p-q} K)(v_{q+1},\cdots, v_p)$$

for any v's $\in V^n$, where ε is the supporting element and $p \geq q \geq 1$.

This is solely a generalization of (12.6) and the proof will be easy by mathematical induction. Applying the above to $L^2/2$, we get

(16.2) $$g(\varepsilon, \varepsilon) = L^2 \qquad (\ g_{ij}(x, y)y^iy^j = \underline{L}^2(x, y)\ ).$$

If we define a Finsler tensor field l of (0, 1)-type by

(16.3) $$l = \nabla^0 L \qquad (\ l_i = \dot\partial_i \underline{L}\ ),$$

and a symmetric Finsler tensor field h of (0, 2)-type by

(16.4) $$h = L(\nabla^0\nabla^0 L), \qquad (\ h_{ij} = \underline{L}(\dot\partial_i\dot\partial_j \underline{L})\ ),$$

the identity

(16.5) $$g(u, v) = h(u, v) + l(u)l(v) \qquad (\ g_{ij} = h_{ij} + l_il_j\ )$$

holds evidently for any $u, v \in V^n$. The symmetric tensor h is called the *angular metric tensor* (cf. Remark 31.1).

PROPOSITION 16.2. *The angular metric tensor field h satisfies $h(\varepsilon) = 0$ ($h_{ij}y^j = 0$). The matrix (h_{ij}) consisting of the components of h is of rank n - 1, provided the space is of dimension n.*

Proof. $h(\varepsilon) = 0$ is obvious from (12.6); so the rank of (h_{ij}) is less than n. If the rank is less than n - 1, then (16.5) shows that the rank of (g_{ij}) is less than n, contradicting to the regularity of L.

We consider a Finsler tensor K of (1, 2)-type, for instance. By the fundamental tensor g a Finsler tensor K_* of (0, 3)-type is constructed such that

(16.6) $$K_*(u, v, w) = g(u, K(v, w)) = (K(v, w), g(u))$$

for any u, v, w $\in V^n$. This construction of K_* from K is called *lowering of index r*; $K_{ijk} = g_{ir}K^r{}_{jk}$. We can also define *raising of index* by g^{-1} to get K* of (2,1)-type; $K^{ij}{}_k = g^{jr}K^i{}_{rk}$. Thus ε_* and 1* are obtained from ε and 1 respectively:

(16.7) $$\varepsilon_*(v) = g(v, \varepsilon) = (\varepsilon, g(v)) \qquad (y_i = g_{ir}y^r),$$

(16.8) $$1^*(v_*) = g^{-1}(v_*, 1) = (g^{-1}(v_*), 1) \qquad (1^i = g^{ir}1_r)$$

for any $v \in V^n$ and any $v_* \in V^n_*$. It is observed that

$$\varepsilon_*(v) = \nabla^0(\nabla^0(L^2/2)v)\varepsilon = \nabla^0(L^2/2)v = L1(v),$$

$$1^*(g(v)) = (g^{-1}(g(v)), 1) = 1(v) = \varepsilon_*(v)/L = (\varepsilon, g(v))/L,$$

that is, we have

(16.9) $$\varepsilon_* = L1 \qquad (y_i = L1_i), \qquad \varepsilon = L1^* \qquad (y^i = L1^i).$$

DEFINITION 16.2. Let M be a Finsler space with a fundamental function L.

(1) The value $\underline{L}(y)$ of $\underline{L}$ at a point $y \in T$ is called the (*absolute*) *length* of the tangent vector $y \in M_x$, $x = \pi_T(y)$.

(2) The *scalar product* $(y_1, y_2)_y$ of $y_1, y_2 \in M_x$ *relative to* $y \in M_x$ is given by $(y_1, y_2)_y = g_u(z^{-1}y_1, z^{-1}y_2)$, $u = (y, z) \in F$.

(3) The value $|y_1|_y = \surd(y_1, y_1)_y$ is called the *length* of y_1 *relative to* y.

It is noted that the scalar product $(y_1, y_2)_y$ does not depend on the choice of a point $z \in \pi_L^{-1}\circ\pi_1(y)$, because from the tensor property of g and (2.14) we have

$$G_{ug}((zg)^{-1}y_1, (zg)^{-1}y_2) = (g^{-1}G_u)(g^{-1}(z^{-1}y_1), g^{-1}(z^{-1}y_2)) = G_u(z^{-1}y_1, z^{-1}y_2)$$

for any $g \in G(n)$, where we use the notation G to denote the fundamental tensor g in order to avoid the confusion.

Let X, Y be two Finsler tensor fields of (1, 0)-type, i.e., Finsler vector fields. We have $X_u, Y_u \in V^n$ at a point $u = (y, z) \in F$, from which tangent vectors zX_u, $zY_u \in M_x$, $x = \pi_T\circ\pi_1(u)$, are derived. The scalar product of them relative to y is given by

$$g_u(z^{-1}(zX_u),\ z^{-1}(zY_u)) = g_u(X_u,\ Y_u).$$

Therefore $g_u(X_u, Y_u)$ is called the *scalar product* of Finsler vector fields X, Y *relative to* y. From this scalar product is defined the concept of relative length of a Finsler vector field.

It follows from (16.2) that the absolute length of y is equal to the root of $g(\varepsilon, \varepsilon)$ and from (16.9) that $1^* = (1^i)$ is a unit vector. This 1^* or $1 = (1_i)$ is called the *normalized supporting element*.

Throughout the following, we consider Finsler spaces only satisfying the assumption that the fundamental function L is (1)p-homogeneous and regular, as it has been mentioned in Definition 16.1. The homogeneity of $\underline{L}(x, y)$ in $y = (y^i)$ is not only necessary for the length integral (16.1) to be invariant by any transformation of parameter t which preserves the orientation of the curve, but also will be essential for the following consideration. On the other hand, the regularity of L is important as it is well-known in Riemannian geometry.

In Finsler geometry it is usual that further assumptions are imposed on L as follows: First

(F 1) L(x, y) is defined and positive for all non-zero y.

This *positiveness* of L, however, seems to arise from our tendency that the length should be positive.

Secondly, it seems that some authors believe the *symmetry property*:

(F 2) $L(x, -y) = L(x, y), \qquad y \neq 0.$

If this assumption holds, the length of a curve does not depend on the choice of its orientation.

Thirdly, because the relative length of a Finsler vector X is defined by $\sqrt{\{g(X, X)\}}$, the following may be desirable:

(F 3) The fundamental tensor g is positive-definite.

Finally, it has made a rule from the first of this book that all the functions we treat are assumed to be differentiable, but the fundamental function L should be differentiable on some restricted domain of y. In fact, if we treat a Riemannian metric $L(x, y) = \sqrt{\{g_{ij}(x)y^i y^j\}}$, then this L is not differentiable at $y = 0$. Thus the following condition is also desirable:

(F 4) L is differentiable at any non-zero y. (cf. Remark 12.1).

However, those additional four conditions are too restrictive and exclude various interesting special Finsler metric, as it will be shown later on. Consequently we do not refer to those four conditions, except the case where some geometrical theory needs them (cf. § 40).

REMARK 16.1. We consider an (r)p-homogeneous function $f(y) = f(y^1, \cdots, y^n)$ which is continuous at $y = 0$. Applying $\lim_{p \to 0}$ to the identity $f(py) = p^r f(y)$, it is seen that (1) if $r > 0$, we have $f(0) = 0$, and (2) if $r = 0$, we have $f(0) = f(y)$. Therefore an (0)p-homogeneous function which is continuous at $y = 0$ must be a constant function. Since the fundamental tensor $g = (g_{ij}(x, y))$ is (0)p-homogeneous, it is conclude that g_{ij} do not depend on y, namely, g is Riemannian, provided g is continuous at $y = 0$.

Next, we consider an (r)p-homogeneous function $f(y)$ which is C^1-differentiable at $y = 0$. We suppose $y = (y^1, y^2)$ for brevity. Then we have

$$\frac{1}{p}\{ f(p, 0) - f(0, 0)\} = \frac{1}{p}\{p^r f(1, 0) - f(0, 0)\}.$$

Suppose $r > 0$. Then the above is equal to $p^{r-1}f(1, 0)$; so (1) if $r > 1$, we get $\{\partial f/\partial y^1\}_0 = 0$, and (2) if $r = 1$, we have $\{\partial f/\partial y^1\}_0 = f(1, 0)(= a_1)$. In the following we treat of the case $r = 1$. Then $\partial f/\partial y^i$ are (0)p-homogeneous, so that $\partial f/\partial y^i = a_i$ (constant). Thus we get $f = (\partial f/\partial y^i)y^i = a_i y^i$, a linear function. Since the fundamental function L is (1)p-homogeneous, it is concluded that L is a linear function $a_i(x)y^i$, provided L is C^1-differentiable at $y = 0$. It is obvious that $L = a_i(x)y^i$ does not satisfy the regularity condition.

REMARK 16.2. In the famous lecture of 1954, Riemann requested the condition (F 1) as well as (F 2). Further, in the famous thesis of 1918, Finsler used (F 1) and (F 3), but not (F 2). On the other hand, (F 1) seems to have first played an essential role when Deicke proved a well-known theorem (Theorem 24.2) in 1953. Similarly (F 2) and (F 4) are assumed in the proof of a well-known theorem due to Brickell (Theorem 24.3), but (F 2) is expected to be removable.

As to the condition (F 2), the author recalls Finsler's letter of 1969 to him:

In der Astronomie misst man die Entfernungen gerne in einem Zeitmass, insbesondere in Lichtjahren. Nimmt man 1 Sekunde als Einheit, dann sind die Einheitsflächen Kugeln mit einem Radius 300,000 km. Zu jedem Punkt im Raum gehört eine solche Einheitsfläche; diese definieren die (in Zeit gemessenen) Entfernungen und damit die Geometrie im Raum, im einfachsten Fall die euklidische. Wenn aber die Lichtstrahlen als ‚kürzeste Linien' in Gravitationsfeldern gekrümmt verlaufen, wird man eine Riemannsche Geometrie erhalten. Ebenso kann in anisotropen Medien die Lichtgeschwindigkeit von der Richtung abhängen, die Einheitsflächen sind dann keine Kugeln mehr. Nun misst man auch im Gelände auf der Erdoberfläche die Entfernungen oft in einem Zeitmass, etwa in Gehminuten (so die Angaben auf Wegzeigern). Die Einheitskurven (für 1 Minute als Einheit) können allgemeine Kurven sein, auch ohne Mittelpunkt (da man in 1 Minute aufwärts weniger weit kommt als abwärts); sie definieren eine ‚allgemeine Geometrie', die ein wohl nicht sehr exaktes, aber doch anschauliches ‚Modell' darstellt. Die ‚kürzesten Linien', auf denen man am schnellsten zum Ziel kommt (etwa zu einer Bergspitze), können kompliziert verlaufen. (See Appendix.)

Finsler calls the indicatrix (Definition 16.3) as Einheitflächen and Finsler geometry as Allgemeine Geometrie. The name "Finsler space" seems to have appeared first in Taylor's paper [3] in 1927.

Further we shall quote a paragraph from Ingarden's monograph [3], concerned with a physical problem. After mentioned the conditions (F 1), (F 2) and (F 3), he writes

The further development of physics shows, however, that for the most important physical applications these assumptions are too restrictive. (F 1) and (F 3) are inconsistent with the pseudo-Euclidean character of physical space-time (existence of the light-cone), (F 2) prohibits, as we shall see, the application to the electromagnetic field theory. The modern presentations of Finsler's geometry are formulated without use of these conditions, and we shall follow this formulation in the present paper.

If (F 1), for instance, does not hold, it is enough for some geometrical treatment to restrict the consideration to a domain where (F 1) holds.

Now, the tangent vector space M_x at a point x of a Finsler space M with a fundamental function L(x, y) is regarded as a Minkowski space with the metric L(x, y) where x is fixed.

DEFINITION 16.3. A *Minkowski space* V is such a real vector space that the length (norm) of a vector $v \in V$ is given by the value L(v) of a function L on V, where L is assumed to satisfy the conditions similar to a Finsler metric. The hypersurface $I = \{v \in V \mid L(v) = 1\}$ is called the *indicatrix*.

Thus we have the indicatrix I_x in each tangent space M_x of a Finsler space M, which is called the *indicatrix at x*. Cf. § 31.

REMARK 16.3. In case of a Riemannian space M with a positive-definite Riemannian metric $L = \sqrt{\{g_{ij}(x)y^i y^j\}}$, each indicatrix I_x is considered as a euclidean unit sphere with the center y = 0, because there is, at each point x, a coordiante system (x^i) such that L^2 is written in the form $L^2 = (y^1)^2 + \cdots + (y^n)^2$ at x. Thus the indicatrices I_x of a Riemannian space are congruent to each other. This situation becomes more complicated in the Finsler case. A Finsler space is called a *Finsler space modelled on a Minkowski space*, if the tangent spaces at arbitrary points are congruent to a unique Minkowski space. See Ichijyo [2, 3, 5, 6].

We consider a Finsler space M with a positive metric L(x, y). There exists always only one point of I_x in any direction y, because the equation $L(x, ay) = 1$, $a > 0$, has a unique solution $a = 1/L(x, y)$ in virtue of the homogeneity. Further, it is known (K.Okubo [2], Rund [18], H.Wang [1]) that I_x is compact, arcwise-connected, simply connected and convex on the conditions (F 1) and (F 3).

REMARK 16.4. Laugwitz' paper [9] gives a short introduction to Minkowski geometry. He shows a geometrical proof of an interesting theorem that, in two-dimensional case, the total length S of the indicatrix (oval) satisfies $6 \leq S \leq 8$, provided (F 1) and (F 2) hold. As to Minkowski geometry, see Barthel [1, 8, 9], Busemann [6], Rund [2], Sumitomo [1], O.Varga [5]. In O.Varga's paper [5] we find an interesting paragraph:

> Der grundlegenden Gedanke, der in dieser Arbeit verwendet wird, ist der folgende: der Minkowskische Raum ist ein affiner Raum, in dem jeder Richtung eine euklidische Massbestimmung zugeordnet ist. Dementsprechend behandeln wir § 1 den Minkowskischen Raum in Bezug auf ein kartesisches Koordinatensystem. Im zweiten Paragraphen führen wir krummlinige Koordinaten ein. Es ist bemerkenswert, dass man dann den ganzen formalen Apparat der Finslerschen Geometrie erhält. Dies ist ein Analogon zur Tatsache, dass der auf krummlinige Koordinaten bezogene Euklidische Raum den formalen Apparat der Riemannschen Geometrie liefert. (See Appendix.)

EXAMPLE 16.1. Let (M, α) be an n-dimensional Riemannian space with a metric $\alpha(x, y) = \sqrt{\{a_{ij}(x)y^i y^j\}}$ and let $\beta(x, dx)$ be a differential one-form $b_i(x)dx^i$ on the manifold M. Then a function

$$L(x, y) = \alpha(x, y) + \beta(x, y)$$

is evidently (1)p-homogeneous and will be regular on some weak assumption (cf. § 30). Thus we have a Finsler space M with this metric L. This interesting metric was first introduced by Randers [1] in 1941 to discuss the general relativity with the electromagnetic field and has been studied by many physicists. See, for instance, Horváth [3, 6], Ingarden [3], Mosharrafa [1, 2], Stephenson [1, 2], Stephenson and Kilmister [1]. This metric is called a *Randers metric*, following Ingarden. Recently we have various geometrical studies of Finsler spaces with a Randers metric. See Eliopoulos [4], Hashiguchi, Hōjō and Matsumoto [1], Matsumoto [17, 21], Matsumoto and Hōjō [1], Shibata, Shimada, Azuma and Yasuda [1], Yasuda [1], Yasuda and Shimada [1], and § 30.

As to Randers metrics, the condition (F 2) holds if and only if $\beta = 0$, while (F 3) will be satisfied on some restriction. (F 4) holds on a domain where α is differentiable. As to (F 1), we consider a two-dimensional Minkowski space V with the Randers metric

$$L = \sqrt{(u^2 + av^2)} + bv,$$

where (u, v) is a coordinate system on V and $a(\neq 0)$, b are constant. We treat the following typical cases:

(i) $a = 2$, $b = 1$. The indicatrix is a circle with the center $(u, v) = (0, -1)$ and the radius $\sqrt{2}$. This L is positive for all non-zero (u, v).

(ii) $a = 1, b = 2$. The indicatrix is a half ($v \le 1/3$) of a hyperbola $3(v - 2/3)^2 - u^2 = 1/3$. This L is negative for $(-\sqrt{3} < u < \sqrt{3}, v = -1)$ and $L = 0$ for $(u = \pm\sqrt{3}, v = -1)$.

(iii) $a = b = 1$. The indicatrix is a parabola $v = -(u^2 - 1)/2$ and $L > 0$, except $(u = 0, v = -1)$.

<u>EXAMPLE 16.2.</u> Let α and β be those as above considered. Then the function

$$L(x, y) = \alpha^2(x, y)/\beta(x, y)$$

is obviously (1)p-homogeneous and will be regular on some weak restriction (cf. § 30). Thus we obtain a Finsler space M with this metric L, which is called a *Kropina metric* (cf. Kropina [1, 2]). If α is assumed to be positive-definite, the algebraic sign of L coincides with that of β. The indicatrix I_x is tangent to a hyperplane $\beta = 0$ at the origin $y = 0$. See Matsumoto [17], Shibata [1].

<u>EXAMPLE 16.3.</u> We are concerned with a two-dimensional euclidean space E^2 with an orthonormal coordinate system (x, y). Let $\overline{PQ}$ be the euclidean length of a segment PQ, and let $\overline{OH}$ be the one of the perpendicular OH from the origin O to the straight line PQ. Then we define the norm $\|PQ\|$ by $\|PQ\| = \overline{PQ}/\overline{OH}$. Let (x, y) and (X, Y) be the coordinates of P and Q respectively. Then we get $\|PQ\| = \{(X - x)^2 + (Y - y)^2\}/\{x(Y - y) - y(X - x)\}$. Consequently, if we put $X = x + \dot{x}$, $Y = y + \dot{y}$, then $\|PQ\|$ is written in the form

$$L(x, y, \dot{x}, \dot{y}) = \frac{\dot{x}^2 + \dot{y}^2}{|x\dot{y} - y\dot{x}|}.$$

This example of two-dimensional Finsler metric $L(x^1, x^2, y^1, y^2)$ is given by Wrona [1] and obviously a kind of Kropina metric. The indicatrix I_P at P(x, y) consists of the two circles of same size, illustrated in Fig.1. In fact, from $\Delta OPK \equiv \Delta PAR$ we have $\overline{OK} = \overline{PR}$, so that $\|PR\| = 1$.

Fig. 1

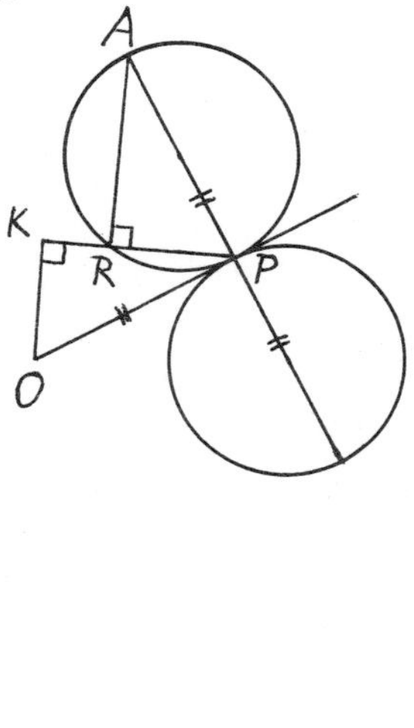

Fig. 2

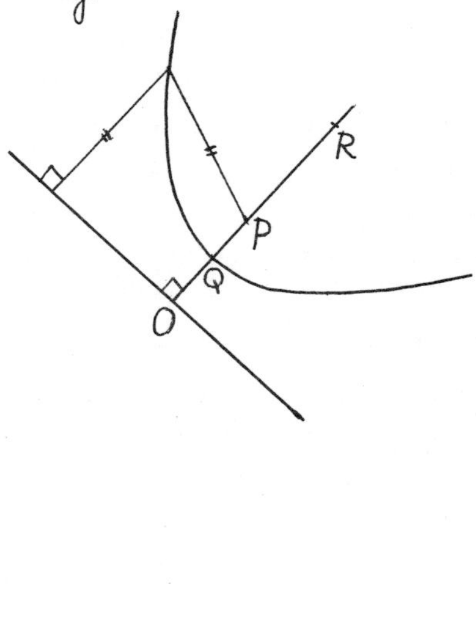

EXAMPLE 16.4. Let $f(x, y) = 0$ be the equation of indicatrix I_x at a point x. Then the fundamental function is given by the implicit function $L(x, y)$ satisfying $f(x, y/L) = 0$. This is a general way to *find L from the equation of indicatrix*. In fact, $f(x, y/L) = 0$ with $L = 1$ reduces to the equation of I_x and $L(x, y)$ is (1)p-homogeneous in y because of $f(x, (ay)/(aL)) = 0$, $a > 0$. By this method, communicated by K.Okubo to the author in 1973, many interesting Finsler metrics will be found.

For instance, we shall find a two-dimensional Finsler metric with parabola as the indicatrix I_P (Fig. 2). Let E^2 be a two-dimensional euclidean space with an orthonormal coordinate system (x, y). We define the indicatrix I_P at a point P(x, y) as a parabola such that the focus is P and the directrix is the perpendicular at the origin O to the straight line OP. In the current coordinates(X, Y), this parabola is written

$$\sqrt{\{(X - x)^2 + (Y - y)^2\}} = \frac{xX + yY}{\sqrt{(x^2 + y^2)}} .$$

If we put $X = x + \dot{x}$, $Y = y + \dot{y}$, the above is rewritten in the form

$$\sqrt{(\dot{x}^2 + \dot{y}^2)} = \frac{x(x + \dot{x}) + y(y + \dot{y})}{\sqrt{(x^2 + y^2)}} .$$

Now, according to Okubo's method, by exchanging $(\dot{x}, \dot{y})$ for $(\dot{x}/L, \dot{y}/L)$, we get

$$L(x, y, \dot{x}, \dot{y}) = \sqrt{\frac{\dot{x}^2 + \dot{y}^2}{x^2 + y^2}} - \frac{x\dot{x} + y\dot{y}}{x^2 + y^2} .$$

This is a kind of Randers metric and it is seen from Hashiguchi, Hōjō and Matsumoto [1] that the Finsler space with this L is a Landsberg space (cf. Definition 25.2). It is interesting to observe that the length of $\overrightarrow{PQ}$ is equal to 1, while the length of $\overleftarrow{PQ}$ is equal to 0, similarly to the fact that $\overrightarrow{PR}$ has zero length, because there is no point of the indicatrix I_P in the direction $\overrightarrow{PR}$. See Hōjō [3].

§ 17. The Cartan connection

Let M be a Finsler space with a Finsler metric L(x, y). We are now in a position to consider a system of axioms satisfied by a Finsler connection FΓ on M by means of which FΓ is uniquely determined from the Finsler metric L. We denote by g the fundamental tensor field of M determined by L.

DEFINITION 17.1. If a Finsler connection FΓ on a Finsler space M satisfies $\nabla^h g = 0$ (resp. $\nabla^v g = 0$), then FΓ is called *h-metrical* (resp. *v-metrical*).

It will be admitted by almost all the mathematicians that the metrical condition as above is essential in the differential geometry of a manifold with a metric (cf. Proposition 5.1). In Proposition 8.2 it has been remarked that each restriction

$\Gamma^v|F(x)$ of the vertical connection Γ^v of $F\Gamma$ is a connection in the principal bundle $F(x)$ and that $F(x)$ is regarded as the linear frame bundle over the fiber $\pi_T^{-1}(x) = M_x$. Now we have the fundamental tensor field g $(g_{ij}(x, y))$, which yields a Riemannian metric on the tangent space M_x, if the point x is fixed. In fact, the Riemannian length s of a curve $t \in [0, 1] \mapsto y(t) \in M_x$ can be defined by the integral

$$s = \int_0^1 \sqrt{\{g_{ij}(x, y(t))\dot{y}^i\dot{y}^j\}}\, dt, \qquad \dot{y}^i = dy^i/dt.$$

Therefore it will be natural to determine the vertical connection Γ^v as the Riemannian connection with respect to the above Riemannian metric. Consequently we introduce first the following two axioms (cf. Proposition 5.1):

(C 1) *$F\Gamma$ is v-metrical.*

(C 2) *The (v)v-torsion tensor field S^1 vanishes.*

Thus the connection coefficients $C_j{}^i{}_k$ of Γ^v are symmetric in j, k from (10.6') and from (5.23) and (9.20) $C_{ijk} = g_{jr}C_i{}^r{}_k$ are given by

$$C_{ijk} = \frac{1}{2}(\dot{\partial}_k g_{ij} + \dot{\partial}_i g_{jk} - \dot{\partial}_j g_{ki}). \tag{17.1}$$

Because g_{ij} is defined by $g_{ij} = \dot{\partial}_i\dot{\partial}_j(L^2/2)$, the above reduces to a simple form

$$C_{ijk} = \frac{1}{2}\dot{\partial}_k g_{ij}. \tag{17.1'}$$

Thus the Cartan tensor field C, namely, the vertical connection Γ^v is completely determined from L.

REMARK 17.1. According to the structure of the Finsler bundle F(M), it may be natural to determine Γ^v as the flat vertical connection F^1, namely, $C_j{}^i{}_k = 0$. The vertical connections due to Berwald as well as Rund are really flat (§ 18).

It may be, however, better to determine Γ^v as the Riemannian connection from such a standpoint that we are concerned with a Finsler metric and the indicatrix I_x is not necessarily a euclidean sphere. By the axiom (C 1) we have some essential merits. For instance, lowering and raising of indices commute with the v-covariant differentiation.

By the two axioms (C 1), (C 2), each tangent space M_x is considered as a Riemannian space equipped with the Riemannian connection Γ^v (cf. § 31). But we can not too emphasize that the Riemannian geometry of each M_x is *affine* in such a sense that a coordinate system (y^i) of $y \in M_x$ is the one induced from a coordinate system (x^i) on the space M, $y^i = \dot{x}^i$ in old-fashioned notations, and obeys only an affine transformation: $\bar{y}^i = (\partial\bar{x}^i/\partial x^j)_x y^j$ with constant coefficients $(\partial\bar{x}^i/\partial x^j)_x$. The tensor property of the Christoffel symbols $C_j{}^i{}_k$ is nothing but a consequence of this affinity.

Therefore a theorem in Riemannian geometry can not be directly applied to the Riemannian space M_x, if this theorem is to be proved by the existence of some special coordinate system. For instance, a well-known theorem "A Riemannian space of zero curvature is flat." is proved by showing the existence of such a coordinate system that all the Christoffel symbols vanish identically. If we carelessly apply this theorem to a two-dimensional Finsler space (cf. Proposition 28.3), the tensor $C_j{}^i{}_k$ vanishes identically, so that any two-dimensional Finsler space reduces to a Riemannian space ! See Theorem 24.1 and the proof of Theorem 24.3.

To study a Finsler space M, we can refer, of course, to any coordinate system in M_x (§ 28), if we pay attention to the above remark.

Now, to determine the V-connection Γ_V and the non-linear connection N of the Finsler connection $F\Gamma = (\Gamma_V, N, \Gamma^v)$ under consideration, we introduce following two axioms, which are analogous to (C 1) and (C 2):

(C 3) *$F\Gamma$ is h-metrical : $\nabla^h g = 0$.*

(C 4) *The (h)h-torsion tensor field T vanishes.*

It follows from (9.18) and (9.19) that (C 3) is expressed as

$$\delta_k g_{ij} = F_{ijk} + F_{jik}, \tag{17.2}$$

where $F_{ijk} = g_{jr}F_i{}^r{}_k$. By the Christoffel process (cf. § 5), from (17.2), (10.12) and (C 4) we get

$$F_{ijk} = \frac{1}{2}(\delta_k g_{ij} + \delta_i g_{jk} - \delta_j g_{ki}). \tag{17.3}$$

The right-hand side still contains unknown coefficients $N^i{}_j$ of the non-linear connection N. As a consequence we have

<u>THEOREM 17.1</u>. *Let a non-linear connection N be given. Then a Finsler connection $F\Gamma=(\Gamma_V, N, \Gamma^v)$ is uniquely determined by the four axioms (C 1), (C 2), (C 3) and (C 4) such that Γ_V and Γ^v are given by (17.3) and (17.1') respectively.*

We consider (17.3) in detail. It is rewritten in a longer form

$$F_{ijk} = \gamma_{ijk} - C_{ijr}N^r{}_k - C_{jkr}N^r{}_i + C_{ikr}N^r{}_j, \tag{17.3'}$$

where (17.1') is used and γ_{ijk} are Christoffel symbols constructed from $g_{ij}(x, y)$ with respect to x^i:

$$\gamma_{ijk} = \frac{1}{2}(\partial_k g_{ij} + \partial_i g_{jk} - \partial_j g_{ki}). \tag{17.4}$$

Contracting (17.3') by y^i and paying attention to the identity $y^iC_{ijk} = 0$, we get

$$y^iF_{ijk} = y^i\gamma_{ijk} - C_{jkr}N^r{}_iy^i.$$

By (9.22) this is rewritten in the form

(17.5) $$D^j_{\ k} + N^j_{\ k} = y^i \gamma_i{}^j{}_k - C_k{}^j{}_r N^r_{\ i} y^i,$$

where we put $\gamma_i{}^j{}_k = g^{jr}\gamma_{irk}$. Contracting (17.5) by y^k, we get

(17.6) $$D^j_{\ k} y^k + N^j_{\ k} y^k = y^i y^k \gamma_i{}^j{}_k.$$

Throughout the following, we shall adopt the index 0 to denote the contraction by supporting element y^i. Thus we have $C_{0jk} = C_{i0k} = C_{ij0} = 0$ and $h_{0j} = h_{i0} = 0$ from Proposition 16.2. Then (17.5) and (17.6) are written respectively as

(17.5') $$D^j_{\ k} + N^j_{\ k} = \gamma_0{}^j{}_k - C_k{}^j{}_r N^r_{\ 0},$$

(17.6') $$D^j_{\ 0} + N^j_{\ 0} = \gamma_0{}^j{}_0.$$

From (17.3'), (17.5') and (17.6') we immediately establish

THEOREM 17.2. *If the deflection tensor field D is given, there exists a unique Finsler connection which satisfies the four axioms (C 1), (C 2), (C 3) and (C 4).*

REMARK 17.2. It should be noted that, contrary to our usage of the index 0, Cartan denotes by the index 0 the contraction by normalized supporting element $l^i = y^i/L$. In our notation, if $K^i_{\ j}(x, y)$ are (r)p-homogeneous, then $K^i_{\ 0}$ are (r + 1)p-homogeneous.

The equation (17.6) is noteworthy. In fact, the *extremal* of the length integral (16.1) is given by the differential equations

$$\frac{d^2x^i}{ds^2} + \gamma_j{}^i{}_k\left(x, \frac{dx}{ds}\right)\frac{dx^j}{ds}\frac{dx^k}{ds} = 0 \qquad \text{(cf. § 40).}$$

On the other hand, the *path*, that is, the autoparallel curve (§ 39) is given by

$$\frac{d^2x^i}{ds^2} + N^i_{\ k}\left(x, \frac{dx}{ds}\right)\frac{dx^k}{ds} = 0.$$

Therefore (17.6) shows that any path coincides with an extremal, if and only if the deflection tensor $D^i_{\ j}$ satisfies $D^i_{\ 0} = 0$. As a consequence, in Theorem 17.2 it seems to be desirable to give $D^i_{\ j}$ such that $D^i_{\ 0} = 0$ holds, although the existence of extremal and path which are different from each other may be allowed. Theorem 17.2 is shown by Hashiguchi [2].

DEFINITION 17.2. The *Cartan connection* CΓ of a Finsler space M with a Finsler metric L is a Finsler connection which is uniquely determined by the system of five axioms:

(C 1) CΓ is v-metrical.

(C 2) The (v)v-torsion tensor field S^1 of $C\Gamma$ vanishes.

(C 3) $C\Gamma$ is h-metrical.

(C 4) The (h)h-torsion tensor field T of $C\Gamma$ vanishes.

(C 5) The deflection tensor field D of $C\Gamma$ vanishes.

The existence and uniqueness of $C\Gamma$ are evident as a special case of Theorem 17.2. The coefficients $(F_j{}^i{}_k, N^i{}_j, C_j{}^i{}_k)$ of $C\Gamma$ are as follows:

(1) $C_j{}^i{}_k = g^{ir}C_{jrk}$ are given by (17.1').

(2) $N^i{}_j$ are given by

$$N^i{}_j = \gamma_0{}^i{}_j - C_j{}^i{}_r\gamma_0{}^r{}_0. \tag{17.7}$$

(3) $F_j{}^i{}_k = g^{ir}F_{jrk}$ are given by (17.3) or (17.3').

<u>REMARK 17.3</u>. Although the aim of Cartan's axioms [4, 5] (cf. Rund [18]) is to determine both the fundamental tensor g and the connection from the Finsler metric, it seems that some of his axioms are rather artificial and are introduced after foreseeing the result. On the other hand, the above axioms (C 1), ···, (C 4) are completely parallel to the axioms of Riemannian connection, and a meaning of (C 5) has been geometrically given by Proposition 13.3. The system of these five axioms is first mentioned in Matsumoto [6]. See Sulanke [2], O.Varga [2, 26].

There are naturally expected some variations from the above system of axioms, and those may be effective to consider some special problems. See Barthel [5], Hashiguchi [2, 4], Hashiguchi and Ichijyo [2], Su [2], Tamássy and Matsumoto [1], Wagner [4] and § 25.

The excellence of the Cartan connection is also verified from the standpoint of variation calculus, as it is shown by Rund and Lovelock [1, p.32].

We are concerned with special behaviors of the torsion and curvature tensors of the Cartan connection $C\Gamma$. It follows from the (1)p-homogeneity of the fundamental function L that $C\Gamma$ is p-homogeneous. Further $C\Gamma$ satisfies, of course, the D-, $C_{\bar{1}}$ and C_2-conditions. Thus, from Theorem 13.2 we have $P^i{}_{j0} = 0$ and $P_h{}^i{}_{j0} = 0$, and from Theorem 13.3

$$R_0{}^i{}_{jk} = R^i{}_{jk}, \qquad P_0{}^i{}_{jk} = P^i{}_{jk}, \qquad S_0{}^i{}_{jk} = 0. \tag{17.8}$$

Next, applying the Ricci identities (10.8'), (10.9') and (10.10') to g_{ij}, from (C 1) and (C 3) we get

$$R_{hijk} = - R_{ihjk}, \qquad P_{hijk} = - P_{ihjk}, \qquad S_{hijk} = - S_{ihjk}, \tag{17.9}$$

where $R_{hijk} = g_{ir}R_h{}^r{}_{jk}$ and so on.

We are concerned with the Bianchi identities. By (C 2)($S^i{}_{jk} = 0$) and (C 4)($T_j{}^i{}_k = 0$), some of them reduce to a little simpler form as follows:

Three identities of the first group are

$$\mathfrak{S}_{(ijk)}\{C_i{}^h{}_r R^r{}_{jk} - R_i{}^h{}_{jk}\} = 0, \tag{17.10}$$

$$\mathfrak{S}_{(ijk)}\{P^h{}_{ir} R^r{}_{jk} + R^h{}_{ij|k}\} = 0, \tag{17.11}$$

$$\mathfrak{S}_{(ijk)}\{P_m{}^h{}_{ir} R^r{}_{jk} + R_m{}^h{}_{ij|k}\} = 0. \tag{17.12}$$

The identity (17.10) does not contain covariant derivatives, and (17.11) is only a consequence of (17.12) by the axiom (C 5)($y^i{}_{|j} = 0$) and (17.8).

Three identities of the second group are

$$\mathfrak{A}_{(ij)}\{C_j{}^h{}_{k|i} + C_i{}^h{}_r P^r{}_{jk} - P_i{}^h{}_{jk}\} = 0, \tag{17.13}$$

$$R^h{}_{ij|k} - R_k{}^h{}_{ij} + \mathfrak{A}_{(ij)}\{R^h{}_{ir} C_j{}^r{}_k + P^h{}_{ir} P^r{}_{jk} + P^h{}_{jk|i}\} = 0, \tag{17.14}$$

$$S_m{}^h{}_{kr} R^r{}_{ij} + R_m{}^h{}_{ij|k} + \mathfrak{A}_{(ij)}\{R_m{}^h{}_{ir} C_j{}^r{}_k + P_m{}^h{}_{ir} P^r{}_{jk} + P_m{}^h{}_{jk|i}\} = 0. \tag{17.15}$$

The identity (17.13) yields an important equation, as it will be shown later on. The identity (17.14) is only a consequence of (17.15), similarly to (17.11).

Two identities of the third group are

$$\mathfrak{A}_{(ij)}\{P^h{}_{ri} C_k{}^r{}_j - P^h{}_{kj|i} + P_i{}^h{}_{kj}\} = 0, \tag{17.16}$$

$$S_m{}^h{}_{ij|k} + \mathfrak{A}_{(ij)}\{P_m{}^h{}_{ri} C_k{}^r{}_j - S_m{}^h{}_{ir} P^r{}_{kj} - P_m{}^h{}_{kj|i}\} = 0. \tag{17.17}$$

The identity (17.16) is also a consequence of (17.17) and yields an important equation, as it will be shown later on.

Two identities of the fourth group are

$$\mathfrak{S}_{(ijk)}\{S_i{}^h{}_{jk}\} = 0, \tag{17.18}$$

$$\mathfrak{S}_{(ijk)}\{S_m{}^h{}_{ij|k}\} = 0. \tag{17.19}$$

These are similar to the identities (5.20') and (5.19') ($\underline{T}_j{}^i{}_k = 0$) which are satisfied by the curvature tensor of a Riemannian connection.

Now we consider the components of the v-curvature tensor S^2. It follows from

(10.7') and (17.1') that

$$S_{hijk} = \mathfrak{A}_{(jk)}\{\dot\partial_k C_{hij} - C_h{}^r{}_j \dot\partial_k g_{ir} + C_h{}^r{}_j C_{rik}\}$$

$$= \mathfrak{A}_{(jk)}\{-\ 2C_h{}^r{}_j C_{rik} + C_h{}^r{}_j C_{rik}\},$$

which implies a surprising simple form of the curvature tensor S^2 of the Riemannian connection $\Gamma^v|F(x)$ as follows:

$$S_{hijk} = \mathfrak{A}_{(jk)}\{C_h{}^r{}_k C_{rij}\}. \tag{17.20}$$

Next we treat the covariant components R_{hijk} of the h-curvature tensor R^2. By (17.9) the Bianchi identity (17.10) is rewritten in the form

$$\mathfrak{A}_{(ijk)}\{C_h{}^r{}_i R_{rjk} + R_{hijk}\} = 0, \tag{17.10'}$$

where $R_{rjk} = g_{ri}R^i{}_{jk} = R_{0rjk}$. Contracting (17.10') by y^h and referring to (17.8), we obtain an important identity

$$\mathfrak{S}_{(ijk)}\{R_{ijk}\} = 0. \tag{17.21}$$

We are next concerned with the hv-curvature tensor P^2. From (17.13) it follows first that

$$\mathfrak{S}_{(ij)}\{C_{hjk|i} + C_h{}^r{}_i P_{rjk} + P_{hijk}\} = 0,$$

where $P_{rjk} = g_{ri}P^i{}_{jk} = P_{0rjk}$. Applying the Christoffel process with respect to h, i, j to the above, we get

$$P_{hijk} = \mathfrak{A}_{(hi)}\{C_{jik|h} + C_h{}^r{}_j P_{rik}\}.$$

Contractions of this by y^h and then by y^j yield

$$P_{ijk} = C_{jik|0} - C_i{}^r{}_j P_{r0k}, \qquad P_{i0k} = 0.$$

Consequently we conclude the interesting expressions of P_{ijk} and P_{hijk} as follows:

$$P_{ijk} = C_{ijk|0}, \tag{17.22}$$

$$P_{hijk} = \mathfrak{A}_{(hi)}\{C_{ijk|h} + C_h{}^r{}_j C_{rik|0}\}. \tag{17.23}$$

The latter immediately yields following identities:

$$\mathfrak{S}_{(hij)}\{P_{hijk}\} = 0, \qquad \mathfrak{S}_{(hik)}\{P_{hijk}\} = 0, \tag{17.24}$$

(17.25) $$\mathfrak{A}_{(jk)}\{P_{hijk}\} = \mathfrak{A}_{(jk)}\{C_h{}^r{}_j P_{rik} - C_i{}^r{}_j P_{rhk}\} = - S_{hijk|0},$$

(17.26) $$P_{hi0k} = P_{hij0} = 0.$$

We return to the identity (17.16), which is written as

$$\mathfrak{A}_{(ij)}\{P_{hir}C_k{}^r{}_j - P_{hkj}|_i + P_{ihkj}\} = 0.$$

Applying the Christoffel process with respect to h, i, j to the above, we get

(17.27) $$P_{hikj} = \mathfrak{A}_{(hi)}\{P_{ijr}C_k{}^r{}_h + P_{ijk}|_h\}$$

$$= \mathfrak{A}_{(hi)}\{P_{khr}C_i{}^r{}_j + \dot{\partial}_h P_{ijk}\}.$$

Finally we shall list up the important miscellaneous formulas for the later use:

(17.28) $$L_{|i} = 0, \qquad L|_i = l_i,$$

(17.29) $$y^i{}_{|j} = 0, \qquad y^i|_j = \delta^i_j,$$

(17.30) $$l_{i|j} = 0, \qquad l_i|_j = \frac{1}{L}h_{ij},$$

(17.31) $$h_{ij|k} = 0, \qquad h_{ij}|_k = -\frac{1}{L}(h_{ik}l_j + h_{jk}l_i).$$

REMARK 17.4. The form (17.20) of S_{hijk} is really simple, because components of the curvature tensor of M_x are written without derivatives of Christoffel symbols $C_j{}^i{}_k$.

In his paper [3], Cartan already gives (17.20), but not (17.23) yet. The latter appears first in his monograph [4]. It seems that (17.27) appears first in Berwald's paper [15], one of his posthumous papers.

§ 18. The Rund and Berwald connections

Let $C\Gamma = (\Gamma^h, \Gamma^v) = (\Gamma_V, N, \Gamma^v)$ be the Cartan connection on a Finsler space M with a Finsler metric L. Throughout the following, the entities of $C\Gamma$ are denoted by the usual notations C_{ijk}, $P^i{}_{jk}$, P_{hijk} etc., which we have used untill now.

The Finsler Γ^h-connection $F\Gamma^h = (\Gamma^h, F^i)$, constructed from $C\Gamma$ by the C-process (Definition 14.2), coincides with the Finsler Γ_V-connection $F\Gamma_V = (\Gamma_V, N, F^i)$, constructed from $C\Gamma$ (Definition 15.3), because the deflection tensor D of $C\Gamma$ vanishes identically (Theorem 15.2).

DEFINITION 18.1. The Finsler Γ_V-connection $R\Gamma$, constructed from the Cartan connection $C\Gamma$, is called the *Rund connection*.

From (14.15) it follows that connection coefficients $F_j{}^i{}_k$ and $N^i{}_j$ of the Rund connection $R\Gamma$ coincide with those of $C\Gamma$, while the Cartan tensor field C of $R\Gamma$ vanishes. In terms of the torsion and curvature tensors of $C\Gamma$, those of $R\Gamma$ are expressed immediately by (14.16). That is, the torsion tensors are such that

$$(18.1)\qquad \begin{aligned} &T = 0, \qquad R^1 = \text{the one of } C\Gamma, \qquad C = 0, \\ &P^1 = \text{the one of } C\Gamma, \qquad S^1 = 0. \end{aligned}$$

The curvature tensors of $R\Gamma$ are as follows:

$$(18.2)\qquad \begin{aligned} &\text{h-curvature } K: \quad K_h{}^i{}_{jk} = R_h{}^i{}_{jk} - C_h{}^i{}_r R^r{}_{jk}, \\ &\text{hv-curvature } F: \quad F_h{}^i{}_{jk} = P_h{}^i{}_{jk} + C_h{}^i{}_{k|j} - C_h{}^i{}_r P^r{}_{jk}, \end{aligned}$$

while the v-curvature S^2 of $R\Gamma$ vanishes identically. In (18.2) it is noted that the h-covariant differentiations $(|)$ of $C\Gamma$ and $R\Gamma$ coincides with each other.

The equations (10.17) and (10.20) give the expressions of components $K_h{}^i{}_{jk}$ and $F_h{}^i{}_{jk}$ in terms of connection coefficients:

$$(18.2')\qquad \begin{aligned} K_h{}^i{}_{jk} &= \mathfrak{A}_{(jk)}\{\delta_k F_h{}^i{}_j + F_h{}^r{}_j F_r{}^i{}_k\}, \\ F_h{}^i{}_{jk} &= \dot\partial_k F_h{}^i{}_j. \end{aligned}$$

Some of the Bianchi identities of $R\Gamma$ reduce to a simple form as follows:

$$(18.3)\qquad \mathfrak{S}_{(ijk)}\{K_i{}^h{}_{jk}\} = 0,$$

$$(18.4)\qquad \mathfrak{S}_{(ijk)}\{P^h{}_{ir} R^r{}_{jk} + R^h{}_{ij|k}\} = 0,$$

$$(18.5)\qquad \mathfrak{S}_{(ijk)}\{F_m{}^h{}_{ir} R^r{}_{jk} + K_m{}^h{}_{ij|k}\} = 0,$$

$$(18.6)\qquad \mathfrak{A}_{(ij)}\{F_i{}^h{}_{jk}\} = 0,$$

$$(18.7)\qquad R^h{}_{ij\cdot k} - K_k{}^h{}_{ij} + \mathfrak{A}_{(ij)}\{P^h{}_{ir} P^r{}_{jk} + P^h{}_{jk|i}\} = 0,$$

$$(18.8)\qquad K_m{}^h{}_{ij\cdot k} + \mathfrak{A}_{(ij)}\{F_m{}^h{}_{ir} P^r{}_{jk} + F_m{}^h{}_{jk|i}\} = 0,$$

$$(18.9)\qquad \mathfrak{A}_{(ij)}\{P^h{}_{kj\cdot i} - F_i{}^h{}_{kj}\} = 0,$$

$$(18.10)\qquad \mathfrak{A}_{(ij)}\{F_m{}^h{}_{kj\cdot i}\} = 0,$$

where we adopt the notation in (9.21).

The Rund connection is obviously h-metrical, but not v-metrical in general. As a consequence the Ricci identities (10.8') and (10.9') applied to g_{ij} yield

$$(18.11)\qquad K_{hijk} + K_{ihjk} + 2C_h{}^r{}_i R_{rjk} = 0,$$

$$(18.12)\qquad F_{hijk} + F_{ihjk} + 2C_h{}^r{}_i P_{rjk} = 2C_{hik|j},$$

where we put $K_{hijk} = g_{ir}K_h{}^r{}_{jk}$, $F_{hijk} = g_{ir}F_h{}^r{}_{jk}$.

REMARK 18.1. In his paper [3], Rund introduces the connection coefficients $P^i{}_{jk}$ by means of which a parallel displacement of a vector field X^i is given by

$$d^*X^i = -P^i{}_{jk}X^j dx^k \qquad \text{(in his notation).}$$

The quantities $P^i{}_{jk}$ are defined by $P^i{}_{jk} = \gamma_j{}^i{}_k - C_j{}^i{}_r\gamma_k{}^r{}_0$. According to his opinion, the idea to introduce the above parallelism is Minkowskian, while Cartan's is euclidean. A.Kawaguchi, the reviewer of Rund's paper [3] in Math. Rev., wrote as follows:

Although the ideas and methods in this paper are interesting and may be a contribution to the theory of Finsler spaces, the introduction of the last one of the four conditions for d^*X^i seems to the reviewer to be incomplete, because the covariant differential $Dg_{ij}(x, x')$ has never been defined, although the author defined that of a tensor $a_{ij}(x)$ whose components depend only on the position x^i but not on the direction x'^i.

In 1954 Rund [11] modified the above $P^i{}_{jk}$ and obtained new connection coefficients $P^{*i}{}_{jk}$. These, however, coincide with $F_j{}^i{}_k$ of CΓ, as indicated by the reviewers Davies in Math. Rev. and Deicke-Süss in Zentralblatt. See Rund [18, p.55], Ohkubo [4], Sulanke [1].

It is noted that in (18.11) and (18.12) the quantities $C_j{}^i{}_k$ are those given by (17.1'); they are not components of the Cartan tensor of RΓ. Throughout the following the notations $C_j{}^i{}_k$ are used to denote the quantities given by (17.1').

DEFINITION 18.2. The Finsler N-connection BΓ, constructed from the Cartan connection CΓ, is called the *Berwald connection*.

Therefore BΓ and CΓ have the common non-linear connection N and it may be said that BΓ is constructed this non-linear connection N only. From Proposition 15.2 it follows that BΓ is obtained from the Rund connection RΓ by the P^1-process. Consequently, in terms of the torsion and curvature tensors of RΓ, those of BΓ are expressed immediately from (15.2) as follows:

First the torsion tensors are such that

(18.13)
$$T = 0, \qquad R^1 = \text{the one of } R\Gamma, \qquad C = 0,$$
$$P^1 = 0, \qquad S^1 = 0.$$

A special character of BΓ is that the (v)hv-torsion tensor P^1 vanishes (Theorem 15.1). Thus (10.14) of BΓ and CΓ and (17.22) show that

(18.14)
$$G_j{}^i{}_k = \dot{\partial}_j N^i{}_k = F_j{}^i{}_k + C_j{}^i{}_{k|0}$$

are connection coefficients of the V-connection Γ_V of BΓ, where $N^i{}_k$ are given by (17.7). Since $T_j{}^i{}_k = G_j{}^i{}_k - G_k{}^i{}_j = 0$, the theory of differential equations shows that there exist locally the functions $N^i(x, y)$ such that $N^i{}_k = \dot{\partial}_k N^i$. In fact, it is observed from (17.4), (17.1') and (17.7) that $y^j y^h \dot{\partial}_k \gamma_{jih} = 0$ and

$$\dot{\partial}_k \gamma_0{}^i{}_0 = (\dot{\partial}_k g^{ir})\gamma_{0r0} + 2\gamma_0{}^i{}_k = -\, 2C^{ir}{}_k \gamma_{0r0} + 2\gamma_0{}^i{}_k = 2N^i{}_k,$$

which implies

(18.15)
$$N^i{}_k = \frac{1}{2}\dot{\partial}_k \gamma_0{}^i{}_0.$$

Secondly, from (15.2) the curvature tensors of BΓ are as follows:

(18.16)
$$\text{h-curvature } H: \quad H_h{}^i{}_{jk} = K_h{}^i{}_{jk} + \mathfrak{A}_{(jk)}\{C_j{}^i{}_{h|0|k} + C_k{}^i{}_{r|0} C_j{}^r{}_{h|0}\},$$
$$\text{hv-curvature } G: \quad G_h{}^i{}_{jk} = F_h{}^i{}_{jk} + C_j{}^i{}_{h|0\cdot k}.$$

The v-curvature tensor S^2 vanishes identically.

We shall show simpler forms of $H_h{}^i{}_{jk}$ and $G_h{}^i{}_{jk}$. It follows immediately from (18. 14) and (10.20) that

(18.17)
$$G_h{}^i{}_{jk} = \dot{\partial}_k G_h{}^i{}_j,$$

similarly to the second of (18.2'). This is also obvious from a general form (10. 16) of the hv-curvature tensor P^2. As to the h-curvature tensor $H_h{}^i{}_{jk}$, we are first concerned with the Bianchi identities of BΓ. Then

(18.18)
$$\mathfrak{S}_{(ijk)}\{H_i{}^h{}_{jk}\} = 0,$$

(18.19)
$$\mathfrak{S}_{(ijk)}\{R^h{}_{ij;k}\} = 0,$$

(18.20)
$$\mathfrak{S}_{(ijk)}\{G_m{}^h{}_{ir} R^r{}_{jk} + H_m{}^h{}_{ij;k}\} = 0,$$

$$(18.21)\qquad H_m{}^h{}_{ij\cdot k} + \mathfrak{A}_{(ij)}\{G_m{}^h{}_{jk;i}\} = 0,$$

$$(18.22)\qquad H_k{}^h{}_{ij} = R^h{}_{ij\cdot k},$$

and further we have three simple identities satisfied by $G_h{}^i{}_{jk}$ which are trivial from (18.14), (18.15) and (18.17), where (;) denotes the h-covariant differentiation with respect to BΓ.

The last identity (18.22) is noteworthy; the h-curvature tensor of BΓ is derived from the (v)h-torsion tensor R^1 of BΓ (the one of CΓ) by partial differentiation in y^i. Furthermore we shall show the interesting equation

$$(18.23)\qquad R^h{}_{ij} = \frac{1}{3}\mathfrak{A}_{(ij)}\{R^h{}_{0j\cdot i}\}.$$

In fact, from Theorem 13.3 or from (18.16) it follows that $H_0{}^i{}_{jk} = R^i{}_{jk}$, so that (18.18) gives

$$R^h{}_{jk} = \mathfrak{A}_{(jk)}\{H_j{}^h{}_{0k}\}.$$

Next, from (18.22) we get

$$\mathfrak{A}_{(ij)}\{R^h{}_{0j\cdot i}\} = \mathfrak{A}_{(ij)}\{\dot\partial_i(y^kR^h{}_{kj})\} = \mathfrak{A}_{(ij)}\{R^h{}_{ij} + y^kR^h{}_{kj\cdot i}\}$$

$$= \mathfrak{A}_{(ij)}\{R^h{}_{ij} + H_i{}^h{}_{0j}\}.$$

These equations yield (18.23) at once.

It is noted here that BΓ is neither h-metrical nor v-metrical in general. Referring to (18.14) we obtain

$$(18.24)\qquad g_{ij;k} = -\ 2C_{ijk|0}.$$

Therefore, with respect to BΓ, the h-covariant derivative of g_{ij} in the direction of supporting element y^k vanishes.

REMARK 18.2. Berwald is the first who introduced the concept of connection in Finsler geometry, although Synge [1] and Taylor [1] are a little noteworthy in this viewpoint. Pinl [2] describes an impressive life of this pioneer in Finsler geometry. Berwald's theory was first published in his paper [2]. Contrary to the case of Cartan, Berwald starts his theory from the equations of geodesics and applies the theory of general paths (Douglas [1]) to define the connection BΓ. The relation (18.14) between BΓ and CΓ is given by Cartan [4]. In his lecture [5], Cartan indicates some week points of BΓ from his standpoint, but it seems that there are several merits of BΓ, in particular, in viewpoint of (18.14), (18.17) and (18.22). From the standpoint of our formulation of general Finsler connection, it may be nonsense to discuss strong and weak points of connections BΓ, CΓ and RΓ, because we have possibilities to introduce other interesting connections from some special standpoints.

It has been observed that $B\Gamma$ is obtained from $R\Gamma$ by the P^1-process and $R\Gamma$ is done from $C\Gamma$ by the C-process. Therefore Proposition 15.2 shows that $B\Gamma$ is obtained by the C-process from a connection which is obtained from $C\Gamma$ by the P^1-process. This leads us to the following definition:

DEFINITION 18.3. A Finsler connection, obtained from the Cartan connection $C\Gamma$ by the P^1-process, is called the *Hashiguchi connection* and denoted by $H\Gamma$.

The quantities of $H\Gamma$ will be given by (15.2) from $C\Gamma$. The special feature of $H\Gamma$ is that the (v)hv-torsion tensor P^1 vanishes, while the (h)hv-torsion tensor C survives. The connection coefficients $(F_j{}^i{}_k, N^i{}_j, C_j{}^i{}_k)$ of $H\Gamma$ are equal to $(G_j{}^i{}_k, N^i{}_j, C_j{}^i{}_k)$ in the notations of the present section. Consequetly the relations of the above typical four Finsler connections are illustrated as follows:

$$\begin{array}{ccc} C\Gamma = (F_j{}^i{}_k, N^i{}_j, C_j{}^i{}_k) & \xrightarrow{\text{C-process}} & R\Gamma = (F_j{}^i{}_k, N^i{}_j, 0) \\ \Big\downarrow P^1\text{-process} & & \Big\downarrow P^1\text{-process} \\ H\Gamma = (G_j{}^i{}_k, N^i{}_j, C_j{}^i{}_k) & \xrightarrow{\text{C-process}} & B\Gamma = (G_j{}^i{}_k, N^i{}_j, 0) \end{array}$$

The non-linear connection N, given by (17.7) or (18.15), is common to all these connections.

REMARK 18.3. The concept of connection $H\Gamma$ was communicated by Hashiguchi to the author in 1969.

A,Kawaguchi [1] showed an interesting idea: $C\Gamma$ is obtained from $B\Gamma$ by a metrization in his sense. (19.14) and (18.24) yield

$$F_j{}^i{}_k = G_j{}^i{}_k + g^{ir} g_{rj;k}/2, \tag{18.25}$$

which might inspire Kawaguchi with this idea. See Miron and Hashiguchi [1].

Recently T.Okada [4] established an important fact as follows: *The Berwald connection $B\Gamma$ is uniquely determined by the following five axioms:*

(1) L-metrical : $\nabla^h L = 0$, (2) deflection tensor $D = 0$,

(3) (v)hv-torsion tensor $P^1 = 0$, (4) (h)h-torsion tensor $T = 0$,

(5) (h)hv-torsion tensor $C = 0$.

CHAPTER IV

DIFFERENTIAL GEOMETRY OF TANGENT BUNDLES

The differential geometry of tangent bundles seems to be originated by S.Sasaki [Tôhoku Math. J. (2) 10 (1958)] and P.Dombrowski [J. Reine Angew. Math. 210 (1962)], and has been studied by many authors from more generalized standpoint. Almost all of them are mainly concerned with linear connections.

We are here interested in this theory from the standpoint of Finsler geometry only. Throughout the present chapter, our theory is based on non-linear connections. By means of a bundle homomorphism of the Finsler bundle $F(M)$ into the linear frame bundle $\bar{L}(T)$ over the tangent bundle $T(M)$ which is defined with respect to a non-linear connection N, a Finsler connection $F\Gamma = (\Gamma, N)$ is thought of as a linear connection in $\bar{L}(T)$ or on $T(M)$. As a consequence three curvature tensors and five torsion tensors of $F\Gamma$ are regarded as fragments of the curvature tensor and torsion tensor of this linear connection in a sense.

§ 19. The N-decompositions of tensors

We first recall the concept of bundle homomorphism of a general principal bundle $P(M) = (P, M, \pi, G)$ into another principal bundle $Q(M) = (Q, M, \tau, H)$ over a common base manifold M. Let β_g (resp. γ_h) be the right translation of P (resp. Q) by $g \in G$ (resp. $h \in H$).

A mapping $r: P \to Q$ is called a *bundle homomorphism* of $P(M)$ into $Q(M)$, if it satisfies the following two conditions:

(1) Corresponding to r, there exists a group homomorphism $s: G \to H$ such that $r \circ \beta_g = \gamma_{s(g)} \circ r$ for any $g \in G$.

(2) The mapping $\underline{r}: M \to M$, induced from r by $\underline{r} \circ \pi = \tau \circ r$, is a transformation of M.

Owing to (1) the mapping r commutes with any right translation, so that the mapping $\underline{r}$ in (2) is uniquely induced by the above equation. If r is a one-to-one mapping of P onto Q, then r is called a *bundle isomorphism*.

Now we are concerned with the Finsler bundle $F(M) = (F, T, \pi_1, G(n))$ of an n-dimensional manifold M; the base space is the total space T of the tangent bundle $T(M)$. Let $\bar{L}(T) = (\bar{L}, T, \bar{\pi}_L, G(2n))$ be the linear frame bundle over T. Its structure group is the general linear group $G(2n)$ of order $2n$ because of $\dim T = 2n$, and the mapping

$$(19.1) \qquad \psi : g = (g^a_b) \in G(n) \mapsto \begin{pmatrix} g^a_b & 0 \\ 0 & g^a_b \end{pmatrix} \in G(2n)$$

is obviously a group homomorphism, which is called the *natural homomorphism*.

DEFINITION 19.1. Assume that a non-linear connection N is given in the tangent bundle T(M). The *N-homomorphism* ϕ_N: $F \to \bar{L}$ is a bundle homomorphism of the Finsler bundle F(M) into the linear frame bundle $\bar{L}(T)$ over the tangent bundle T(M) such that

$$\phi_N : u = (y, z) \in F \mapsto \{l_y(z), l_y^v(z)\} \in \bar{L},$$

where $l_y(z)$ is the set of lifts $l_y(z_a)$, $z = \{z_a\}$, to the point $y \in T$ with respect to the given N, and $l_y^v(z)$ is the set of vertical lifts $l_y^v(z_a)$ to y.

Thus $\phi_N(u)$, $u = (y, z)$, is the linear frame $\{l_y(z_1),\cdots, l_y(z_n), l_y^v(z_1),\cdots, l_y^v(z_n)\}$ of T with the origin y. It is evident that the mapping ϕ_N is a bundle homomorphism with the natural homomorphism ψ, namely, we have

$$(19.2) \qquad \phi_N \circ \beta_g = \gamma_{\psi(g)} \circ \phi_N,$$

where γ_h, $h \in G(2n)$, is a right translation of $\bar{L}(T)$. The mapping $\underline{\phi}_N$, induced from ϕ_N into the base space T, is clearly the identity mapping.

It is remarked that ϕ_N is not a bundle isomorphism, because $\phi_N(u)$ is such a special frame of T that the first n vectors $l_y(z_a)$ are horizontal with respect to N and the second n vectors $l_y^v(z_a)$ are vertical.

We consider the tangent bundle $\bar{T}(T) = (\bar{T}, T, \bar{\pi}_T, V^{2n}, G(2n))$ over the tangent bundle T(M). The standard fiber of $\bar{T}(T)$ is a 2n-dimensional vector space V^{2n} which is regarded as the product $V^n \times V^n$, where V^n is the standard fiber of T(M). From the fixed base $B = \{e_a\}$ of V^n (cf. Remark 2.3), a base $\bar{B} = \{\bar{e}_\alpha\}$, $\alpha = 1,\cdots, 2n$, of V^{2n} is derived by

$$\bar{e}_a = (e_a, 0), \qquad \bar{e}_{(a)} = (0, e_a),$$

where $(a) = n + a$. Throughout the following we denote n + i, n + a by (i), (a) respectively.

The action $\bar{\xi}$ of G(2n) on V^{2n} is defined based on this $\bar{B}$, similarly to the action ξ of G(n) on V^n (cf. (2.12)). Thus we have

$$(19.3) \qquad \{\psi(g)\}\bar{v} = (g_b^a \bar{v}^b e_a, g_b^a \bar{v}^{(b)} e_a) \in V^n \times V^n,$$

$$g = (g_b^a), \qquad \bar{v} = \bar{v}^\alpha \bar{e}_\alpha = \bar{v}^a \bar{e}_a + \bar{v}^{(a)} \bar{e}_{(a)}.$$

Next, we are concerned with an admissible mapping $\bar{z}\bar{\alpha} : \bar{v} \in V^{2n} \mapsto \bar{z}\bar{v} \in \bar{T}$, $\bar{z} \in \bar{L}$, of $\bar{T}$. If we take points $u = (y, z) \in F$, $\bar{z} = \phi_N(u) \in \bar{T}$ and $\bar{v} = (v_1, v_2) \in V^n \times V^n$, then we have

$$(19.4) \qquad \bar{z}\bar{\alpha}(\bar{v}) = l_y(zv_1) + l_y^v(zv_2),$$

from which the inverse $(_{\bar{z}}\bar{\alpha})^{-1}: \bar{T} \to V^{2n}$ is given by

$$(19.5) \qquad (_{\bar{z}}\bar{\alpha})^{-1}\bar{X} = (z^{-1}(\pi_T'\bar{X}),\ z^{-1}((1_y^v)^{-1}(v_N\bar{X}))), \quad \bar{z} = \phi_N(y, z),\ \bar{X} \in T_y,$$

where $v_N\bar{X}$ is the vertical component of $\bar{X}$ with respect to N (cf. (8.2)).

Since $\bar{L}(T)$ is a linear frame bundle, the basic form $\bar{\theta}$ on $\bar{L}(T)$ is defined by Definition 3.6.: $\bar{\theta}_{\bar{z}} = (_{\bar{z}}\bar{\alpha})^{-1}\circ\bar{\pi}_L'$. In particular, at a point of $\phi_N(F) \subset \bar{L}$, we have

$$(19.6) \qquad \bar{\theta}\circ\phi_N' = (\theta^h, \theta^v),$$

where θ^h is the h-basic form on F(M) and θ^v is the v-basic form with respect to N under consideration. (19.6) is easily shown from (19.5), (6.10) and (8.2).

A fundamental vector field $\bar{Z}(\bar{A})$ is also defined on $\bar{L}(T)$, corresponding to $\bar{A} \in G'(2n)$, by Definition 2.5. At a point of $\phi_N(F)$ we have

$$(19.7) \qquad \phi_N'Z(A) = \bar{Z}\begin{pmatrix} A & 0 \\ 0 & A \end{pmatrix}.$$

In fact, from (6.4) and (19.2) we observe

$$\phi_N'(Z(A)_u) = \phi_N'\circ{}_u\beta'(A_e) = {}_{\phi_N(u)}\gamma'\circ\psi'(A_e).$$

From $\psi'(A_e) = \begin{pmatrix} A_e & 0 \\ 0 & A_e \end{pmatrix}$, the above shows (19.7).

We now introduce an important concept. Consider a tensor field K on T of (1, 1)-type, for instance. K is regarded as a $(V^{2n})_1^1$-valued function on $\bar{L}$ which satisfies $K\circ\gamma_h = h^{-1}K$, $h \in G(2n)$, (cf. Definition 3.7). From K we get four mappings K_1^1, K_2^1, K_1^2, K_2^2: $F \to (V^n)_1^1$ as follows: For $v_* \in V_*^n$ and $v \in V^n$ we have $(v_*, 0)$, $(0, v_*) \in V_*^n\times V_*^n = V_*^{2n}$ and $(v, 0)$, $(0, v) \in V^n\times V^n = V^{2n}$. Then

$$(19.8) \qquad \begin{aligned} K_1^1(v_*, v) &= (K\circ\phi_N)((v_*, 0), (v, 0)), \\ K_2^1(v_*, v) &= (K\circ\phi_N)((v_*, 0), (0, v)), \\ K_1^2(v_*, v) &= (K\circ\phi_N)((0, v_*), (v, 0)), \\ K_2^2(v_*, v) &= (K\circ\phi_N)((0, v_*), (0, v)). \end{aligned}$$

These K_1^1, K_2^1, K_1^2, and K_2^2 are Finsler tensor fields of (1, 1)-type. In fact, we consider K_2^1, for instance. For $u \in F$ and $g \in G(n)$ it is seen from (19.2) that

$$K\circ\phi_N(ug) = K\circ\gamma_{\psi(g)}\circ\phi_N(u) = \psi(g)^{-1}(K\circ\phi_N(u)),$$

and from (19.3) and (2.14) that

$$(K^1_2 \circ \beta_g)(v_*, v) = \{\psi(g)^{-1}(K \circ \phi_N)\}\{(v_*, 0), (0, v)\}$$

$$= (K \circ \phi_N)\{\psi(g)(v_*, 0), \psi(g)(0, v)\} = (K \circ \phi_N)\{(gv_*, 0), (0, gv)\}$$

$$= K^1_2(gv_*, gv) = (g^{-1}K^1_2)(v_*, v).$$

Consequently we are led to

DEFINITION 19.2. The four Finsler tensor fields K^1_1, K^1_2, K^2_1, K^2_2, defined by (19.8), are called the *N-decompositions* of a tensor field K on T with respect to a non-linear connection N.

In general the N-decompositions of a tensor field of (r, s)-type are defined in a similar way, and those are 2^{r+s} Finsler tensor fields of the same type.

The converse will be obvious. That is, a collection of any 2^{r+s} Finsler tensor fields of (r, s)-type constitutes the N-decompositions of a tensor field of (r, s)-type on T. It is, however, remarked that, for instance, from two Finsler vector fields Y, Z we obtain two tangent vector fields on T, one having the N-decompositions $(X^1, X^2) = (Y, Z)$ and the other having $(X^1, X^2) = (Z, Y)$.

We shall often be concerned with tensor fields of (1, 2)-type and (1, 3)-type on T later on. For instance, we deal with a tensor field K of (1, 2)-type on T. For $\bar{v}_1$, $\bar{v}_2 \in V^{2n}$ we have $K(\bar{v}_1, \bar{v}_2) \in V^{2n} = V^n \times V^n$. Then the following formulas are immediately verified by Definition 19.2:

$$(19.9)\quad \begin{aligned} K((v_1, 0), (v_2, 0)) &= (K^1_{11}(v_1, v_2), K^2_{11}(v_1, v_2)), \\ K((v_1, 0), (0, v_2)) &= (K^1_{12}(v_1, v_2), K^2_{12}(v_1, v_2)), \\ K((0, v_1), (v_2, 0)) &= (K^1_{21}(v_1, v_2), K^2_{21}(v_1, v_2)), \\ K((0, v_1), (0, v_2)) &= (K^1_{22}(v_1, v_2), K^2_{22}(v_1, v_2)). \end{aligned}$$

We consider the components of N-decompositions. Let K be a tensor field of (1, 1)-type, for instance, and put $K_{\bar{z}} = \{K^\alpha_\beta\}\bar{e}^{-\beta}_\alpha$ at a point $\bar{z} = \phi_N(u)$, $u = (y, z)$. Then

$$(K^1_2)_u(v_*, v) = \{K^\alpha_\beta\}\bar{e}^{-\beta}_\alpha\{(v_a e^a, 0), (0, v^b e_b)\}$$

$$= K^\alpha_\beta(v_a \delta^a_\alpha)(v^b \delta^\beta_{(b)}) = K^a_{(b)} v_a v^b,$$

which implies $(K^1_2)_u = K^a_{(b)} e^b_a$. Namely, if K^α_β are components of K with respect to the frame $\bar{z} = \phi_N(y, z)$, then $K^a_{(b)}$ are components of K^1_2 with respect to the frame z. Similarly, K^a_b, $K^{(a)}_b$, $K^{(a)}_{(b)}$ are components of K^1_1, K^2_1, K^2_2 respectively with respect to

z.

Next, we consider the components of N-decompositions with respect to a coordinate system (x^i, y^i, z^i_a). Let K^λ_μ be the components of K with respect to the coordinate system $(x^i, y^i) = (\bar{x}^\lambda)$. Then (3.14') gives $K^\alpha_\beta = (\bar{z}^{-1})^\alpha_\lambda K^\lambda_\mu \bar{z}^\mu_\beta$. It follows from Definition 19.1, (9.1) and (3.18) that

$$\bar{z}_a = l_y(z_a) = z^i_a \left(\frac{\partial}{\partial x^i} - N^j{}_i \frac{\partial}{\partial y^j} \right),$$

$$\bar{z}_{(a)} = l^v_y(z_a) = z^i_a \frac{\partial}{\partial y^i} .$$

Therefore the N-homomorphism ϕ_N is written in the form

$$\phi_N : (x^i, y^i, z^i_a) \mapsto (\bar{x}^\lambda, \bar{z}^\lambda_\alpha), \qquad \bar{x}^i = x^i, \quad \bar{x}^{(i)} = y^i,$$

(19.10)
$$\bar{z}^i_a = z^i_a, \qquad \bar{z}^{(i)}_a = - z^j_a N^i{}_j,$$

$$\bar{z}^i_{(a)} = 0, \qquad \bar{z}^{(i)}_{(a)} = z^i_a .$$

The elements $(\bar{z}^{-1})^\alpha_\lambda$ of the matrix $\bar{z}^{-1}$ are given by

$$(\bar{z}^{-1})^a_i = (z^{-1})^a_i, \qquad (\bar{z}^{-1})^a_{(i)} = 0,$$

(19.10')
$$(\bar{z}^{-1})^{(a)}_i = (z^{-1})^a_j N^j{}_i, \qquad (\bar{z}^{-1})^{(a)}_{(i)} = (z^{-1})^a_i .$$

Accordingly we have, for instance,

$$K^a_b = (\bar{z}^{-1})^a_\lambda K^\lambda_\mu \bar{z}^\mu_b = (z^{-1})^a_i K^i_j z^j_b - (z^{-1})^a_i K^i_{(r)} z^j_b N^r{}_j ,$$

which gives the first of the following (19.11). In like manner the components of N-decompositions are given by

$$(K^1_1)^i_j = K^i_j - K^i_{(r)} N^r{}_j ,$$

$$(K^1_2)^i_j = K^i_{(j)} ,$$

(19.11)
$$(K^2_1)^i_j = K^{(i)}_j + K^r_j N^i{}_r - K^s_{(r)} N^i{}_s N^r{}_j - K^{(i)}_{(r)} N^r{}_j ,$$

$$(K^2_2)^i_j = K^{(i)}_{(j)} + K^r_{(j)} N^i{}_r .$$

REMARK 19.1. From (19.11) the following fact is observed: Among $(2n)^2$ components $(K^i_j, K^i_{(j)}, K^{(i)}_j, K^{(i)}_{(j)})$ of K with respect to the coordinate system (x^i, y^i), the second $K^i_{(j)}$ constitute the components of the Finsler tensor K^1_2, while each one of the other does not so. Although this fact will be verified by coordinate transformation, the computation will be very complicated.

In particular, we are concerned with a tensor field $\bar{X}$ of (1, 0)-type, i.e., a tangent vector field of T, considered as the mapping $\bar{L} \to V^{2n}$. From $\bar{X}$ we obtain its N-decompositions $\bar{X}^1$, $\bar{X}^2$ which are Finsler vector fields.

<u>PROPOSITION 19.1.</u> *The N-decompositions $\bar{X}^1$, $\bar{X}^2$ of a tangent vector field $\bar{X}$ of T are given by*

$$(\bar{X}^1)_u = z^{-1}(\pi_T'\bar{X}), \qquad (\bar{X}^2)_u = z^{-1}((l_y^v)^{-1}(v_N\bar{X}))$$

at a point $u = (y, z) \in F$.

The proof is obvious from (19.5). In terms of components $(X^i, X^{(i)})$ of $\bar{X}$ in an induced coordinate system $(\bar{x}^i, \bar{x}^{(i)})$, $\bar{x}^i = x^i$, $\bar{x}^{(i)} = y^i$, the N-decompositions are written as

$$(\bar{X}^1)^i = X^i, \qquad (\bar{X}^2)^i = X^{(i)} + X^j N^i_{\ j}.$$

We observe

$$\bar{X} = X^i \frac{\partial}{\partial x^i} + X^{(i)} \frac{\partial}{\partial y^i} = X^i\left(\frac{\partial}{\partial x^i} - N^j_{\ i} \frac{\partial}{\partial y^j}\right) + (X^{(i)} + X^j N^i_{\ j}) \frac{\partial}{\partial y^i} .$$

Thus it may be said that $\bar{X}^1$ and $\bar{X}^2$ are horizontal and vertical components of $\bar{X}$ respectively. Futher it follows from (19.4) that

$$\bar{X}_y = l_y(z\bar{X}^1_u) + l^v_y(z\bar{X}^2_u), \qquad u = (y, z). \tag{19.12}$$

From a tangent vector field X of the base space M, we can construct the horizontal lift X^h with respect to the non-linear connection N under consideration and the vertical lift X^v. Applying Proposition 19.1 to X^h and X^v, we get

$$(X^h)^1_u = z^{-1}X, \qquad (X^h)^2 = 0, \qquad (X^v)^1 = 0, \qquad (X^v)^2_u = z^{-1}X, \qquad u = (y, z), \tag{19.13}$$

If we specially consider the intrinsic-vertical vector field l^v (Definition 3.11), it follows from Definition 6.3 that

$$(l^v)^1 = 0, \qquad (l^v)^2 = \varepsilon. \tag{19.14}$$

Further, with respect to the non-linear connection N, the *intrinsic-horizontal vector field* l^h is defined by $(l^h)_y = l_y(y)$. Then we have

$$(l^h)^1 = \varepsilon, \qquad (l^h)^2 = 0. \tag{19.15}$$

§ 20. Linear connections of Finsler type

The concept of bundle homomorphism $r: P \to Q$ is interesting from the standpoint of the theory of connections, because it induces a connection $\Gamma' = r(\Gamma)$ in Q from a connection Γ in P as follows: Let q be a point of $r(P) \subset Q$. Then we put $\Gamma'_q = r'(\Gamma_p)$. Because Γ is invariant by any right translation and r commutes with it, Γ'_q is well-defined by the above equation, independently on the choice of p such that $r(p) = q$. Next, since the induced mapping $\underline{r}$ from r is a transformation of the base space M, there exists a point $q_1 = \gamma_h(q) \in r(P)$ on the fiber through any point $q \in Q$ which is not contained in r(P), so that Γ' at q is defined by $\Gamma'_q = (\gamma_h^{-1})'\Gamma'_{q_1}$. Thus we get a distribution Γ' all over Q, and it is evident that Γ' is a connection in Q.

DEFINITION 20.1. Assume that a Finsler connection $F\Gamma = (\Gamma, N)$ is given in the Finsler bundle F(M). Then we obtain the N-homomorphism $\phi_N: F \to \bar{L}$ by N. The induced connection $\bar{\Gamma} = \phi_N(\Gamma)$ from Γ is called the *linear connection of Finsler type.*

Thus, from an ordinary connection Γ in F and a non-linear connection N in T, we can construct a kind of linear connection $\bar{\Gamma}$ in $\bar{L}$. The connection form $\bar{\omega}$ of $\bar{\Gamma}$ has the value in the Lie algebra G'(2n) and we obtain a basic vector field $\bar{B}(\bar{v})$, corresponding to $\bar{v} \in V^{2n}$, with respect to $\bar{\Gamma}$.

PROPOSITION 20.1. *Let $\bar{\omega}$ be the connection form of a linear connection of Finsler type $\bar{\Gamma} = \phi_N(\Gamma)$ and $\bar{B}(\bar{v})$ be a basic vector field with respect to $\bar{\Gamma}$. Then*

$$(1) \quad \bar{\omega}\circ\phi'_N = \begin{pmatrix} \omega & 0 \\ 0 & \omega \end{pmatrix}, \qquad (2) \quad \begin{aligned} \phi'_N B^h(v) &= \bar{B}((v, 0)), \\ \phi'_N B^v(v) &= \bar{B}((0, v)), \end{aligned}$$

where ω is the connection form of Γ and $B^h(v)$ (resp. $B^v(v)$) is an h-(resp. v)-basic vector field with respect to the Finsler connection $F\Gamma = (\Gamma, N)$.

Proof. If we consider a G'(2n)-valued form $\alpha = \bar{\omega}\circ\phi'_N - \begin{pmatrix} \omega & 0 \\ 0 & \omega \end{pmatrix}$ on F, then $\alpha Z(A) = 0$ follows from (19.7) immediately. Further both $\phi'_N B^h(v)$ and $\phi'_N B^v(v)$ are contained in $\bar{\Gamma}$ by Definition 20.1, so that $\alpha B^h(v) = \alpha B^v(v) = 0$ are evident. Therefore we have $\alpha = 0$. Next we shall show (2). It follows from (19.6) that $\bar{\theta}\circ\phi'_N B^h(v) = (v, 0)$ and $\bar{\theta}\circ\phi'_N B^v(v) = (0, v)$, so that (2) is verified by Definition 4.6.

We denote by $\bar{\nabla}K$ the covariant derivative of a tensor field K on T with respect to the connection $\bar{\Gamma} = \phi_N(\Gamma)$. The relation between N-decompositions of $\bar{\nabla}K$ and h-and v-covariant derivatives of the N-decompositions of K is given by

THEOREM 20.1. *Let K^α_β and $(\bar{\nabla}K)^\alpha_{\beta\gamma}$, $\alpha, \beta, \gamma = 1, 2$, be respective N-decompositions* *of a tensor field K of (1, 1)-type, for instance, and the covariant derivative $\bar{\nabla}K$ of K with respect to the linear connection of Finsler type $\bar{\Gamma} = \phi_N(\Gamma)$. Then we have*

$$(\bar{\nabla}K)^{\alpha}_{\beta 1} = \nabla^h(K^{\alpha}_{\beta}), \qquad (\bar{\nabla}K)^{\alpha}_{\beta 2} = \nabla^v(K^{\alpha}_{\beta}),$$

where ∇^h (resp. ∇^v) is the h- (resp. v-)covariant differentiation with respect to the Finsler connection $F\Gamma = (\Gamma, N)$.

Proof. We are concerned with $(\bar{\nabla}K)^1_{21}$, for instance. From Proposition 20.1 we observe

$$(\bar{\nabla}K)^1_{21}(v_*, v_1, v_2) = \bar{\nabla}K((v_*, 0), (0, v_1), (v_2, 0))$$

$$= \bar{B}((v_2, 0))K((v_*, 0), (0, v_1))) = B^h(v_2)K^1_2(v_*, v_1),$$

which compeletes the proof.

The above theorem distinctly shows an effect of introducing N-decompositions and linear connection of Finsler type.

Let X, Y be two tangent vector fields of T. The *covariant derivative* $\bar{\nabla}_X Y$ of Y *along* X or *in the direction* X is a tangent vector field defined by

$$\bar{\nabla}_X Y(\bar{v}_*) = \bar{\nabla}Y(\bar{v}_*, \tilde{X}), \tag{20.1}$$

where $\bar{v}_* \in V^{2n}_*$ and $\tilde{X}$ is the associated tensor function $\bar{L} \to V^{2n}$ with X. It follows from Proposition 19.1 that (19.5) is written

$$\tilde{X} = \bar{z}^{-1}X = (X^1, X^2). \tag{20.2}$$

We consider the N-decompositions of $\bar{\nabla}_X Y$. From (19.8) and Theorem 20.1 we obtain, for instance,

$$(\bar{\nabla}_X Y)^1(v_*) = \bar{\nabla}Y((v_*, 0), (X^1, X^2)) = (\bar{\nabla}Y)^1_1(v_*, X^1) + (\bar{\nabla}Y)^1_2(v_*, X^2)$$

$$= \nabla^h Y^1(v_*, X^1) + \nabla^v Y^1(v_*, X^2).$$

In like manner we get

$$(\bar{\nabla}_X Y)^1 = \nabla^h Y^1(X^1) + \nabla^v Y^1(X^2),$$

$$(\bar{\nabla}_X Y)^2 = \nabla^h Y^2(X^1) + \nabla^v Y^2(X^2). \tag{20.3}$$

In particular, we treat of the horizontal lift X^h and vertical lift X^v of a tangent vector field X of the base space M. As to $\bar{\nabla}_{X^h} Y^h$, it follows from (20.3) and (19.13) that

$$(\bar{\nabla}_{X^h} Y^h)^1 = \nabla^h(z^{-1}Y)(z^{-1}X), \qquad (\bar{\nabla}_{X^h} Y^h)^2 = 0.$$

Therefore, denoting by $\bar{X}$ the Finsler vector field induced from X (cf. Example 6.1), we obtain $(\bar\nabla_{X^h}Y^h)^1 = \nabla^h\bar Y(\bar X)$ and $(\bar\nabla_{X^h}Y^h)^2 = 0$. Similarly we get

$$
(20.4)\qquad
\begin{aligned}
&(\bar\nabla_{X^h}Y^h)^1 = \nabla^h\bar Y(\bar X), && (\bar\nabla_{X^h}Y^h)^2 = 0,\\
&(\bar\nabla_{X^h}Y^v)^1 = 0, && (\bar\nabla_{X^h}Y^v)^2 = \nabla^h\bar Y(\bar X),\\
&(\bar\nabla_{X^v}Y^h)^1 = \nabla^v\bar Y(\bar X), && (\bar\nabla_{X^v}Y^h)^2 = 0,\\
&(\bar\nabla_{X^v}Y^v)^1 = 0, && (\bar\nabla_{X^v}Y^v)^2 = \nabla^v\bar Y(\bar X).
\end{aligned}
$$

Here we write the connection coefficients $\bar\Gamma_\mu{}^\lambda{}_\nu$ of the linear connection of Finsler type $\bar\Gamma = \phi_N(\Gamma)$ in terms of the connection coefficients $(F_j{}^i{}_k, N^i{}_j, C_j{}^i{}_k)$ of the Finsler connection $F\Gamma = (\Gamma, N)$. Referring to the quantities Γ's appeared in (9.3), we have

$$
(20.5)\qquad
\begin{aligned}
&\bar\Gamma_j{}^i{}_k = \Gamma_j{}^i{}_k, && \bar\Gamma_j{}^{(i)}{}_k = \partial_k N^i{}_j + N^r{}_j\Gamma_r{}^i{}_k - N^i{}_r\Gamma_j{}^r{}_k,\\
&\bar\Gamma_{(j)}{}^i{}_k = 0, && \bar\Gamma_{(j)}{}^{(i)}{}_k = \Gamma_j{}^i{}_k,\\
&\bar\Gamma_j{}^i{}_{(k)} = C_j{}^i{}_k, && \bar\Gamma_j{}^{(i)}{}_{(k)} = \dot\partial_k N^i{}_j + N^r{}_j C_r{}^i{}_k - N^i{}_r C_j{}^r{}_k,\\
&\bar\Gamma_{(j)}{}^i{}_{(k)} = 0, && \bar\Gamma_{(j)}{}^{(i)}{}_{(k)} = C_j{}^i{}_k.
\end{aligned}
$$

These are easily obtained from (2) of Proposition 20.1 and (19.10).

We shall here show an interesting application of the theory in the present section. Let $\underline{\nabla}$ be the covariant differentiation with respect to a linear connection $\underline{\Gamma}$ in L(M). It is well-known that a tangent vector field X of M is called *concurrent with respect to* $\underline{\Gamma}$, if $\underline{\nabla}X = -\delta$ holds, where δ is the tensor field with Kronecker deltas as the components. Tachibana [1] generalizes this notion to a Finsler space. He calls a tangent vector field X of a Finsler space M concurrent with respect to the Cartan connection $C\Gamma$, if

$$
(20.6)\qquad X^i{}_{|j} = -\delta^i_j, \qquad X^i C_i{}^j{}_k = 0
$$

are satisfied. The first equation of (20.6) is a natural but formal generalization of the case of a linear connection, and any origin of the second equation is not shown by Tachibana. Further, it is unsatisfactory for the author that X is assumed to be a tangent vector field of M, not a Finsler vector field.

In the following, we examine a reasonable definition of the concept of concurrency of a Finsler vector field. We already have the linear connection of Finsler type $\bar{\Gamma} = \phi_N(\Gamma)$ induced from a Finsler connection $F\Gamma = (\Gamma, N)$ into the linear frame bundle $\bar{L}(T)$. Therefore, according to the ordinary definition, a tangent vector field X of T is said to be concurrent with respect to $\bar{\Gamma}$ if $\bar{\nabla}X = -\delta$ holds, i.e, $\bar{\nabla}_Y X = -Y$ for any Y. Then (20.3) shows that the concurrency of X is written as

$$(20.7) \qquad (1) \quad \nabla^h X^1 = -\delta, \qquad (2) \quad \nabla^v X^1 = 0,$$

$$(20.8) \qquad (1) \quad \nabla^h X^2 = 0, \qquad (2) \quad \nabla^v X^2 = -\delta.$$

If the Finsler connection $F\Gamma$ satisfies the D-and C_1-conditions, (20.8) holds for $X^2 = -\varepsilon$ (cf. (9.14) and (9.16)). Consequently we are naturally led to

DEFINITION 20.2. With respect to a Finsler connection $F\Gamma = (\Gamma, N)$ satisfying the D- and C_1-conditions, a Finsler vector field X is called *concurrent*, if $(X^1, X^2) = (X, -\varepsilon)$ are the N-decompositions of a concurrent vector field of the tangent bundle T(M) with respect to the linear connection of Finsler type $\phi_N(\Gamma)$, i.e., $\nabla^h X = -\delta$ and $\nabla^v X = 0$ hold.

If $F\Gamma$ further satisfies the C_2-condition, the equation $\nabla^v X(\varepsilon) = 0$, obtained from (2) of (20.7), reduces to $\nabla^0 X(\varepsilon) = 0$, so that X is (0)p-homogeneous.

We shall restrict our consideration to the Cartan connection $C\Gamma$, which satisfies the D-, C_1- and C_2-conditions. Then (20.7) and the Ricci identity (10.9') give

$$(20.9) \qquad X^r P_{rijk} + C_{ijk} = 0.$$

Since P_{rijk} are skew-symmetric in r, i, we get $X^i C_{ijk} = 0$ from (20.9), so that (2) of (20.7) reduces to $\nabla^0 X = 0$, i.e., the components X^i of X are functions of position x only. Thus X is solely a tangent vector field of the Finsler space M. Consequently we have returned to Tachibana's definition (20.6). Furthermore, it is observed that

$$\dot{\partial}_j X_i = \dot{\partial}_j (g_{ir} X^r) = (g_{ir} X^r)|_j + g_{sr} X^r C_i{}^s{}_j = 0$$

from $X^r|_j = 0$.

Summarizing up the above, we have

THEOREM 20.2. *A concurrent vector field X with respect to the Cartan connection $C\Gamma$ is characterized by (20.6). Both contravariant and covariant components of X depend on a position alone.*

REMARK 20.1. Matsumoto and Eguchi [1] treats Finsler spaces admitting a concurrent vector field. What is a real geometrical meaning of such a strange property that both contravariant and covariant components of X depend on a position alone ? Tachibana [1] shows that a Finsler space admits a concurrent vector field $v^i(x)$ if and only if there exists a coordinate system (x^i) with reference to which ds^2 of the space is written in the form

$$ds^2 = (dx^n)^2 + (x^n)^2 H(x^\alpha, dx^\beta),$$

where $H(x^\alpha, dx^\beta)$ is the square of line-element of an arbitrary (n - 1)-dimensional Finsler space. This is just a generalization of the well-known theorem in Riemannian case. Although the vector field $(0,\cdots, 0, -x^n)$ is certainly concurrent and Tachibana's condition is sufficient, the author can not follow his proof. It seems to the author that Tachibana treats a Riemannian metric obtained from the Finsler metric by specifying the supporting element as the concurrent vector field under consideration, similarly to the concept of so-called *osculating Riemannian space*. See Maebashi [1], O.Varga [2, 4].

The greatest care must be taken in dealing with the supporting element. Relating to this remark, the author must write here that the augument of O.Varga [20] may be false, apart from his conclusion. Because the distribution he gives there unfortunately depends on a field of supporting element, so that we can not arrive at his result in his way, although it is analogous to the Riemannian case.

It is an interesting idea for studying Finsler geometry to *bring some concepts in Riemannian geometry into Finsler geometry*, but we must get rid of our prejudice which comes from Riemannian geometry. For instance, the result of Shing [1] is probably false; he finds interesting conditions $R_{hijk|m} = 0$ and $C_{hij|k} = 0$ for a Finsler space to be *locally symmetric*, similarly to the Riemannian case (cf. Soós [3]), but he refers to the mistaken identities satisfied by the curvature tensor which are analogous to the Riemannian case. Nevertheless Shing's idea is more interesting than only a formal generalization appearing in Misra [12, 13] and his successors.

We shall find the torsion tensor $\bar{T}$ and the curvature tensor $\bar{R}$ of the linear connection of Finsler type $\bar{\Gamma} = \phi_N(\Gamma)$. These appear in the structure equation

$$[\bar{B}(\bar{v}_1), \bar{B}(\bar{v}_2)] = \bar{B}(\bar{T}(\bar{v}_1, \bar{v}_2)) + \bar{Z}(\bar{R}(\bar{v}_1, \bar{v}_2)).$$

If we take $\bar{v}_1 = (v_1, 0)$, $\bar{v}_2 = (v_2, 0)$, then it follows from Proposition 20.1 that

$$[\bar{B}((v_1, 0)), \bar{B}((v_2, 0))] = \phi'_N[B^h(v_1), B^h(v_2)],$$

so that (10.3) and (19.7) yield

$$\bar{B}(\bar{T}((v_1, 0), (v_2, 0))) + \bar{Z}(\bar{R}((v_1, 0), (v_2, 0)))$$

$$= \bar{B}((T(v_1, v_2), 0)) + \bar{B}((0, R^1(v_1, v_2))) + \bar{Z}\begin{pmatrix} R^2(v_1, v_2) & 0 \\ 0 & R^2(v_1, v_2) \end{pmatrix},$$

which implies

$$\bar{T}((v_1, 0), (v_2, 0)) = (T(v_1, v_2), R^1(v_1, v_2)),$$

$$\bar{R}((v_1, 0), (v_2, 0)) = \begin{pmatrix} R^2(v_1, v_2) & 0 \\ 0 & R^2(v_1, v_2) \end{pmatrix}.$$

From (19.9), in terms of N-decompositions of $\bar{T}$ and $\bar{R}$, the above equations show that $\bar{T}_1{}^1{}_1(v_1, v_2) = T(v_1, v_2)$, $\bar{T}_1{}^2{}_1(v_1, v_2) = R^1(v_1, v_2)$ and $\bar{R}_1{}^1{}_{11}(v_1, v_2) = R^2(v_1, v_2)$, $\bar{R}_2{}^1{}_{11} = \bar{R}_1{}^2{}_{11} = 0$, $\bar{R}_2{}^2{}_{11}(v_1, v_2) = R^2(v_1, v_2)$. In the similar way the following theorem is established:

THEOREM 20.3. *The N-decompositions* $\bar{T}_\beta{}^\alpha{}_\gamma$, $\bar{R}_\beta{}^\alpha{}_{\gamma\delta}$, $\alpha, \beta, \gamma, \delta = 1, 2$, *of the torsion and curvature tensors* $\bar{T}$, $\bar{R}$ *of the linear connection of Finsler type* $\bar{\Gamma} = \phi_N(\Gamma)$ *are written, in terms of the Finsler connection* $F\Gamma = (\Gamma, N)$, *as*

$$\bar{T}_1{}^1{}_1(u, v) = T(u, v), \qquad \bar{T}_1{}^2{}_1(u, v) = R^1(u, v),$$

$$\bar{T}_1{}^1{}_2(u, v) = C(u, v), \qquad \bar{T}_1{}^2{}_2(u, v) = P^1(u, v),$$

$$\bar{T}_2{}^1{}_1(u, v) = -\ C(v, u), \qquad \bar{T}_2{}^2{}_1(u, v) = -\ P^1(v, u),$$

$$\bar{T}_2{}^1{}_2(u, v) = 0, \qquad \bar{T}_2{}^2{}_2(u, v) = S^1(u, v),$$

$$\bar{R}_\beta{}^\alpha{}_{11}(u, v) = \delta^\alpha_\beta R^2(u, v), \qquad \bar{R}_\beta{}^\alpha{}_{12}(u, v) = \delta^\alpha_\beta P^2(u, v),$$

$$\bar{R}_\beta{}^\alpha{}_{21}(u, v) = -\ \delta^\alpha_\beta P^2(v, u), \qquad \bar{R}_\beta{}^\alpha{}_{22}(u, v) = \delta^\alpha_\beta S^2(u, v),$$

where u, v *are any vectors of* V^n.

REMARK 20.2. It may be said that Theorem 20.3 is the most fruitful result arising from introducing the concept of linear connection of Finsler type. As to the linear connection $\bar{\Gamma} = \phi_N(\Gamma)$, we have, of course, only one torsion tensor $\bar{T}$ and only one curvature tensor $\bar{R}$, while there are five torsion tensors and three curvature tensors as to the Finsler connection $F\Gamma = (\Gamma, N)$. In the viewpoint of Theorem 20.3, it is observed that these torsion and curvature tensors of $F\Gamma$ may be regarded solely as *fragments* of those of $\bar{\Gamma}$.

In the usual classical treatment of Finsler geometry, the author must agree with Busemann. He began his interesting lecture [4] with the following suggestive and cynical words:

The term "Finsler space" evokes in most mathematicians the picture of an impenetrable forest whose entire vegetation consists of tensors. The purpose of the present lecture is to show that the association of tensors (or differential forms) with Finsler spaces is due to an historical accident, and ···.

By the term "historical accident" he may signify the theories due to Berwald and

Cartan. Even Rund's book [18] is really a forest of tensors.

Moreover, three Ricci identities and eleven Bianchi identities of the Finsler connection $F\Gamma = (\Gamma, N)$ are fragments of only one Ricci identity and only two Bianchi identities of the linear connection of Finsler type $\bar{\Gamma} = \phi_N(\Gamma)$.

It seems to the author that Deicke's papers [2, 3] are written from our standpoint in this chapter, although his expression is not so clear.

It should be noted that the equations of Theorem 20.3 elegantly give the N-decompositions of $\bar{T}$ and $\bar{R}$, but the components of $\bar{T}$ and $\bar{R}$ in an induced coordinate system (x^i, y^i) are of complicated form. For instance, we write the components $\bar{T}_\mu{}^\lambda{}_\nu$ of $\bar{T}$; those are written as

$$\bar{T}_\mu{}^\lambda{}_\nu = \bar{z}^\lambda_\alpha \bar{T}_\beta{}^\alpha{}_\gamma (\bar{z}^{-1})^\beta_\mu (\bar{z}^{-1})^\gamma_\nu,$$

where $\bar{T}_\beta{}^\alpha{}_\gamma$ are components of $\bar{T}$ with respect to the frame $\bar{z} = \phi_N(u)$, $u = (y, z)$. Theorem 20.3 shows that $\bar{T}_\beta{}^\alpha{}_\gamma$ are given by

$$\bar{T}_b{}^a{}_c = T_b{}^a{}_c, \qquad \bar{T}_b{}^{(a)}{}_c = R^a{}_{bc}, \qquad \bar{T}_b{}^a{}_{(c)} = C_b{}^a{}_c,$$

$$\bar{T}_b{}^{(a)}{}_{(c)} = P^a{}_{bc}, \qquad \bar{T}_{(b)}{}^a{}_{(c)} = 0, \qquad \bar{T}_{(b)}{}^{(a)}{}_{(c)} = S^a{}_{bc}.$$

The components $T_b{}^a{}_c$ of T, for instance, with respect to the frame z are written as $T_b{}^a{}_c = (z^{-1})^a_i T_j{}^i{}_k z^j_b z^k_c$ in terms of components $T_j{}^i{}_k$ in (x^i, y^i). Consequently (19.10) and (19.10') yield

$$\begin{aligned}
\bar{T}_j{}^i{}_k &= \bar{z}^i_\alpha \bar{T}_\beta{}^\alpha{}_\gamma (\bar{z}^{-1})^\beta_j (\bar{z}^{-1})^\gamma_k \\
&= z^i_a\{\bar{T}_b{}^a{}_c (z^{-1})^b_j (z^{-1})^c_k + \bar{T}_b{}^a{}_{(c)} (z^{-1})^b_j (z^{-1})^c_r N^r{}_k + \bar{T}_{(b)}{}^a{}_c (z^{-1})^b_r N^r{}_j (z^{-1})^c_k\} \\
&= T_j{}^i{}_k + C_j{}^i{}_r N^r{}_k - C_k{}^i{}_r N^r{}_j,
\end{aligned}$$

$$\begin{aligned}
\bar{T}_j{}^{(i)}{}_{(k)} &= \bar{z}^{(i)}{}_\alpha \bar{T}_\beta{}^\alpha{}_\gamma (\bar{z}^{-1})^\beta_j (\bar{z}^{-1})^\gamma_{(k)} \\
&= \{- z^r_a N^i{}_r \bar{T}_b{}^a{}_{(c)} (z^{-1})^b_j + z^i_a \bar{T}_b{}^{(a)}{}_{(c)} (z^{-1})^b_j + z^i_a \bar{T}_{(b)}{}^{(a)}{}_{(c)} (z^{-1})^b_r N^r{}_j\} (z^{-1})^c_k \\
&= - N^i{}_r C_j{}^r{}_k + P^i{}_{jk} + S^i{}_{rk} N^r{}_j.
\end{aligned}$$

The other components will be similarly obtained.

As above seen, the linear connection of Finsler type $\bar{\Gamma}$ is not symmetric ($\bar{T} \neq 0$) in general. We shall derive a symmetric connection from $\bar{\Gamma}$. By (4.14) any linear connection $\bar{\Gamma}'$ in $\bar{L}(T)$ is given by

$$\bar{B}'(\bar{v}) = \bar{B}(\bar{v}) + \bar{Z}(\bar{D}(\bar{v})), \tag{20.10}$$

where $\bar{D}$ is the difference tensor field of $\bar{\Gamma}'$ from $\bar{\Gamma}$. Then the torsion tensor field $\bar{T}'$ of $\bar{\Gamma}'$ is written

$$\bar{T}'(\bar{v}_1, \bar{v}_2) = \bar{T}(\bar{v}_1, \bar{v}_2) - \bar{D}(\bar{v}_1, \bar{v}_2) + \bar{D}(\bar{v}_2, \bar{v}_1).$$

In particular, if we put $\bar{D} = \bar{T}/2$, we obtain $\bar{T}' = 0$, so that $\bar{\Gamma}'$ becomes symmetric.

DEFINITION 20.3. A symmetric connection $\bar{\Gamma}'$ whose difference tensor $\bar{D}$ from a linear connection of Finsler type $\bar{\Gamma} = \phi_N(\Gamma)$ is equal to $\bar{T}/2$ is called a *linear symmetric connection of Finsler type*.

The basic vector field of the above $\bar{\Gamma}'$ is written

$$\bar{B}'(\bar{v}) = \bar{B}(\bar{v}) + \frac{1}{2}\bar{Z}(\bar{T}(\bar{v})),$$

so that the covariant derivative $\bar{\nabla}'K$ of a tensor field K with respect to $\bar{\Gamma}'$ is given by

$$\bar{\nabla}'K(\bar{v}) = \bar{\nabla}K(\bar{v}) - \frac{1}{2}\bar{T}(\bar{v})\cdot K.$$

In particular, from Theorems 20.1 and 20.3, the N-decompositions of covariant derivative $\bar{\nabla}'X$ of a tangent vector field X of T are given by

$$\begin{aligned} (\bar{\nabla}'X)^1_1 &= \nabla^h X^1 - \frac{1}{2}\{T(X^1, \delta) - C(\delta, X^2)\}, \\ (\bar{\nabla}'X)^1_2 &= \nabla^v X^1 - \frac{1}{2} C(X^1, \delta), \\ (\bar{\nabla}'X)^2_1 &= \nabla^h X^2 - \frac{1}{2}\{R^1(X^1, \delta) - P^1(\delta, X^2)\}, \\ (\bar{\nabla}'X)^2_2 &= \nabla^v X^2 - \frac{1}{2}\{P^1(X^1, \delta) - S^1(\delta, X^2)\}. \end{aligned} \tag{20.11}$$

Let X, Y be two tangent vector fields of the base space M. Then, referring to the induced Finsler vector fields $\bar{X}$, $\bar{Y}$ from X, Y, it follows from (19.13) and (20.11) that for $v_* \in V_*^n$ we have

$$(\bar{\nabla}'_{X^h} Y^h)^1(v_*) = \bar{\nabla}'Y^h((v_*, 0), X^h) = (\bar{\nabla}'Y^h)^1_1(v_*, \bar{X}) = \nabla^h\bar{Y}(v_*, \bar{X}) - \frac{1}{2}T(\bar{Y}, v_*, \bar{X}),$$

$$(\bar{\nabla}'_{X^h} Y^h)^2(v_*) = \bar{\nabla}'Y^h((0, v_*), X^h) = (\bar{\nabla}'Y^h)^2_1(v_*, \bar{X}) = -\frac{1}{2}R^1(v_*, \bar{Y}, \bar{X}).$$

Thus the first two equations of the following (20.12) are shown. Similarly we get

$$\left\{ (\bar{\nabla}'_{X^h} Y^h)^1 = \nabla^h\bar{Y}(\bar{X}) - \frac{1}{2}T(\bar{Y}, \bar{X}), \qquad (\bar{\nabla}'_{X^h} Y^h)^2 = -\frac{1}{2}R^1(\bar{Y}, \bar{X}), \right.$$

$$(20.12)\quad \begin{cases} (\bar{\nabla}'_{X^h} Y^v)^1 = \frac{1}{2}\, C(\bar{X}, \bar{Y}), & (\bar{\nabla}'_{X^h} Y^v)^2 = \nabla^h \bar{Y}(\bar{X}) + \frac{1}{2}\, P^1(\bar{X}, \bar{Y}), \\ (\bar{\nabla}'_{X^v} Y^h)^1 = \nabla^v \bar{Y}(\bar{X}) - \frac{1}{2}\, C(\bar{Y}, \bar{X}), & (\bar{\nabla}'_{X^v} Y^h)^2 = -\frac{1}{2}\, P^1(\bar{Y}, \bar{X}), \\ (\bar{\nabla}'_{X^v} Y^v)^1 = 0, & (\bar{\nabla}'_{X^v} Y^v)^2 = \nabla^v \bar{Y}(\bar{X}) + \frac{1}{2}\, S^1(\bar{X}, \bar{Y}). \end{cases}$$

REMARK 20.3. In 1964, Yano and Ledger [J. London Math. Soc. 40 (1964)] gave a symmetric connection on the tangent bundle T(M) which is derived from a symmetric linear connection on the base space M. Comparing (20.12) with the equations given by them, it is seen that our symmetric connection $\bar{\Gamma}'$ is a direct generalization of their connection to the Finsler case.

§ 21. Lifts of Finsler metrics

Assume that the total space T of the tangent bundle T(M) is a Riemannian space with a fundamental tensor field $\bar{g}$. Then we obtain the N-decompositions $\bar{g}_{11}$, $\bar{g}_{12}$ $(= \bar{g}_{21})$, $\bar{g}_{22}$ of $\bar{g}$ with respect to a given non-linear connection N, and the scalar product $\bar{g}(X, Y)$ of tangent vectors X, Y of T is written

$$(21.1)\qquad \bar{g}(X, Y) = \bar{g}_{11}(X^1, Y^1) + \bar{g}_{12}(X^1, Y^2) + \bar{g}_{21}(X^2, Y^1) + \bar{g}_{22}(X^2, Y^2),$$

where (X^1, X^2) (resp. (Y^1, Y^2)) are N-decompositions of X (resp. Y). These N-decompositions $\bar{g}_{\alpha\beta}$, $\alpha, \beta = 1, 2$, are Finsler tensor fields of (0, 2)-type, and $\bar{g}_{11}$ and $\bar{g}_{22}$ are evidently symmetric tensors.

DEFINITION 21.1. We consider a Finsler space M with a fundamental tensor field g and a Finsler connection $F\Gamma = (\Gamma, N)$. The α-*lift* $\bar{g}$ of g to the tangent bundle T(M) is a tensor field of (0, 2)-type of T whose N-decompositions are given by

$$\bar{g}_{11} = \bar{g}_{22} = g, \qquad \bar{g}_{12} = \bar{g}_{21} = \alpha g,$$

where α is a Finsler scalar field.

By means of the α-lift $\bar{g}$, we introduce the scalar product $\bar{g}(X, Y)$ of tangent vectors X, Y of T:

$$(21.2)\qquad \bar{g}(X, Y) = g(X^1, Y^1) + g(X^2, Y^2) + \alpha\{g(X^1, Y^2) + g(X^2, Y^1)\}.$$

The tangent bundle T over a Finsler space M now becomes a (quasi-)Riemannian space with the fundamental tensor field $\bar{g}$. The determinant consisting of components of $\bar{g}$ is obviously equal to $(1 - \alpha^2)|g|^2$, where $|g|$ is the determinant consisting of components of g. Therefore α may be assumed to satisfy $-1 < \alpha < 1$.

Let X, Y be tangent vectors of M. In terms of the induced Finsler vectors $\bar{X}$, $\bar{Y}$, from (21.2) we get

$$\bar{g}(X^h, Y^h) = \bar{g}(X^v, Y^v) = g(\bar{X}, \bar{Y}),$$

$$\bar{g}(X^h, Y^v) = \bar{g}(X^v, Y^h) = \alpha g(\bar{X}, \bar{Y}).$$

$g(\bar{X}, \bar{Y})$ is, of course, equal to $g(X, Y)$. Therefore

PROPOSITION 21.1. *With respect to an α-lift $\bar{g}$ of a Finsler fundamental tensor field g,*

(1) the scalar product of $l_y X$ and $l_y Y$ or $l^v_y X$ and $l^v_y Y$ is equal to the scalar product of X, Y relative to y,

(2) the horizontal subspace N_y is orthogonal to the fiber through y with respect to the 0-lift.

We find the components $\bar{g}_{\lambda\mu}$ of α-lift $\bar{g}$ with respect to an induced coordinate system (x^i, y^i). From (19.10') we have

$$\begin{aligned}\bar{g}_{ij} = \bar{g}_{\alpha\beta}(\bar{z}^{-1})^\alpha_i(\bar{z}^{-1})^\beta_j &= (z^{-1})^a_i\{\bar{g}_{ab}(z^{-1})^b_j + \bar{g}_{a(b)}(z^{-1})^b_r N^r{}_j\} \\ &+ (z^{-1})^a_r N^r{}_i\{\bar{g}_{(a)b}(z^{-1})^b_j + \bar{g}_{(a)(b)}(z^{-1})^b_s N^s{}_j\} \\ &= g_{ij} + \alpha g_{ir}N^r{}_j + \alpha g_{rj}N^r{}_i + g_{rs}N^r{}_i N^s{}_j.\end{aligned}$$

In the similar way we obtain

$$\bar{g}_{ij} = g_{ij} + \alpha(g_{ir}N^r{}_j + g_{rj}N^r{}_i) + g_{rs}N^r{}_i N^s{}_j,$$

(21.2')

$$\bar{g}_{i(j)} = \bar{g}_{(j)i} = \alpha g_{ij} + g_{rj}N^r{}_i, \qquad \bar{g}_{(i)(j)} = g_{ij}.$$

If we put $\delta y^i = dy^i + N^i{}_r dx^r$, then (21.2') shows

$$\bar{g}_{\lambda\mu}d\bar{x}^\lambda d\bar{x}^\mu = g_{ij}dx^i dx^j + 2\alpha g_{ij}dx^i\delta y^j + g_{ij}\delta y^i \delta y^j.$$

REMARK 21.1. From (21.2') it is seen that the 0-lift is nothing but the lift due to Sasaki [Tôhoku Math. J. (2) 10 (1958)] in Riemannian case. On the other hand, the *complete lift* due to Yano and Kobayashi [J. Math. Soc. Japan 18 (1966)] is such that its N-decompositions are given by $\bar{g}_{11} = \bar{g}_{22} = 0$, $\bar{g}_{12} = \bar{g}_{21} = g$.

We are concerned with a Finsler space M with a fundamental tensor field g and a Finsler connection $F\Gamma = (\Gamma, N)$. Then an α-lift $\bar{g}$ of g with respect to N and the linear connection of Finsler type $\bar{\Gamma} = \phi_N(\Gamma)$ are introduced. Therefore we have the covariant derivative $\bar{\nabla}\bar{g}$ of $\bar{g}$ with respect to $\bar{\Gamma}$. From Theorem 20.1 the N-decompositions of $\bar{\nabla}\bar{g}$ are given by

$$(21.3)\qquad \begin{aligned} &(\bar{\nabla}\bar{g})_{111} = (\bar{\nabla}\bar{g})_{221} = \nabla^h g, \qquad &(\bar{\nabla}\bar{g})_{112} = (\bar{\nabla}\bar{g})_{222} = \nabla^v g,\\ &(\bar{\nabla}\bar{g})_{121} = (\bar{\nabla}\bar{g})_{211} = \nabla^h(\alpha g), \qquad &(\bar{\nabla}\bar{g})_{122} = (\bar{\nabla}\bar{g})_{212} = \nabla^v(\alpha g). \end{aligned}$$

From this the following theorem is obvious:

THEOREM 21.1. *A linear connection of Finsler type $\bar{\Gamma} = \phi_N(\Gamma)$ is metrical with respect to an α-lift $\bar{g}$ of g, if and only if the Finsler connection $F\Gamma = (\Gamma, N)$ is h-and v-metrical with respect to g and the scalar α is constant.*

This is very convenient situation: For instance, the linear connection of Finsler type which is derived from the Cartan connection $C\Gamma$ is metrical with respect to any α-lift, provided α is constant. In the following we shall restrict our discussion to the Cartan connection $C\Gamma$ and an α-lift $\bar{g}$ with a constant α.

Now, from a given Finsler metric $L(x, y)$, we first get the Finsler fundamental tensor g and the Cartan connection $C\Gamma = (\Gamma, N)$. Then the linear connection of Finsler type $\bar{\Gamma} = \phi_N(\Gamma)$ is derived from $C\Gamma$. On the other hand, an α-lift $\bar{g}$ of g with a constant α is constructed with respect to N of $C\Gamma$ and the Riemannian connection $\Gamma(\bar{g})$ is obtained from $\bar{g}$.

$$\begin{array}{ccccccc} & & & \nearrow & C\Gamma = (\Gamma, N) & \longrightarrow & \bar{\Gamma} = \phi_N(\Gamma) \\ L & \longrightarrow & g & & & & \\ & & & \searrow & \bar{g} & \longrightarrow & \Gamma(\bar{g}) \end{array}$$

Consequently we have two linear connection $\bar{\Gamma}$ and $\Gamma(\bar{g})$ on $T(M)$.

DEFINITION 21.2. The difference tensor field S of the Riemannian connection $\Gamma(\bar{g})$ from the linear connection of Finsler type $\bar{\Gamma}$, derived from the Cartan connection $C\Gamma$, is called the *strain tensor field* of $C\Gamma$.

By (4.14) a basic vector field $B^r(\bar{v})$ of the Riemannian connection $\Gamma(\bar{g})$ is written

$$(21.4)\qquad B^r(\bar{v}) = \bar{B}(\bar{v}) + \bar{Z}(S(\bar{v})),$$

where $\bar{B}(\bar{v})$ is the one of $\bar{\Gamma}$.

REMARK 21.2. The notion of strain tensor field can be similarly defined for any Finsler connection, provided that a fundamental function L is given. However, from the viewpoint of Theorem 21.1 and the fact that the connection $\Gamma(\bar{g})$ is metrical, it seems that the strain tensor field of a non-metrical Finsler connection is rather nonsense.

There remains an arbitrariness in constructing a lift $\bar{g}$ of the fundamental tensor g, arising from the choice of α, but the strain tensor S of $C\Gamma$ may be useful to consider the property of $C\Gamma$ in such a sense that S expresses the "strain" of $C\Gamma$ from the well-known Riemannian connection $\Gamma(\bar{g})$.

The torsion tensor $T^r (= 0)$ and the curvature tensor R^r of the Riemannian connection $\Gamma(\bar{g})$ are defined by the structure equation

$$[B^r(\bar{v}_1), B^r(\bar{v}_2)] = B^r(T^r(\bar{v}_1, \bar{v}_2)) + \bar{Z}(R^r(\bar{v}_1, \bar{v}_2)).$$

Substituting from (21.4), the above immediately gives

$$(21.5) \quad \begin{aligned} T^r(1, 2) &= \bar{T}(1, 2) - \mathfrak{A}_{(12)}\{S(1, 2)\}, \\ R^r(1, 2) &= \bar{R}(1, 2) - \mathfrak{A}_{(12)}\{\bar{\nabla}S(1, 2) - S(S(1), 2) + S(S(1, 2))\}, \end{aligned}$$

where $S(1, 2) = S(\bar{v}_2)\cdot\bar{v}_1$.

Next, from (21.4) it follows that

$$(21.6) \quad \nabla^r\bar{g}(1, 2, 3) = \bar{\nabla}\bar{g}(1, 2, 3) + \bar{g}(S(1, 3), 2) + \bar{g}(S(2, 3), 1),$$

where ∇^r is the covariant differentiation with respect to $\Gamma(\bar{g})$.

Now the Riemannian connection $\Gamma(\bar{g})$ is uniquely determined by the conditions $\nabla^r\bar{g} = 0$ and $T^r = 0$, and $C\Gamma$ is metrical, so that (21.6) and (21.5) give

$$(21.7) \quad \bar{g}(S(1, 3), 2) + \bar{g}(S(2, 3), 1) = 0,$$

$$(21.8) \quad \bar{T}(1, 2) - S(1, 2) + S(2, 1) = 0.$$

If we rewrite these equations in terms of covariant strain tensor S_* and covariant torsion tensor $\bar{T}_*$ (cf. (16.6)) such that

$$S_*(1, 2, 3) = S(1, \bar{g}(2), 3), \qquad \bar{T}_*(1, 2, 3) = \bar{T}(1, \bar{g}(2), 3),$$

then S_* is easily found as follows:

$$(21.9) \quad 2S_*(1, 2, 3) = \bar{T}_*(1, 2, 3) - \bar{T}_*(2, 3, 1) + \bar{T}_*(3, 1, 2).$$

To find the N-decompositions $S_{*\alpha\beta\gamma}$, $\alpha, \beta, \gamma = 1, 2$, of S_*, we first find the N-decompositions $\bar{T}_{*\alpha\beta\gamma}$ of $\bar{T}_*$. From Definition 21.1 it follows that

$$\bar{T}_{*\beta 1\gamma}(1, 2, 3) = \bar{T}_{\beta}{}^{1}{}_{\gamma}(1, g(2), 3) + \bar{T}_{\beta}{}^{2}{}_{\gamma}(1, \alpha g(2), 3),$$

$$\bar{T}_{*\beta 2\gamma}(1, 2, 3) = \bar{T}_{\beta}{}^{1}{}_{\gamma}(1, \alpha g(2), 3) + \bar{T}_{\beta}{}^{2}{}_{\gamma}(1, g(2), 3).$$

Consequently Theorem 20.3 ($T = S^1 = 0$) leads us to

$$\bar{T}_{*111}(1, 2, 3) = \alpha R^1_*(2, 1, 3), \qquad \bar{T}_{*121}(1, 2, 3) = R^1_*(2, 1, 3),$$

$$(21.10)\quad \begin{cases} \bar{T}_{*112}(1,\ 2,\ 3) = -\ \bar{T}_{*211}(3,\ 2,\ 1) = C_*(1,\ 2,\ 3) + \alpha P^1_*(2,\ 1,\ 3), \\ \bar{T}_{*122}(1,\ 2,\ 3) = -\ \bar{T}_{*221}(3,\ 2,\ 1) = \alpha C_*(1,\ 2,\ 3) + P^1_*(2,\ 1,\ 3), \\ \bar{T}_{*212} = \bar{T}_{*222} = 0, \end{cases}$$

where the components of covariant torsion tensors R^1_*, C_*, P^1_* are respectively defined by $R_{ijk} = g_{ir}R^r{}_{jk}$, $C_{ijk} = g_{jr}C_i{}^r{}_k$, $P_{ijk} = g_{ir}P^r{}_{jk}$.

Then, from (21.9) and (21.10) we obtain, for instance,

$$2S_{*112}(1,\ 2,\ 3) = \bar{T}_{*112}(1,\ 2,\ 3) - \bar{T}_{*121}(2,\ 3,\ 1) + \bar{T}_{*211}(3,\ 1,\ 2)$$

$$= C_*(1,\ 2,\ 3) + \alpha P^1_*(2,\ 1,\ 3) - R^1_*(3,\ 2,\ 1) - C_*(2,\ 1,\ 3) - \alpha P^1_*(1,\ 2,\ 3).$$

Since C_* and P^1_* are completely symmetric for $C\Gamma$, we obtain $2S_{*112}(1,\ 2,\ 3) = R^1_*(3,\ 1,\ 2)$. In like manner we conclude

<u>PROPOSITION 21.2</u>. *The N-decompositions $S_{*\alpha\beta\gamma}$, $\alpha, \beta, \gamma = 1, 2$, of covariant strain tensor field S_* of the Cartan connection $C\Gamma$ are given by*

$$S_{*111}(1,\ 2,\ 3) = \alpha R^1_*(3,\ 1,\ 2), \qquad S_{*112}(1,\ 2,\ 3) = \frac{1}{2}R^1_*(3,\ 1,\ 2),$$

$$S_{*121}(1,\ 2,\ 3) = -\ S_{*211}(2,\ 1,\ 3)$$
$$= \frac{1}{2}R^1_*(2,\ 1,\ 3) + C_*(1,\ 2,\ 3) + \alpha P^1_*(1,\ 2,\ 3),$$

$$S_{*122}(1,\ 2,\ 3) = -\ S_{*212}(1,\ 2,\ 3) = P^1_*(1,\ 2,\ 3) + \alpha C_*(1,\ 2,\ 3),$$

$$S_{*221} = S_{*222} = 0.$$

The N-decompositions $S_\beta{}^\alpha{}_\gamma$ of strain tensor S itself are found as follows:

$$S_{*\beta 1\gamma}(1,\ 2,\ 3) = S_\beta{}^1{}_\gamma(1,\ g(2),\ 3) + S_\beta{}^2{}_\gamma(1,\ \alpha g(2),\ 3),$$

$$S_{*\beta 2\gamma}(1,\ 2,\ 3) = S_\beta{}^1{}_\gamma(1,\ \alpha g(2),\ 3) + S_\beta{}^2{}_\gamma(1,\ g(2),\ 3).$$

Solving the above equations with respect to $S_\beta{}^1{}_\gamma$, $S_\beta{}^2{}_\gamma$, we get

$$(1 - \alpha^2)S_\beta{}^1{}_\gamma(1,\ 2) = S_{*\beta 1\gamma}(1,\ g^{-1},\ 2) - \alpha S_{*\beta 2\gamma}(1,\ g^{-1},\ 2),$$

$$(1 - \alpha^2)S_\beta{}^2{}_\gamma(1,\ 2) = S_{*\beta 2\gamma}(1,\ g^{-1},\ 2) - \alpha S_{*\beta 1\gamma}(1,\ g^{-1},\ 2).$$

Thus, from Proposition 21.2 we obtain the N-decompositions of strain tensor S of $C\Gamma$ as follows: Putting $\bar{\alpha} = 1 - \alpha^2$, we have

$$
\begin{aligned}
\bar{\alpha} S_{1}{}^{1}{}_{1}(u, v) &= \alpha\{R^{1}_{*}(v, u, g^{-1}) - C(u, v) - \tfrac{1}{2} R^{1}(u, v) - \alpha P^{1}(u, v)\}, \\
\bar{\alpha} S_{1}{}^{2}{}_{1}(u, v) &= \tfrac{1}{2} R^{1}(u, v) + C(u, v) + \alpha\{P^{1}(u, v) - \alpha R^{1}_{*}(v, u, g^{-1})\}, \\
\bar{\alpha} S_{1}{}^{1}{}_{2}(u, v) &= \tfrac{1}{2} R^{1}_{*}(v, u, g^{-1}) - \alpha\{P^{1}(u, v) + \alpha C(u, v)\}, \\
\bar{\alpha} S_{1}{}^{2}{}_{2}(u, v) &= P^{1}(u, v) + \alpha\{C(u, v) - \tfrac{1}{2} R^{1}_{*}(v, u, g^{-1})\}, \\
\bar{\alpha} S_{2}{}^{1}{}_{1}(u, v) &= \tfrac{1}{2} R^{1}_{*}(u, v, g^{-1}) - C(u, v) - \alpha P^{1}(u, v), \\
\bar{\alpha} S_{2}{}^{2}{}_{1}(u, v) &= \alpha\{C(u, v) - \tfrac{1}{2} R^{1}_{*}(u, v, g^{-1}) + \alpha P^{1}(u, v)\}, \\
\bar{\alpha} S_{2}{}^{1}{}_{2}(u, v) &= - P^{1}(u, v) - \alpha C(u, v), \\
\bar{\alpha} S_{2}{}^{2}{}_{2}(u, v) &= \alpha\{P^{1}(u, v) + \alpha C(u, v)\}.
\end{aligned}
\tag{21.11}
$$

<u>REMARK 21.3</u>. Even in the simplest case $\alpha = 0$, there are still five surviving N-decompositions

$$
S_{1}{}^{2}{}_{1}(u, v) = \tfrac{1}{2} R^{1}(u, v) + C(u, v), \qquad S_{1}{}^{1}{}_{2}(u, v) = \tfrac{1}{2} R^{1}_{*}(v, u, g^{-1}),
$$

$$
S_{1}{}^{2}{}_{2}(u, v) = P^{1}(u, v), \qquad S_{2}{}^{1}{}_{1}(u, v) = \tfrac{1}{2} R^{1}_{*}(u, v, g^{-1}) - C(u, v),
$$

$$
S_{2}{}^{1}{}_{2}(u, v) = - P^{1}(u, v)
$$

of the strain tensor S of $C\Gamma$. It seems to the author that the concept of 0-lift may not be suitable for studying a p-homogeneous Finsler connection, because the Finsler tensor field $S_{1}{}^{2}{}_{1}$, for instance, is not p-homogeneous. Thus it may be said that to find a lift of g to T(M) which is effective in a sense for studying Finsler connections is still an open problem.

Next, from (21.5) the covariant curvature tensor field R^{r}_{*} is written

$$
R^{r}_{*}(1, 2, 3, 4) = \bar{R}_{*}(1, 2, 3, 4) - \mathfrak{A}_{(34)}\{\bar{\nabla} S_{*}(1, 2, 3, 4) - S_{*}(S(1, 3), 2, 4) + S_{*}(1, 2, S(3, 4))\}.
\tag{21.12}
$$

It follows further from Theorem 20.3 that the N-decompositions $\bar{R}_{*\alpha\beta\gamma\delta}$, $\alpha, \beta, \gamma, \delta = 1, 2$, of covariant curvature tensor $\bar{R}_*$ are written as

$$
\begin{aligned}
&\bar{R}_{*1111} = \bar{R}_{*2211} = R^2_*, \qquad && \bar{R}_{*1211} = \bar{R}_{*2111} = \alpha R^2_*, \\
&\bar{R}_{*1112} = \bar{R}_{*2212} = P^2_*, && \bar{R}_{*1212} = \bar{R}_{*2112} = \alpha P^2_*, \\
&\bar{R}_{*1122} = \bar{R}_{*2222} = S^2_*, && \bar{R}_{*1222} = \bar{R}_{*2122} = \alpha S^2_*.
\end{aligned} \tag{21.13}
$$

Thus we have made all the preparations for finding the curvature tensor R^r of the Riemannian connection $\Gamma(\bar{g})$ in terms of torsion and curvature tensors of the Cartan connection $C\Gamma$, that is, (21.12) together with Theorem 20.1, Proposition 21.2, (21.11) and (21.13). For instance, from (21.12) and so on we have

$$
\begin{aligned}
R^r_{*1211}(1, 2, 3, 4) = \alpha R^2_*(1, 2, 3, 4) - \mathfrak{A}_{(34)}\{\nabla^h S_{*121}(1, 2, 3, 4) \\
- S_{*121}(S_1{}^1{}_1(1, 3), 2, 4) - S_{*221}(S_1{}^2{}_1(1, 3), 2, 4) \\
+ S_{*121}(1, 2, S_1{}^1{}_1(3, 4)) + S_{*122}(1, 2, S_1{}^2{}_1(3, 4))\}.
\end{aligned}
$$

We do not, however, write here the expressions of all the N-decompositions of R^r_*, because those are too complicated and we do not need all of them in the following.

REMARK 21.4. If we are concerned with a Riemannian space M and the 0-lift, then the expressions of N-decompositions of R^r_* will become somewhat simpler. But, if we need the components of R^r_* in an induced coordinate system (x^i, y^i), those will be quite complicated and long, as it will be conjectured from (19.11). See Matsumoto [8, 12] and Akbar-Zadeh and Wegrzynowska [2].

§ 22. Isotropy of Finsler spaces

We consider the Riemannian space T, namely, the total space of the tangent bundle T(M) which is equipped with an α-lift $\bar{g}$ (α = constant) of the Finsler fundamental tensor g with respect to the non-linear connection N of the Cartan connection $C\Gamma = (\Gamma, N)$. The covariant curvature tensor R^r_* of this T is given by (21.12).

The concept of isotropy of a Riemannian space is defined as to be of constant curvature. Thus the Riemannian space T is isotropic, if and only if there exists a scalar field K such that the curvature tensor R^r_* is written

$$R^r_*(1, 2, 3, 4) = K\{\bar{g}(1, 3)\bar{g}(2, 4) - \bar{g}(1, 4)\bar{g}(2, 3)\}. \tag{22.1}$$

To examine this condition in detail, we shall first find an N-decomposition R^r_{*2222} of R^r_*. From (21.12) it is written

$$R^r_{*2222}(1, 2, 3, 4) = \bar{R}_{*2222}(1, 2, 3, 4) - \mathfrak{A}_{(34)}\{\nabla^v S_{*222}(1, 2, 3)$$

$$- S_{*122}(S_2{}^1{}_2(1, 3), 2, 4) - S_{*222}(S_2{}^2{}_2(1, 3), 2, 4)$$

$$+ S_{*221}(1, 2, S_2{}^1{}_2(3, 4)) + S_{*222}(1, 2, S_2{}^2{}_2(3, 4))\}.$$

From (21.13) and Proposition 21.2 this is rewritten as

$$R^r_{*2222}(1, 2, 3, 4) = S^2_*(1, 2, 3, 4)$$

$$+ \mathfrak{A}_{(34)}\{P^1_*(S_2{}^1{}_2(1, 3), 2, 4) + \alpha C_*(S_2{}^1{}_2(1, 3), 2, 4)\}.$$

From (21.11) and (17.22) we have $S_2{}^1{}_2(v, \varepsilon) = 0$ for any $v \in V^n$, so that (17.20) and (17.22) lead us to $R^r_{*2222}(1, 2, 3, \varepsilon) = 0$. As a consequence, from (19.14) we have

PROPOSITION 22.1. *Let T be the total space of the tangent bundle $T(M)$ which is regarded as a Riemannian space with the fundamental tensor $\bar{g}$, where $\bar{g}$ is an α-lift (α = constant) of a Finsler fundamental tensor g with respect to the non-linear connection N of the Cartan connection $C\Gamma = (\Gamma, N)$. Then the sectional curvature of a two-section spanned by the intrinsic-vertical vector l^v and any vertical vector is equal to zero.*

Therefore, if the Riemannian space T as above is isotropic, the scalar curvature K must vanish, i.e., T must be locally flat. Then we find the condition for T to be locally flat.

We first examine an N-decomposition R^r_{*1122} of R^r_*:

$$R^r_{*1122}(1, 2, 3, 4) = \bar{R}_{*1122}(1, 2, 3, 4) - \mathfrak{A}_{(34)}\{\nabla^v S_{*112}(1, 2, 3, 4)$$

$$- S_{*112}(S_1{}^1{}_2(1, 3), 2, 4) - S_{*212}(S_1{}^2{}_2(1, 3), 2, 4)$$

$$+ S_{*111}(1, 2, S_2{}^1{}_2(3, 4)) + S_{*112}(1, 2, S_2{}^2{}_2(3, 4))\}$$

$$= S^2_*(1, 2, 3, 4) - \mathfrak{A}_{(34)}\{ \tfrac{1}{2} \nabla^v R^1_*(3, 1, 2, 4)$$

$$- \tfrac{1}{2} R^1_*(4, S_1{}^1{}_2(1, 3), 2) + P^1_*(S_1{}^2{}_2(1, 3), 2, 4)$$

$$+ \alpha C_*(S_1{}^2{}_2(1, 3), 2, 4) + \alpha R^1_*(S_2{}^1{}_2(3, 4), 1, 2)$$

$$+ \tfrac{1}{2} R^1_*(S_2{}^2{}_2(3, 4), 1, 2)\}.$$

Put $v_4 = \varepsilon$ and consider $R^r_{*1122}(1, 2, 3, \varepsilon)$. Because

$$S_{1\ 2}^{\ 1}(v, \varepsilon) = S_{1\ 2}^{\ 2}(v, \varepsilon) = S_{2\ 2}^{\ 1}(v, \varepsilon) = S_{2\ 2}^{\ 1}(\varepsilon, v) = S_{2\ 2}^{\ 2}(v, \varepsilon) = S_{2\ 2}^{\ 2}(\varepsilon, v) = 0$$

for any $v \in V^n$, we obtain

$$R^r_{*1122}(1, 2, 3, \varepsilon) = -\frac{1}{2}\{\nabla^v R^1_*(3, 1, 2, \varepsilon) - \nabla^v R^1_*(\varepsilon, 1, 2, 3)\}$$

$$= -R^1_*(3, 1, 2) = 0,$$

so that the (v)h-torsion tensor R^1 of $C\Gamma$ should vanish.

Secondly, we examine R^r_{*1112}. From $R^1 = 0$ it is written as

$$R^r_{*1112}(1, 2, 3, 4) = P^2_*(1, 2, 3, 4) - \frac{1}{1 - \alpha^2}\{P^1_*\big(C(1, 3) + \alpha P^1(1, 3), 2, 4\big)$$

$$+ \alpha C_*\big(C(1, 3) + \alpha P^1(1, 3), 2, 4\big) - C_*\big(P^1(1, 4)$$

$$+ \alpha C(1, 4), 2, 3\big) - \alpha P^1_*\big(P^1(1, 4) + \alpha C(1, 4), 2, 3\big)\}.$$

Putting $v_1 = \varepsilon$ in the above, we immediately have $R^r_{*1112}(\varepsilon, 2, 3, 4) = P^1_*(2, 3, 4) = 0$, so that the (v)hv-torsion tensor P^1 should vanish and $P^2 = 0$ from (17.27). Then we have

$$R^r_{*1112}(1, 2, 3, 4) = \frac{\alpha}{1 - \alpha^2}\{C_*(C(1, 4), 2, 3) - C_*(C(1, 3), 2, 4)\} = 0,$$

which is equal to $\{\alpha/(1 - \alpha^2)\}S^2_*(1, 2, 3, 4) = 0$ from (17.20), so that we get $\alpha S^2 = 0$.

Thirdly an N-decomposition R^r_{*1212} is now written as

$$R^r_{*1212}(1, 2, 3, 4) = \alpha\nabla^h C_*(1, 2, 4, 3) - \nabla^v C_*(1, 2, 3, 4) - C_*(1, 2, C(4, 3)),$$

so that $R^r_{*1212}(1, 2, 3, \varepsilon) = 0$ leads us finally to $C_*(1, 2, 3) = 0$. Consequently the metric tensor g must be Riemannian (Theorem 24.1), so that $R^1 = 0$ means the flatness of the Riemannian space M with the metric g.

Summarizing up the above all, we establish

THEOREM 22.1. *The Riemannian space T in Proposition 22.1 is isotropic, if and only if the Finsler fundamental function L is flat Riemannian. Then T is also flat.*

REMARK 22.1. In the author's monograph [12] the above theorem is proved in case of $\alpha = 0$. Honestly speaking, the above consideration was a trial to find a generalization of the concept of constant curvature to Finsler case. That is, the author tried to define a Finsler space M of constant curvature as such that the Riemannian space T with the metric $\bar{g}$ is of constant curvature. In consequence of Theorem 22.1, it must be said that this trial was unsuccessful.

We shall show another trial to generalize the concept of constant curvature to Finsler case. We have the linear connection of Finsler type $\bar{\Gamma} = \phi_N(\Gamma)$, which is derived from the Cartan connection $C\Gamma = (\Gamma, N)$, and an α-lift $\bar{g}$ (α = constant) of the fundamental tensor g. Although $\bar{\Gamma}$ is not Riemannian, it is metrical with respect to $\bar{g}$.

Let $\bar{R}$ be the curvature tensor of $\bar{\Gamma}$ and $\bar{R}_*$ be the covariant one. Instead of (21.1), we are now concerned with the equation

$$\bar{R}_*(1, 2, 1, 2) = K\{ \bar{g}(1, 1)\bar{g}(2, 2) - \bar{g}(1, 2)^2 \}, \tag{22.2}$$

where K is a scalar on the space T and $\bar{v}_1$, $\bar{v}_2$ (denoted by their indices only) are arbitrary vectors of V^{2n}. This equation is analogous to the case where the sectional curvature does not depend on the section.

We first consider (22.2) such that $\bar{v}_1 = (0, v_1)$ and $\bar{v}_2 = (0, v_2)$, namely, the two-section is vertical. It then follows from (21.13) and Definition 21.1 that (22.2) in this case is written as

$$S^2_*(1, 2, 1, 2) = K\{ g(1, 1)g(2, 2) - g(1, 2)^2 \}. \tag{22.3}$$

This leads, however, immediately to a trivial result K = 0, because $S^2_*(v, \varepsilon, v, \varepsilon) = 0$ for any $v \in V^n$.

REMARK 22.2. Since the components S_{hijk} of S^2_* satisfy the identities similar to (5.24), the condition (22.3) immediately gives

$$S_{hijk} = K(g_{hj}g_{ik} - g_{hk}g_{ij}),$$

which implies K = 0 by contraction by y^h.

Secondly, we deal with (22.2) such that $\bar{v}_1 = (v_1, 0)$ and $\bar{v}_2 = (0, v_2)$, namely, the two-section being spanned by a horizontal vector and a vertical vector. Then we similarly have

$$\alpha P^2_*(1, 2, 1, 2) = K\{g(1, 1)g(2, 2) - \alpha^2 g(1, 2)^2\}. \tag{22.4}$$

In terms of components it is written as

$$\alpha(P_{hijk} + P_{hkji} + P_{jihk} + P_{jkhi}) = K\{ 4g_{hj}g_{ik} - 2\alpha^2(g_{hi}g_{jk} + g_{hk}g_{ij})\}. \tag{22.4'}$$

Contracting this by $y^h y^j y^k$ and paying attention to (17.26), we get K = 0, so that (22.4') gives $\alpha = 0$ or

$$P_{hijk} + P_{hkji} + P_{jihk} + P_{jkhi} = 0.$$

If we contract this by y^h, it follows from (17.8) and (17.26) that $P_{ijk} = 0$, so that (17.27) gives $P_{hijk} = 0$.

Consequently we are led to only trivial results as follows:

PROPOSITION 22.2. *(1) If the equation (22.3) holds for any v_1, $v_2 \in V^n$, then the v-curvature tensor S^2 vanishes. (2) If the equation (22.4) holds for any v_1, $v_2 \in V^n$, then $K = \alpha = 0$ or $K = 0$ and $P^2 = 0$.*

Thirdly we consider (22.2) for a horizontal two-section, namely, $\bar{v}_1 = (v_1, 0)$ and $\bar{v}_2 = (v_2, 0)$. Similarly we have

$$R^2_*(1, 2, 1, 2) = K\{ g(1, 1)g(2, 2) - g(1, 2)^2 \}. \tag{22.5}$$

In terms of components it is written

$$R_{hijk} + R_{hkji} + R_{jihk} + R_{jkhi} = K\{4g_{hj}g_{ik} - 2(g_{hi}g_{jk} + g_{hk}g_{ij})\}. \tag{22.5'}$$

It is well-known that in Riemannian case the equation $R_{hijk} = K(g_{hj}g_{ik} - g_{hk}g_{ij})$ is easily derived from (22.2') in virtue of (5.24), in particular (5.26). While in Finsler case we have not such identities, but we show that an identity similar to (5.26) is derived and in consequence an equation similar to the Riemannian case is concluded by somewhat complicated computation as follows.

The components R_{hijk} satisfy $R_{hijk} = - R_{ihjk} = - R_{hikj}$ and (17.10'). The last one is written in the form

$$\mathfrak{S}_{(ijk)}\{ R_{hijk}\} = M_{hijk}, \tag{22.6}$$

where we put

$$M_{hijk} = - \mathfrak{S}_{(ijk)}\{C_h{}^r{}_i R_{rjk}\}. \tag{22.6'}$$

Then we observe

$$\begin{aligned} R_{jkhi} &= - R_{jhik} - R_{jikh} + M_{jkhi} = R_{hjik} + R_{ijkh} + M_{jkhi} \\ &= (- R_{hikj} - R_{hkji} + M_{hjik}) + (- R_{ikhj} - R_{ihjk} + M_{ijkh}) + M_{jkhi} \\ &= 2R_{hijk} + (R_{khji} + R_{kihj}) + M_{hjik} + M_{ijkh} + M_{jkhi} \\ &= 2R_{hijk} + (- R_{kjih} + M_{khji}) + M_{hjik} + M_{ijkh} + M_{jkhi}, \end{aligned}$$

which finally implies

$$R_{jkhi} = R_{hijk} + N_{hijk}, \tag{22.7}$$

putting

$$N_{hijk} = \frac{1}{2}(M_{khji} + M_{hjik} + M_{ijkh} + M_{jkhi}) = \mathfrak{A}_{(jk)}\{C_i{}^r{}_j R_{rhk} - C_h{}^r{}_j R_{rik}\}. \tag{22.8}$$

The identity (22.7) is a generalization of (5.26).

Now, by means of (22.6) and (22.7), the left-hand side of (22.5') is rewritten as

$$R_{hijk} + (- R_{hjik} - R_{hikj} + M_{hkji}) + (R_{hkji} + N_{hkji}) + (R_{hijk} + N_{hijk})$$

$$= 3R_{hijk} - R_{hjik} + (- R_{hjik} - R_{hikj} + M_{hkji}) + M_{hkji} + N_{hkji} + N_{hijk}.$$

Therefore (22.5') is rewritten in the form

$$(22.5'')\qquad \begin{aligned} &4R_{hijk} - 2R_{hjik} + 2M_{hkji} + N_{hkji} + N_{hijk} \\ &= K\{4g_{hj}g_{ik} - 2(g_{hi}g_{jk} + g_{hk}g_{ij})\}. \end{aligned}$$

Construct an equation from (22.5") by interchanging indices i, j, and dividing by 2. Then, summing (22.5") and the equation as thus obtained, we get

$$(22.9)\qquad \begin{aligned} &3R_{hijk} + 2M_{hkji} + M_{hkij} + N_{hkji} + N_{hijk} + \frac{1}{2}(N_{hkij} + N_{hjik}) \\ &= 3K(g_{hj}g_{ik} - g_{hk}g_{ij}). \end{aligned}$$

Contracting this by y^h and paying attention to $M_{0ijk} = 0$ and $N_{0ijk} = \mathfrak{A}_{(jk)}\{C_i{}^r{}_j R_{r0k}\}$, we obtain

$$(22.10)\qquad R_{ijk} + \frac{1}{2}(C_j{}^r{}_i R_{r0k} - C_k{}^r{}_i R_{r0j}) = K(y_j g_{ik} - y_k g_{ij}).$$

Once more contraction by y^j gives $R_{i0k} = K(L^2 g_{ik} - y_i y_k)$. Therefore (22.10) reduces to $R_{ijk} = K(y_j g_{ik} - y_k g_{ij})$, so that $M_{hijk} = N_{hijk} = 0$. Then (22.9) finally yields

$$(22.11)\qquad R_{hijk} = K(g_{hj}g_{ik} - g_{hk}g_{ij}).$$

Consequently, any horizontal two-section is isotropic, if and only if (22.11) holds. This leads us to

DEFINITION 22.1. If any horizontal two-section is isotropic with respect to the linear connection of Finsler type, derived from CΓ, namely, (22.11) holds, then the Finsler space with CΓ is called *h-isotropic*.

REMARK 22.3. The identity (22.7) will be useful in some problems concerned with Cartan's R_{hijk}.

The concept of isotropy in the above sense is introduced by Akbar-Zadeh [5]. He gives (22.3) and (22.5), while (22.4) is given by him only in the case $\alpha = 1$. A geometrical meaning of the isotropy is first shown by Matsumoto [11]. See O.Varga [7, p.377].

THEOREM 22.2. *If an n (> 2)-dimensional Finsler space with CΓ is h-isotropic, the scalar curvature K in (22.11) is constant. Further, if K does not vanish, the hv-curvature tensor* P^2 *satisfies* $P_h{}^i{}_{jk} = P_h{}^i{}_{kj}$ *and the v-curvature tensor* S^2 *vanishes identically.*

The first half of Theorem 22.2 is a generalization of the well-known Schur's theorem in Riemannian geometry.

The proof is a little long. If we first contract (17.15) by $y^m y^j$ and pay attention to (17.26), we get

$$R_{mhij|k} y^m y^j = R_{h0r} C_i{}^r{}_k + P_{hik|0}.$$

It follows from (17.22) that the members of the right-hand side of the above are symmetric in i, k, so that we have

$$(R_{mhij|k} - R_{mhkj|i}) y^m y^j = 0.$$

By substitution from (22.11), the above is rewritten as

$$K|_i h_{hk} - K|_k h_{hi} = 0,$$

where h_{hk} is the angular metric tensor. The scalar curvature K is obviously (0)p-homogeneous, so that $K|_i y^i = 0$. From (16.5) we have $h_{hk} g^{hk} = n - 1$. Thus the contraction of the above by g^{hk} yields $(n - 2)K|_i = 0$, which implies K to be a function K(x) of position x alone, provided n > 2.

Next, substitution from (22.11) into (17.12) yields

$$\mathfrak{S}_{(ijk)}\{K_{|k}(g_{mi}g_{hj} - g_{mj}g_{hi}) - K(P_{mhjk} - P_{mhkj})y_i\} = 0. \tag{22.12}$$

Contracting this by $g^{mk} y^h y^i$ and referring to (17.8), (17.9) and (17.26), we get

$$(n - 2)(L^2 K_{|j} - K_{|0} y_j) = 0. \tag{22.13}$$

If we differentiate (22.13) v-covariantly, we get

$$2K_{|j} y_k + L^2 K_{|j}|_k - (K_{|r}|_k y^r + K_{|k}) y_j - K_{|0} g_{jk} = 0.$$

The Ricci identity (10.9') yields $K_{|i}|_j = - K_{|r} C_i{}^r{}_j$, so that the above may be rewritten in the form

$$2K_{|j} y_k - L^2 K_{|r} C_j{}^r{}_k - K_{|k} y_j - K_{|0} g_{jk} = 0.$$

This and (22.13) lead to $K_{|j} = 0$, so that K is finally a constant.

Now the equation (22.12) reduces to

$$K\, \mathfrak{S}_{(ijk)}\{(P_{mhjk} - P_{mhkj}) y_i\} = 0.$$

Contraction by y^i and (17.26) yield $P_{mhjk} = P_{mhkj}$ easily, provided $K \neq 0$.

Finally, from (17.15) we obtain

$$\mathfrak{S}_{(ijk)}\{S_{mhkr} R^r{}_{ij} + R_{mhij|k}\} = 0.$$

Substitution from (22.11) gives $2K\, \mathfrak{S}_{(ijk)}\{S_{mhjk} y_i\} = 0$, which immediately implies $S_{mhjk} = 0$.

REMARK 22.4. Theorem 22.2 is proved by Akbar-Zadeh [7]. In an exceptional case $n = 2$, the equation (22.11) always holds (cf. (28.8')). The symmetry of P^2 is remarkable from some standpoints (cf. (17.25) and Proposition 28.4).

The fact that the beautiful equation (22.11) ($K \neq 0$) causes $S^2 = 0$ really surprises us because of Brickell's theorem (Theorem 24.3). Thus it must be again said that Akbar-Zadeh's trial to define a Finsler space of constant curvature is unsuccessful. This unsuccess suggests us that it does not always give a satisfactory result to bring into Finsler geometry some notions which are fruitful in Riemannian gometry, without any modification. Compare the ideas in the present section with Berwald's idea in § 26.

§ 23. The almost complex N-structures

In general an almost complex structure J of a manifold M is defined as a tensor field of (1, 1)-type of M which satisfies $J^2 = -1$, namely, if we consider the associated tensor function $\tilde{J}$ with J, then $\tilde{J}(\tilde{J}(v)) = -v$ for any $v \in V^n$. The theory of manifolds with an almost complex structure has been studied by many authors.

DEFINITION 23.1. Assume that a non-linear connection N is given in the tangent bundle T(M). The *almost complex N-structure* J is a tensor field of (1, 1)-type of T such that the N-decompositions J^α_β, $\alpha, \beta = 1, 2$, of J with respect to N are given by

$$J^1_1 = J^2_2 = 0, \qquad J^2_1 = - J^1_2 = \delta,$$

where δ is the Kronecker delta.

If we take $\bar{v} = (v_1, v_2) \in V^{2n}$, from (19.9) the almost complex N-structure J is such that

$$\begin{aligned} J((v_1, v_2)) &= J((v_1, 0)) + J((0, v_2)) \\ &= (J^1_1(v_1), J^2_1(v_1)) + (J^1_2(v_2), J^2_2(v_2)) = (0, v_1) + (-v_2, 0). \end{aligned}$$

Thus we have a simple expression of this J:

$$J((v_1, v_2)) = (-v_2, v_1), \tag{23.1}$$

from which $J(J(v_1, v_2)) = (-v_1, -v_2)$ is immediately shown.

In the above treatment, we regard J as the associated tensor function defined on $\bar{L}(T)$ with the value in $(V^{2n})^1_1$. On the other hand, if we regard J as a tensor field of T in the ordinary sense, it follows from (19.13) that

$$J(X^h) = X^v, \qquad J(X^v) = -X^h \tag{23.2}$$

for any tangent vector X of the base space M.

The components J^{α}_{β}, α, $\beta = 1, \cdots, 2n$, of J with respect to a frame $\bar{z} = \phi_N(u)$, $u = (y, z) \in F$, are obviously given by

(23.3) $$J^a_b = J^{(a)}_{(b)} = 0, \qquad J^{(a)}_{b} = - J^{a}_{(b)} = \delta^a_b.$$

On the other hand, the components J^{λ}_{μ}, λ, $\mu = 1, \cdots, 2n$, of J in an induced coordinate system (x^i, y^i) are given by $J^{\lambda}_{\mu} = \bar{z}^{\lambda}_{\alpha} J^{\alpha}_{\beta} (\bar{z}^{-1})^{\beta}_{\mu}$, so that (19.10) and (19.10') yield, for instance,

$$J^i_j = z^i_a \{ J^a_b (z^{-1})^b_j + J^{a}_{(b)} (z^{-1})^b_r N^r_j \} = - N^i_j.$$

Similarly the components of J in (x^i, y^i) are

(23.3') $$J^i_j = - N^i_j, \qquad J^{i}_{(j)} = - \delta^i_j.$$
$$J^{(i)}_{j} = \delta^i_j + N^i_r N^r_j, \qquad J^{(i)}_{(j)} = N^i_j.$$

Assume now that we have a Finsler connection $F\Gamma = (\Gamma, N)$ of the manifold M. Corresponding to the almost complex N-structure J with respect to N of $F\Gamma$, a linear operator J_F of the tangent bundle over the Finsler bundle F(M) is introduced by

(23.4) $$J_F(Z(A)) = Z(A), \qquad J_F(B^h(v)) = B^v(v), \qquad J_F(B^v(v)) = - B^h(v),$$

where $B^h(v)$ (resp. $B^v(v)$) is an h- (resp. v-) basic vector field with respect to $F\Gamma$. Thus Γ^h and Γ^v of $F\Gamma = (\Gamma^h, \Gamma^v)$ change places with each other by J_F and we have $\pi'_1 \circ J_F = J \circ \pi'_1$. From (19.7) and Proposition 20.1 we see

$$\phi'_N \circ J_F(Z(A)) = \bar{Z}\begin{pmatrix} A & 0 \\ 0 & A \end{pmatrix}, \qquad \phi'_N \circ J_F(B^h(v)) = \bar{B}((0, v)),$$
$$\phi'_N \circ J_F(B^v(v)) = - \bar{B}((v, 0)).$$

Therefore, if we introduce from J a linear operator $\bar{J}$ of the tangent bundle over $\bar{L}(T)$ such that

(23.5) $$\bar{J}(\bar{Z}(\bar{A})) = \bar{Z}(\bar{A}), \qquad \bar{J}(\bar{B}(v_1, v_2)) = \bar{B}((- v_2, v_1)),$$

we obtain

(23.6) $$\phi'_N \circ J_F = \bar{J} \circ \phi'_N .$$

From (23.6) it is seen that $\bar{J}$ and J_F are regarded as lifts of J to $\bar{L}(T)$ and F(M) respectively in a sense.

It is well-known that an almost complex structure J of a manifold M is *integrable*, if and only if the tensor field E of (1, 2)-type of M defined as follows vanishes identically.

$$E(X, Y) = [X, Y] + J([J(X), Y]) + J([X, J(Y)]) - [J(X), J(Y)]$$

for any tangent vector fields X, Y of M.

We shall examine the integrability condition of an almost complex N-structure J. Corresponding to the tensor field E defined from J, we introduce an operator E_F by

$$E_F(X, Y) = [X, Y] + J_F([J_F(X), Y]) + J_F([X, J_F(Y)]) - [J_F(X), J_F(Y)], \tag{23.7}$$

where X, Y are any tangent vector fields of the Finsler bundle F. From $\pi_1'\circ J_F = J\circ\pi_1'$ it is evident that

$$\pi_1'(E_F(1_uX, 1_uY)) = E(X, Y), \tag{23.8}$$

where 1_uX is the lift of a tangent vector field X of T with respect to the Finsler connection $F\Gamma$.

We consider E_F by means of the structure equations of $F\Gamma$. Since the almost complex N-structure J is defined with respect to the non-linear connection N only, it is enough to be concerned with the Finsler N-connection FN (Definition 15.1) which is determined by N only. In this case we have $C = P^1 = 0$ from Theorem 15.1, so that Lie bracket $[B^h(v_1), B^v(v_2)]$ is vertical and $[B^v(v_1), B^v(v_2)] = 0$. Then

$$E_F(B^h(1), B^h(2)) = B^h(T(1, 2)) + B^v(R^1(1, 2)) + J_F([B^v(1), B^h(2)])$$
$$+ J_F([B^h(1), B^v(2)]) - [B^v(1), B^v(2)] + Z(\cdots).$$

It is not necessary to write the vertical component $Z(\cdots)$ explicitly in the viewpoint of (23.8). Similarly we obtain

$$E_F(B^h(1), B^h(2)) = B^h(T(1, 2)) + B^v(R^1(1, 2)) + Z(\cdots),$$

$$E_F(B^h(1), B^v(2)) = B^h(R^1(1, 2)) - B^v(T(1, 2)) + Z(\cdots), \tag{23.9}$$

$$E_F(B^v(1), B^v(2)) = - B^h(T(1, 2)) - B^v(R^1(1, 2)) + Z(\cdots).$$

Consequently (23.8) and (23.9) lead to

<u>THEOREM 23.1</u>. *The almost complex N-structure J is integrable, if and only if the (h)h-torsion tensor T and (v)h-torsion tensor R^1 of the Finsler N-connection FN vanish identically.*

<u>REMARK 23.1</u>. In terms of connection coefficients $(F_j{}^i{}_k = \dot\partial_k N^i{}_j, N^i{}_j, 0)$ of FN the above condition is written as

$$\dot\partial_k N^i{}_j = \dot\partial_j N^i{}_k, \qquad \delta_k N^i{}_j = \delta_j N^i{}_k.$$

A systematical description of almost complex structures is seen in the book "Differential geometry on complex and almost complex spaces" by Yano, Pergamon 1965.

In a case where N is the associated connection with a symmetric linear connection $\underline{\Gamma}$, the condition in Theorem 23.1 reduces to $y^h \underline{R}_h{}^i{}_{jk} = 0$ only, so that we get $\underline{R}_h{}^i{}_{jk} = 0$.

In general, a pair (g, J) of any Riemannian fundamental tensor field g and an almost complex structure J is called *almost Hermitian*, if J preserves invariant the scalar product g(X, Y) of any tangent vectors X, Y, namely, g(J(X), J(Y)) = g(X, Y).

We have an α-lift $\bar{g}$ of a Finsler fundamental tensor field g, which is a Riemannian fundamental tensor field of the tangent bundle T(M), so that the pair $(\bar{g}, J)$ of $\bar{g}$ and the almost complex N-structure J is obtained; both are defined with respect to a same non-linear connection N.

THEOREM 23.2. *A pair $(\bar{g}, J)$ of an α-lift $\bar{g}$ and the almost complex N-structure J with respect to a non-linear connection N is almost Hermitian, if and only if α is equal to zero.*

This shows the importance of 0-lift. We shall prove the theorem. Corresponding to a tangent vector $X \in T_y$, there exist $X_1, X_2 \in M_x$, $x = \pi_T(y)$, such that $X = X_1^h + X_2^v$ with respect to the N in the notations of (19.13). Then (19.13) gives respective N-decompositions X^β, $J(X)^\beta$, $\beta = 1, 2$, of X, J(X) such that

$$X^1 = z^{-1}X_1, \qquad X^2 = z^{-1}X_2,$$

$$J(X)^1 = -z^{-1}X_2, \qquad J(X)^2 = z^{-1}X_1.$$

Thus (21.2) leads to

$$\bar{g}(J(X), J(Y)) = g(-X_2, -Y_2) + g(X_1, Y_1) + \alpha\{g(-X_2, Y_1) + g(X_1, -Y_2)\}.$$

Accordingly, comparing this with (21.2), the proof is completed.

It is well-known that there exists a symmetric connection with respect to which a given almost complex structure J is covariant constant, if and only if J is integrable. Although the almost complex N-structure J is not integrable in general, it may be possible to be covariant constant with respect to a non-symmetric connection. From this viewpoint the following theorem is interesting.

THEOREM 23.3. *An almost complex N-structure is covariant constant with respect to the linear connection of Finsler type $\bar{\Gamma} = \phi_N(\Gamma)$ for any Γ.*

The proof will be obvious from Theorem 20.1 and the fact that the N-decompositions of J have constant components 0 or δ.

We consider the above almost Hermitian structure $(\bar{g}, J)$ where $\bar{g}$ is the 0-lift. Let J_* be the covariant almost complex N-structure, that is, $J_*(u, v) = J(u, \bar{g}(v))$. From Theorem 23.2 it follows that

$$J_*(u, v) = \bar{g}(J(u), v) = \bar{g}(J(J(u)), J(v)) = -\bar{g}(u, J(v)) = -J_*(v, u),$$

so that the tensor J_* of (0, 2)-type is skew-symmetric.

The almost Hermitian structure $(\bar{g}, J)$ is called *almost Kaehler*, if the differential two-form with the components J_* as its coefficients is closed, or

$$\mathfrak{S}_{(123)}\{\nabla^r J_*(1, 2, 3)\} = 0, \tag{23.10}$$

in the notation ∇^r of (21.6). From (21.4) and Theorem 23.3 we have

$$\nabla^r J_*(1, 2, 3) = -S_*(1, J(2), 3) + S_*(2, J(1), 3).$$

Therefore, by (21.9) $(T^r = 0)$ the condition (23.10) is written as

$$\mathfrak{S}_{(123)}\{\bar{T}_*(1, J(2), 3)\} = 0.$$

From (21.10) $(\alpha = 0)$ and Definition 23.1 the N-decompositions of the above equation are given by

$$\mathfrak{S}_{(123)}\{R^1_*(1, 2, 3)\} = 0, \qquad P^1_*(1, 2, 3) = P^1_*(2, 1, 3),$$

$$C_*(1, 2, 3) = C_*(1, 3, 2).$$

These conditions are all satisfied by the Cartan connection $C\Gamma$ from (17.21) and (17.22). Consequently

THEOREM 23.4. *The almost complex Hermitian structure $(\bar{g}, J)$ of Theorem 23.2 is almost Kaehler with respect to the Cartan connection $C\Gamma$.*

REMARK 23.2. In the above theorem the metrical condition of $C\Gamma$ is very effective. We shall be, however, able to generalize the result to the case of a non-metrical Finsler connection, for instance, the Berwald connection $B\Gamma$. See Matsumoto [12].

Finally we shall cite some papers relating to almost complex and complex structures of Finsler spaces: Heil [1], Ichijyo [1], Prakash [2], Prakash and Dhawan [1], Rani [2], Rizza [1, 2, 3], Sanini [2, 4]. In particular, we finds in Heil [1] an interesting theorem: If a Finsler space M^{2n} admits a complex structure which constructs generalized almost Hermitian structure with the given Finsler metric g, then g is a Riemannian metric and M^{2n} is an Hermitian space.

CHAPTER V

SPECIAL FINSLER SPACES

In Riemannian geometry we have many interesting theorems such that if a Riemannian space is assumed to have special geometrical properties, or to satisfy special tensor equations, or to admit special tensor fields, or so on, then the space reduces to one of well-known space-forms, for instance, euclidean spaces, spheres, topological spheres, projective spaces, and so on.

On the other hand, in Finsler geometry we have hardly had such space-forms. We have special Finsler spaces, namely, Riemannian spaces and Minkowski spaces, but there are various kinds of Riemannian spaces and Minkowski spaces. As a consequence, we have an important problem to classify all the Minkowski spaces, but this is a completely open problem. Moreover, it is easy to write down concrete forms of fundamental functions $L(x, y)$ which are interesting as a function, for instance, a Randers metric. But, if we try to find special properties of torsion and curvature tensors of a Finsler space with such a fundamental function, then we must come soon to the deadlock of enormous computation.

It may be said that to find interesting Finsler space-forms has just begun. Thus it is essential for the progress of Finsler geometry to find Finsler spaces which are quite analogous to Riemannian spaces, but not Riemannian, and Minkowski spaces which are quite analogous to flat space, but not flat.

In the present chapter we are mainly concerned with special tensor equations satisfied by torsion, curvature and other important tensors. Almost all of them are well-known, but some of them are concepts which have been introduced by our contemporaries. In spite of their ceaseless efforts, they can hardly arrive at complete solutions, namely, finding fundamental functions of such special Finsler spaces, except several cases.

It seems to the author that there exist some essential tensors in Finsler spaces, other than torsion and curvature tensors, which vanish in Riemannian case and are not satisfactorily studied yet.

§ 24. Riemannian spaces and locally Minkowski spaces

Let M be a Riemannian space with a Riemannian metric $L(x, y)$, that is,

$$L(x, y)^2 = g_{ij}(x)y^iy^j.$$

The components of Finsler fundamental tensor field g of this space are above coefficients $g_{ij}(x)$, so that (17.1') gives $C_{ijk} = 0$. Conversely, if $C_{ijk} = 0$, then $L(x, y)$ is obviously Riemannian. Therefore we have

THEOREM 24.1. *Among Finsler spaces, the class of all the Riemannian spaces is characterized by* $C_{ijk} = 0$, *where* C_{ijk} *is given by (17.1').*

The condition $C_{ijk} = 0$ means that the vertical connection Γ^v of the Cartan connection $C\Gamma$ is flat, namely, $\Gamma^v = F^1$.

There will be many authors having bitter experiences that on some desirable and interesting conditions the Finsler space all too soon reduced to a Riemannian space. Almost all of such simple results must have not been published. The following is an typical example of such results which are not worth writing.

PROPOSITION 24.1. *If the tensor* C_{ijk} *is v-recurrent, i.e., there exists a Finsler vector field* X_i *such that the v-covariant derivative* $C_{hij}|_k$ *of* C_{hij} *with respect to* $C\Gamma$ *is of the form* $C_{hij}|_k = C_{hij}X_k$, *the space is Riemannian.*

Proof. From $C_{hij}y^h = 0$ and the assumption we get

$$C_{hij}|_k y^h = 0 = (C_{hij}y^h)|_k - C_{kij} = - C_{kij},$$

so that the space is Riemannian.

REMARK 24.1. It may be a little interesting to generalize the notion of recurrency of a tensor field in a Riemannian space to a Finsler space. But we then get only a formal and non-geometrical problem. We recently have a lot of papers concerned with this subject. See, for example, Matsumoto [13], Mishra and Pande [3], Misra [10], Misra and Meher [3], Moór [13, 15, 18], Sen [4], Sinha and Singh [1, 2]. In these papers long and troublesome tensor calculus occupies almost all the pages, and meaningless and compulsory conditions are sometimes imposed for getting some trivial results.

We have, however, two important theorems which assert the space to reduce to a Riemannian space. These are never *all too soon*, but quite astonishing theorems. Because the proofs of them exceed the contents of our treatment, we shall show only an outline of the proof.

The first is so-called *Deicke's theorem* :

THEOREM 24.2. *If the torsion vector* $A_i = LC_i$, $C_i = C_i{}^r{}_r$, *vanishes identically, the space is Riemannian, provided that the fundamental function* $L(x, y)$ *is positive and* C^4*-differentiable for any non-zero* y^i.

REMARK 24.2. This theorem was first proved by Deicke [1] in 1953 by applying the affine differential geometry, and Brickell [1] gave another proof in 1965 based on an inequality between geometric and arithmetic means and E.Hopf's maximum principle for elliptic differential operators.

A geometrical meaning of C_i is as follows: Let g be the determinant consisting of components $g_{ij}(x, y)$ of fundamental tensor. Then

$$\dot{\partial}_i\sqrt{g} = \frac{g}{2\sqrt{g}}\, g^{jk}\dot{\partial}_i g_{jk} = \sqrt{g}\, g^{jk}C_{jki},$$

so that C_i are given by

$$C_i = \dot{\partial}_i(\log\sqrt{g}). \tag{24.1}$$

As a consequence, if $A_i = 0$, then g is a function of position only, and the concept of volume-element can be introduced by $\sqrt{g}\, dx^1\cdots dx^n$, as in the Riemannian case.

We shall sketch the proof of Theorem 24.2 given by Brickell. He first remarks that the matrix $(g_{ij}(x, y))$ is positive-definite, which is shown by Deicke from positiveness of L and constancy of g. Then the characteristic roots of $(g_{ij}(x, y_2))$ with respect to $(g_{ij}(x, y_1))$ are all positive and these roots are also characteristic roots of the matrix $(g_{ij}(x, y_1))^{-1}(g_{ij}(x, y_2))$. Therefore, based on an inequality between geometric and arithmetic means, we obtain

$$\mathrm{Tr}\{(g_{ij}(x, y_1))^{-1}(g_{ij}(x, y_2))\} \geq n\{\det(g_{ij}(x, y_1))^{-1}(g_{ij}(x, y_2))\}^{1/n} = n.$$

Next, the elliptic operator $\Delta = g^{ij}\dot{\partial}_i\dot{\partial}_j$ is introduced and it is shown that the matrix $\Delta(g_{ij}(x, y))$ is positively semi-definite. Then Hopf's maximum principle leads us to the conclusion.

REMARK 24.3. Importance of the class of Finsler spaces with $A_i = 0$ had been noticed by many authors. See Barthel [5], Su [2], Wegener [1]. In 1934 Cartan spent about three pages of his monograph [4] to treat such spaces and showed a close relation between them and hyperareal spaces due to himself. In 1952 Moór [7] discussed a duality of such Finsler spaces and Cartan spaces (hyperareal spaces) with $A^i = 0$.

Berwald already stated in one of his posthumous [17, p.769] that any two-dimensional Finsler space with $A_i = 0$ is Riemannian (see § 28), although it seems that he had already noticed it when he introduced the main scalar in 1927 [5].

PROPOSITION 24.2. *Let M be a Finsler space of n dimensions with the Finsler metric*

$$L(x, y) = c(x)(y^1y^2\cdots y^n)^{1/n},$$

where $c(x)$ is an arbitrary non-zero function of position x alone. The torsion vector A_i of M vanishes identically.

Proof. The fundamental tensor g of this space M is given by

$$g_{ij} = \frac{c^2L^2}{ny^iy^j}\left(\frac{2}{n} - \delta^i_j\right),$$

where i, j are not summation indices. Therefore we have the $\det(g_{\dot{i}j}) = c^{2n}(-1)^{n-1}n^{-n}$ and so $\dot{\partial}_k \det(g_{ij}) = 0$.

REMARK 24.4. Soon after Deicke's paper [1] was published, Moór wrote a paper [9] as a supplementary note to a previous paper [7]. In this note he showed the interesting example given in Proposition 24.2 due to Berwald [Über Finslersche und Cartansche Geometrie II. Compositio Math. 7 (1939), 141-176]. This example is not a counterexample of Theorem 24.2, but asserts the necessity of positiveness of L in Theorem 24.2; if n is odd, then the $\det(g_{ij})$ is positive, but L is not always positive.

We shall turn to the second important theorem, called *Brickell's theorem*. As already mentioned in Remark 17.1, vanishing of the v-curvature tensor S^2 of CΓ does not directly imply the flatness of tangent space M_x, which is regarded as a Riemannian space with the fundamental tensor $g_{ij}(x, y)$, where x is fixed. In fact, it will be seen in § 28 that S^2 vanishes for any two-dimensional Finsler space.

In 1946 Lichnerowicz [2, 3] showed a generalization of the Gauss-Bonnet formula in a Riemannian space to a class of Finsler spaces, called the Berwald space by him, which are characterized by $S^2 = 0$ of CΓ. (We do not, however, use the name "Berwald space" for such a Finsler space. Cf. § 25.)

THEOREM 24.3. *If the v-curvature tensor S^2 of the Cartan connection CΓ on an n (≥ 3)-dimensional Finsler space vanishes identically, the space is Riemannian, provided that the fundamental function L(x, y) satisfies the conditions (F 1, 2, 3, 4) of § 16.*

This astonishing theorem was shown by Brickell [3] in 1967, and Schneider [1] gave another proof from an affine-geometric standpoint. In the following we shall sketch Brickell's proof. First the four lemmas are given :

(1) Let f(y) be a real-valued function of class C^1 on R^n_0, where R^n_0 denotes R^n with the origin removed. f(y) is (0)p-homogeneous, if and only if $(\dot{\partial}_i f)y^i = 0$.

(2) Let f(y) be a real-valued function of class C^1 on R^n_0 and $\dot{\partial}_i f$ are (0)p-homogeneous. Then f = g + c, where g is (1)p-homogeneous and c is a constant.

(3) Let H^n_0 be a Riemannian space defined on R^n_0 with a positive-definite Riemannian metric $g_{ij}(y)dy^idy^j$, where $g_{ij} = \dot{\partial}_i\dot{\partial}_j(L^2/2)$. We also give R^n and R^n_0 the euclidean metric $(dy^1)^2 + \cdots + (dy^n)^2$ and denotes these spaces by E^n, E^n_0 respectively.

Suppose that H_0^n has zero curvature and $n \geq 3$. Then there is an isometry $y \mapsto Y(y) = (Y^i(y))$ of H_0^n onto E_0^n which is differentiable of class C^3, (1)p-homogeneous and such that Y^i satisfy $Y^i(\partial_j \partial_k Y^i) = 0$.

(4) Suppose that L^2 is (2)p-homogeneous. Then the isometry Y given in (3) is (1)-homogeneous.

It is noted here that (1) and (2) are rather trivial, and (3) is proved, based on the well-known isometric immersion of H_0^n into E_0^n.

Next, he deals with the Laplace operator Δ on E^n and shows

(5) Suppose that Y^i are real-valued functions of E_0^n of class C^2, (1)-homogeneous and satisfy $Y^i \Delta Y^i = 0$. Then they are homogeneous linear functions.

Finally he considers the functions Y^i given in (3). From (4) it is seen that they are (1)p-homogeneous and thus satisfy the hypotheses of (5). Therefore they are homogeneous *linear* functions, Consequently the proof has been completed.

It should be emphasized that the most essential point of above proof is to show the linearity of Y^i. In fact, (Y^i) is then regarded as an induced coordinate system of the tangent space M_x. See Remark 17.1.

REMARK 24.5. Another proof of Theorem 24.3 due to Schneider is deduced from an interesting isoperimetric inequality.

In 1956 A.Kawaguchi [3] already concluded that a Finsler space with $S^2 = 0$ is Riemannian, although it seems to the author that his viewpoint is rather intuitive. In 1959 Laugwitz [10] proved that S^2 vanishes for any two-dimensional Finsler space and in 1965 [14] stated the conjecture that a Finsler space of n ($\geq$ 3) dimensions with $S^2 = 0$ will be Riemannian.

It is noticed by careful examination of Brickell's proof that the condition (F 2), i.e., $L(x, -y) = L(x, y)$ is not made full use. In fact, in § 30 we shall show an example to cause the problem "Whether or not the symmetry assumption can be completely removed ?" See Brickell [6, p.27, note on Theorem 3].

It should be, however, remarked that the conditions (F 1, 4) can not be removed. This is confirmed by an interesting example due to Kikuchi [3] as follows:

EXAMPLE 24.1. Let M^n, $n = 2k + r$, be a Finsler space of n dimensions with a fundamental function L(x, y) such that

$$L(x, y)^2 = \{L_1(x, y^1, y^2)\}^2 + \cdots + \{L_k(x, y^{2k-1}, y^{2k})\}^2 + \sum_{p,q=2k+1}^{n} g_{pq}(x) y^p y^q,$$

where $L_a(x, y^{2a-1}, y^{2a})$, $a = 1, \cdots, k$, are (1)p-homogeneous functions in y^{2a-1}, y^{2a} and g_{pq} are functions of x^i alone. The v-curvature tensor S^2 (of $C\Gamma$) of this M^n vanishes identically, which is evident by the fact that $S^2 = 0$ for two-dimensional case. As a special case of the above, we have a direct product of k Finsler spaces of two dimensions and a Riemannian space of n - 2k dimensions. As to the above L, (F 1, 4) are not satisfied at a point $(y^i) = (0, 0, y^3, \cdots, y^n)$, for instance.

It was often mentioned that each tangent space M_x of a Finsler space M with the connection $C\Gamma$ is regarded as a Riemannian space with the fundamental tensor $g_{ij}(x, y)$, where x is fixed. This viewpoint yields various interesting studies and results of Finsler spaces. See § 31 and O.Varga [18].

On the other hand, the tangent space M_x is naturally regarded as a Minkowski space (Definition 16.3) with the Minkowski metric L(x, y), where x is fixed. As a consequence, if the fundamental function L(x, y) reduces to a function of y^i only, i.e., $\partial_i L = 0$, in some coordinate system (x^i), it may be said that all the tangent spaces of the coordinate neighborhood of (x^i) have a common Minkowski geometry. It is remarked that $\partial_i L$ are not components of a Finsler tensor field. Thus we are led to the following definition:

<u>DEFINITION 24.1</u>. A Finsler space with a fundamental function L(x, y) is called *locally Minkowski*, if there exists a coordinate system (x^i) in which L is a function of y^i only. Such a coordinate system (x^i) is called *adapted*.

It is well-known in Riemannian geometry that a normal coordinate system undergoes affine transformation. We have a similar situation as to adapted coordinate systems as follows:

<u>PROPOSITION 24.3</u>. *A transformation between any two adapted coordinate systems of a locally Minkowski space M is affine. Conversely, a coordinate system obtained from an adapted coordinate system of M by an affine transformation is also adapted.*

Proof. If we consider two coordinate systems (x^i), $(\bar{x}^a)$, then we have

$$\frac{\partial y^i}{\partial \bar{x}^a} = \frac{\partial}{\partial \bar{x}^a}\left(\bar{y}^b \frac{\partial x^i}{\partial \bar{x}^b}\right) = \bar{y}^b \frac{\partial^2 x^i}{\partial \bar{x}^a \partial \bar{x}^b},$$

which implies

$$\frac{\partial L}{\partial \bar{x}^a} = \frac{\partial L}{\partial x^i}\frac{\partial x^i}{\partial \bar{x}^a} + \frac{\partial L}{\partial y^i}\bar{y}^b \frac{\partial^2 x^i}{\partial \bar{x}^a \partial \bar{x}^b}.$$

If (x^i) is adapted and the transformation $(x^i) \to (\bar{x}^a)$ is affine (with constant coefficients), the above gives $\partial L/\partial \bar{x}^a = 0$, so that $(\bar{x}^a)$ is adapted also.

Conversely, if both (x^i) and $(\bar{x}^a)$ are adapted, components g_{ij} of the fundamental tensor field g do not depend on x^i, so that connection coefficients $F_j{}^i{}_k$ of $C\Gamma$ vanish from (17.3) and (17.7). Accordingly coefficients $\bar{F}_b{}^a{}_c$ in $(\bar{x}^a)$ are written in the form

$$\bar{F}_{b\ c}^{\ a} = \frac{\partial^2 x^i}{\partial\bar{x}^b \partial\bar{x}^c}\frac{\partial\bar{x}^a}{\partial x^i}.$$

Thus, $\bar{F}_{b\ c}^{\ a} = 0$ implies $\partial^2 x^i/\partial\bar{x}^b\partial\bar{x}^c = 0$, so that $(x^i) \to (\bar{x}^a)$ is affine.

REMARK 24.6. It seems to the author that Proposition 24.3 has been known by some authors since the early period of Finsler geometry, because in 1943 O.Varga mentioned it in his paper [5, p.151]. So far as the author knows, the proof was first published by Heil [2] in 1966. See Tamássy and Matsumoto [1].

THEOREM 24.4. *A Finsler space is locally Minkowski, if and only if*

(1) $R^2 = \nabla^h C = 0$ *in* $C\Gamma$: $R_{i\ jk}^{\ h} = C_{hij|k} = 0.$

(2) $K = \nabla^h C = 0$, *or* $K = F = 0$ *in* $R\Gamma$: $K_{i\ jk}^{\ h} = C_{hij|k} = 0$, *or* $K_{i\ jk}^{\ h} = F_{i\ jk}^{\ h} = 0.$

(3) $H = G = 0$ *in* $B\Gamma$: $H_{i\ jk}^{\ h} = G_{i\ jk}^{\ h} = 0.$

Proof. The connection coefficients $F_{j\ k}^{\ i}$, $N^i_{\ j}$ of $C\Gamma$ vanish in an adapted coordinate system, so that (10.13) and (10.15) give $R^2 = 0$. Next we have

$$C_{hij|k} = \partial_k C_{hij} = \frac{1}{2}\dot{\partial}_j(\partial_k g_{hi}) = 0.$$

Conversely, if $R^2 = 0$ and $\nabla^h C = 0$, then $P^2 = 0$ and $P^1 = 0$ from (17.22) and (17.23), so that $F_{h\ jk}^{\ i} = \dot{\partial}_k F_{h\ j}^{\ i} = 0$ from (10.19'). Consequently $F_{h\ j}^{\ i}$ are functions of position only, and (10.15) gives

$$\mathfrak{A}_{(jk)}\{\partial_k F_{h\ j}^{\ i} + F_{h\ j}^{\ r} F_{r\ k}^{\ i}\} = 0.$$

According to these facts, in similar way to the case of a Riemannian space with zero curvature, there exists a coordinate system $(\bar{x}^a)$ such that $\bar{F}_{b\ c}^{\ a} = 0$, hence $\bar{N}^a_{\ b} = 0$ also, so that (17.2) gives $\bar{g}_{ab} = \bar{g}_{ab}(\bar{y})$. Consequently (1) is proved.

Next, (17.8) and (18.2) show that $R^2 = 0$ is equivalent to $K = 0$. It follows from (18.16) that $K = \nabla^h C = 0$ imply $H = 0$ and $H = \nabla^h C = 0$ imply $K = 0$. Further Theorem 25.2 and Proposition 25.1 show that $\nabla^h C = 0$ is equivalent to $G = 0$ or $F = 0$. Consequently the condition (2) or (3) is equivalent to (1).

REMARK 24.7. Theorem 24.4 was mentioned by Berwald [7] in 1928 and by Cartan [4] in 1934, but both of them did not show any proof. Rund gave a proof in his book [18, p.136]. See also O.Varga [5, p.162], Wagner [12, p.126]. The above proof is due to Matsumoto [14].

It is remarked that the above conditions (2) and (3) state vanishing of the h-and hv-curvature tensors in $R\Gamma$ and $B\Gamma$ respectively, but (1) does not so. $R^2 = P^2 = 0$ in $C\Gamma$ is not sufficient for the space to be locally Minkowski. But, is it true ? See Remark 30.7.

THEOREM 24.5. *A Finsler space is locally Minkowski, if and only if* $R^1 = 0$ $(R^i{}_{jk} = 0)$ *and* $\nabla^h C = 0$ $(C_{hij|k} = 0)$ *in* $C\Gamma$.

Proof. $R^1 = 0$ follows from $R^2 = 0$ by (17.8). Conversely, $R^1 = 0$ yields $K = R^2$ from (18.2) and $H = 0$ from (18.22). Further $\nabla^h C = 0$ yields $H = K$ from (18.16), so that we get $R^2 = 0$.

REMARK 24.8. Theorem 24.5 is a simple remark given by Soós [4]. Those characterizations of locally Minkowski spaces are remarkable in such a point of view that a problem whether some special coordinate systems exist or not is solved by an invariant way, that is, by tensor equations. While independence of components of a tensor field T from y^i is obviously equivalent to the tensor equation $\nabla^0 T = 0$, independence from x^i leads us to such a problem, which will be generally rather difficult to solve. For instance, what is an invariant expression of $\partial_k C_{hij} = 0$? Pay attention to an interesting method given in Tamássy and Matsumoto [1].

§ 25. Berwald spaces and Landsberg spaces

In 1928 Berwald had a lecture [7], in which he enumerated some special classes of Finsler spaces. Among them a class of *affinely connected Finsler spaces* which had been introduced in his previous paper [3] was characterized by a tensor equation. Such a Finsler space belonging to this class is defined as follows:

DEFINITION 25.1. If the connection coefficients $G_j{}^i{}_k$ of the Berwald connection $B\Gamma$ given by (18.14) are functions of position alone, the space is called a *Berwald space*.

Thus the Berwald connection $B\Gamma$ of a Berwald space is similar to a linear connection as to the parallel displacement of vector fields. Although the name *"affinely connected space"* due to Berwald himself is popular, we shall adopt the name *"Berwald space"* following Wagner [3]. According to (18.17), one of the tensorial characterizations of such a space is immediately obtained:

THEOREM 25.1. *A Finsler space is a Berwald space, if and only if the hv-curvature tensor G* $(G_i{}^h{}_{jk})$ *of the Berwald connection* $B\Gamma$ *vanishes identically.*

REMARK 25.1. It seems that the tensor property of $\dot\partial_k G_i{}^h{}_j$ $(= G_i{}^h{}_{jk})$ has been known in the early period, but the recognition that $G_i{}^h{}_{jk}$ are components of the hv-curvature tensor of $B\Gamma$, corresponding to $P_i{}^h{}_{jk}$ of $C\Gamma$, may be a fruit of our theory based on fiber bundles.

From (18.16) it follows that the condition $G_i{}^h{}_{jk} = 0$ is equivalent to

$$F_h{}^i{}_{jk} + C_j{}^i{}_{h|0\cdot k} = 0.$$

By (17.22) this may be written as

$$F_{hijk} + P_{hij|k} - C_i{}^r{}_k P_{rjh} + C_j{}^r{}_k P_{rhi} + C_h{}^r{}_k P_{rij} = 0. \tag{25.1}$$

Accordingly the identity (18.12) is written

$$C_{hik|j} + P_{hij|k} + C_j{}^r{}_k P_{rhi} - C_h{}^r{}_i P_{rjk} = 0.$$

Contracting this by y^h, we obtain $P_{kij} = 0$, so that the above reduces to $C_{hik|j} = 0$. Conversely, it follows from $C_{hik|j} = 0$, (17.22) and (17.23) that $P_{ijk} = 0$ and $P_{hijk} = 0$. Therefore (18. 2) gives $F_h{}^i{}_{jk} = 0$ and then (25.1) holds. Consequently

<u>THEOREM 25.2</u>. *In terms of the Cartan connection $C\Gamma$ a Berwald space is characterized by $\nabla^h C = 0$ ($C_{hij|k} = 0$).*

<u>REMARK 25.2</u>. Cartan states the above theorem in his monograph [4. p.38, footnote] without proof. Berwald gave a proof in one of his posthumous papers [15].

From (18.14) it is observed that $F_j{}^i{}_k$ of a Berwald space coincide with $G_j{}^i{}_k$, so that $F_j{}^i{}_k$ are also functions of position only. The converse is true. In fact, (18.2) gives

$$P_h{}^i{}_{jk} + C_h{}^i{}_{k|j} - C_h{}^i{}_r P^r{}_{jk} = 0.$$

Contracting this by y^h, we get $P^i{}_{jk} = 0$, so that (17.27) yields $P_{hijk} = 0$. Consequently the above reduces to $C_h{}^i{}_{k|j} = 0$. Therefore

<u>PROPOSITION 25.1</u>. *A Finsler space is a Berwald space, if and only if the connection coefficients $F_j{}^i{}_k$ of the Cartan connection $C\Gamma$ are functions of position only, namely, the hv-curvature tensor F ($F_i{}^h{}_{jk}$) of the Rund connection $R\Gamma$ vanishes identically.*

<u>REMARK 25.3</u>. Berwald [5, 14] finds fundamental functions of all the two-dimensional Berwald spaces which are not locally Minkowski. In § 28 we shall see his beautiful theory to find them. According to Berwald's result, the fundamental function L(x, dx) of any two-dimensional Berwald space which is not locally Minkowski is a (1)p-homogeneous function of two linearly independent differential one-forms. Therefore we have an interesting problem to study n-dimensional Finsler spaces the fundamental functions L(x, dx) of which are (1)p-homogeneous functions of n linearly independent differential one-forms. See Ichijyo [2, 6], Matsumoto and Shimada [2, 3].

Theorem 29.6 gives some informations on three-dimensional Berwald spaces. Recently we have noteworthy results on Berwald spaces due to Szabó [5].

Among special classes of Finsler spaces which were enumerated by Berwald in his lecture, quoted at the beginning of the present section, we find the following generalization of special two-dimensional spaces which had been considered by Landsberg [1, 2, 3]:

DEFINITION 25.2. A Finsler space is called a *Landsberg space*, if the Berwald connection $B\Gamma$ is h-metrical.

REMARK 25.4. Landsberg is concerned with the notion of curvature of extremal field on two-dimensional variation problem. He notices the importance of such a space that the curvature depends on position only, because the well-known Gauss-Bonnet formula holds in such a space. He gives such an example that the integrand is given by

$$L(x, y) = \{a_i(x)y^i\}^r\{b_j(x)y^j\}^s, \qquad r + s = 1.$$

It will be easily verified that the space with this L as its fundamental function is really a Berwald space. See (28.28 (i)).

In 1926 Berwald [3] indicated that the "*Streckenkrümmung*" (stretch curvature) due to himself vanishes in the space considered by Landsberg. See Cartan [1, p.124, footnote], Moór [11, p.95], Rund [18, p.262].

THEOREM 25.3. *In terms of the Cartan connection $C\Gamma$, a Landsberg space is characterized by (1) the (v)hv-torsion tensor P^1 ($P^i{}_{jk}$) vanishes identically, namely, $\nabla^h C(\varepsilon) = 0$ ($C_{hij|0} = 0$), or (2) the hv-curvature tensor P^2 ($P_i{}^h{}_{jk}$) vanishes identically.*

Proof. (1) follows at once from (18.24) and (17.22). (2) is also obvious from (17.8) and (17.27).

REMARK 25.5. Cartan notices the importance of spaces with $\nabla^h C(\varepsilon) = 0$ in a viewpoint of the parallel displacement. See Cartan [4, p.37]. Berwald [15] indicates (1) of Theorem 25.3.

The concept of Landsberg space is noteworthy from a standpoint of the Berwald connection $B\Gamma$, because it is h-metrical in such a space. Furthermore, from the above (1) and (18.14) we see

PROPOSITION 25.2. *A Finsler space is a Landsberg space, if and only if the hor-*

izontal connections Γ^h *of* $C\Gamma$ *and* $B\Gamma$ *coincide with each other.*

REMARK 25.6. As the concept of Finsler space is a generalization of Riemannian space and we have many interesting concepts and results in Riemannian geometry, it is naturally important to bring these concepts into Finsler geometry and to intend to establish the similar results. We have indeed had various interesting results from this standpoint. It may be, however, more important for future progress of Finsler geometry to consider properties of tensors which reduce to zero in the Riemannian case, such as P^1, P^2, S^2 and so on of $C\Gamma$. From this point of view, the class of Landsberg spaces must be very important, but it is not satisfactorily studied yet. We have few papers concerned with Landsberg spaces. See Hashiguchi, Hōjō and Matsumoto [1], Ichijyo [4], H.Kawaguchi [1].

In two-dimensional case the (v)hv-torsion tensor P^1 and the hv-curvature tensor P^2 have special forms (28.12'), (28.13') respectively (Proposition 28.4). The following theorems give answers of problems arising from those special forms; the proofs are easy.

THEOREM 25.4. *If the hv-curvature tensor* $P_{h\,i\,j\,k}$ *is symmetric in* j, k, *and the (v)hv-torsion tensor* $P_{i\,j\,k}$ *is proportional to the (h)hv-torsion tensor* $C_{i\,j\,k}$, *then* $P_{hijk} = 0$ *(Landsberg space), or the v-curvature tensor* $S_{hijk} = 0$.

THEOREM 25.5. *If there exists a Finsler vector field* X_i *such that*

$$P_{hijk} = X_h C_{ijk} - X_i C_{hjk},$$

then $P_{hijk} = 0$ *(Landsberg space), or* $S_{hijk} = 0$. *In every case we have* $S_{hijk|m} = 0$.

REMARK 25.7. As to those theorems, see Hashiguchi [3], Matsumoto [15]. According to Theorem 24.3, those theorems assert that on respective assumptions the space of dimension $n \geq 3$ is Landsberg, provided that the assumptions in Theorem 24.3 is satisfied, because a Riemannian space is, of course, a Landsberg space.

Now the Cartan connection $C\Gamma$ has three curvature tensors R^2, P^2 and S^2. Theorem 25.3 shows

$$\text{hv-curvature } P^2 = 0 \quad \rightarrow \quad \text{Landsberg space},$$

and, roughly speaking, Theorem 24.3 asserts

$$\text{v-curvature } S^2 = 0 \quad \rightarrow \quad \text{Riemannian space}.$$

Then the problem to examine the condition $R^2 = 0$ naturally arises. It is noted that the case "$R^2 = 0$" is exceptional in Theorem 22.2. As for this interesting problem, only the following is now known:

THEOREM 25.6. *If the h-curvature tensor R^2 $(R_i{}^h{}_{jk})$ of the Cartan connection $C\Gamma$ vanishes identically, (1) $P_{hij|k}$ and $C_{hij|k|0}$ are completely symmetric, and (2) the tensor*

$$Q_{hijk} = \mathfrak{A}_{(jk)}\{P^r{}_{hk}P_{rij}\}$$

vanishes identically.

Proof. One of the Bianchi identities (17.14) now reduces to

$$\mathfrak{A}_{(ij)}\{P_{hir}P^r{}_{jk} + P_{hjk|i}\} = 0. \tag{25.2}$$

Cyclic permutation of indices h, j, k in (25.2) and summation yield

$$3P_{hjk|i} = P_{hik|j} + P_{jih|k} + P_{kij|h},$$

which easily implies $P_{hjk|i} = P_{hji|k}$; this is written as $C_{hjk|0|i} = C_{hji|0|k}$. In case of $R^2 = 0$, the Ricci identity (10.8') gives $C_{hjk|m|i} = C_{hjk|i|m}$, so that we get $C_{hjk|0|i} = C_{hjk|i|0}$. Then (1) is proved. As a consequence of (25.2) and (1), we get (2).

REMARK 25.8. The results of Theorem 25.6 are a little similar to those of Theorem 22.2, because (1) corresponds to the symmetry $P_{hijk} = P_{hikj}$ and the tensor Q_{hijk} appearing in (2) has a quite analogous form to S_{hijk}, according to (17.20) and (17.22).

In 1943 Wagner [4] gave the notion of generalized Berwald space. According to his definition, a Finsler space is said to be a *generalized Berwald space*, if it is possible to introduce a generalized Cartan connection with the (h)h-torsion tensor T different from zero, in such a way that the connection coefficients $F_j{}^i{}_k$ would be functions of position alone. He obtained an invariant characterization of only two-dimensional case in terms of the main scalar (§ 28).

We shall be concerned with this interesting but vague notion due to Wagner and give an exact formulation of generalized Cartan connection and generalized Berwald space, which is studied by Hashiguchi [4].

We first recall Theorems 17.1, 17.2 and Definition 17.2. Changing the axiom (C 4), we shall introduce a system of axioms as follows:

PROPOSITION 25.3. *There exists a unique Finsler connection $C\Gamma T$ which satisfies the following five axioms:*

(C 1) $C\Gamma T$ is v-metrical.

(C 2) The (v)v-torsion tensor S^1 of $C\Gamma T$ vanishes.

(C 3) $C\Gamma T$ is h-metrical.

(C 4') The (h)h-torsion tensor T of $C\Gamma T$ is a given skew-symmetric Finsler ten-

sor field of (1, 2)-type.

(C 5) *The deflection tensor D of CΓT vanishes.*

Proof. The (h)hv-torsion tensor C is, of course, given by (17.1'). Next, from (17.2) we obtain an equation similar to (17.3'):

(25.3) $$F_{ijk} = \gamma_{ijk} - C_{ijr}N^r_{\ k} - C_{jkr}N^r_{\ i} + C_{ikr}N^r_{\ j} + A_{ijk},$$

where the additional term A_{ijk} arising from (C 4') is defined by

(25.4) $$2A_{ijk} = T_{ijk} - T_{jki} + T_{kij}, \qquad T_{ijk} = g_{jr}T_i{}^r{}_k.$$

From (25.3) the proof is easily finished in similar way to the proof of Theorem 17.2.

Furthermore, we consider a condition for a Finsler space with the connection CΓT to be such that the connection coefficients $F_j{}^i{}_k$ be functions of position alone (cf. Definition 25.1). It follows first from (10.19') that the condition is equivalent to

$$P_{hijk} = - C_{hik|j} + C_{hir}P^r_{\ jk}.$$

Since the connection CΓT is metrical, P_{hijk} are skew-symmetric in h, i, similarly to the case of CΓ, while the right-hand members of the above equation are symmetric in h, i. Thus we get

$$P_{hijk} = 0, \qquad C_{hik|j} = C_{hir}P^r_{\ jk}.$$

From Theorem 13.3 and $P_{hijk} = 0$ it follows that $P_{0ijk} = P_{ijk} = 0$, so that the above reduces to

(25.5) $$P_{hijk} = 0, \qquad C_{hik|j} = 0.$$

It is remarked that the former of (25.5) is a consequence of the latter from (17.23) in case of CΓ and that (17.23) is obtained from the Bianchi identity (11.4'). In the Finsler connection CΓT we shall thus return to (11.4') again, where $T_j{}^i{}_k$ survive. In a similar way by means of which (17.23) is derived from (11.4'), we get a generalization of (17.23):

(25.6) $$2P_{hijk} = 2\,\mathfrak{A}_{(hi)}\{C_{ijk|h} + C_{hjr}P^r_{\ ik}\} + (g_{ir}T_h{}^r{}_{j\cdot k} - g_{hr}T_i{}^r{}_{j\cdot k} - g_{jr}T_i{}^r{}_{h\cdot k}).$$

Here appear three aditional terms; the first being skew-symmetric in h, j, while the remaining two terms being symmetric in h, j. Therefore (25.5) implies

(25.7) $$T_h{}^r{}_{j\cdot k} = 0.$$

Conversely, if $C_{hij|k} = 0$ and (25.7) are satisfied, then $P_{hijk} = 0$ is a consequence of (25.6), as it will be easily verified. Therefore

THEOREM 25.7. *A necessary and sufficient condition for a Finsler connection $C\Gamma T$, given by Proposition 25.3, to be such that the connection coefficients $F_j{}^i{}_k$ do not depend on y^i is that $C_{hij|k} = 0$ and (25.7) hold.*

Next, following Wagner, we suppose that the (h)h-torsion tensor T is so-called *semi-symmetric*, namely, there exists a Finsler covariant vector field s_i such that

$$T_j{}^i{}_k = \delta^i_j s_k - \delta^i_k s_j. \tag{25.8}$$

In this case the condition (25.7) implies that s_i are functions of position only; they are components of an ordinary covariant vector field on the space M.

Summarizing up all the above, we are led to the following definitions and theorem:

DEFINITION 25.3. Let $s_i(x)$ be components of a covariant vector field on a Finsler space M. The *Wagner connection* $W\Gamma$ of the Finsler space M is a Finsler connection which is uniquely determined by the following five axioms:

(C 1) $W\Gamma$ is v-metrical.

(C 2) The (v)v-torsion tensor S^1 of $W\Gamma$ vanishes.

(C 3) $W\Gamma$ is h-metrical.

(C 4s) The (h)h-torsion tensor T of $W\Gamma$ is given by (25.8).

(C 5) The deflection tensor D of $W\Gamma$ vanishes.

REMARK 25.9. Corresponding to each covariant vector field s_i on the space M, we can construct a unique Wagner connection $W\Gamma$ from the fundamental function L. Therefore, pay attention to the expression of the following definition.

DEFINITION 25.4. A Finsler space M is called a *Wagner space*, if there exists a covariant vector field $s_i(x)$ such that the connection coefficients $F_j{}^i{}_k$ of the Wagner connection $W\Gamma$, corresponding to s_i, are functions of position only.

THEOREM 25.8. *A Finsler space is a Wagner space, if and only if there exists a covariant vector field $s_i(x)$ such that $\nabla^h C = 0$ holds with respect to the Wagner connection $W\Gamma$, corresponding to $s_i(x)$.*

REMARK 25.10. It will be shown in § 28 that a two-dimensional Finsler space is a Berwald space, if and only if the main scalar I is constant. On the other hand, Wagner [4] shows a beautiful theorem that a two-dimensional Finsler space is a Wagner space, if and only if $\partial I/\partial\theta$ is a function of I, where θ is the Landsberg angle (§ 28). He gives an interesting example of a two-dimensional Wagner space whose fundamental function L(x, y) is a so-called *cubic metric*

$$L(x, y) = \{a_{ijk}(x)y^i y^j y^k\}^{1/3}.$$

As to this space, the main scalar I satisfies $\partial I/\partial\theta = -3I^2 - 3/2$. See Matsumoto and Numata [1], Wagner [2].

The n-dimensional Finsler spaces with a cubic metric have been studied by several authors. For example, Liber [1], Tonooka [1, 2, 3, 4, 5], Wegener [2, 3]. Further Shimada [3].

The theory of Finsler connections with a surviving (h)h-torsion tensor T has been studied by several authors. Matsumoto [6] treats a semi-symmetric torsion. Moór [16] makes use of the torsion tensor to construct a natural frame field in four-dimensional Finsler spaces.

The following theorem due to Hashiguchi and Ichijyo [2] may be the most fruitful result arising from the introduction of the Wagner connection.

THEOREM 25.9. *(1) A Wagner space remains to be a Wagner space by any conformal change of metric. (2) The condition that a Finsler space be conformal to a Berwald space is that the spaces becomes a Wagner space with respect to a gradient vector field $s_i(x)$. (3) The condition that a Finsler space be conformal to a locally Minkowski space is that the space becomes a Wagner space with respect to a gradient vector field $s_i(x)$ and its h-curvature tensor (in the sense of the Wagner space) vanishes identically.* (The proof is omitted.)

REMARK 25.11. As to Wagner connections and Wagner spaces, see Hashiguchi and Ichijyo [1], Hashiguchi and Varga [1], Tamássy and Matsumoto [1].

§ 26. Finsler spaces of scalar curvature

In Riemannian geometry the concept of constant curvature is important, because the class of Riemannian spaces of constant curvature is the best typical model of space-forms and is regarded as a generalization of the so-called non-euclidean planes. The canonical forms of ds^2 of such spaces are well-known.

We have considered two trials of generalization of this concept to Finsler geometry in § 22 with respect to the Cartan connection CΓ. It may be, however, said that both trials are unsuccessful in the viewpoint of Theorems 22.1 and 22.2 together with Theorem 24.3.

The first purpose of the present section is to generalize this concept to Finsler geometry with respect to the Berwald connection BΓ.

DEFINITION 26.1. Let $X = (X^i)$ be a tangent vector of a Finsler space M at a point $x = (x^i)$. The quantity $K(x, y, X)$ at a supporting element (x, y) given by

$$H_{hijk}y^h X^i y^j X^k = K(g_{hj}g_{ik} - g_{hk}g_{ij})y^h X^i y^j X^k$$

is called the *curvature at (x, y) with respect to X*, where H_{hijk} are components of the h-curvature tensor H of the Berwald connection BΓ.

Thus the curvature $K(x, y, X)$ is defined similarly to the Riemannian case. It may be regarded as the sectional curvature of a two-section spanned by y and X, but it should be emphasized that H_{hijk} and g_{ij} are evaluated at this supporting element (x, y), so that $K(x, y + X, X)$, for instance, is different from $K(x, y, X)$ in general, although $y + X$ and X span the same two-section with the one spanned by y and X.

It is observed from (18.2) and (18.16) that

$$H_{hijk}y^h y^j = K_{hijk}y^h y^j = R_{hijk}y^h y^j ,$$

so that in so far as $K(x, y, X)$ alone is concerned, it is also written in terms of the h-curvature tensors of CΓ as well as RΓ. In the following, however, it will be seen that certain special properties of H_{hijk} play an important role.

From (17.8), (16.2) and (16.5) it follows that the curvature $K(x, y, X)$ is given by

(26.1) $$R_{i0k}X^i X^k = KL^2 h_{ik}X^i X^k.$$

DEFINITION 26.2. If the curvature $K(x, y, X)$ does not depend on X at every supporting element (x, y), the space is said to be *of scalar curvature K(x, y)*.

In this case the scalar curvature K is a (0)p-homogeneous Finsler scalar field, and (26.1) reduces to

(26.2) $$R_{i0k} = KL^2 h_{ik},$$

because of $R_{i0k} = R_{k0i}$ from (17.21).

We shall derive more convenient equation from (26.2). Paying attention to (17.1'), the equation (18.23) is rewritten in the form

(26.3) $$R_{ijk} = \frac{1}{3}\mathfrak{A}_{(jk)}\{R_{i0k|j} - R_{r0k}C_i{}^r{}_j\}.$$

Substituting from (26.2) and paying attention to (17.31), we immediately get

(26.4) $$R_{ijk} = h_{ik}K_j - h_{ij}K_k,$$

where we put

(26.5) $$K_j = \frac{L^2}{3}K|_j + KLl_j.$$

Conversely, (26.2) is a consequence of (26.4) by Proposition 16.2. In fact, contracting (26.4) by y^j, we get $R_{i0k} = h_{ik}K_0$, and, if we put $K = K_0/L^2$, the above procedure from (26.2) to (26.4) yields (26.5). Therefore

THEOREM 26.1. *A Finsler space is of scalar curvature, if and only if (1) R_{i0k} is proportional to h_{ik}, that is, (26.2) holds, or (2) there exists a Finsler covariant vector field K_j such that R_{ijk} is written in the form (26.4).*

We shall find the form of h-curvature tensor H_{hijk} in case of scalar curvature. From (18.22) and (17.31) we have

$$\dot\partial_h R_{ijk} = H_{hijk} + 2C_{ihr}R^r{}_{jk}, \qquad \dot\partial_k h_{ij} = 2C_{ijk} - \frac{1}{L}(h_{ik}l_j + h_{jk}l_i).$$

Then, from (26.4) we immediately get

$$\text{(26.6)} \qquad H_{hijk} = \mathfrak{A}_{(jk)}\{h_{ik}K_{j\cdot h} + \frac{1}{L}(h_{ih}l_j + h_{jh}l_i)K_k\}.$$

Now we are concerned with a special case where the scalar curvature K is a function of position only. Then we have $K_j = KLl_j$, so that (26.4) and (26.6) reduce respectively to

$$\text{(26.7)} \qquad R_{ijk} = KL(h_{ik}l_j - h_{ij}l_k) = KL(g_{ik}l_j - g_{ij}l_k),$$

$$\text{(26.8)} \qquad H_{hijk} = K(g_{hj}g_{ik} - g_{hk}g_{ij}).$$

Evidently the former is a consequence of the latter.

PROPOSITION 26.1. *If the scalar curvature K of a Finsler space of scalar curvature is a function of position alone, then K is a constant, provided the dimension is more than two.*

This is a generalization of Schur's theorem in Riemannian geometry and similar to the result of Theorem 22.2. The proof is as follows: From (26.7) we get

$$R_{ijk|h} = K_{|h}(y_j g_{ik} - y_k g_{ij}).$$

From this and (26.7) the Bianchi identity (17.11) is written as

$$\mathfrak{S}_{(ijk)}\{K_{|k}(y_i g_{hj} - y_j g_{hi})\} = 0.$$

Contraction of the above by $g^{hj}y^k$ leads to $(n-2)(LK_{|i} - K_{|0}l_i) = 0$, so that we get $LK_{|i} = K_{|0}l_i$ in case of $n > 2$. We differentiate the result v-covariantly. At this time the Ricci identity (10.9') shows

$$K_{|i|j} - K_{|j|i} = - K|_r C_i{}^r{}_j - K|_r P^r{}_{ij},$$

that is, $K_{|i|j} = 0$ in this case. Thus we get $K_{|0} = 0$, so that $K_{|i} = 0$. From the assumption $K = K(x)$, we see $K_{|i} = \partial_i K = 0$.

DEFINITION 26.3. If the scalar curvature K of a Finsler space of scalar curvature is a constant, the space is said to be *of constant curvature K*.

THEOREM 26.2. *A Finsler space of n (≥ 3) dimensions is of constant curvature K, if and only if the equation (26.8) holds.*

Proof. We have to show the sufficiency only. From (26.8) we get (26.7). Then

$$R_{ijk\cdot h} = K_{\cdot h}(y_j g_{ik} - y_k g_{ij}) + K(g_{jh}g_{ik} - g_{kh}g_{ij}) + 2K(y_j C_{ikh} - y_k C_{ijh}).$$

Substituting from (26.8) and paying attention to (18.22), we get $K_{\cdot h}(y_j g_{ik} - y_k g_{ij}) = 0$, which implies $K_{\cdot h} = 0$. Then Proposition 26.1 completes the proof.

REMARK 26.1. The equation (26.8) is of a quite similar form to the curvature tensor of a Riemannian space of constant curvature, and shows the importance of the Berwald connection BΓ. The concept of a Finsler space of scalar curvature is introduced by Berwald [17] before 1942, because he died in 1942.

The equation (22.11) with respect to CΓ holds in any two-dimensional Finsler space because of $R_{hijk} = - R_{ihjk} = - R_{hikj}$. While (26.8) is not satisfied generally in such spaces, because we have not the identity $H_{hijk} = - H_{ihjk}$ by means of the fact that BΓ is not h-metrical. See (28.8') and (28.17).

We shall recall the proof of the second part of Theorem 22.2, in which the Bianchi identities (17.12) and (17.15) play the essential role. We shall apply the similar procedure of the proof to BΓ. The Bianchi identities of BΓ, corresponding to them, are (18.20) and (18.21) respectively. Substituting into them from (26.7) and (26.8), we obtain two equations. The first is trivial because of the identity $G_h{}^i{}_{jk} = G_h{}^i{}_{kj}$. The second is written

$$\mathfrak{A}_{(ij)}\{2KC_{mik}g_{hj} + g_{hr}G_m{}^r{}_{jk;i}\} = 0.$$

Similarly to the proof of Theorem 22.2, we apply the process of cyclic permutation of indices i, j, k and summation, but only a trivial equation is derived. Consequently it seems that we have not a theorem as to BΓ similar to Theorem 22.2.

We have really non-trivial examples of spaces of scalar curvature and constant curvature. See Numata [3], Yasuda and Shimada [1].

As for the h-curvature tensor K_{hijk} of the Rund connection RΓ, the analogous equation

$$K_{hijk} = K(g_{hj}g_{ik} - g_{hk}g_{ij})$$

leads us to a trivial result immediately. In fact, in this case we get $R_{ijk} = K(y_jg_{ik} - y_kg_{ij})$ and (18.11) reduces to

$$K(y_jC_{hik} - y_kC_{hji}) = 0,$$

which immediately implies $K = 0$ or $C_{hik} = 0$.

We now return to consideration of Finsler spaces of scalar curvature. From (17.14), by contracting by y^j and summing with respect to $h = k$, we obtain

$$R^h{}_{ij|h}y^j - R^h{}_{0r}C_i{}^r{}_h - P^h{}_{ih|0} = 0.$$

Substituting from (26.2), (26.4) and (17.22), the above is rewritten as

$$\text{(26.9)} \qquad \frac{n+1}{3} L^2K\big|_i + L^2KC_i + C_{i|0|0} = 0.$$

From this we get

<u>PROPOSITION 26.2.</u>

(1) If the torsion vector C_i of an $n(\geq 3)$-dimensional Finsler space of scalar curvature vanishes, the space is of constant curvature.

(2) If a Finsler space is of constant curvature, we have $C_{i|0|0} = - L^2KC_i$.

(3) If a Finsler space of non-zero scalar curvature K is a Landsberg space, K is given by $|g|^{-3/2(n+1)}\phi$, where $|g|$ is the determinant of components of the fundamental tensor g and ϕ is a function of position only.

Proof of (3). From (26.9) and Theorem 25.3 we get $(n + 1)K\big|_i/3 + KC_i = 0$, which by (24.1) is rewritten as

$$\frac{1}{K}\frac{\partial K}{\partial y^i} + \frac{3}{n+1}\frac{\partial \log\sqrt{|g|}}{\partial y^i} = 0.$$

Then integration completes the proof.

<u>REMARK 26.2.</u> In his paper [17, p.777, footnote], Berwald states that in two-dimensional case (26.9) is one of Bianchi identities. It is, however, noticed that in general-dimensional case it is also derived from the Bianchi identity (17.14).

The result (1) of Proposition 26.2 is mentioned by Berwald, but we should recall Theorem 24.2.

Next, (2) is effectively used by Ingarden in his interesting monograph [3, p.45] concerned with a problem of theoretical physics.

Finally, (3) is shown by Matsumoto [18] in three-dimensional case, provided that the space is a Berwald space, instead of a Landsberg space, but we must refer to

Theorem 30.6, proved recently. Similarly Theorem 30.7 has surpassed an interesting result due to Soós [4]: If $C^rR_{rjk} = 0$ holds in a Berwald space of scalar curvature, the space is locally Minkowski, provided $C_i \neq 0$.

We shall consider (26.4) again. Putting $R_j = R^r{}_{jr}$ $(= R_0{}^r{}_{jr})$, it follows that

$$3R_j = (n - 2)L^2K|_j + 3(n - 1)Ky_j.$$

Differentiating this v-covariantly, we get

$$3R_j|_i = 2(n - 2)K|_j y_i + 3(n - 1)K|_i y_j + (n - 2)L^2K|_j|_i + 3(n - 1)Kg_{ji},$$

which implies

$$(26.10) \qquad 3(R_j|_i - R_i|_j) = (n + 1)(y_jK|_i - y_iK|_j).$$

From Proposition 26.1 and (26.10) we easily conclude

THEOREM 26.3. *A necessary and sufficient condition for a Finsler space of scalar curvature K and dimension more than two to be of constant curvature is that the v-covariant derivative* $R_j|_i$ *of the tensor* $R_j = R_0{}^r{}_{jr}$ *is symmetric. In this case we have* $R_j|_i = (n - 1)Kg_{ji}$.

REMARK 26.3. Theorem 26.3 expresses merely the simple notion "constant curvature" in a complicated condition, but it seems to the author that it might be a result of Berwald's final effort [17, p.775], before he was deported to Poland in 1941, which generalizes the concept of Einstein space to Finsler geometry. In fact, if the space is Riemannian with the curvature tensor $R_h{}^i{}_{jk}$, the above R_j is equal to $R_j = y^hR_h{}^r{}_{jr} = y^hR_{hj}$, where R_{hj} is the so-called Ricci tensor, so that $R_{j\cdot i}$ are nothing but R_{ji}. It is noted that $R_j|_i - R_i|_j = R_{j\cdot i} - R_{i\cdot j}$, and that the last equation given in Theorem 26.3 is written as $R_{ji} = (n - 1)Kg_{ji}$; this is just the equation by means of which the concept of Einstein space is defined.

So far as the author knows, we have no paper yet which shows a trial to generalize the notion of Einstein space to Finsler geometry, except Eguchi and Ichijyo [1], Rund [21], Sharma and Trivedi [1]. See also Matsumoto and Shimada [1]: The Einstein tensor of a Riemannian space is defined by

$$G_{ij} = R_{ij} - \frac{R}{2} g_{ij}, \qquad R = g^{ij}R_{ij}.$$

This is symmetric and satisfies the *conservation law* $G^r{}_{j|r} = 0$ which is important in physics. To generalize this concept to Finsler geometry, we must find such a tensor as G_{ij}. It is, however, obvious that the contracted h-curvature tensors of BΓ, CΓ and RΓ are all not symmetric in general and $2R^r{}_{j|r} = R_{|j}$ does not hold with respect to a general CΓ, for instance.

In the following we shall see Rund's trial to generalize the notion of Einstein space to Finsler geometry.

It has been already remarked that the curvature K(x, y, X) with respect to X is also written in the form

$$K_{hijk}y^hX^iy^jX^k = K(g_{hj}g_{ik} - g_{hk}g_{ij})y^hX^iy^jX^k$$

in terms of RΓ. We now consider an orthonormal frame z^i_a, $a = 1,\cdots, n$, at a point x of an n-dimensional Finsler space M and introduce the quantities

$$\mu_{ab}(x, y) = K_{hijk}z^h_a z^i_b z^j_a z^k_b,$$

from which we have

$$\mu_a(x, y) = \sum_{b=1}^{n} \mu_{ab}(x, y) = K_{hj}z^h_a z^j_a ,$$

where $K_{hj} = K_h{}^r{}_{jr}$, the Ricci tensor of RΓ. Therefore the *mean curvature* $\mu_a(x, y)$ with respect to z^i_a, due to Rund, does not depend on the choice of z^i_b, $b = 1,\cdots, n, \neq a$.

We now take the direction of z^i_a as that of the supporting element y^i. Then the mean curvature with respect to y^i is given by

(26.11) $$\mu(x, y) = K_{hj}l^hl^j = L^{-2}K_{hj}y^hy^j.$$

The equation $K_{hj}(x, y)y^hy^j = 1$ with fixed x gives a hypersurface of the tangent space M_x, which is called the *K-hypersurface* at x.

DEFINITION 26.4. A *direction of the principal curvature* at x is such that the radius of the K-hypersurface at x has an extremal length in that direction.

Thus a direction of the principal curvature is obtained by

$$\dot\partial_k\{g_{ij}y^iy^j - \lambda(K_{ij}y^iy^j - 1)\} = 0$$

for the Lagrange multiplier λ, that is,

$$\{2g_{ik} - \lambda(K_{ik} + K_{ki} + y^jK_{ij\cdot k})\}y^i = 0.$$

Contracting this by y^k, we get $\lambda = \mu^{-1}$ by (26.11), so that the above is written

(26.12) $$(K_{ik} + K_{ki} + y^jK_{ij\cdot k})y^i = 2\mu y_k.$$

As a consequence we are led to the following definition and proposition:

DEFINITION 26.5. A Finsler space is called an *Einstein-Finsler space*, if it is homogeneous at every point with respect to the mean curvature, namely, any direction

y^i is a direction of the principal curvature.

PROPOSITION 26.3. *A Finsler space is Einstein-Finsler, if and only if there exists a Finsler scalar field* μ *which satisfies (26.12) for any* y^i.

It is evident that if an Einstein-Finsler space in the above sense is Riemannian, the space is Einstein in the Riemannian sense.

We first write (26.12) in terms of $C\Gamma$. It follows from (18.2) that

$$K_{hj} = R_{hj} - C_h{}^s{}_r R^r{}_{js}, \qquad K_{0j} = R_{0j}, \qquad K_{h0} = R_{h0} - C_h{}^s{}_r R^r{}_{0s}.$$

and from (18.2) and (18.8)

$$y^m y^i K_{mi\cdot k} = - P^r{}_{rk|0} = - C_{k|0|0}.$$

Therefore (26.12) is written in the form

$$R_{0j} + R_{j0} - C_j{}^s{}_r R^r{}_{0s} - C_{j|0|0} = 2\mu y_j. \tag{26.13}$$

Further, from (22.7) and (22.8) we have

$$R_{j0} = R_{0j} + 2(C_j{}^s{}_r R^r{}_{0s} - C^r R_{r0j}),$$

so that (26.13) may be written as

$$2R_{0j} + C_j{}^s{}_r R^r{}_{0s} - 2C^r R_{r0j} - C_{j|0|0} = 2\mu y_j. \tag{26.14}$$

Contraction by y^j yields $\mu = L^{-2} R_{00}$.

Assume that an Einstein-Finsler space is of scalar curvature K. Then, from (26.2) and (26.4), the equation (26.14) becomes

$$(n - 1)L^2 K|_j + 2\{(n - 1)K - \mu\}y_j = 0,$$

from which we easily get $\mu = (n - 1)K$ and $K|_j = 0$. Therefore Proposition 26.1 leads to

PROPOSITION 26.4. *An* n (≥ 3)*-dimensional Finsler space of scalar curvature* K *is Einstein-Finsler, if and only if it is of constant curvature. In this case* $\mu = (n - 1)K$.

Next, we are concerned with the conservation law in an Einstein-Finsler space of constant curvature K. We start from the Bianchi identity (17.12), which, from (26.7) and (17.25), is now written as

$$\mathfrak{S}_{(ijk)}\{KS_{mhij|0}y_k + R_{mhij|k}\} = 0.$$

Contracting this by $g^{mk}g^{hj}$, we get

$$- KS_{|0}y_i + 2R^r{}_{i|r} - R_{|i} = 0,$$

where $S = S_h{}^r{}_{jr}g^{hj}$. This is written in the form

$$(- KSy_iy^r + 2R^r{}_i - R\delta^r_j)_{|r} = 0.$$

Consequently we have

THEOREM 26.4. *In a Finsler space of constant curvature K, the tensor field*

$$G^j{}_i = R^j{}_i - \frac{1}{2}(R\delta^j_i + KSy^jy_i)$$

satisfies the conservation law $G^j{}_{i|j} = 0$*, where* $R_{ij} = R_i{}^r{}_{jr}$*,* $R = R^r{}_r$ *and* $S = S_h{}^r{}_{jr}g^{hj}$ *in terms of* $C\Gamma$.

REMARK 26.4. In Riemannian case, the above $G^j{}_i$ is nothing but the Einstein tensor, so that Rund's generalization may be natural, although the restriction to Finsler spaces of constant curvature is a weak point.

In Finsler case, we have three curvature tensors of h-, hv- and v-type, and it is natural that an unknown Einstein tensor will be defined mainly in terms of the h-curvature tensor $R_i{}^h{}_{jk}$. As to the above $G^j{}_i$, it is just so, but the author thinks it a little strange that the v-curvature tensor $S_i{}^h{}_{jk}$, but not the hv-curvature tensor $P_i{}^h{}_{jk}$, appears in an additional term.

In Eguchi and Ichijyo [1] the equation

$$K_{ij}y^j = -(n-1)g_{ij}y^j$$

is given as a necessary condition for a Finsler space M to be such that the holonomy group of a generalized normal projective connection in a projective vector bundle over M leaves an F-hypersurface invariant. They write that a Finsler space satisfying the above equation may be called a *Finsler-Einstein space*. In Riemannian geometry it is well-known that there are close relations between Einstein spaces and special properties of holonomy groups.

Recently Shibata [2] showed a theorem: In a Finsler space of scalar curvature K, if the scalar curvature K is h-covariant constant ($K_{|i} = 0$), the tensor field

$$*G^j{}_i = R^j{}_i - \frac{1}{2}(R\delta^j_i + SB_iy^j) + B_rS^r{}_iy^j$$

satisfies the conservation law $*G^j{}_i|_j = 0$, where $B_i = (KL^3)_{\cdot i}/3L$. We also find the same result in Ishikawa's paper [2]. It is easy to show that $*G^j{}_i$ reduces to Rund's $G^j{}_i$ in case of constant curvature. But the matter has been settled by the following theorem due to Matsumoto and Tamássy [1]:

THEOREM 26.5. *Assume that a Finsler space M with a fundamental function L is of non-zero scalar curvature K. If an (r)p-homogeneous scalar field S on M is h-covariant constant, then S is necessarily equal to cL^r with a constant c.*

COROLLARY 26.1. *A Finsler space of scalar curvature K is of constant curvature, if and only if K is h-covariant constant.*

In Theorem 26.5 the assumption $S_{|i} = 0$ does not depend on the choice of the connections BΓ, CΓ and RΓ, as it is clear from (9.19) and $S_{|i} = \delta_i S$. We shall prove Theorem 26.5. The Ricci identity (10.8') and (26.4) lead to

$$S_{|i|j} - S_{|j|i} = - S_{\cdot k}(h^k{}_j K_i - h^k{}_i K_j) = 0.$$

From the homogeneity assumption on S, we have $S_{\cdot k}h^k{}_j = S_{\cdot j} - rSl_j/L$, so that the above gives $S_{\cdot j} - rSl_j/L = sK_j$ with some scalar s. Contracting this by y^j and paying attention to (26.5), we get $s = 0$, so that $S_{\cdot j} = rSl_j/L$, that is, $(\dot\partial_j S)/S = r(\dot\partial_j L)/L$. Thus integration yields $S = L^r f(x)$ with some function $f(x)$. Finally $S_{|i} = 0 = L^r f_{|i}$ shows f = constant.

REMARK 26.5. It is noted that Theorem 26.5 holds even for two-dimensional case, and further any two-dimensional Finsler space is of scalar curvature.

§ 27. Intrinsic fields of orthonormal frame

To the theory of two-dimensional Finsler spaces makes Berwald a great contribution by his papers [3, 4, 5, 13, 14]. His theory is developed based on the intrinsic field of orthonormal frame which consists of the normalized supporting element l^i and the unit vector orthogonal to l^i.

Following Berwald's idea on two-dimensional case, Moór [11] introduces in a three-dimensional Finsler space the intrinsic field of orthonormal frame which consists of l^i, the normalized torsion vector C^i/C (cf. Theorem 24.2) and the unit vector orthogonal to them, and develops a theory of three-dimensional Finsler spaces. By applying Moór's frame, Matsumoto [18, 20] shows various interesting results on three-dimensional Finsler spaces.

In the present section we give a theory of intrinsic orthonormal frame fields on n-dimensional Finsler spaces as a generalization of Berwald's and Moór's ideas.

Now we consider an n-dimensional Finsler space M with a fundamental function L(x, y). Let $\delta^{ij\cdots k}_{pq\cdots r}$ be generalized Kronecker deltas and put

$$\gamma_{ij\cdots k} = \delta^{12\cdots n}_{ij\cdots k}, \qquad \gamma^{ij\cdots k} = \delta^{ij\cdots k}_{12\cdots n}.$$

Then the so-called ε-tensors are defined by

$$\varepsilon_{ij\cdots k} = \sqrt{|g|}\,\gamma_{ij\cdots k}, \qquad \varepsilon^{ij\cdots k} = (\sqrt{|g|})^{-1}\gamma^{ij\cdots k},$$

where $|g|$ is the determinant consisting of the components of the fundamental tensor $g = (g_{ij})$. If we have to consider the case where $|g|$ is negative, $-|g|$ may be used, instead of $|g|$.

If we have n - 1 linearly independent vectors $e^i_{\alpha)}$, $\alpha = 1,\cdots, n-1$, in M, then the vector

$$(27.1) \qquad e^k_{n)} = \varepsilon^{hi\cdots jk} e_{1)h} e_{2)i} \cdots e_{n-1)j}$$

is orthogonal to $e^i_{\alpha)}$, $\alpha = 1,\cdots, n-1$. Further, if these $e^i_{\alpha)}$ have the unit length and orthogonal to each other, then $e^i_{\alpha)}$, $\alpha = 1,\cdots, n$, as thus obtained, constitute an orthonormal frame.

Now, from the fundamental function L(x, y) we introduce Finsler tensor fields of $(0, 2\alpha - 1)$-type respectively by

$$(27.2) \qquad L_{i_1 i_2 \cdots i_{2\alpha-1}} = \frac{1}{2^\alpha} \dot\partial_{i_1} \dot\partial_{i_2} \cdots \dot\partial_{i_{2\alpha-1}} L^2 .$$

It is noted that $L_{i_1 i_2 \cdots i_{2\alpha-1}}$ are completely symmetric and, in particular, $L_i = Ll_i = y_i$ and $L_{ijk} = C_{ijk}$. Then we get a sequence of covariant vectors

$$(27.3) \qquad L_{1)i} = L_i, \qquad L_{\alpha)i} = L_{ij_1 j_2 \cdots j_{2\alpha-3} j_{2\alpha-2}} g^{j_1 j_2} \cdots g^{j_{2\alpha-3} j_{2\alpha-2}} .$$

The second vector $L_{2)i}$ is obviously equal to $C_i = C_{ijk} g^{jk}$.

<u>DEFINITION 27.1</u>. If n - 1 covariant vectors $L_{\alpha)i}$, $\alpha = 1,\cdots, n-1$, as above given, are linearly independent, the Finsler space is called *strongly non-Riemannian*.

In the following, we denote by $L_{n)i}$ the one defined by (27.3), if $L_{\alpha)i}$, $\alpha = 1,\cdots, n$, are linearly independent. While, if $L_{n)i}$ defined by (27.3) is a linear combination of $L_{\alpha)i}$, $\alpha = 1,\cdots, n-1$, then we denote by $L_{n)i}$ the covector whose components are constructed from $L_{\alpha)i}$, $\alpha = 1,\cdots, n-1$, by (27.1). In either case we get n covectors $L_{\alpha)i}$, $\alpha = 1,\cdots, n$, on every coordinate neighborhood of a strongly non-Riemannian Finsler space.

<u>REMARK 27.1</u>. According to Definition 27.1, any two-dimensional Finsler space is strongly non-Riemannian, because $L_{1)i} = y_i$ does not vanish. Next, any Finslar space of three dimensions with non-zero C_i is strongly non-Riemannian (cf. Theorem 24.2). It is known that there exist non-Riemannian Finsler spaces of general dimensions which is not strongly non-Riemannian. See Matsumoto [26].

We show how to construct an orthonormal frame from the above linearly independent n covectors $L_{\alpha)i}$. First put

$$e_{1)}^{i} = \frac{1}{L} L_{1)}^{i} = 1^{i}.$$

Here and in the following, we shall freely use raising and lowering of indices; $L_{1)}^{i} = g^{ij}L_{1)j}$. If we put

$$N_{1)ij} = g_{ij} - e_{1)i}e_{1)j},$$

then $N_{1)ij}$ is nothing but the angular metric tensor h_{ij} and the matrix $N_{1)} = (N_{1)ij})$ is of rank n - 1 from Proposition 16.2. Because of $N_{1)}{}^{i}{}_{j}e_{1)}{}^{j} = 0$, the vector $e_{1)}$ is an eigenvector, corresponding to an eigenvalue 0 of $N_{1)}$. It is obvious that any vector X orthogonal to $e_{1)}$ satisfies $N_{1)}{}^{i}{}_{j}X^{j} = X^{i}$, so that other eigenvalues of $N_{1)}$ are all equal to 1. In particular, we have $N_{1)}{}^{i}{}_{j}L_{2)}{}^{j} = L_{2)}{}^{i}$ from $C_{i}1^{i} = 0$. Thus the second vector $e_{2)}$ is introduced by

$$e_{2)}{}^{i} = \frac{1}{L_2} L_{2)}{}^{i},$$

where L_2 is the length of $L_{2)}{}^{i}$ relative to y^{i}.

Next, we put

$$(27.4) \qquad N_{2)ij} = N_{1)ij} - e_{2)i}e_{2)j}, \qquad E_{3)}{}^{i} = N_{2)}{}^{i}{}_{j}L_{3)}{}^{j}.$$

It is easily verified that the matrix $N_{2)} = (N_{2)ij})$ has the rank n - 2, $e_{1)}$ and $e_{2)}$ are eigenvectors, corresponding to the eigenvalue 0 of $N_{2)}$, and other eigenvalues are all equal to 1. Thus $E_{3)}$ is orthogonal to $e_{1)}$ and $e_{2)}$, and the third vector is defined by

$$e_{3)}{}^{i} = \frac{1}{E_3} E_{3)}{}^{i},$$

where E_3 is the length of $E_{3)}{}^{i}$ relative to y^{i}.

The orthonormal frame $\{e_{\alpha)}\}$, $\alpha = 1, \cdots, n$, is now inductively constructed by repetition of the above process. That is, let $N_{r)}{}^{i}{}_{j}$ be such that $N_{r)ij}$ are symmetric and

(1) the rank of $N_{r)} = (N_{r)ij})$ is equal to n - r,

(2) the orthonormal r vectors $e_{1)}, \cdots, e_{r)}$ span the eigenspace, corresponding to an eigenvalue 0 of $N_{r)}$,

(3) other eigenvalues of $N_{r)}$ are all equal to 1.

Then we put

$$(27.5) \qquad E_{r+1)}{}^{i} = N_{r)}{}^{i}{}_{j}L_{r+1)}{}^{j}, \qquad e_{r+1)}{}^{i} = \frac{1}{E_{r+1}} E_{r+1)}{}^{i},$$

where E_{r+1} is the length of $E_{r+1)}{}^{i}$ relative to y^{i}, and define

$$(27.6) \qquad N_{r+1)ij} = N_{r)ij} - e_{r+1)i}e_{r+1)j}.$$

DEFINITION 27.2. The orthonormal frame $\{e_{\alpha)}\}$, $\alpha = 1, \cdots, n$, as above defined in every coordinate neighborhood of a strongly non-Riemannian Finsler space is called the *Miron frame*.

REMARK 27.2. The idea to construct the Miron frame was communicated from Miron to the author in 1974. See Matsumoto and Miron [1]. While the tensors $L_{i_1 i_2 \cdots i_{2\alpha-1}}$ given by (27.2) are $(3 - 2\alpha)$p-homogeneous, $e_{\alpha)}{}^i$ are (0)p-homogeneous, because these are obtained from $E_{\alpha)}{}^i$ by dividing by their lengths.

It seems that an essential point, constructing the Miron frame, is the contraction by g^{ij} such as in (27.3). On the other hand, there will be some other ways to introduce such tensors $L_{i_1 i_2 \cdots i_{2\alpha-1}}$ than (27.2). For instance, we may consider

$$\dot{\partial}_{i_1} \cdots \dot{\partial}_{i_{2\alpha-1}} L^{2\alpha-1}.$$

These have a merit to be (0)p-homogeneous. Anyhow it will be desirable that such quantities are completely symmetric, because $L_{\alpha)i}$ are uniquely defined by the way such as (27.3), independently on the choice of pair (h, k) of indices with respect to which the contraction g^{hk} is done.

We consider the Miron frame $\{e_{\alpha)}\}$. If a tensor $T^i{}_j$ of (1, 1)-type, for instance, is given, we define scalars

$$T_{\alpha\beta} = T^i{}_j e_{\alpha)i} e_{\beta)}{}^j, \tag{27.7}$$

then $T^i{}_j$ are written in the form

$$T^i{}_j = T_{\alpha\beta} e_{\alpha)}{}^i e_{\beta)j}, \tag{27.7'}$$

where the summation convention is also applied to Greek indices. These $T_{\alpha\beta}$ are called the *scalar components* of $T^i{}_j$ with respect to the Miron frame.

DEFINITION 27.3. Let $H_{\alpha)\beta\gamma}$ (resp. $V_{\alpha)\beta\gamma}/L$) be scalar components of the h- (resp. v-)covariant derivatives $e_{\alpha)}{}^i{}_{|j}$ (resp. $e_{\alpha)}{}^i|_j$) of the vectors $e_{\alpha)}$ belonging to the Miron frame $\{e_{\alpha)}\}$. $H_{\alpha)\beta\gamma}$ (resp. $V_{\alpha)\beta\gamma}$) are called *h-* (resp. *v-*)*connection scalars*.

Therefore $H_{\alpha)\beta\gamma}$ and $V_{\alpha)\beta\gamma}$ are (0)p-homogeneous, provided that the Finsler connection under consideration is p-homogeneous, and we have

$$\begin{aligned} e_{\alpha)}{}^i{}_{|j} &= H_{\alpha)\beta\gamma} e_{\beta)}{}^i e_{\gamma)j}, \\ L e_{\alpha)}{}^i|_j &= V_{\alpha)\beta\gamma} e_{\beta)}{}^i e_{\gamma)j}. \end{aligned} \tag{27.8}$$

These equations are called the *movement equations* of the Miron frame. The orthogonality of this frame field yields

$$(27.9)\qquad H_{\alpha)\beta\gamma} = -H_{\beta)\alpha\gamma}, \qquad V_{\alpha)\beta\gamma} = -V_{\beta)\alpha\gamma}.$$

In the following we are concerned with the Cartan connection $C\Gamma$ only. The first vector $e_{1)}{}^{i}$ of the Miron frame is equal to the normalized supporting element l^i, so that (17.30) gives

$$(27.10)\qquad H_{1)\beta\gamma} = 0, \qquad V_{1)\beta\gamma} = \delta^2_\beta\delta^2_\gamma + \cdots + \delta^n_\beta\delta^n_\gamma = \delta_{\beta\gamma} - \delta^1_\beta\delta^1_\gamma.$$

Let $C_{\alpha\beta\gamma}/L$ be scalar components of the (h)hv-torsion tensor $C_j{}^i{}_k$:

$$(27.11)\qquad LC_j{}^i{}_k = C_{\alpha\beta\gamma}e_{\alpha)j}e_{\beta)}{}^{i}e_{\gamma)k}.$$

Contraction by $e_{1)}{}^{j} = y^j/L$ gives $0 = C_{1\beta\gamma}e_{\beta)}{}^{i}e_{\gamma)k}$ and contraction by $g^{jk} = e_{\mu)}{}^{j}e_{\mu)}{}^{k}$ does $LC^i = C_{\mu\beta\mu}e_{\beta)}{}^{i}$. For $n \geq 3$ we have $e_{2)}{}^{i} = C^i/C$, where C is the length of C^i, $C^2 = g_{ij}C^iC^j$, so that we have

<u>PROPOSITION 27.1</u>. *With respect to the Miron frame the scalar components $C_{\alpha\beta\gamma}$ of LC_{ijk} are such that*

(1) $C_{\alpha\beta\gamma}$ *are completely symmetric,* *(2)* $C_{1\beta\gamma} = 0$,

(3) $C_{2\mu\mu} = LC$, $C_{3\mu\mu} = \cdots = C_{n\mu\mu} = 0$ *for* $n \geq 3$, *where* C *is the length of* C^i.

We consider covariant derivatives of (27.7'). It is easily shown that $T^i{}_{j|k}$ is written

$$T^i{}_{j|k} = (\delta_k T_{\alpha\beta})e_{\alpha)}{}^{i}e_{\beta)j} + T_{\alpha\beta}e_{\alpha)}{}^{i}{}_{|k}e_{\beta)j} + T_{\alpha\beta}e_{\alpha)}{}^{i}e_{\beta)j|k}.$$

Therefore, let $T_{\alpha\beta,\gamma}$ be scalar components of $T^i{}_{j|k}$:

$$(27.12)\qquad T^i{}_{j|k} = T_{\alpha\beta,\gamma}e_{\alpha)}{}^{i}e_{\beta)j}e_{\gamma)k}.$$

Then we obtain

$$(27.12')\qquad T_{\alpha\beta,\gamma} = (\delta_k T_{\alpha\beta})e_{\gamma)}{}^{k} + T_{\mu\beta}H_{\mu)\alpha\gamma} + T_{\alpha\mu}H_{\mu)\beta\gamma}.$$

Similarly, if we put

$$(27.13)\qquad LT^i{}_j\big|_k = T_{\alpha\beta;\gamma}e_{\alpha)}{}^{i}e_{\beta)j}e_{\gamma)k},$$

then scalar components $T_{\alpha\beta;\gamma}$ of $LT^i{}_j\big|_k$ are given by

$$(27.13')\qquad T_{\alpha\beta;\gamma} = L(\dot\partial_k T_{\alpha\beta})e_{\gamma)}{}^{k} + T_{\mu\beta}V_{\mu)\alpha\gamma} + T_{\alpha\mu}V_{\mu)\beta\gamma}.$$

The scalar components $T_{\alpha\beta,\gamma}$ (resp. $T_{\alpha\beta;\gamma}$) are called *h-* (resp. *v*)*-scalar deriv-*

atives of $T_{\alpha\beta}$.

We are concerned with the Ricci identities. From (27.12) the Ricci identity (10.8') is written as

$$T^h{}_{i|j|k} - T^h{}_{i|k|j} = (T_{\alpha\beta,\gamma,\delta} - T_{\alpha\beta,\delta,\gamma})e_{\alpha)}{}^h e_{\beta)i} e_{\gamma)j} e_{\delta)k}$$

$$= T_{\mu\beta} e_{\mu)}{}^r e_{\beta)i} R_r{}^h{}_{jk} - T_{\alpha\mu} e_{\alpha)}{}^h e_{\mu)r} R_i{}^r{}_{jk} - \frac{1}{L} T_{\alpha\beta;\mu} e_{\alpha)}{}^h e_{\beta)i} e_{\mu)r} R^r{}_{jk}.$$

Introduce the scalar components

$$R_{\alpha\beta\gamma\delta} = R_h{}^i{}_{jk} e_{\alpha)}{}^h e_{\beta)i} e_{\gamma)}{}^j e_{\delta)}{}^k$$

of the h-curvature tensor $R_h{}^i{}_{jk}$. Then scalar components $LR_{\beta\gamma\delta}$ of the (v)h-torsion tensor $R^i{}_{jk}$ are given by

$$LR_{1\beta\gamma\delta} = R^i{}_{jk} e_{\beta)i} e_{\gamma)}{}^j e_{\delta)}{}^k.$$

Consequently we have one of Ricci identities of scalar form

$$T_{\alpha\beta,\gamma,\delta} - T_{\alpha\beta,\delta,\gamma} = T_{\mu\beta} R_{\mu\alpha\gamma\delta} + T_{\alpha\mu} R_{\mu\beta\gamma\delta} - T_{\alpha\beta;\mu} R_{1\mu\gamma\delta}. \tag{27.14}$$

Secondly, the Ricci identity (10.9') is similarly written in the form

$$T_{\alpha\beta,\gamma;\delta} - T_{\alpha\beta;\delta,\gamma} = T_{\mu\beta} P_{\mu\alpha\gamma\delta} + T_{\alpha\mu} P_{\mu\beta\gamma\delta} - T_{\alpha\beta,\mu} C_{\mu\gamma\delta} - T_{\alpha\beta;\mu} P_{1\mu\gamma\delta}, \tag{27.15}$$

where $P_{\alpha\beta\gamma\delta}/L$ are scalar components of the hv-curvature tensor $P_h{}^i{}_{jk}$:

$$P_{\alpha\beta\gamma\delta} = LP_h{}^i{}_{jk} e_{\alpha)}{}^h e_{\beta)i} e_{\gamma)}{}^j e_{\delta)}{}^k.$$

Thirdly we consider the Ricci identity (10.10'). From (27.13) and (17.28) we see

$$T_{\alpha\beta;\gamma;\delta} e_{\alpha)}{}^h e_{\beta)i} e_{\gamma)j} e_{\delta)k} = L(LT^h{}_i|_j)|_k = L(e_{1)k} T^h{}_i|_j + LT^h{}_i|_j|_k),$$

which implies

$$(T_{\alpha\beta;\gamma;\delta} - T_{\alpha\beta;\delta;\gamma}) e_{\alpha)}{}^h e_{\beta)i} e_{\gamma)j} e_{\delta)k}$$

$$= e_{1)k} T_{\alpha\beta;\gamma} e_{\alpha)}{}^h e_{\beta)i} e_{\gamma)j} - e_{1)j} T_{\alpha\beta;\delta} e_{\alpha)}{}^h e_{\beta)i} e_{\delta)k} + L^2 (T^r{}_i S_r{}^h{}_{jk} - T^h{}_r S_i{}^r{}_{jk}).$$

Consequently the scalar form of (10.10') is given by

$$T_{\alpha\beta;\gamma;\delta} - T_{\alpha\beta;\delta;\gamma} = T_{\mu\beta} S_{\mu\alpha\gamma\delta} + T_{\alpha\mu} S_{\mu\beta\gamma\delta} + T_{\alpha\beta;\gamma}\delta^1_\delta - T_{\alpha\beta;\delta}\delta^1_\gamma, \tag{27.16}$$

where $S_{\alpha\beta\gamma\delta}/L^2$ are scalar components of the v-curvature tensor $S_h{}^i{}_{jk}$:

$$S_{\alpha\beta\gamma\delta} = L^2 S_h{}^i{}_{jk} e_{\alpha)}{}^h e_{\beta)i} e_{\gamma)}{}^j e_{\delta)}{}^k.$$

We apply these Ricci identities of scalar form to the vectors $e_{\alpha)}$. Because scalar components of $e_{\alpha)}$ with respect to the Miron frame are equal to $\delta_{\alpha\beta}$. In the following, however, we denote them by $\delta_{\alpha)\beta}$ in order to avoid the confusion, while scalar components of the fundamental tensor g are $\delta_{\alpha\beta}$. Then (27.14) leads us to

$$\delta_{\alpha)\beta,\gamma,\delta} - \delta_{\alpha)\beta,\delta,\gamma} = \delta_{\alpha)\mu}R_{\mu\beta\gamma\delta} - \delta_{\alpha)\beta;\mu}R_{1\mu\gamma\delta}.$$

The equation (27.8) shows

$$(27.17) \qquad \delta_{\alpha)\beta,\gamma} = H_{\alpha)\beta\gamma}, \qquad \delta_{\alpha)\beta;\gamma} = V_{\alpha)\beta\gamma}.$$

Accordingly we obtain

$$(27.18) \qquad H_{\alpha)\beta\gamma,\delta} - H_{\alpha)\beta\delta,\gamma} = R_{\alpha\beta\gamma\delta} - V_{\alpha)\beta\mu}R_{1\mu\gamma\delta}.$$

REMARK 27.3. We should notice the difference between $\delta_{\alpha\beta,\gamma}$ and $\delta_{\alpha)\beta,\gamma}$ such that

$$\delta_{\alpha\beta,\gamma} = (\delta_k\delta_{\alpha\beta})e_{\gamma)}{}^k + \delta_{\mu\beta}H_{\mu)\alpha\gamma} + \delta_{\alpha\mu}H_{\mu)\beta\gamma} = H_{\beta)\alpha\gamma} + H_{\alpha)\beta\gamma} = 0,$$

$$\delta_{\alpha)\beta,\gamma} = (\delta_k\delta_{\alpha)\beta})e_{\gamma)}{}^k + \delta_{\alpha)\mu}H_{\mu)\beta\gamma} = H_{\alpha)\beta\gamma}.$$

Similarly the following formulas are derived from (27.15) and (27.16) respectively:

$$(27.19) \qquad H_{\alpha)\beta\gamma;\delta} - V_{\alpha)\beta\delta,\gamma} = P_{\alpha\beta\gamma\delta} - H_{\alpha)\beta\mu}C_{\mu\gamma\delta} - V_{\alpha)\beta\mu}P_{1\mu\gamma\delta},$$

$$(27.20) \qquad V_{\alpha)\beta\gamma;\delta} - V_{\alpha)\beta\delta;\gamma} = S_{\alpha\beta\gamma\delta} + V_{\alpha)\beta\gamma}\delta^1_\delta - V_{\alpha)\beta\delta}\delta^1_\gamma.$$

These equations (27.18), (27.19) and (27.20) are called the *fundamental equations* of the Miron frame.

REMARK 27.4. The Miron frame is a generalization of Berwald's frame in two-dimensional case and Moór's frame in three-dimensional case which will play various remarkable roles in the following two sections, but it is a future problem to discuss Finsler spaces, for instance, of four dimensions with reference to the Miron frame.

§ 28. Two-dimensional Finsler spaces

It is natural to expect that Finsler spaces of low dimension have some special properties, as it is so in Riemannian case. In the present section we consider two-dimensional Finsler spaces in detail.

The Miron frame $\{e_{1)}, e_{2)}\}$ of this case is called the *Berwald frame*. The first vector $e_{1)}{}^i$ is the normalized supporting element $l^i = y^i/L(x, y)$ and the second $e_{2)}{}^i$

$= m^i$ is the unit vector orthogonal to l^i relative to y^i, which is constructed by (27. 1), that is,

$$(28.1) \qquad (m^1, m^2) = \frac{1}{\sqrt{|g|}}(-l_2, l_1), \qquad (m_1, m_2) = \sqrt{|g|}\,(-l^2, l^1).$$

If C^i has non-zero length C, then $m^i = \pm C^i/C$.

From (27.9) and (27.10) it follows that the connection scalars $H_{\alpha)\beta\gamma}$ and $V_{\alpha)\beta\gamma}$ of two-dimensional case are such that

$$H_{\alpha)\beta\gamma} = 0, \qquad V_{\alpha)\beta 1} = 0, \qquad V_{\alpha)\beta 2} = \gamma_{\alpha\beta}, \text{ (cf. p.176)}$$

which implies

$$(28.2) \qquad l^i{}_{|j} = 0, \qquad L l^i|_j = m^i m_j, \qquad m^i{}_{|j} = 0, \qquad L m^i|_j = -l^i m_j.$$

From Proposition 27.1 it is seen that surviving scalar component of LC_{ijk} is C_{222} only. Therefore, putting $I = C_{222}$, we get

$$(28.3) \qquad LC_{ijk} = I m_i m_j m_k.$$

<u>DEFINITION 28.1</u>. The scalar field I appearing in (28.3) is called the *main scalar* of a two-dimensional Finsler space.

<u>REMARK 28.1</u>. The main scalar I plays various important roles in the theory of two-dimensional Finsler spaces due to Berwald. As it was mentioned at the beginning of the last section, Berwald published various interesting papers on this subject. Among them the posthumous paper [14] is written in the most elegant form. The main scalar I is defined in his paper [5].

From (28.3) we have $LC_i = Im_i$, so that $C_i = 0$ implies $C_{ijk} = 0$; so the space is Riemannian, as it is indicated by Berwald [17, p.769] (cf. Theorem 24.2).

The covariant components l_i of the first vector of Berwald frame are defined by (16.3), $l_i = \dot\partial_i L$. On the other hand, as to the covariant components m_i of the second vector, we consider differential equations

$$(28.4) \qquad \dot\partial_i \theta = \frac{1}{L} m_i.$$

From (28.2) and (28.3) it follows that

$$\dot\partial_j(\dot\partial_i\theta) = -\frac{1}{L^2}(l_i m_j + l_j m_i) + \frac{I}{L^2} m_i m_j,$$

which are symmetric in i, j. Therefore the integrability condition of (28.4) is satisfied, so that we have a solution θ which is determined up to an additional function of position only.

DEFINITION 28.2. The quantity θ which is defined by (28.4) is called the *Landsberg angle* of a two-dimensional Finsler space.

From $\dot{\partial}_i L = l_i$ and (28.4) it is observed that the Jacobian matrix $\partial(L, \theta)/\partial(y^1, y^2)$ is non-singular, so that (L, θ) may be regarded as a coordinate system on each tangent plane M_x with the origin removed. Thus a Finsler scalar field $S(x^1, x^2, y^1, y^2)$ is thought of as a function of variables x^1, x^2, L, θ. It is easy to show that

$$\text{(28.5)} \qquad \frac{\partial y^i}{\partial L} = l^i, \qquad \frac{\partial y^i}{\partial \theta} = Lm^i,$$

from which we have

$$\text{(28.6)} \qquad \frac{\partial S}{\partial L} = (\dot{\partial}_i S)l^i = \frac{1}{L} S_{;1}, \qquad \frac{\partial S}{\partial \theta} = L(\dot{\partial}_i S)m^i = S_{;2}$$

in terms of v-scalar derivatives $S_{;\alpha}$.

It follows from (27.12) and (27.13) that we have

$$\text{(28.7)} \qquad S_{|i} = S_{,1}l_i + S_{,2}m_i, \qquad LS|_i = S_{;1}l_i + S_{;2}m_i.$$

PROPOSITION 28.1.

(1) If a Finsler scalar field S is (r)p-homogeneous, then $S_{;1} = rS$.

(2) If S is (0)p-homogeneous, then $S_{;1} = 0$, $LS|_i = S_{;2}m_i$ *and* $S(x^1, x^2, L, \theta)$ *does not depend on L.*

In the following we mainly deal with (0)p-homogeneous scalar fields which are scalar components of Finsler tensor fields with respect to the Berwald frame. Therefore the above (2) is important.

REMARK 28.2. The Landsberg angle θ was introduced by Landsberg [1, 2, 3] in 1908 and fully used by Berwald later on. The coordinate system (L, θ) in a tangent plane M_x is regarded as a generalization of a polar coordinate system (r, ϕ) of a euclidean plane. In fact, putting

$$z^1 = r\cos\phi, \qquad z^2 = r\sin\phi,$$

we obtain

$$\frac{\partial\phi}{\partial z^1} = \frac{1}{r}\left(-\frac{z^2}{r}\right), \qquad \frac{\partial\phi}{\partial z^2} = \frac{1}{r}\left(\frac{z^1}{r}\right).$$

These are analogous to (28.4) by (28.1). It is noted that the coordinate system (L, θ) on M_x is not an induced one. Cf. Remark 17.1.

We are now concerned with the Cartan connection $C\Gamma$ and find scalar components of torsion and curvature tensors with respect to the Berwald frame. To do so, we first state

PROPOSITION 28.2. *Let T_{ij} be a tensor of (0,2)-type of a two-dimensional Finsler space and $T_{\alpha\beta}$ be scalar components of T_{ij} with respect to the Berwald frame:*

$$T_{ij} = T_{11}l_i l_j + T_{12}l_i m_j + T_{21}m_i l_j + T_{22}m_i m_j.$$

(1) *If T_{ij} is symmetric, we have $T_{12} = T_{21}$.*

(2) *If T_{ij} is skew-symmetric, we have $T_{11} = T_{22} = 0$, $T_{12} = - T_{21}$ and there exists a scalar T such that $T_{ij} = T(l_i m_j - l_j m_i)$.*

(3) *If T_{ij} satisfies T_{0j} (resp. T_{i0}) $= 0$, then $T_{11} = T_{12} = 0$ (resp. $T_{11} = T_{21} = 0$).*

(4) *If T_{ij} is skew-symmetric and $T_{0j} = 0$, then $T_{ij} = 0$.*

First we consider the h-curvature tensor R_{hijk}, which is skew-symmetric in h, i and j, k, so that (2) of Proposition 28.2 yields the simple form

$$R_{hijk} = R(l_h m_i - l_i m_h)(l_j m_k - l_k m_j), \tag{28.8}$$

where R is a scalar, called the *h-scalar curvature*. From (28.8) and (17.8) the (v)h-torsion tensor R_{ijk} is written

$$R_{ijk} = LRm_i(l_j m_k - l_k m_j). \tag{28.9}$$

It is noted here that $l_h m_i - l_i m_h$ is, of course, h-covariant constant, and further v-covariant constant from (28.2):

$$(l_h m_i - l_i m_h)_{|j} = 0, \qquad (l_h m_i - l_i m_h)|_j = 0. \tag{28.10}$$

Secondly, we deal with the hv-curvature tensor P_{hijk}. Its components satisfy (17.9) and (17.26). Therefore (2) and (3) of Proposition 28.2 lead to

$$LP_{hijk} = P(l_h m_i - l_i m_h)m_j m_k,$$

where P is a scalar. This P is expressed by the main scalar I as follows: From (28.3) and (28.7) we get

$$LC_{hij|k} = (I_{,1}l_k + I_{,2}m_k)m_h m_i m_j, \tag{28.11}$$

so that from (17.22) the (v)hv-torsion tensor P_{ijk} is written as

$$P_{ijk} = C_{ijk|0} = I_{,1}m_i m_j m_k. \tag{28.12}$$

Consequently (17.23) yields

$$LP_{hijk} = I_{,1}(l_h m_i - l_i m_h)m_j m_k, \tag{28.13}$$

which implies $P = I_{,1}$.

Thirdly, the v-curvature tensor S_{hijk} has a conspicuous property, namely, from (4) of Proposition 28.2 we have

PROPOSITION 28.3. *The v-curvature tensor S_{hijk} of $C\Gamma$ of any two-dimensional Finsler space vanishes identically.*

From the above special forms of torsion and curvature tensors, certain interesting relations between important tensors are derived. First, from (28.3), the equations (28.12) and (28.13) may be written in the form

$$P_{ijk} = PC_{ijk}, \qquad P = LI_{,1}/I, \tag{28.12'}$$

$$LP_{hijk} = P(l_h C_{ijk} - l_i C_{hjk}). \tag{28.13'}$$

Next we consider the tensor $g_{hijk} = g_{hj}g_{ik} - g_{hk}g_{ij}$. This is skew-symmetric in h, i and j, k, and also $g_{hijk}l^h m^i l^j m^k = 1$, so that Proposition 28.2 leads to

$$g_{hijk} = (l_h m_i - l_i m_h)(l_j m_k - l_k m_j).$$

Therefore (28.8) may be written in the distinctive form

$$R_{hijk} = R(g_{hj}g_{ik} - g_{hk}g_{ij}). \tag{28.8'}$$

Summarizing up the above results as for P_{hijk}, P_{ijk} and R_{hijk}, we have

PROPOSITION 28.4. *In any two-dimensional Finsler space,*

(1) The h-curvature tensor R_{hijk} is written in the forms (28.8) and (28.8').

(2) The hv-curvature tensor P_{hijk} is symmetric in j, k and written in the forms (28.13) and (28.13').

(3) The (v)hv-torsion tensor P_{ijk} is written in the forms (28.12) and (28.12').

REMARK 28.3. Proposition 28.3 is perhaps first shown by Laugwitz [10], while it seems that the special form (28.8') of R_{hijk} has been noticed by many authors, because we have a model in Riemannian geometry. The symmetry property of P_{hijk}, stated in (2) of Proposition 28.4, will be remarkable. The special form (28.13') of P_{hijk} will be first noticed by Matsumoto [15]. The following special form of the hv-curvature tensor G_{hijk} of $B\Gamma$ is given by Berwald [14]. Cf. Theorems 25.4 and 25.5, and Definition 22.1.

Next we consider the Bianchi identities. In two-dimensional case the five identities (17.10), (17.11), (17.12), (17.18) and (17.19) obviously reduce to trivial identities. In general (17.14) and (17.16) are solely consequence of (17.15) and

(17.17) respectively. The identity (17.13) is nothing but the equation (17.23). Therefore it is enough to treat of two identities (17.15) and (17.17) only.

Paying attention to (28.10) and Proposition 28.1, it follows from (28.8) that

$$LR_{hijk}|_m = R_{;2}(l_h m_i - l_i m_h)(l_j m_k - l_k m_j)m_m,$$

and from (28.13) that

$$LP_{hijk}|_m = (I_{,1,1}l_m + I_{,1,2}m_m)(l_h m_i - l_i m_h)m_j m_k.$$

Thus (17.15) is written in the simple scalar form

$$R_{;2} + RI + I_{,1,1} = 0. \tag{28.14}$$

Next we turn to consideration of (17.17). According to Proposition 28.3, it reduces to

$$\mathfrak{A}_{(ij)}\{P_{mhri}C_k{}^r{}_j - P_{mhkj}|_i\} = 0.$$

From (28.13) and Proposition 28.1 we obtain

$$L^2P_{hijk}|_m = (l_h m_i - l_i m_h)\{(I_{,1;2}m_m - I_{,1}l_m)m_j m_k - I_{,1}(l_j m_k + l_k m_j)m_m\}.$$

Accordingly (17.17) reduces to a trivial identity, as it is easily seen.

REMARK 28.4. The Bianchi identity (28.14) is given by Berwald [5, p.206], and it seems that the singleness of Bianchi identity in two-dimensional case is also shown there. See Berwald [14, p.92].

We are also interested in special forms of curvature tensors of BΓ and RΓ, although they have not such simple forms as those of CΓ.

From (18.2) we easily obtain the following forms of h- and hv-curvature tensors K_{hijk}, F_{hijk} of RΓ:

$$K_{hijk} = R(l_h m_i - l_i m_h - I m_h m_i)(l_j m_k - l_k m_j), \tag{28.15}$$

$$LF_{hijk} = \{I_{,1}(l_h m_j + l_j m_h)m_i - I_{,1}m_h m_j l_i - (II_{,1} - I_{,2})m_h m_j m_i\}m_k. \tag{28.16}$$

From (18.16) we similarly obtain the following forms of h- and hv-curvature tensors H_{hijk}, G_{hijk} of BΓ:

$$H_{hijk} = \{R(l_h m_i - l_i m_h) + R_{;2}m_h m_i\}(l_j m_k - l_k m_j), \tag{28.17}$$

$$LG_{hijk} = \{-2I_{,1}l_i + (I_{,2} + I_{,1;2})m_i\}m_h m_j m_k. \tag{28.18}$$

In the above, the Bianchi identity (28.14) was used to derive (28.17). Conversely, if we first find H_{hijk} from (18.22) and compare it with the expression of H_{hijk} obtained from (18.16), we are led to the identity (28.14).

We shall begin to consider some special Finsler spaces of two dimensions. First, from (28.12) and Theorem 25.3 we get

THEOREM 28.1. *A two-dimensional Finsler space is a Landsberg space, if and only if the main scalar I satisfies* $I_{,1} = (\delta_i I) l^i = 0$.

REMARK 28.5. Two-dimensional Landsberg spaces with a special metric is discussed by Hashiguchi, Hōjō and Matsumoto [1]. Recently F.Ikeda made detailed studies of two-dimensional Landsberg spaces.

Next we consider the condition $C_{hij|k} = 0$; from (28.11) it follows that this condition is equivalent to $I_{,1} = I_{,2} = 0$. Applying Ricci identity (27.14) to the main scalar I, we get

$$I_{,1,2} - I_{,2,1} = - I_{;2}R.$$

Therefore $C_{hij|k} = 0$ yields $I_{;2}R = 0$. Since $I_{;1} = 0$ identically, $I_{;2} = 0$ implies that I is a constant. Consequently, from Theorems 24.4 and 25.2 we have

THEOREM 28.2. *Assume that a two-dimensional Finsler space M is not locally Minkowski. Then M is a Berwald space, if and only if the main scalar I is constant.*

We shall find a weaker condition for I than the above. Differentiating (28.3) v-covariantly, we get

$$Ll_h C_{ijk} + L^2 C_{ijk}|_h = I_{;2} m_h m_i m_j m_k - I(l_i m_j m_k + l_j m_k m_i + l_k m_i m_j) m_h.$$

From (28.3) this may be written as

$$(28.19) \qquad LT_{ijkh} = I_{;2} m_i m_j m_k m_h,$$

where the tensor field T_{ijkh} is defined by

$$(28.20) \qquad T_{ijkh} = LC_{ijk}|_h + l_i C_{jkh} + l_j C_{ikh} + l_k C_{ijh} + l_h C_{ijk}.$$

It is evident that this tensor T_{ijkh} is also defined by (28.20) in case of general dimension and is completely symmetric. From (28.19) we get the following interesting theorem:

THEOREM 28.3. *The tensor field T_{ijkh} of a two-dimensional Finsler space M vanishes, if and only if the main scalar I of M is a function of position only.*

REMARK 28.6. We can not too much emphasize the importance of tensor T_{ijkh}, as it will be understood in the following (§§ 29, 30, 31). This tensor was first introduced in 1972 by Matsumoto [16, p.510], when he considered a theory of transformations of Finsler spaces (cf. Theorem 37.4). Theorem 28.3 was communicated by Hashiguchi to the author in 1972. Independently and almost simultaneously H.Kawaguchi [1] noticed the importance of T_{ijkh} from a standpoint of Landsberg spaces. He adopts the letter "T" to denote this tensor; so we call this tensor *T-tensor*. Recently Hashiguchi [5] proves that $T_{ijkh} = 0$ is necessary and sufficient for a Landsberg space of general dimension to be such that it is still Landsberg under any conformal change of metric.

The remainder of the present section is devoted to showing an interesting application of the theory of two-dimensional Finsler spaces, namely, to finding all the two-dimensional Berwald space which are not locally Minkowski; it was performed by Berwald [5, 14].

We first introduce the quantity

$$B^p = \frac{L^2}{2|g|^{1/p}}, \qquad |g| = \det(g_{ij}), \tag{28.21}$$

and put

$$B^p_i = \frac{|g|^{1/p}}{L}\dot{\partial}_i B^p, \qquad B^p_{ij} = |g|^{1/p}\dot{\partial}_i\dot{\partial}_j B^p, \qquad B^p_{ijk} = L|g|^{1/p}\dot{\partial}_i\dot{\partial}_j\dot{\partial}_k B^p.$$

What we are going to examine is the condition $B^p_{ijk} = 0$ for B^p to be a quadratic form of y^i.

From (16.3) and (24.1) we get

$$B^p_i = l_i - \frac{1}{p}A_i, \qquad A_i = LC_i. \tag{28.22}$$

Then, from (28.22) and (17.30) we can derive

$$B^p_{ij} = g_{ij} - \frac{2}{p}(l_iA_j + l_jA_i) + \frac{2}{p^2}A_iA_j - \frac{1}{p}D_{ij}, \qquad D_{ij} = L^2(C_i|_j + C_{ijr}C^r). \tag{28.23}$$

Finally, paying attention to

$$T_{ij} = T_{ijkh}g^{kh} = LC_i|_j + l_iC_j + l_jC_i,$$

we obtain

$$B^p_{ijk} = B^{pT}_{ijk} + B^{pA}_{ijk}. \tag{28.24}$$

Here two terms of the right-hand side are given by

$$B^{pT}_{ijk} = -\frac{L}{p}(LT_{ij}|_k + A^rT_{rijk}) + \frac{L}{p}\mathfrak{S}_{(ijk)}\{\frac{2}{p}A_iT_{jk} - l_iT_{jk} - A_i{}^r{}_jT_{rk}\},$$

$$B^{pA}_{ijk} = 2A_{ijk} - \frac{1}{p}(A_kh_{ij} - A_i{}^r{}_jA_{rk} + \frac{4}{p^2}A_iA_jA_k)$$

$$+ \frac{1}{p} \mathfrak{S}_{(ijk)} \{\frac{2}{p} A_{ij}A_k - A_i h_{jk} - A_i{}^r{}_j A_{rk}\},$$

where $A_{ijk} = LC_{ijk}$ and $A_{ij} = A_{ijr}A^r$.

If $T_{ijkh} = 0$, then $B^{pT}_{ijk} = 0$. As a consequence, if the main scalar I is a function of position alone, we have $B^p_{ijk} = B^{pA}_{ijk}$.

From (28.3) and $h_{ij} = m_i m_j$, the expression of B^{pA}_{ijk} in two-dimensional case is easily obtained:

$$(28.25) \qquad p^3 \, B^{pA}_{ijk} = 2(p - 2)\{p^2 - (p - 1)I^2\}Im_i m_j m_k.$$

Therefore we obtain

PROPOSITION 28.5. *If the T-tensor T_{ijkh} of a two-dimensional Finsler space vanishes identically, then $B^2 = L^2/2\sqrt{|g|}$ is a quadratic form in y^i.*

REMARK 28.6. It is noted that if p = 2, the factor $p^2 - (p - 1)I^2$ of the right-hand side of (28.25) becomes $4 - I^2$ and its algebraic sign will be important in the following discussion.

We now treat of a two-dimensional Finsler space with the constant main scalar I. Then B^2 is written in the quadratic form $B^2 = \{b_{ij}(x)y^i y^j\}/2$ in y^i. Since $L^2C_i|_j = - I(l_i m_j + l_j m_i)$, the equation (28.23) is rewritten as

$$b_{ij}(x) = l_i h_j + m_i k_j, \qquad h_j = \frac{1}{\sqrt{|g|}} (l_j - \frac{I}{2} m_j), \qquad k_j = \frac{1}{\sqrt{|g|}} (m_j - \frac{I}{2} l_j).$$

The determinant b consisting of b_{ij} is equal to $(l_1 m_2 - l_2 m_1)(h_1 k_2 - h_2 k_1)$. Here $4|g|(h_1 k_2 - h_2 k_1) = (4 - I^2)(l_1 m_2 - l_2 m_1)$ is easily seen, so that we get $4|g|b = (4 - I^2)(l_1 m_2 - l_2 m_1)$. It is obvious that $|g| = (l_1 m_2 - l_2 m_1)^2$, so that we have $b = (4 - I^2)/4$.

Because the constant b is the determinant of quadratic form $2B^2$, its algebraic sign dominates the behavior of B^2 as follows:

(i) $I^2 > 4$. In this case the form $2B^2$ is written as

$$2B^2 = \beta\gamma, \qquad \beta = p_i(x)y^i, \qquad \gamma = q_i(x)y^i,$$

where two linear forms β, γ are linearly independent, and

$$I^2 - 4 = (p_1 q_2 - p_2 q_1)^2.$$

(ii) $I^2 < 4$. In this case the form $2B^2$ is written as

$$2B^2 = \beta^2 + \gamma^2, \qquad \beta = p_i(x)y^i, \qquad \gamma = q_i(x)y^i,$$

where two linear forms β, γ are linearly independent, and

$$4 - I^2 = 4(p_1q_2 - p_2q_1)^2.$$

(iii) $I^2 = 4$. In this case we have

$$2B^2 = \beta^2, \qquad \beta = p_i(x)y^i.$$

In either case, it follows from (24.1) and (28.4) that

$$\dot{\partial}_i(\log\sqrt{|g|}) = \frac{I}{L} m_i = I\dot{\partial}_i\theta,$$

which implies $\sqrt{|g|} = J(x)\exp(I\theta)$, where $J(x)$ is an arbitrary function. Thus (28.21) yields

(28.26) $$L^2 = 2B^2J(x)\exp(I\theta).$$

Next, we examine the Landsberg angle θ. In the first two cases (i), (ii), the pair (β, γ) can be regarded as variables, instead of (y^1, y^2), and we obtain

$$\frac{\partial y^1}{\partial\beta} = \frac{q_2}{r}, \quad \frac{\partial y^1}{\partial\gamma} = -\frac{p_2}{r}, \quad \frac{\partial y^2}{\partial\beta} = -\frac{q_1}{r}, \quad \frac{\partial y^2}{\partial\gamma} = \frac{p_1}{r}. \qquad r = p_1q_2 - p_2q_1,$$

Therefore the equations (28.4) are written as

$$\frac{\partial\theta}{\partial\beta} = \frac{1}{Lr}(m_1q_2 - m_2q_1), \qquad \frac{\partial\theta}{\partial\gamma} = -\frac{1}{Lr}(m_1p_2 - m_2p_1).$$

According to (28.1) the above are rewritten in the form

(28.27) $$\frac{\partial\theta}{\partial\beta} = -\frac{\gamma}{2B^2r}, \qquad \frac{\partial\theta}{\partial\gamma} = \frac{\beta}{2B^2r}.$$

Now we consider the case (i). Paying attention to $2B^2 = \beta\gamma$ and $r^2 = I^2 - 4$, the integration of (28.27) yields $\theta = (I^2 - 4)^{-1/2}\log(\gamma/\beta)$, and (28.26) leads to

(28.28(i)) $$L^2 = \beta\gamma\left\{\frac{\gamma}{\beta}\right\}^{I/\sqrt{(I^2-4)}}.$$

In the above and the following, arbitrary functions of x^i which appear as integrating (28.27), together with $J(x)$, are supposed to be contained in β, γ.

In the case (ii) we similarly obtain

(28.28(ii)) $$L^2 = (\beta^2 + \gamma^2)\exp\left\{\frac{2I}{\sqrt{(4 - I^2)}}\operatorname{Arctan}\frac{\gamma}{\beta}\right\}.$$

The remaining case (iii) is in a somewhat special situation. Put $\gamma = p_2(x)y^2$, provided $p_1p_2 \neq 0$. Then γ is linearly independent of β, and (28.27) similarly holds. Consequently we obtain $\theta = \gamma/(r\beta)$, and finally

(28.28(iii)) $$L^2 = \beta^2\exp\left\{\frac{I\gamma}{\beta}\right\}.$$

Summarizing up all the above, we establish the theorem due to Berwald as follows:

THEOREM 28.4. *All the two-dimensional Finsler spaces with a constant main scalar I are classified, according as (i) $I^2 > 4$, (ii) $I^2 < 4$ or (iii) $I^2 = 4$, into the groups such that the fundamental function L(x, y) is given by (i), (ii) or (iii) of (28.28) respectively.*

REMARK 28.7. In his paper [5], Berwald further finds all the two-dimensional Finsler spaces with $I = I(x)$, namely, $T_{ijkh} = 0$. Moreover he treats of two-dimensional Finsler spaces with rectilinear extremals in his paper [14].

The quantity B^2 given by (28.21) is introduced by Berwald, and the general B^p is considered by Matsumoto [20] to find special three-dimensional Finsler spaces with $T_{ijkh} = 0$. See Matsumoto and Shimada [2].

§ 29. Three-dimensional Finsler spaces

The Miron frame of a three-dimensional Finsler space is called the *Moór frame*, because Moór [11] first introduced it to study such spaces. The first vector $e_{1)}{}^i$ of Moór frame $\{e_{1)}, e_{2)}, e_{3)}\}$ is the normalized supporting element l^i, the second $e_{2)}{}^i$ is the normalized torsion vector $m^i = C^i/C$ and the third $e_{3)}{}^i = n^i$ is constructed from them by the method mentioned in § 27. Throughout the present section, we suppose that the length C of C^i does not vanish, namely, the space is strongly non-Riemannian.

We first consider the connection scalars $H_{\alpha)\beta\gamma}$ and $V_{\alpha)\beta\gamma}$ of the Moór frame. These satisfy the identities (27.9) and (27.10). The equations (27.8) give

$$e_{2)}{}^i{}_{|j} = H_{2)3\gamma}e_{3)}{}^i e_{\gamma)j}, \qquad e_{3)}{}^i{}_{|j} = - H_{2)3\gamma}e_{2)}{}^i e_{\gamma)j}.$$

We now define a Finsler vector field

$$h_j = H_{2)3\gamma}e_{\gamma)j}. \tag{29.1}$$

Then, making use of skew-symmetric quantities $\gamma_{\alpha\beta\gamma}$ $(= \delta^{123}_{\alpha\beta\gamma})$, we get

$$e_{\alpha)}{}^i{}_{|j} = \gamma_{1\alpha\beta}e_{\beta)}{}^i h_j, \tag{29.2}$$

or explicitly

$$l^i{}_{|j} = 0, \qquad m^i{}_{|j} = n^i h_j, \qquad n^i{}_{|j} = - m^i h_j. \tag{29.2'}$$

On the other hand, from (27.9) and (27.10) it follows that $V_{1)2\gamma} = \delta^2_\gamma$, and $V_{1)3\gamma} = \delta^3_\gamma$, so that (27.8) yields

$$Le_{1)}{}^i|_j = e_{2)}{}^i e_{2)j} + e_{3)}{}^i e_{3)j}, \qquad Le_{2)}{}^i|_j = - e_{1)}{}^i e_{2)j} + V_{2)3\gamma}e_{3)}{}^i e_{\gamma)j},$$

$$Le_{3)}{}^{i}\big|_{j} = - e_{1)}{}^{i} e_{3)_j} - V_{2)3\gamma} e_{2)}{}^{i} e_{\gamma)j}.$$

Therefore, if we define a Finsler vector field

(29.3) $$v_j = V_{2)3\gamma} e_{\gamma)j} ,$$

we obtain

(29.4) $$Le_{\alpha)}{}^{i}\big|_{j} = \delta^1_\alpha \delta^i_j - e_{1)}{}^{i} e_{\alpha)j} + \gamma_{1\alpha\beta} e_{\beta)}{}^{i} v_j ,$$

or explicitly

(29.4') $$Ll^i\big|_j = \delta^i_j - l^i l_j = h^i{}_j ,$$
$$Lm^i\big|_j = - l^i m_j + n^i v_j , \qquad Ln^i\big|_j = - l^i n_j - m^i v_j .$$

DEFINITION 29.1. The Finsler vector field h_j (resp. v_j) given by (29.1) (resp. (29.3)) is called the *h-*(resp. *v-*) *connection vector*.

The scalars $H_{2)3\gamma}$ (resp. $V_{2)3\gamma}$) are considered as the scalar components h_γ (resp. v_γ) of the h (resp. v)-connection vector with respect to the Moór frame, and skew-symmetric matrices $(H_{\alpha)\beta\gamma})$ and $(V_{\alpha)\beta\gamma})$, γ being fixed, are respectively written

(29.5) $$(H_{\alpha)\beta\gamma}) = \begin{pmatrix} 0 & 0 & 0 \\ 0 & 0 & h_\gamma \\ 0 & -h_\gamma & 0 \end{pmatrix}, \qquad (V_{\alpha)\beta\gamma}) = \begin{pmatrix} 0 & \delta^2_\gamma & \delta^3_\gamma \\ -\delta^2_\gamma & 0 & v_\gamma \\ -\delta^3_\gamma & -v_\gamma & 0 \end{pmatrix}.$$

In three-dimensional case a similar fact to Proposition 28.1 holds also. Further it is seen from (29.4') that $Lm^i\big|_j l^j = 0 = (v_j l^j) n^i$, so that $v_1 = v_j l^j = 0$. Consequently

PROPOSITION 29.1.

(1) If a Finsler scalar field S of a three-dimensional Finsler space is (0)p-homogeneous, then we have $S_{;1} = 0$.

(2) The first scalar component v_1 *of v-connection vector* v_i *vanishes identially.*

We shall generalize the notion of Berwald's main scalar to three-dimensional case. From Proposition 27.1 we have $C_{222} + C_{233} = LC$, $C_{322} + C_{333} = 0$. Thus, putting

(29.6) $$H = C_{222}, \qquad I = C_{233}, \qquad J = C_{333} = - C_{223}, \qquad (H + I = LC),$$

we get

(29.6') $$LC_{ijk} = Hm_im_jm_k - J\,\mathfrak{S}_{(ijk)}\{m_im_jn_k\} + I\,\mathfrak{S}_{(ijk)}\{m_in_jn_k\} + Jn_in_jn_k.$$

<u>DEFINITION 29.2</u>. Three scalar fields H, I, J given by (29.6) are called the *main scalars* of a three-dimensional Finsler space.

<u>PROPOSITION 29.2</u>. *Let T_{ij} be a skew-symmetric Finsler tensor of a three-dimensional Finsler space.*

(1) If we put $*T^i = \varepsilon^{ijk}T_{jk}/2$, then we obtain $T_{jk} = \varepsilon_{jki}{*T^i}$.

(2) If T_{ij} satisfies $T_{i0} = 0$, there exists a scalar T such that $T_{ij} = T(m_in_j - m_jn_i)$.

We shall show (2) only. Because $\gamma_{\alpha\beta\gamma}$ are scalar components of ε_{ijk}, it follows from (1) that $T_{\alpha\beta} = \gamma_{\alpha\beta\mu}{*T_\mu}$, where $T_{\alpha\beta}$ and $*T_\mu$ are respective scalar components of T_{ij} and $*T^i$. The assumption $T_{i0} = 0$ is equivalent to $T_{\alpha 1} = 0$, so that surviving scalar components $T_{\alpha\beta}$ are T_{23}, T_{32} $(= -T_{23})$ only. Thus, putting $T = 2T_{23}$, the proof is completed.

We find scalar components of torsion and curvature tensors of $C\Gamma$. Because the v-curvature tensor S_{hijk} is skew-symmetric in h, i as well as j, k, and $S_{0ijk} = S_{hi0k} = 0$, Proposition 29.2 implies

(29.7) $$L^2S_{hijk} = S(m_hn_i - m_in_h)(m_jn_k - m_kn_j),$$

where S is a (0)p-homogeneous scalar.

Consider the tensor $h_{hijk} = h_{hj}h_{ik} - h_{hk}h_{ij}$. This satisfies the identities similar to those of S_{hijk}, and further $h_{hijk}m^hn^im^jn^k = 1$ holds because of $h_{ij} = m_im_j + n_in_j$. Therefore h_{hijk} is written as

$$h_{hijk} = (m_hn_i - m_in_h)(m_jn_k - m_kn_j).$$

Consequently we obtain

(29.7') $$L^2S_{hijk} = S(h_{hj}h_{ik} - h_{hk}h_{ij}).$$

From (27.11) and (17.20), the scalar components $S_{\alpha\beta\gamma\delta}$ of L^2S_{hijk} are also written as

(29.8) $$S_{\alpha\beta\gamma\delta} = C_{\alpha\delta\mu}C_{\mu\beta\gamma} - C_{\alpha\gamma\mu}C_{\mu\beta\delta}.$$

Then (29.7) shows $S = S_{2323}$. Therefore, from (29.8) and (29.6) it follows that

(29.7") $$S = 2J^2 + I^2 - HI.$$

The scalar S is called the *v-curvature*.

Summarizing up the above results on S_{hijk}, we have

THEOREM 29.1. *The v-curvature tensor S_{hijk} of any three-dimensional Finsler space is written in the form (29.7'). The v-curvature S is given by (29.7") in terms of main scalars H, I, J.*

REMARK 29.1. The conspicuous form (29.7') of S_{hijk} in three-dimensional case, a little similar to (28.8'), is shown by Matsumoto [15]. Another proof of (29.7') is given in § 31. The importance of the form (29.7') in case of general dimension is indicated by O.Varga [18], Kikuchi [2]. Cf. Theorem 31.6.

Next we consider the hv-curvature tensor P_{hijk}. Because P_{hijk} is skew-symmetric in h, i, from Proposition 29.2 it is written

(29.9) $$LP_{hijk} = \varepsilon_{hir}{}^{*}P^{r}{}_{jk}.$$

The scalar components $P_{\alpha\beta\gamma\delta}$ of LP_{hijk} are given by

(29.9') $$P_{\alpha\beta\gamma\delta} = \gamma_{\alpha\beta\mu}{}^{*}P_{\mu\gamma\delta},$$

where ${}^{*}P_{\mu\gamma\delta}$ are scalar components of ${}^{*}P^{r}{}_{jk}$.

From (29.9) the (v)hv-torsion tensor $P_{ijk} = P_{0ijk}$ is written

(29.10) $$P_{ijk} = l^{h}\varepsilon_{hir}{}^{*}P^{r}{}_{jk},$$

and its scalar components $P_{\beta\gamma\delta}$ are

(29.10') $$P_{\beta\gamma\delta} = \gamma_{1\beta\mu}{}^{*}P_{\mu\gamma\delta}.$$

From (17.26) it follows that $P_{\alpha\beta1\delta} = P_{\alpha\beta\gamma1} = 0$, which implies

(29.11) $${}^{*}P_{\alpha1\gamma} = {}^{*}P_{\alpha\beta1} = 0.$$

We are concerned with equations (17.22) and (17.23). From (27.11) it follows that

$$LC_{hij|k} = C_{\alpha\beta\gamma,\delta}e_{\alpha)h}e_{\beta)i}e_{\gamma)j}e_{\delta)k}.$$

From (29.6) the explicit form of $C_{\alpha\beta\gamma,\delta}$ is immediately obtained as follows:

$$C_{1\beta\gamma,\delta} = 0, \qquad C_{222,\delta} = H_{,\delta} + 3Jh_{\delta},$$

(29.13) $$C_{223,\delta} = -J_{,\delta} + (H - 2I)h_{\delta},$$

$$C_{233,\delta} = I_{,\delta} - 3Jh_{\delta}, \qquad C_{333,\delta} = J_{,\delta} + 3Ih_{\delta},$$

where $H_{,\delta}$, for instance, is the h-scalar derivative of the single scalar H, namely,

$H_{,\delta} = (\delta_i H) e_{\delta)}{}^i$.

REMARK 29.2. As we put $H = C_{222}$, we should notice the difference between $H_{,\delta}$ and $C_{222,\delta}$. According to the formula (27.12'), we see

$$C_{222,\delta} = (\delta_k C_{222}) e_{\delta)}{}^k + 3C_{\alpha 22} H_{\alpha) 2\delta}$$

$$= (\delta_k H) e_{\delta)}{}^k + 3C_{322} H_{3) 2\delta} = H_{,\delta} + 3(-J)(-h_\delta).$$

Thus we get the second equation of (29.13).

From (17.22) we obtain $P_{\alpha\beta\gamma} = C_{\alpha\beta\gamma,1}$. Accordingly from (29.10') we have

(29.11') $$*P_{2\gamma\delta} = -C_{3\gamma\delta,1}, \qquad *P_{3\gamma\delta} = C_{2\gamma\delta,1}.$$

From (17.23) it follows immediately that

(29.9") $$P_{\alpha\beta\gamma\delta} = \mathfrak{A}_{(\alpha\beta)}\{C_{\beta\gamma\delta,\alpha} + C_{\alpha\gamma\mu}C_{\mu\beta\delta,1}\},$$

so that (29.9') implies

(29.11") $$*P_{1\gamma\delta} = \mathfrak{A}_{(23)}\{C_{3\gamma\delta,2} + C_{2\gamma\mu}C_{\mu 3\delta,1}\}.$$

Here we consider the condition for P_{hijk} to be symmetric in j, k (cf. Theorem 22. 2 and Proposition 28.4). From (17.25) this is equivalent to $S_{hijk|0} = 0$. Therefore (29.7') leads to $S_{|0} = 0$, so that we get

THEOREM 29.2. *In a three-dimensional Finsler space the hv-curvature tensor* P_{hijk} *of* $C\Gamma$ *is symmetric in j, k, if and only if the v-curvature S satisfies* $S_{,1} = (\delta_i S) l^i = 0$.

We turn to the consideration of h-curvature tensor R_{hijk} and (v)h-torsion tensor R_{ijk}. Similarly to the case of v-curvature tensor, it follows from Proposition 29.2 that there exists a tensor $*R^{rs}$ such that

(29.14) $$R_{hijk} = \varepsilon_{hir}\varepsilon_{jks} *R^{rs},$$

and scalar components $R_{\alpha\beta\gamma\delta}$ of R_{hijk} are written as

(29.14') $$R_{\alpha\beta\gamma\delta} = \gamma_{\alpha\beta\mu}\gamma_{\gamma\delta\tau} *R_{\mu\tau} = \delta^{\alpha\beta\mu}_{\gamma\delta\tau} *R_{\mu\tau}$$

in terms of scalar components $*R_{\mu\tau}$ of $*R^{rs}$. Then the (v)h-torsion tensor R_{ijk} is written as

(29.15) $$\frac{1}{L} R_{ijk} = l^h \varepsilon_{hir}\varepsilon_{jks} *R^{rs},$$

and scalar components $R_{\beta\gamma\delta}$ of R_{ijk}/L are given by

(29.15') $$R_{\beta\gamma\delta} = \gamma_{1\beta\mu}\gamma_{\gamma\delta\tau}{}^*R_{\mu\tau}.$$

<u>THEOREM 29.3</u>. *The h-curvature tensor R_{hijk} of $C\Gamma$ of any three-dimensional Finsler space is written in the form*

$$R_{hijk} = \mathfrak{A}_{(jk)}\{g_{hj}L_{ik} + g_{ik}L_{hj}\},$$

where L_{ij} is defined by (29.17).

Proof. We introduce the *h-Ricci tensor* $R_{ij} = R_{hikj}g^{hk}$. Because R_{hijk} satisfies $R_{hijk} = R_{ihkj}$, the h-Ricci tensor R_{ij} is also defined by $R_{ij} = R_{i\ jh}^{\ h}$. It is, however, noted that R_{ij} is not necessarily symmetric (cf. (22.7)). Now, from (29.14') the scalar components $R_{\alpha\beta}$ of R_{ij} are given by

(29.16) $$R_{\alpha\beta} = \gamma_{\pi\alpha\mu}\gamma_{\pi\beta\tau}{}^*R_{\mu\tau} = \delta_{\alpha\beta}{}^*R_{\mu\mu} - {}^*R_{\beta\alpha},$$

which implies that the *h-scalar curvature* $R = R_{ij}g^{ij}$ is given by $R = 2{}^*R_{\mu\mu}$. If we introduce the tensor

(29.17) $$L_{ij} = R_{ij} - \frac{R}{4}g_{ij},$$

its scalar components $L_{\alpha\beta}$ are given by

(29.17') $$L_{\alpha\beta} = \frac{1}{2}\delta_{\alpha\beta}{}^*R_{\mu\mu} - {}^*R_{\beta\alpha}.$$

It then follows from (29.14') that

$$\begin{aligned} R_{\alpha\beta\gamma\delta} &= \delta^{\alpha\beta\mu}_{\gamma\delta\tau}{}^*R_{\mu\tau} = \delta^{\alpha\beta}_{\gamma\delta}{}^*R_{\mu\mu} - \delta^{\alpha\beta}_{\gamma\tau}{}^*R_{\delta\tau} + \delta^{\alpha\beta}_{\delta\tau}{}^*R_{\gamma\tau} \\ &= (\delta^\alpha_\gamma\delta^\beta_\delta - \delta^\alpha_\delta\delta^\beta_\gamma){}^*R_{\mu\mu} - (\delta^\alpha_\gamma{}^*R_{\delta\beta} - \delta^\beta_\gamma{}^*R_{\delta\alpha}) + (\delta^\alpha_\delta{}^*R_{\gamma\beta} - \delta^\beta_\delta{}^*R_{\gamma\alpha}) \\ &= \delta_{\alpha\gamma}(\frac{1}{2}\delta_{\beta\delta}{}^*R_{\mu\mu} - {}^*R_{\delta\beta}) + \delta_{\beta\delta}(\frac{1}{2}\delta_{\alpha\gamma}{}^*R_{\mu\mu} - {}^*R_{\gamma\alpha}) \\ &\quad - \delta_{\alpha\delta}(\frac{1}{2}\delta_{\beta\gamma}{}^*R_{\mu\mu} - {}^*R_{\gamma\beta}) - \delta_{\beta\gamma}(\frac{1}{2}\delta_{\alpha\delta}{}^*R_{\mu\mu} - {}^*R_{\delta\alpha}). \end{aligned}$$

This is nothing but the result which we wish to prove, written in terms of scalar components.

<u>REMARK 29.2</u>. The above remarkable form of R_{hijk}, shown first by Matsumoto [18], is conjectured from the observation that the curvature tensor of a three-dimensional

Riemannian space has such a form as to show vanishing of the Weyl conformal curvature tensor. Cf. Theorem 31.2.

We consider the identity (17.10); appearance of $C_{j\ k}^{\ i}$ in it is a reason why the h-Ricci tensor R_{ij} is not necessarily symmetric. From (29.14') and (29.15') the identity (17.10') is written in the form

$$\mathfrak{S}_{(\beta\gamma\delta)}\{\gamma_{\alpha\beta\mu}\gamma_{\gamma\delta\tau}{}^{*}R_{\mu\tau} + C_{\alpha\beta\pi}\gamma_{1\pi\mu}\gamma_{\gamma\delta\tau}{}^{*}R_{\mu\tau}\} = 0.$$

This essentially gives only three identities where $\alpha = 1, 2, 3$ and $\beta = 1$, $\gamma = 2$, $\delta = 3$. Thus we obtain

$$*R_{23} - *R_{32} = 0,$$

$$(29.18) \qquad *R_{13} - *R_{31} + (H - I)*R_{23} + J(*R_{22} - *R_{33}) = 0,$$

$$*R_{12} - *R_{21} + 2J*R_{23} + I(*R_{22} - *R_{33}) = 0.$$

Assume that the h-Ricci tensor R_{ij} is symmetric, namely, $*R_{\alpha\beta}$ is symmetric. Then The last two equations of (29.18) reduce to

$$(H - I)*R_{23} + J(*R_{22} - *R_{33}) = 0,$$

$$2J*R_{23} + I(*R_{22} - *R_{33}) = 0.$$

If $(H - I)I - 2J^2$ $(= - S)$ does not vanish, the above yields $*R_{23} = 0$, $*R_{22} = *R_{33}$. Therefore

THEOREM 29.4. *Assume that the v-curvature tensor S_{hijk} of a three-dimensional Finsler space does not vanish. Then the h-Ricci tensor R_{ij} is symmetric, if and only if $*R_{\alpha\beta}$ are symmetric and $*R_{23} = 0$, $*R_{22} = *R_{33}$.*

Next, we are concerned with the tensor $C_{hij|k}$. From (27.11) it follows that

$$L^2C_{hij|k} + LC_{hij}{}^1{}_k = C_{\alpha\beta\gamma;\delta}e_{\alpha)h}e_{\beta)i}e_{\gamma)j}e_{\delta)k},$$

which implies

$$(29.19) \qquad L^2C_{hij|k} = (C_{\alpha\beta\gamma;\delta} - C_{\alpha\beta\gamma}\delta^1_\delta)e_{\alpha)h}e_{\beta)i}e_{\gamma)j}e_{\delta)k}.$$

The explicit form of $C_{\alpha\beta\gamma;\delta}$ is easily obtained:

$$C_{1\beta\gamma;\delta} = - C_{\beta\gamma\delta}, \qquad C_{222;\delta} = H_{;\delta} + 3Jv_\delta,$$

$$(29.19') \qquad C_{223;\delta} = - J_{;\delta} + (H - 2I)v_\delta,$$

$$C_{233;\delta} = I_{;\delta} - 3Jv_\delta, \qquad C_{333;\delta} = J_{;\delta} + 3Iv_\delta,$$

where $H_{;\delta}$, for instance, is the v-scalar derivative of the single scalar H, namely, $H_{;\delta} = L(\dot{\partial}_i H)e_{\delta)}{}^i$.

REMARK 29.3. As we put $-J = C_{223}$, according to the formula (27.13'), we have

$$C_{223;\delta} = L(\dot{\partial}_k C_{223})e_{\delta)}{}^k + 2C_{\alpha 23}V_{\alpha)2\delta} + C_{22\alpha}V_{\alpha)3\delta}$$

$$= -L(\dot{\partial}_k J)e_{\delta)}{}^k + 2C_{323}V_{3)2\delta} + C_{222}V_{2)3\delta} = -J_{;\delta} + 2I(-v_\delta) + Hv_\delta,$$

which shows the third of (29.19'), the difference between $C_{223;\delta}$ and $-J_{;\delta}$.

Obviously the tensor $C_{hij|k}$ is completely symmetric. Accordingly (29.19) yields

(29.20) $$C_{\alpha\beta\gamma;\delta} - C_{\alpha\beta\delta;\gamma} = C_{\alpha\beta\gamma}\delta^1_\delta - C_{\alpha\beta\delta}\delta^1_\gamma.$$

This is explicitly written as

$$(H - 2I)v_2 - 3Jv_3 = J_{;2} + H_{;3},$$

(29.20') $$3Jv_2 + (H - 2I)v_3 = I_{;2} + J_{;3},$$

$$3Iv_2 + 3Jv_3 = -J_{;2} + I_{;3}.$$

From (29.6') and the first and third equations of (29.20') it follows that

(29.21) $$Cv_2 = C_{;3}.$$

Thus scalar components of v_i are such that $v_1 = 0$, $v_2 = C_{;3}/C$.

We consider the tensor T_{hijk} which is defined by (28.20). From (29.19) the scalar components $T_{\alpha\beta\gamma\delta}$ of LT_{hijk} are given by

(29.22) $$T_{\alpha\beta\gamma\delta} = C_{\alpha\beta\gamma;\delta} + \delta^1_\alpha C_{\beta\gamma\delta} + \delta^1_\beta C_{\alpha\gamma\delta} + \delta^1_\gamma C_{\alpha\beta\delta}.$$

From $T_{hijk}1^k = 0$ we have $T_{\alpha\beta\gamma 1} = 0$, accordingly the surviving components $T_{\alpha\beta\gamma\delta}$ are only

(29.22') $$T_{\alpha\beta\gamma\delta} = C_{\alpha\beta\gamma;\delta}, \qquad \alpha, \beta, \gamma, \delta = 2, 3.$$

We shall be concerned with the condition $T_{hijk} = 0$. From (29.19') and (29.6') it follows first that

$$C_{223;\delta} + C_{333;\delta} = LCv_\delta,$$

so that (29.22') yields $v_i = 0$ from $C \neq 0$. Thus we have $H_{;\delta} = I_{;\delta} = J_{;\delta} = 0$ from (29.19'). Consequently

THEOREM 29.5. *The T-tensor T_{hijk} of a three-dimensional Finsler space vanishes identically, if and only if the v-connection vector v_i vanishes and all the main scalars are functions of position only.*

REMARK 29.4. Theorem 29.5 is a generalization of Theorem 28.3. According to Proposition 29.3, shown later on, $T_{hijk} = 0$ implies that the main scalars $H(x)$, $I(x)$, $J(x)$ satisfy $S = 2J^2 + I^2 - HI = -1$.

Now, by (29.7") the v-curvature S is expressed in terms of main scalars. We shall show another expression of S in terms of v-scalar derivative of scalar components of the v-connection vector v_i. From (27.13') we have

$$V_{\alpha)\beta\gamma;\delta} = L(\dot{\partial}_i V_{\alpha)\beta\gamma})e_{\delta)}{}^i + V_{\alpha)\mu\gamma}V_{\mu)\beta\delta} + V_{\alpha)\beta\mu}V_{\mu)\gamma\delta}.$$

From (27.9) and (27.10) the first of the following (29.23) is immediately obtained. While we see

$$V_{2)3\gamma;\delta} = L(\dot{\partial}_i v_\gamma)e_{\delta)}{}^i + v_\mu V_{\mu)\gamma\delta} - V_{1)2\gamma}V_{1)3\delta}.$$

Since the first two terms of right-hand side constitute $v_{\gamma;\delta}$, we obtain the second of (29.23). Thus we get

$$\text{(29.23)} \qquad \begin{aligned} V_{1)\beta\gamma;\delta} &= -\delta^1_\beta\delta_{\gamma\delta} - \delta^1_\gamma\delta_{\beta\delta}, \\ V_{2)3\gamma;\delta} &= v_{\gamma;\delta} - \delta^2_\gamma\delta^3_\delta. \end{aligned}$$

We are concerned with (27.20); it reduces trivial for the case $\alpha = 1$, as it will be seen from (29.23). In case of $\alpha = 2$, $\beta = 3$, (27.20) is written as

$$v_{\gamma;\delta} - v_{\delta;\gamma} - \delta^2_\gamma\delta^3_\delta + \delta^2_\delta\delta^3_\gamma = S\gamma_{1\gamma\delta} + v_\gamma\delta^1_\delta - v_\delta\delta^1_\gamma.$$

If we put $\gamma = 1$, the above reduces trivial. Thus the above leads to the only one equation

$$\text{(29.24)} \qquad S = v_{2;3} - v_{3;2} - 1.$$

PROPOSITION 29.3. *If the v-connection vector v_i of a three-dimensional Finsler space is v-covariant constant, the v-curvature S is equal to - 1.*

Consequently, if $T_{hijk} = 0$, Theorem 29.5 shows $S = -1$.
Vanishing of the v-connection vector v_i is really in close relation to possibil-

ity of a generalization of the Landsberg angle θ to three-dimensional case. In fact, to generalize the Landsberg angle, it seems natural from (28.4) that we introduce the system of differential equations of the form

$$(29.25) \qquad \dot{\partial}_i\theta = \pi m_i, \qquad \dot{\partial}_i\phi = \tau n_i,$$

where π, τ are functions which should be chosen such that (29.25) is integrable. From (29.4') we have

$$\dot{\partial}_j(\dot{\partial}_i\theta) = (\dot{\partial}_j\pi)m_i + \frac{\pi}{L}(-\,l_i m_j + n_i v_j + m_r C_i{}^r{}_j),$$

$$\dot{\partial}_j(\dot{\partial}_i\phi) = (\dot{\partial}_j\tau)n_i + \frac{\tau}{L}(-\,l_i n_j - m_i v_j + n_r C_i{}^r{}_j).$$

Thus the integrability condition is written as

$$(29.26) \quad \begin{cases} (1) \quad \pi_{;1} + \pi = 0, \qquad \tau_{;1} + \tau = 0, \\ (2) \quad \pi_{;3} - \pi v_2 = 0, \qquad \tau_{;2} + \tau v_3 = 0. \end{cases}$$

Assume that π, τ are (-1)p-homogeneous, because θ and ϕ are desirable to be (0)p-homogeneous. Then (1) of (29.26) holds automatically. Hence, putting $\pi = \tau = 1/L$ as in (28.4), the condition (2) holds also, provided $v_2 = v_3 = 0$, namely, $v_i = 0$. Consequently

PROPOSITION 29.4. *If the v-connection vector v_i of a three-dimensional Finsler space vanishes identically, there exist two scalar fields θ, ϕ satisfying the differential equations*

$$\dot{\partial}_i\theta = \frac{1}{L}m_i, \qquad \dot{\partial}_i\phi = \frac{1}{L}n_i.$$

These scalars θ, ϕ are defined up to additional functions of position only and may be called the *Landsberg angles* of such a special three-dimensional Finsler space.

Now we shall turn our consideration to three-dimensional Berwald spaces. From Theorem 25.2 and (29.13) we obtain $C_{223,\delta} + C_{333,\delta} = LCh_\delta = 0$ for such a space, which implies $h_i = 0$. Thus (29.13) yields that all the main scalars are h-covariant constant. Conversely, it is obvious that these conditions are sufficient for a Finsler space to be a Berwald space.

We discuss such spaces in detail. It is first observed from (27.9) and (27.10) that

$$H_{2)3\gamma,\delta} = (\delta_i H_{2)3\gamma})e_{\delta)}{}^i + H_{2)\mu\gamma}H_{\mu)3\delta} + H_{2)3\mu}H_{\mu)\gamma\delta}$$

$$= (\delta_i h_\gamma) e_{\delta)}{}^i + h_\mu H_{\mu)\gamma\delta} = h_{\gamma,\delta} \; .$$

Therefore (27.18) in case of $\alpha = 2$, $\beta = 3$ are written as

$$h_{\gamma,\delta} - h_{\delta,\gamma} = ({}^*R_{1\mu} - v_2{}^*R_{3\mu} + v_3{}^*R_{2\mu})\gamma_{\mu\gamma\delta}. \tag{29.27}$$

Similarly from (27.19) we have

$$h_{\gamma;\delta} - v_{\delta,\gamma} = {}^*P_{1\gamma\delta} - h_\mu C_{\mu\gamma\delta} - v_\mu \gamma_{1\mu\tau}{}^*P_{\tau\gamma\delta}. \tag{29.28}$$

Next, applying (27.14) to v_i, we have

$$v_{\alpha,\beta,\gamma} - v_{\alpha,\gamma,\beta} = (v_\mu \gamma_{\mu\alpha\tau} - v_{\alpha;\mu}\gamma_{1\mu\tau})\gamma_{\beta\gamma\pi}{}^*R_{\tau\pi}. \tag{29.29}$$

As to Berwald spaces, we have $h_i = 0$ as above and $P_{hijk} = 0$. Therefore (29.28) yields $v_{\delta,\gamma} = 0$, so that v_i is h-covariant constant. On the other hand, (29.27) reduces to

$${}^*R_{1\mu} - v_2{}^*R_{3\mu} + v_3{}^*R_{2\mu} = 0. \tag{29.30}$$

Further (29.29) is simply written as

$$(v_\mu \gamma_{\mu\alpha\tau} - v_{\alpha;\mu}\gamma_{1\mu\tau}){}^*R_{\tau\pi} = 0.$$

Because of $v_{1;\mu} = - v_\mu$, the above is trivial in case $\alpha = 1$, and thus from the above we obtain only

$$v_3{}^*R_{1\pi} + v_{2;2}{}^*R_{3\pi} - v_{2;3}{}^*R_{2\pi} = 0,$$

$$v_2{}^*R_{1\pi} - v_{3;2}{}^*R_{3\pi} + v_{3;3}{}^*R_{2\pi} = 0.$$

Substituting from (29.30), these equations are written in the form

$$\begin{aligned} &\{v_{2;3} + (v_3)^2\}{}^*R_{2\pi} - (v_{2;2} + v_2 v_3){}^*R_{3\pi} = 0, \\ &(v_{3;3} - v_2 v_3){}^*R_{2\pi} - \{v_{3;2} - (v_2)^2\}{}^*R_{3\pi} = 0. \end{aligned} \tag{29.31}$$

Finally, applying (27.14) to the main scalars which are h-covariant constant, we obtain

$$H^{(p)}{}_{;3}{}^*R_{2\tau} - H^{(p)}{}_{;2}{}^*R_{3\tau} = 0, \qquad p = 1, 2, 3, \tag{29.32}$$

where we put $(H^{(1)}, H^{(2)}, H^{(3)}) = (H, I, J)$.

We now discuss Berwald spaces of three dimensions, according as the rank ρ of matrix $({}^*R_{\alpha\beta})$. The rank ρ is less than three from (29.30).

(i) $\rho = 0$. This means $R_{hijk} = 0$, so that the space is locally Minkowski by Theorem 24.4.

(ii) $\rho = 1$. It then follows from (29.31) and (29.32) that we have $H^{(p)}{}_{;3} : H^{(p)}{}_{;2} = v_{23} : v_{22} = v_{33} : v_{32}$, $p = 1, 2, 3$, where we put

$$v_{\alpha\beta} = v_{\alpha;\beta} + \gamma_{1\alpha\mu}v_\beta v_\mu, \qquad \alpha, \beta = 2, 3. \tag{29.33}$$

(iii) $\rho = 2$. It then follows similarly that $v_{\alpha\beta} = 0$, $\alpha, \beta = 2, 3$, and $H^{(p)}{}_{;2} = H^{(p)}{}_{;3} = 0$, so that all the main scalars are v-covariant constant; so they are constant. In this case, from (29.24) we get $S + (v)^2 + 1 = 0$, where v is the length of v_i.

Summarizing up the above all, we conclude

THEOREM 29.6. *A necessary and sufficient condition for a three-dimensional Finsler space to be a Berwald space is that the h-connection vector field h_i vanishes and all the main scalars H, I, J are h-covariant constant.*

In this case, the v-connection vector field v_i is h-covariant constant. Let ρ be the rank of matric (R_{hijk}) where (hi) and (jk) show the number of the row and column respectively. Then ρ is less than three and

(i) if $\rho = 0$, the space is locally Minkowski,

(ii) if $\rho = 1$, we have $H^{(p)}{}_{;3} : H^{(p)}{}_{;2} = v_{23} : v_{22} = v_{33} : v_{32}$, $p = 1, 2, 3$,

(iii) if $\rho = 2$, all the main scalars are constant, $v_{\alpha\beta} = 0$, $\alpha, \beta = 2, 3$, and there is a relation $S + (v)^2 + 1 = 0$ between v-curvature S and length v of v-connection vector v_i.

REMARK 29.5. This Theorem 29.6 is shown by Matsumoto [18] and regarded as a generalization of Theorem 28.2, although we have an exceptional case (ii). It is noted that the rank of matrix (R_{hijk}) is equal to that of $(R_{\alpha\beta\gamma\delta})$ or $({}^*R_{\alpha\beta})$ by (29.14').

Finally we consider three-dimensional Finsler spaces of scalar curvature. In terms of scalar components the equation (26.4) is written

$$3\gamma_{1\beta\mu}\gamma_{\gamma\delta\tau}{}^*R_{\mu\tau} = \mathfrak{A}_{(\gamma\delta)}\{(K_{;\gamma} + 3K\delta^1_\gamma)(\delta_{\beta\delta} - \delta^1_\beta\delta^1_\delta)\}.$$

This immediately leads us to

$${}^*R_{23} = {}^*R_{32} = 0, \qquad {}^*R_{22} = {}^*R_{33} = K,$$

$$3{}^*R_{31} = -K_{;3}, \qquad 3{}^*R_{21} = -K_{;2}.$$

It then follows from (29.18) that ${}^*R_{\alpha\beta}$ are symmetric, so that the h-Ricci tensor R_{ij} is symmetric. Consequently we have

THEOREM 29.7. *A three-dimensional Finsler space is of scalar curvature K, if and only if we have*

$$ {}^*R_{12} = {}^*R_{21} = -\frac{1}{3} K_{;2}, \qquad {}^*R_{13} = {}^*R_{31} = -\frac{1}{3} K_{;3}, $$

$$ {}^*R_{22} = {}^*R_{33} = K, \qquad {}^*R_{23} = {}^*R_{32} = 0. $$

In this case the h-Ricci tensor R_{ij} is symmetric.

REMARK 29.6. In the present section the Ricci identities are fully used, but not the Bianchi identities. Although the latter identities will have very complicated forms, it may be expected that we shall get more detailed informations on three-dimensional Finsler spaces from the Bianchi identities. In fact, Matsumoto gave a more detailed result on three-dimensional Berwald spaces of scalar curvature in his paper [18], but such spaces are really Riemannian or locally Minkowski, as it will be shown in the next section owing to Bianchi identities.

§ 30. C-reducibility of Finsler spaces

In § 16, two special Finsler metrics are presented; one is a Randers metric and the other is a Kropina metric. Generalizing these metrics, we introduce

DEFINITION 30.1. A Finsler metric L is called an *(α, β)-metric*, when L is a (1) p-homogeneous function of two variables

$$ \alpha(x, y) = \sqrt{\{a_{ij}(x)y^i y^j\}}, \qquad \beta(x, y) = b_i(x)y^i, $$

where $a_{ij} = a_{ji}$ and $\det(a_{ij})$ does not vanish.

REMARK 30.1. In the above definition $\alpha^2(x, dx)$ is a quasi-Riemannian metric and $\beta(x, dx)$ is a differential one-form. We shall confine our following discussion to a domain D of the tangent bundle T(M) over a Finsler space M with an (α, β)-metric L(α, β), which is suitably restricted according as the behavior of α, β and the form of the function L(α, β). For instance, if we consider a Randers metric L = α + β, the domain D is such that α^2 is non-negative, and if a Kropina metric $L = \alpha^2/\beta$ is concerned, then D is taken such that β does not vanish.

A Finsler metric $L = \alpha^p\beta^q$, p + q = 1, is an interesting generalization of Kropina type, but $L = (\alpha^2/\beta) + \beta$ is thought of as a kind of Kropina metric. Because

$$ \frac{\alpha^2}{\beta} + \beta = \frac{1}{\beta}\{a_{ij}(x) + b_i(x)b_j(x)\}y^i y^j, $$

so that, if we put $\bar{\alpha} = \sqrt{\{(a_{ij} + b_i b_j)y^i y^j\}}$, then L is written as $L = \bar{\alpha}^2/\beta$.

Throughout the present section, following notations are adopted:

$$Y_i = a_{ij}y^j, \qquad B^i = a^{ij}b_j,$$

where (a^{ij}) is the inverse matrix of (a_{ij}). Further, subscripts α, β denote partial differentiations by α, β respectively.

As for an (α, β)-metric $L(\alpha, \beta)$, we first obtain

$$(30.1) \qquad y_i = g_{ij}y^j = L(\dot{\partial}_i L) = pY_i + LL_\beta b_i,$$

$$(30.2) \qquad h_{ij} = g_{ij} - l_i l_j = L(\dot{\partial}_i\dot{\partial}_j L)$$

$$= pa_{ij} + q_0 b_i b_j + q_{-1}(b_i Y_j + b_j Y_i) + q_{-2} Y_i Y_j,$$

where coefficients are given by

$$(30.2') \qquad p = \frac{L}{\alpha} L_\alpha, \qquad q_0 = LL_{\beta\beta}, \qquad q_{-1} = \frac{L}{\alpha} L_{\alpha\beta},$$
$$q_{-2} = \frac{L}{\alpha^2}\left(L_{\alpha\alpha} - \frac{L_\alpha}{\alpha}\right).$$

Owing to the homogeneity or $h_{ij}y^j = 0$, we have two identities

$$(30.3) \qquad p + q_{-1}\beta + q_{-2}\alpha^2 = 0, \qquad q_0\beta + q_{-1}\alpha^2 = 0.$$

REMARK 30.2. In (30.2) the subscripts of coefficients q_0, q_{-1}, q_{-2} are used to indicate respective degrees of homogeneity. In the following such usage of subscripts is also adopted. A coefficients without subscript, for instance, p in (30.2), are (0)p-homogeneous.

From (30.1) and (30.2) the fundamental tensor g_{ij} is given by

$$(30.4) \qquad g_{ij} = pa_{ij} + p_0 b_i b_j + p_{-1}(b_i Y_j + b_j Y_i) + p_{-2} Y_i Y_j,$$

where coefficients are given by

$$p_0 = q_0 + (L_\beta)^2 = LL_{\beta\beta} + (L_\beta)^2,$$

$$(30.4') \qquad p_{-1} = q_{-1} + \frac{p}{L} L_\beta = \frac{1}{\alpha}(LL_{\alpha\beta} + L_\alpha L_\beta),$$

$$p_{-2} = q_{-2} + \left(\frac{p}{L}\right)^2 = \frac{1}{\alpha^2}\{LL_{\alpha\alpha} + (L_\alpha)^2 - \frac{L}{\alpha} L_\alpha\}.$$

From (30.3) we get identities

(30.5) $$p_0\beta + p_{-1}\alpha^2 = LL_\beta, \qquad p_{-1}\beta + p_{-2}\alpha^2 = 0.$$

The inverse matrix (g^{ij}) of (g_{ij}) has to be found for the later use. The following proposition gives a useful method to find such a matrix:

PROPOSITION 30.1. *Let a non-singular symmetric n-matrix (A_{ij}) and n quantities c_i be given, and put $B_{ij} = A_{ij} + c_i c_j$. The inverse matrix (B^{ij}) of (B_{ij}) and the $\det(B_{ij})$ are given by*

$$B^{ij} = A^{ij} - \frac{1}{1 + c^2} c^i c^j, \qquad \det(B_{ij}) = A(1 + c^2),$$

where (A^{ij}) is the inverse matrix of (A_{ij}), $A = \det(A_{ij})$, $c^i = A^{ij}c_j$ and $c^2 = c^i c_i$.

The proof is easily obtained by verifying $B_{ij}B^{ik} = \delta^k_j$ directly. It is seen that (B_{ij}) is also non-singular if $1 + c^2 \neq 0$.

From (30.4) the components g_{ij} may be written as

$$g_{ij} = pa_{ij} + c_i c_j + d_i d_j,$$

where we put

$$c_i = \pi b_i, \qquad d_i = \pi_0 b_i + \pi_{-1} Y_i,$$

$$\pi^2 + \pi_0^2 = p_0, \qquad \pi_0\pi_{-1} = p_{-1}, \qquad \pi_{-1}^2 = p_{-2}.$$

Then, putting

$$B_{ij} = pa_{ij} + c_i c_j, \qquad g_{ij} = B_{ij} + d_i d_j,$$

and applying the method of Proposition 30.1 twice, we obtain $|g| = \det(g_{ij})$ as

(30.6) $$|g| = p^{n-2} a\tau,$$

where $a = \det(a_{ij})$ and τ is defined by

(30.6') $$\tau = p(p + p_0 b^2 + p_{-1}\beta) + (p_0 p_{-2} - p_{-1}^2)(\alpha^2 b^2 - \beta^2),$$
$$b^2 = a^{ij} b_i b_j.$$

The inverse (g^{ij}) is now given by

(30.7) $$g^{ij} = \frac{1}{p} a^{ij} - s_0 B^i B^j - s_{-1}(B^i y^j + B^j y^i) - s_{-2} y^i y^j,$$

where coefficients are defined by

(30.7') $$\begin{cases} s_0 = \dfrac{1}{\tau p}\{pp_0 + (p_0 p_{-2} - p_{-1}^2)\alpha^2\}, \\ s_{-1} = \dfrac{1}{\tau p}\{pp_{-1} + (p_0 p_{-2} - p_{-1}^2)\beta\}, \end{cases}$$

$$s_{-2} = \frac{1}{\tau p}\{pp_{-2} + (p_0 p_{-2} - p_{-1}^2)b^2\}.$$

Now, differentiating (30.4) by y^k and paying attention to $\dot\partial_i Y_k = a_{ik}$, we easily get

$$(30.8)\qquad 2pC_{ijk} = r_{-1}b_i b_j b_k + \mathfrak{S}_{(ijk)}\{h_{ij}P_k + r_{-2}b_i b_j Y_k + r_{-3}b_i Y_j Y_k\} + r_{-4}Y_i Y_j Y_k,$$

where P_k is defined by

$$(30.9)\qquad P_k = p_{-1}b_k + p_{-2}Y_k,$$

and coefficients are

$$(30.8')\qquad \begin{aligned} r_{-1} &= pp_{0\beta} - 3p_{-1}q_0, & r_{-2} &= pp_{-1\beta} - p_{-2}q_0 - 2p_{-1}q_{-1}, \\ r_{-3} &= pp_{-2\beta} - p_{-1}q_{-2} - 2p_{-2}q_{-1}, & r_{-4} &= \frac{1}{\alpha}pp_{-2\alpha} - 3p_{-2}q_{-2}. \end{aligned}$$

From (30.3) and (30.5) we obtain important relations

$$(30.10)\qquad r_{-\mu}\beta + r_{-\mu-1}\alpha^2 = 0, \qquad \mu = 1, 2, 3.$$

The form (30.8) of C_{ijk} may be written in a simpler form as follows: From (30.5) and (30.10) we have

$$p_{-2} = \gamma p_{-1}, \qquad r_{-\mu-1} = \gamma^\mu r_{-1}, \qquad \mu = 1, 2, 3,$$

where $\gamma = -\beta/\alpha^2$. Then it is easy to show

$$(30.11)\qquad 2pC_{ijk} = \mathfrak{S}_{(ijk)}\{H_{ij}P_k\},$$

where we put

$$(30.12)\qquad H_{ij} = h_{ij} + \frac{r_{-1}}{3p_{-1}^3}P_i P_j.$$

Further, by direct computation from (30.7) and (30.8), we obtain an expression of $C_i = C_{ijk}g^{jk}$ as

$$C_i = C_{-1}b_i + C_{-2}Y_i.$$

It is not necessary to write explicit forms of the coefficients C_{-1}, C_{-2}. From $C_i y^i = 0$ it follows that $C_{-1}\beta + C_{-2}\alpha^2 = 0$, which implies

$$(30.13)\qquad C_i = \frac{C_{-1}}{p_{-1}}P_i.$$

It is noted that we may assume here $p_{-1} \neq 0$; $p_{-1} = 0$ implies $L^2 = u\alpha^2 + v\beta^2$ with some constants u, v and this L^2 is essentially Riemannian.

From (30.11) and (30.13) it is seen that there exists a Finsler tensor field A_{ij} such that

(30.14) $$C_{ijk} = \mathfrak{S}_{(ijk)}\{A_{ij}C_k\},$$

and A_{ij} are symmetric and satisfy $A_{i0} = 0$.

The above simple and distinctive form of C_{ijk} leads us to

DEFINITION 30.2. A Finsler space of dimension n more than two is called *quasi-C-reducible*, if there exists a symmetric Finsler tensor field A_{ij}, satisfying $A_{i0} = 0$, in terms of which C_{ijk} is written in the form (30.14).

REMARK 30.3. As it will be easily seen from (28.3), the equation (30.14) is not a restriction for two-dimensional spaces.

To study Finsler spaces with some important tensors of special form is an interesting and valuable problem. So far as the author knows, a simple form of C_{ijk}, except (30.14), has been presented by Moór [6]: Generalizing the special form (28.3) in two-dimensional case, he obtains some interesting results, but Theorem 24.3 shows that, roughly speaking, he essentially treats of Riemannian spaces.

On the other hand, Matsumoto [17] presents a problem to study Finsler spaces with C_{ijk} of the form

$$C_{ijk} = \mathfrak{S}_{(ijk)}\{A_{ij}B_k\},$$

where A_{ij} is assumed to be symmetric and satisfy $A_{i0} = 0$, and B_i to satisfy $B_0 = 0$. It seems, however, that further restriction should be imposed on A_{ij} or B_i to obtain interesting results on such spaces. As examples of such A_{ij} and B_i, we know h_{ij} and C_i, and in concequence we arrive at the concept of quasi-C-reducibility as above ($B_i = C_i$). It is easily shown that if $A_{ij} = h_{ij}$, we necessarily have $B_i = C_i/(n + 1)$. Cf. Definition 30.3. See also A.Kawaguchi [3, (6.16)].

THEOREM 30.1.

(1) Any Finsler spaces with an (α, β)-metric is quasi-C-reducible.

(2) A three-dimensional Finsler space is quasi-C-reducible, if and only if J among the main scalars vanishes.

(3) The main scalar J of any three-dimensional Finsler spaces with an (α, β)-metric vanishes.

Proof. We have to show (2) only. If a three-dimensional Finsler space is quasi-C-reducible, in terms of the Moór frame (l^i, m^i, n^i) we obtain $J = C_{333} = LC_{ijk}n^in^jn^k = 0$ because of $C_in^i = 0$ and (30.14). Conversely, if J ($= C_{333} = - C_{223}$) = 0, from (29.6') it follows that

$$LC_{ijk} = Hm_im_jm_k + I\,\mathfrak{S}_{(ijk)}\{m_in_jn_k\},$$

which is rewritten as

$$LC_{ijk} = \mathfrak{S}_{(ijk)} \{ (\frac{H}{3} m_i m_j + I n_i n_j) m_k \}. \tag{30.15}$$

From $m_k = c_k/c$ we have the form (30.14).

In particular, we are concerned with a Randers metric $L = \alpha + \beta$. In this case (30.4) and (30.7) yield

$$g_{ij} = \frac{L}{\alpha} a_{ij} + b_i b_j + \frac{1}{\alpha}(b_i Y_j + b_j Y_i) - \frac{\beta}{\alpha^3} Y_i Y_j, \tag{30.16}$$

$$g^{ij} = \frac{\alpha}{L} a^{ij} - \frac{\alpha}{L^2}(B^i y^j + B^j y^i) + \frac{\alpha b^2 + \beta}{L^3} y^i y^j. \tag{30.17}$$

It is easily shown that $r_{-\mu} = 0$, accordingly (30.8) may be written as

$$2LC_{ijk} = \mathfrak{S}_{(ijk)} \{ h_{ij} (b_k - \frac{\beta}{\alpha^2} Y_k) \}. \tag{30.18}$$

It is observed that (30.18) has such a form as

$$C_{ijk} = \mathfrak{S}_{(ijk)} \{ h_{ij} B_k \}.$$

In this case $B_i = C_i/(n + 1)$ is necessarily obtained, which is immediately verified by contracting by g^{jk}. Therefore we are led to

DEFINITION 30.3. A Finsler space of dimension n more than two is called *C-reducible*, if C_{ijk} is written in the form

$$C_{ijk} = \frac{1}{n+1} \mathfrak{S}_{(ijk)} \{ h_{ij} C_k \}.$$

REMARK 30.4. The C-reducibility is not a restriction for any Riemannian space and for any two-dimensional Finsler space. A Finsler space with a Randers metric is an example of C-reducible space, as above shown.

If $C_i = 0$ for a C-reducible Finsler space, then $C_{ijk} = 0$ immediately, so that the space is concluded to be Riemannian, without reference to Theorem 24.2.

The following theorem characterizes Randers metrics and Kropina metrics among (α, β)-metrics:

THEOREM 30.2. *An n (≥ 3)-dimensional Finsler space with an (α, β)-metric is C-reducible, if and only if the metric is of Randers type or of Kropina type.*

Proof. We consider a C-reducible non-Riemannian Finsler space with an (α, β)-metric. From $C_i \neq 0$ and (30.13) we have $C_{-1} \neq 0$. Therefore (30.12) is written

$$H_{ij} = h_{ij} + \frac{r_{-1}}{3p_{-1}(C_{-1})^2} C_i C_j.$$

Substituting this and (30.13) into (30.11) and using the equation in Definition 30. 3, we obtain

$$P \mathfrak{S}_{(ijk)}\{h_{ij}C_k\} = QC_iC_jC_k, \tag{30.19}$$

where we put

$$P = \frac{2p}{n+1} - \frac{p_{-1}}{C_{-1}}, \qquad Q = \frac{r_{-1}}{(C_{-1})^3}.$$

Contracting (30.19) by C^k, we obtain $PC^2 h_{ij} = (QC^2 - 2P)C_iC_j$. From this it follows that $PC \neq 0$ leads us to a contradiction such that the rank of matrix (h_{ij}) is less than two (cf. Proposition 16.2), and that $C = 0$ implies $P = 0$. Thus we get $P = 0$ in any case, so that (30.19) gives $Q = 0$. Consequently we have $r_{-1} = 0$ together with $r_{-2} = r_{-3} = r_{-4} = 0$ from (30.10).

Conversely, if $r_{-\mu} = 0$, $\mu = 1, \cdots 4$, then (30.8) reduces to the equation which shows the C-reducibility.

We shall integrate the differential equation

$$r_{-4} = \frac{L^2}{\alpha^4}(L_\alpha L_{\alpha\alpha\alpha} + \frac{3}{\alpha} L_\alpha L_{\alpha\alpha} - 3L^2_{\alpha\alpha}) = 0.$$

If $L_{\alpha\alpha} = 0$, we are instantly led to a Randers metric. If $L_{\alpha\alpha} \neq 0$, the above is written as

$$\frac{L_{\alpha\alpha\alpha}}{L_{\alpha\alpha}} + \frac{3}{\alpha} - \frac{3L_{\alpha\alpha}}{L_\alpha} = 0.$$

Paying attention to homogeneity, the integration of the above gives $\alpha^3 L_{\alpha\alpha} = c_1\beta^2 L^3_\alpha$ with a constant c_1. Further we integrate to obtain

$$L^2_\alpha = \frac{\alpha^2}{c_2\alpha^2 + c_1\beta^2}$$

with another constant c_2. Then, according as $c_2 \neq 0$ or $c_2 = 0$, we obtain

$$L(\alpha, \beta) = \pm \frac{1}{c_2}\sqrt{(c_2\alpha^2 + c_1\beta^2)} + c_3\beta,$$

$$L(\alpha, \beta) = \pm \frac{\alpha^2}{2c_1\beta} + c_3\beta,$$

where c_3 is a constant. In essential the former is of Randers type and the latter is of Kropina type.

REMARK 30.5. As to a Randers metric, see Example 16.1 and also Ingarden [3]. We shall quote some lines from Ingarden's monograph:

We shall call it a Randers' space, since Randers (1941) seems to have been the first to consider this kind of spaces, although he regarded them not as Finsler spaces but as 'affinely connected Riemannian spaces', which is a rather confusing notion. Randers could not use, therefore, the methods of Finsler's geometry and tried to reduce the study to a sort of 5-dimensional Kaluza-Klein geometry where Riemannian methods plus a method of special projecting of tensors are used. Spaces with the Randers metric were also considered by Stephenson and Kilmister in 1953, but in investigations of these spaces they simply use pure Riemannian methods, which is obviously erroneous.

Eliopoulos [4] finds the equations of motion in a Finsler space with a Randers metric. Further he [5] treats of a generalized Randers metric such that $\beta(x, y)$ is (1)p-homogeneous in y^i. Hashiguchi, Hōjō and Matsumoto [1] consider two-dimensional Finsler spaces with an (α, β)-metric to find concrete form of fundamental function L in case of Randers metric and $L = \alpha^{m+1}/\beta^m$, $m \neq 0, -1$, such that Finsler spaces with those L be Landsberg. Yasuda [1] finds the connection coefficients $F_j{}^i{}_k$ of $C\Gamma$ of a space with a Randers metric, and recently Matsumoto [21] discusses the torsion and curvature tensors of such spaces. Further Numata [1] shows that if the v-Ricci tensor $S_{ij} = S_i{}^r{}_{jr}$ or the tensor $T_{ij} = T_i{}^r{}_{jr}$ of a Finsler space with an (α, β)-metric vanishes, the space is Riemannian. Since an (α, β)-metric is not necessarily symmetric, it seems that the fact "$S_h{}^i{}_{jk} = 0 \rightarrow$ Riemannian" shows a possibility of justification of Theorem 24.3 without the assumption of symmetry.

The best fruits of introducing the concept of C-reducibility will be such that we can notice special interesting forms of (v)hv-torsion tensor and hv- and v-curvature tensors of $C\Gamma$, as it will be shown in the following. We have already had only one such form (29.7') of the v-curvature tensor, but it will be too restrictive. Any model of special forms of P_{ijk} and P_{hijk} can not be found from the analogy of Riemannian geometry, because these tensors are equal to zero in Riemannian case.

First, from (17.20) and the equation in Definition 30.3 we obtain a special form of the v-curvature tensor:

$$S_{hijk} = \mathfrak{A}_{(jk)}\{h_{hj}M_{ik} + h_{ik}M_{hj}\}, \tag{30.20}$$

where the symmetric tensor M_{ij} is given by

$$M_{ij} = -\frac{1}{(n+1)^2}\left(\frac{C^2}{2} h_{ij} + C_i C_j\right). \tag{30.20'}$$

It is noteworthy that the form (30.20) is similar to the one of R_{hijk} in Theorem 29. 3 and more general than (29.7').

<u>THEOREM 30.3</u>. *There is no C-reducible non-Riemannian Finsler space of dimension more than three, whose v-curvature tensor S_{hijk} has the form (29.7').*

Proof. From (29.7') and (30.20) we get

$$\mathfrak{A}_{(jk)}\{h_{hj}\bar{M}_{ik} + h_{ik}\bar{M}_{hj}\} = 0,$$

where $\bar{M}_{ik} = L^2M_{ik} - Sh_{ik}/2$. Contraction by g^{hj} gives

$$(n - 3)\bar{M}_{ik} + \bar{M}h_{ik} = 0, \qquad \bar{M} = \bar{M}_{ik}g^{ik}.$$

From this we get $2(n - 2)\bar{M} = 0$, so that $\bar{M}_{ik} = 0$. Thus, from (30.20') the equation of the form $ph_{ij} = C_iC_j$ is derived, hence $C_i = 0$. Since the C-reducibility is assumed, $C_i = 0$ implies $C_{ijk} = 0$ instantly.

Secondly, from (17.22), (17.31) and the C-reducibility we obtain the special form of the (v)hv-torsion tensor P_{ijk}:

(30.21) $$P_{ijk} = \frac{1}{n + 1}\,\mathfrak{S}_{(ijk)}\{h_{ij}C_{k|0}\},$$

which is of a similar form to the one of C_{ijk}.

Thirdly, from (17.23) and the C-reducibility we obtain the special form of the hv-curvature tensor P_{hijk}:

(30.22) $$P_{hijk} = N_{hi}h_{jk} + \mathfrak{A}_{(hi)}\{h_{ij}N^{(1)}_{kh} + h_{ik}N^{(2)}_{jh}\},$$

where we put

$$N^{(1)}_{ij} = \frac{1}{n + 1}C_{i|j} - \frac{1}{(n + 1)^2}(C_iC_{j|0} + \tfrac{1}{2}C^rC_{r|0}h_{ij}),$$

(30.22') $$N^{(2)}_{ij} = \frac{1}{n + 1}C_{i|j} + \frac{1}{(n + 1)^2}(C_{i|0}C_j + \tfrac{1}{2}C^rC_{r|0}h_{ij}),$$

$$N_{ij} = - N^{(1)}_{ij} + N^{(1)}_{ji} = - N^{(2)}_{ij} + N^{(2)}_{ji}.$$

<u>THEOREM 30.4</u>. *If a Finsler space is C-reducible and a Landsberg space, it is a Berwald space.*

Proof. From (17.22), (17.23) and Theorem 25.3 we have $C_{ijk|h} - C_{hjk|i} = 0$, so that $C_{i|h} - C_{h|i} = 0$ and $C_{i|0} = 0$. In this case (30.22') gives

$$N^{(1)}_{ij} = N^{(2)}_{ij} = \frac{1}{n + 1}C_{i|j}, \qquad N_{ij} = 0.$$

Therefore (30.22) reduces to

$$\mathfrak{A}_{(hi)}\{h_{ij}C_{k|h} + h_{ik}C_{j|h}\} = 0,$$

from which the contraction by g^{hk} easily yields

$$C_{i|j} = \mu h_{ij}, \tag{30.23}$$

where μ is a scalar. From (30.20'), by substituting from (30.23) and using the C-reducibility, we obtain

$$M_{ij|k} = -\frac{\mu}{n+1} C_{ijk}. \tag{30.24}$$

The Bianchi identity (17.17) is now reduced to $S_{mhij|k} = 0$, namely, from (30.20)

$$\mathfrak{A}_{(ij)}\{h_{mi}M_{hj|k} + h_{hj}M_{mi|k}\} = 0.$$

Substituting from (30.24), this is written in the form

$$\mathfrak{A}_{(ij)}\{h_{mi}C_{hjk} + h_{hj}C_{mik}\} = 0,$$

provided $\mu \neq 0$. By contracting by $g^{mi}g^{hj}$, the above gives $C_i = 0$. Accordingly the space is Riemannian. If $\mu = 0$, then $C_{i|j} = 0$ from (30.23), so that $C_{hij|k} = 0$ because of the C-reducibility. Consequently the space is a Berwald space.

<u>THEOREM 30.5</u>. *If the hv-curvature tensor P_{hijk} of an n (≥ 4)-dimensional Finsler space is written as (30.22) together with (30.22') and symmetric in j, k, then C_i or $C_{i|0}$ vanishes, and the space is a Landsberg space.*

We omit the proof of Theorem 30.5, because it will be easily shown from $P_{hijk} - P_{hikj} = 0$ by contraction.

Finally we consider the T-tensor T_{hijk} which is defined by (28.20). Paying attention to (17.31), from the equation in Definition 30.3 we get

$$C_{hij|k} = \frac{1}{n+1}\mathfrak{S}_{(hij)}\{h_{hi}C_{j|k} - \frac{1}{L}(C_i l_j + C_j l_i)h_{hk}\}. \tag{30.25}$$

The identity $C_{hij|k} - C_{hik|j} = 0$ must be taken into account:

$$\mathfrak{A}_{(jk)}\{h_{ij}C_{hk} + h_{hj}C_{ik}\} = 0, \tag{30.26}$$

where we put

$$C_{ij} = C_{i|j} + \frac{1}{L}(C_i l_j + C_j l_i). \tag{30.26'}$$

Paying attention to $C_{0j} = 0$, contraction of (30.26) by g^{hk} yields

$$C_{ij} = \frac{C^*}{n-1} h_{ij}, \tag{30.27}$$

where $C^* = g^{ij}C_{i|j}$. Hence (30.25) is written in the form

$$C_{hij|k} = \frac{C^*}{n^2 - 1} \mathfrak{S}_{(hij)}\{h_{hi}h_{jk}\} - \frac{1}{L}(l_h C_{ijk} + l_i C_{hjk} + l_j C_{hik} + l_k C_{hij}),$$

which is nothing but the equation

$$T_{hijk} = \frac{LC^*}{n^2 - 1} \mathfrak{S}_{(hij)}\{h_{hi}h_{jk}\}. \tag{30.28}$$

This is a conspicuous form of T_{hijk} and the importance of this tensor may be understood by this special form. It is noted that $LC_{ij} = T_{ij}$ $(= T_{ijkh}g^{kh})$.

Summarizing up the above all, we have

<u>PROPOSITION 30.2</u>. *The (v)hv-torsion tensor $P_{i\,\cdot\,k}^{\;\,j}$, hv-curvature tensor $P_{h\,i\,\cdot\,k}^{\;\;\;j}$, v-curvature tensor $S_{h\,i\,\cdot\,k}^{\;\;\;j}$ and the T-tensor $T_{h\,i\,\cdot\,k}^{\;\;\;j}$ of a C-reducible Finsler space are written in the special forms (30.21), (30.22), (30.20) and (30.28) respectively.*

<u>REMARK 30.6</u>. The above noteworthy special forms of important tensors are indicated by Matsumoto [21]. It is noted that the angular metric tensor h_{ij}, not the fundamental tensor g_{ij}, appears always in each form of such tensors that becomes zero by contraction by y^i. Cf. Theorem 29.3 and see § 31.

Recently we have many papers concerned with C-reducible Finsler spaces. See the references of Matsumoto and Hōjō [1], further F.Ikeda [2, 3], Watanabe and Ikeda [2].

It will be an interesting problem to consider Finsler spaces which satisfy some well-known two conditions simultaneously. In the following we are concerned with such problems:

<u>THEOREM 30.6</u>. *If an n (≥ 3)-dimensional Finsler space is a Landsberg space and of non-zero scalar curvature K, the space is a Riemannian space of constant curvature K.*

<u>THEOREM 30.7</u>. *If an n (≥ 3)-dimensional Finsler space is a Berwald space and of scalar curvature K, the space is Riemannian or locally Minkowski, according as $K \neq 0$ or $K = 0$.*

To prove Theorem 30.6, we shall first show

<u>LEMMA 30.1</u>. *An n (≥ 3)-dimensional Landsberg space of non-zero scalar curvature K is C-reducible and the torsion vector C_i is equal to $-(n + 1)K_{\cdot i}/3K$.*

Then it follows from Lemma 30.1 and Theorem 30.4 that such a Finsler space as mentioned in Theorem 30.6 is a Berwald space, so that Theorem 30.7 leads us to the proof of Theorem 30.6. Consequently it is enough for the proof of Theorem 30.6 to show Lemma 30.1 and Theorem 30.7.

We are first concerned with the proof of Lemma 30.1. For a Landsberg space $g_{ij;k} = 0$ (cf. (18.24)) holds by Definition 25.2, so that the Bianchi identity (18.21) may be written as

$$H_{hijk\cdot m} - 2C_i{}^r{}_m H_{hrjk} + \mathfrak{A}_{(jk)}\{G_{hikm;j}\} = 0. \tag{30.29}$$

Similarly the Ricci identity (10.9') for the Berwald connection BΓ yields

$$G_{hijk} + G_{ihjk} = 2C_{hik;j},$$

which implies $G_{h0jk} = 0$, because $G_{0hjk} = 0$ holds in general from (18.14) and (18.17). Therefore it follows from (30.29) that $H_{hijk\cdot m}y^i = 0$. We consider $H_{hijk\cdot m}y^i y^k = 0$, which is written as

$$(H_{hijk}y^i y^k)_{\cdot m} - H_{hmj0} - H_{h0jm} = 0. \tag{30.30}$$

From (26.5) and (26.6) we get

$$K_0 = L^2 K, \qquad K_{r\cdot i}y^r = LKl_i + \frac{2L^2}{3}K_{\cdot i},$$

$$H_{hmj0} = LK(h_{mh}l_j + h_{jh}l_m) - h_{mj}K_{r\cdot h}y^r - h_{mh}K_j,$$

$$H_{h0jm} = h_{jh}K_m - h_{mh}K_j.$$

Substituting from these equations, (30.30) is rewritten as

$$3KC_{hmj} + \mathfrak{S}_{(hmj)}\{h_{hm}K_{\cdot j}\} = 0. \tag{30.32}$$

Since $K \neq 0$ is assumed, the above shows that the space is C-reducible and, comparing with the equation in Definition 30.3, we get $C_i = -(n+1)K_{\cdot i}/3K$. Thus the proof of Lemma 30.1 is completed.

We shall prove Theorem 30.7. It is enough for it to show that the space under consideration is Riemannian, provided $K \neq 0$. Since $G_{hijk} = 0$ for the Berwald space from Theorem 25.1, the equation (30.29) gives

$$(H_{hijk}y^k)_{\cdot m} - H_{hijm} - 2C_i{}^r{}_m H_{hrj0} = 0. \tag{30.33}$$

We rewrite (30.33) by substituting from (26.6). At that time we have to pay attention to

$$h_{ij\cdot k} = 2C_{ijk} - \frac{1}{L}(h_{ik}l_j + h_{jk}l_i)$$

$$= -\frac{2}{3K}h_{ij}K_{\cdot k} - h_{jk}\left(\frac{2}{3K}K_{\cdot i} + \frac{1}{L}l_i\right) - h_{ki}\left(\frac{2}{3K}K_{\cdot j} + \frac{1}{L}l_j\right),$$

which is obtained from (30.32). Then, by long but trivial computation, (30.33) is rewritten in the simple form

$$\mathfrak{S}_{(hmj)}\{h_{ih}K_{mj}\} = 0, \qquad K_{mj} = L^2K_{\cdot m\cdot j} + L(K_{\cdot m}l_j + K_{\cdot j}l_m),$$

which instantly gives $K_{mj} = 0$. Owing to $C_i = -(n+1)K_{\cdot i}/3K$, this K_{mj} may be written as

$$K_{mj} = \frac{3LK}{n+1}\left(\frac{3L}{n+1}C_mC_j - C_ml_j - C_jl_m - LC_{m\cdot j}\right) = 0.$$

On the other hand, from (30.27) with (30.26') we get

$$\frac{LC^*}{n-1}h_{mj} = C_ml_j + C_jl_m + LC_{m\cdot j} - \frac{L}{n+1}(C^2h_{mj} + 2C_mC_j).$$

Eliminating $C_{m\cdot j}$ from these two equations, we get

$$\left(\frac{C^2}{n+1} + \frac{C^*}{n-1}\right)h_{mj} = \frac{1}{n+1}C_mC_j.$$

If the coefficients of h_{mj} does not vanish, the above shows that the rank of (h_{mj}) is at most one, which is a contradiction to the assumption on dimension. Thus we have $C_i = 0$, so that the space is Riemannian.

REMARK 30.7. These interesting Theorems 30.6 amd 30.7 are proved by Numata [2]. T.Varga [1] also gives another proof of Theorem 30.7. The above Numata's proof shows great importance of Bianchi identities. Although Matsumoto [18] showed an interesting result on three-dimensional Berwald spaces of scalar curvature with reference to the Moór frame, it is only a special case of above Lemma 30.1.

We observe that a problem remains still in Theorem 30.6; a Landsberg space with zero scalar curvature is exceptional. Thus, *consider a Finsler space with* $R_{hijk} = P_{hijk} = 0$. This condition is obviously weaker than that for locally Minkowski spaces, but it is conjectured that such a Finsler space may be locally Minkowski.

In a Finsler space with an (α, β)-metric, we have a linear connection $(\Gamma_j{}^i{}_k(x))$, that is, the Riemannian connection of the Riemannian metric α. Therefore we obtain a Finsler connection $(\Gamma_j{}^i{}_k, N^i{}_j = y^k\Gamma_k{}^i{}_j, C_j{}^i{}_k)$, where $C_j{}^i{}_k$ are usual quantities. Making use of this Finsler connection, Hashiguchi and Ichijyo [1] develop an interesting theory and find examples of Berwald spaces and Wagner spaces among Finsler spaces with an (α, β)-metric.

§ 31. Indicatrices as Riemannian spaces

It has been seen that the tangent space M_x at each point x of an n-dimensional Finsler space with a fundamental function L(x, y) is regarded as an n-dimensional Riemannian space with the fundamental tensor $g_{ij}(x, y)$, where $x = (x^i)$ is fixed. Then, in terms of the Cartan connection $C\Gamma$ of M, components $C_j{}^i{}_k$ of the (h)hv-torsion tensor are Christoffel symbols of M_x (cf. (17.1)) and the v-curvature tensor $S_h{}^i{}_{jk}$ is nothing but the Riemannian curvature tensor of M_x (cf. (10.7')). Consequently the Riemannian covariant differentiation coincides with the v-covariant one of M.

The indicatrix I_x at a point x is a hypersurface of the Riemannian space M_x which is defined by the equation $L(x, y) = 1$, where x is fixed. Consequently I_x is regarded as an $(n - 1)$-dimensional Riemannian space with the Riemannian metric induced from the one of M_x. In the present section we shall consider the Riemannian space I_x from this point of view. Throughout the present section, the dimension of the space M is assumed to be more than two.

We apply the theory of hypersurfaces of Riemannian spaces to I_x in M_x. Let $y^i = y^i(u^\alpha)$, $\alpha = 1,\cdots, n - 1$, be parametric equations of I_x in M_x; we have

$$(31.1) \qquad L(x, y(u)) = 1.$$

Differentiation of (31.1) by u^α yields

$$(31.2) \qquad l_i Y^i_\alpha = 0, \qquad Y^i_\alpha = \frac{\partial y^i}{\partial u^\alpha}.$$

The quantities Y^i_α are so-called projection operators and regarded as components of $n - 1$ linearly independent vectors Y_α tangent to I_x. Therefore (31.2) shows that the normalized supporting element l^i is the unit normal of I_x. Since we then have the field of frame $(Y^i_1,\cdots, Y^i_{n-1}, l^i)$ of M_x defined along I_x, quantities Y^α_i are uniquely defined along I_x by

$$(31.3) \qquad Y^i_\alpha Y^\beta_i = \delta^\beta_\alpha, \qquad l^i Y^\alpha_i = 0, \qquad Y^i_\alpha Y^\alpha_j + l^i l_j = \delta^i_j,$$

and $(Y^1_i,\cdots, Y^{n-1}_i, l_i)$ is the dual coframe of (Y^i_α, l^i).

The induced Riemannian fundamental tensor $g_{\alpha\beta}(u)$ of I_x is defined by $g_{\alpha\beta} = g_{ij}(x, y(u))Y^i_\alpha Y^j_\beta$. From (31.2) and (16.5) $g_{\alpha\beta}$ is written in the form

$$(31.4) \qquad g_{\alpha\beta} = h_{ij} Y^i_\alpha Y^j_\beta,$$

where h_{ij} is the angular metric tensor of M.

<u>REMARK 31.1</u>. From (31.4) it is observed that h_{ij} gives the length on I_x, analogous to the unit sphere, so that h_{ij} is called the angular metric tensor.

The so-called Gauss and Weingarten formulas of the hypersurface I_x are written as

$$(31.5_1) \qquad \frac{\partial Y^i_\alpha}{\partial u^\beta} + C_j{}^i{}_k Y^j_\alpha Y^k_\beta = \Gamma_\alpha{}^\gamma{}_\beta Y^i_\gamma + \Gamma_{\alpha\beta} l^i,$$

$$(31.5_2) \qquad \frac{\partial l^i}{\partial u^\beta} + C_j{}^i{}_k l^j Y^k_\beta = - \Gamma^\gamma_\beta Y^i_\gamma,$$

where $\Gamma_\alpha{}^\gamma{}_\beta$ are Christoffel symbols constructed from $g_{\alpha\beta}$ with respect to u^α, $\Gamma_{\alpha\beta}$ components of the second fundamental tensor of I_x and $\Gamma^\gamma_\beta = g^{\gamma\alpha}\Gamma_{\alpha\beta}$. The second term of the left-hand side of (31.5_2) is evidently equal to zero. Differentiating (31.2) by u^β, from (17.30) we obtain

$$h_{ij}Y^i_\alpha Y^j_\beta + l_i \frac{\partial Y^i_\alpha}{\partial u^\beta} = 0.$$

From this together with (31.4) and (31.5_1) it is observed that $\Gamma_{\alpha\beta}$ are equal to $-g_{\alpha\beta}$. Thus, according to the theory of hypersurfaces of Riemannian spaces, we get

PROPOSITION 31.1. *The indicatrix I_x is a totally umbilical hypersurface with the mean curvature 1 of the tangent Riemannian space M_x, provided that M is of dimension more than two.*

Accordingly (31.5) are rewritten in respective definitive forms

$$(31.6_1) \qquad \frac{\partial Y^i_\alpha}{\partial u^\beta} + C_j{}^i{}_k Y^j_\alpha Y^k_\beta = \Gamma_\alpha{}^\gamma{}_\beta Y^i_\gamma - g_{\alpha\beta} l^i,$$

$$(31.6_2) \qquad \frac{\partial l^i}{\partial u^\beta} = Y^i_\beta.$$

It is noted that y^i coincides with l^i on I_x $(L = 1)$.

We are concerned with the so-called Gauss and Codazzi equations which express the integrability conditions of (31.6). The Gauss equation is written

$$(31.7) \qquad S_{hijk}Y^h_\alpha Y^i_\beta Y^j_\gamma Y^k_\delta = R_{\alpha\beta\gamma\delta} - (g_{\alpha\gamma}g_{\beta\delta} - g_{\alpha\delta}g_{\beta\gamma}),$$

where $R_{\alpha\beta\gamma\delta}$ is the Riemannian curvature tensor of I_x. While the Codazzi equation is trivial, because $\Gamma_{\alpha\beta} = -g_{\alpha\beta}$ and $S_{hijk}l^h = 0$.

Let $C_{\alpha\beta\gamma}$ and $S_{\alpha\beta\gamma\delta}$ be respective projections of C_{ijk} and S_{hijk} onto I_x, namely,

$$C_{\alpha\beta\gamma} = C_{ijk}Y^i_\alpha Y^j_\beta Y^k_\gamma, \qquad S_{\alpha\beta\gamma\delta} = S_{hijk}Y^h_\alpha Y^i_\beta Y^j_\gamma Y^k_\delta.$$

From (17.20) we have

$$(31.8) \qquad S_{\alpha\beta\gamma\delta} = C_{\alpha\delta\mu}C_\beta{}^\mu{}_\gamma - C_{\alpha\gamma\mu}C_\beta{}^\mu{}_\delta.$$

From (31.7) the Ricci tensor $R_{\alpha\beta} = R_\alpha{}^\mu{}_{\beta\mu}$ and the scalar curvature $R = R_{\alpha\beta}g^{\alpha\beta}$ of I_x are given by

$$(31.9) \qquad R_{\alpha\beta} = S_{\alpha\beta} + (n-2)g_{\alpha\beta}, \qquad R = S + (n-1)(n-2),$$

where $S_{\alpha\beta} = S_{ij}Y^i_\alpha Y^j_\beta$ is the projection of the v-Ricci tensor $S_{ij} = S_i{}^r{}_{jr}$ onto I_x and S is the v-scalar curvature $S = S_{ij}g^{ij}$.

REMARK 31.2. Such a distinctive form of the Gauss equation as (31.7) is found by many authors, for example, A.Kawaguchi [3, p.176], Kikuchi [2], Laugwitz [10, p. 33], O.Varga [18, 21].

DEFINITION 31.1. A Finsler tensor T is called *indicatory*, if its components $T^{i\cdots}_{j\cdots}$ satisfy $T^{0\cdots}_{j\cdots} = \cdots = T^{i\cdots}_{0\cdots} = \cdots = 0$.

The familiar tensors h_{ij}, C_{ijk}, P_{ijk}, S_{hijk} and T_{hijk} are obviously indicatory. Importance of the indicatory property for studying indicatrix is that, roughly speaking, an indicatory tensor is thought of as a tensor on I_x as follows:

PROPOSITION 31.2. *If $T^i_{\ j}$, for instance, is an indicatory tensor, it is written in the form*

$$T^i_{\ j} = T^\alpha_{\ \beta} Y^i_\alpha Y^\beta_j$$

on the indicatrix I_x, where $T^\alpha_{\ \beta}$ is the projection of $T^i_{\ j}$ onto I_x.

Proof. From (31.3) we have

$$T^\alpha_{\ \beta} Y^i_\alpha Y^\beta_j = (T^h_{\ k} Y^\alpha_h Y^k_\beta) Y^i_\alpha Y^\beta_j = T^h_{\ k}(\delta^i_h - l^i l_h)(\delta^k_j - l^k l_j),$$

which is equal to $T^i_{\ j}$ from the indicatory property.

The above proposition shows that, as to an indicatory tensor, its behavior is completely governed by that of its projection onto I_x. In fact, suppose that we have two indicatory tensors $H_{ij}(x, y)$ and $K_{ij}(x, y)$ such that the respective projections $H_{\alpha\beta}$ and $K_{\alpha\beta}$ onto I_x have a relation $H_{\alpha\beta} = \mu K_{\alpha\beta}$ with some quantity μ on I_x. It then follows from Proposition 31.2 that $H_{ij} = \mu K_{ij}$ on I_x. Then, if H_{ij} (resp. K_{ij}) is p-homogeneous of degree s (resp. t), then by extending μ all over M_x as a (0)-homogeneous function on M we observe

$$H_{ij}(x, y) = L^s H_{ij}(x, y/L) = L^s \mu K_{ij}(x, y/L) = L^{s-t} \mu K_{ij}(x, y),$$

because y/L is on I_x. Thus we obtain $L^{-s} H_{ij}(x, y) = L^{-t} \mu K_{ij}(x, y)$ on M_x.

DEFINITION 31.2. Let H_{ij} (resp. K_{ij}) be (s) (resp. (t))p-homogeneous indicatory Finsler tensor field, and suppose that $H_{\alpha\beta} = \mu K_{\alpha\beta}$ holds on every indicatrix. Then we have $L^{-s} H_{ij} = L^{-t} \mu K_{ij}$ in the Finsler space, where (0)p-homogeneous scalar μ is obtained as the extension of μ on I_x. The process to get the above relation from $H_{\alpha\beta} = \mu K_{\alpha\beta}$ is called the *homogeneous extension*.

It should be remarked that the homogeneous extension can be applied to indicatory tensors only.

As an illustrative application of this process, we shall show another proof of the equation (29.7') which holds in any three-dimensional Finsler space. In this case the indicatrix I_x is a two-dimensional Riemannian space, so that the curvature tensor $R_{\alpha\beta\gamma\delta}$ of I_x is written in the well-known form

$$R_{\alpha\beta\gamma\delta} = R(g_{\alpha\gamma}g_{\beta\delta} - g_{\alpha\delta}g_{\beta\gamma}),$$

where R is the Gauss curvature of I_x. It then follows from (31.7) that

$$S_{\alpha\beta\gamma\delta} = (R - 1)(g_{\alpha\gamma}g_{\beta\delta} - g_{\alpha\delta}g_{\beta\gamma}).$$

Now, applying the homogeneous extension to indicatory tensors S_{hijk} and $h_{hj}h_{ik} - h_{hk}h_{ij}$ and paying attention to (31.4), we get

$$L^2 S_{hijk} = (R - 1)(h_{hj}h_{ik} - h_{hk}h_{ij}),$$

which shows (29.7'). Further the above shows that S_{hijk} vanishes, if and only if the Gauss curvature R is equal to 1. More generally, we have

<u>THEOREM 31.1</u>. *The v-curvature tensor $S_{h\cdot ijk}$ vanishes at a point x, if and only if the indicatrix I_x is of constant curvature 1, provided that the space is of dimension more than two.*

Proof. Since S_{hijk} is indicatory, it vanishes, if and only if the projection $S_{\alpha\beta\gamma\delta}$ onto I_x vanishes. Thus (31.7) leads to the proof at once.

<u>REMARK 31.3</u>. According to Theorem 31.1, we have a conjecture that if S_{hijk} vanishes identically, the Finsler space is Riemannian. In fact, roughly speaking, "I_x is of constant curvature 1" implies "I_x is a unit sphere", so that the metric is Riemannian. See Theorem 24.3 and A.Kawaguchi [3, p.177].

Moreover, generalizing (29.7'), we shall show the following as to four-dimensional Finsler spaces.

<u>THEOREM 31.2</u>. *The v-curvature tensor S_{hijk} of any four-dimensional Finsler space is written in the form*

$$S_{hijk} = \mathfrak{A}_{(jk)}\{h_{hj}M_{ik} + h_{ik}M_{hj}\},$$

where $M_{ij} = S_{ij} - Sh_{ij}/4$.

This form of S_{hijk} is quite similar to the form (30.20) and to the form of R_{hijk} in three-dimensional case which is given by Theorem 29.3. The proof of Theorem 31.2 is as follows: If we put

$$S'_{hijk} = \mathfrak{A}_{(jk)}\{h_{hj}M_{ik} + h_{hk}M_{ij}\},$$

both S_{hijk} and S'_{hijk} are indicatory and (-2)p-homogeneous. Their projections onto I_x are respectively given by $S_{\alpha\beta\gamma\delta}$ and

$$S'_{\alpha\beta\gamma\delta} = \mathfrak{A}_{(\gamma\delta)}\{g_{\alpha\gamma}M_{\beta\delta} + g_{\beta\delta}M_{\alpha\gamma}\},$$

where $M_{\beta\delta} = S_{\beta\delta} - Sg_{\beta\delta}/4$. It follows from (31.9) that $S_{\alpha\beta} = R_{\alpha\beta} - 2g_{\alpha\beta}$ and $S = R - 6$, so that we have

$$M_{\alpha\beta} = N_{\alpha\beta} - \frac{1}{2}\, g_{\alpha\beta}, \qquad N_{\alpha\beta} = R_{\alpha\beta} - \frac{R}{4}\, g_{\alpha\beta}.$$

It is well-known that the conformal curvature tensor of three-dimensional Riemannian space I_x vanishes identically, that is, we have

$$R_{\alpha\beta\gamma\delta} = \mathfrak{A}_{(\gamma\delta)}\{g_{\alpha\gamma}N_{\beta\delta} + g_{\beta\delta}N_{\alpha\gamma}\}.$$

Thus (31.7) easily leads to $S_{\alpha\beta\gamma\delta} = S'_{\alpha\beta\gamma\delta}$, and we get the proof by homogeneous extension.

We consider the condition for the Riemannian space I_x to be Einstein, namely, $R_{\alpha\beta} = Rg_{\alpha\beta}/(n - 1)$. In this case (31.9) shows $S_{\alpha\beta} = Sg_{\alpha\beta}/(n - 1)$ and we get $S_{ij} = Sh_{ij}/(n - 1)$ by homogeneous extension.

THEOREM 31.3.

(1) *The indicatrix* I_x *is an Einstein space as a Riemannian space, if and only if* $S_{ij} = Sh_{ij}/(n - 1)$, *where* $S_{ij} = S_i{}^r{}_{jr}$ *and* $S = S_{ij}g^{ij}$.

(2) *If the indicatrix* I_x *of a four-dimensional Finsler space is an Einstein space, i.e.,* $S_{ij} = Sh_{ij}/3$, *the v-curvature tensor* S_{hijk} *is written in the form*

$$S_{hijk} = \frac{S}{6}\,(h_{hj}h_{ik} - h_{hk}h_{ij}).$$

(3) *If the v-Ricci tensor* S_{ij} *of a four-dimensional Finsler space vanishes, the v-curvature tensor* S_{hijk} *vanishes.*

Proof of (2). From $S_{ij} = Sh_{ij}/3$ and $M_{ij} = S_{ij} - Sh_{ij}/4$ in Theorem 31.2 we get $M_{ij} = Sh_{ij}/12$. Then Theorem 31.2 leads instantly to the proof.

REMARK 31.4. Theorems 31.2 and 31.3 may be important from the standpoint of theoretical physics. Y.Takano [5] considers the so-called variational principle in Finsler spaces and obtains interesting equations, similar to Einstein's field equation in Riemannian space. Theorems 31.2 and 31.3 are proved by Matsumoto [22], complying with Takano's request to find four-dimensional Finsler spaces such that $S_{hijk} \neq 0$ and $S_{ij} = 0$. According to (3) of Theorem 31.3 and Theorem 24.3, it seems that a four-dimensional Finsler space with $S_{ij} = 0$ reduces to a Riemannian space. For physical applications, however, the assumptions of Brickell's theorem are, of course, too restrictive, so that we can not immediately assert such conclusion.

In the above treatment the indicatory property of h_{ij} and S_{hijk} plays the essential role. We shall consider a general process by means of which any tensor can be converted into an indicatory tensor. Let T_{ij} be an arbitrary Finsler tensor and $T_{\alpha\beta}$ be its projection onto I_x. It is observed that

$$T_{\alpha\beta}Y^\alpha_i = (T_{jk}Y^j_\alpha Y^k_\beta)Y^\alpha_i = T_{jk}(\delta^j_i - l^j l_i)Y^k_\beta = T_{jk}h^j{}_i Y^k_\beta = (T_{ik} - l_i T_{0k})Y^k_\beta.$$

This observation leads us to

<u>DEFINITION 31.3.</u> Let $T^i{}_j$ be a Finsler tensor, for instance, of (1, 1)-type. The Finsler tensor

$$T^{(i)}{}_j = T^r{}_j h^i{}_r = T^i{}_j - \frac{1}{L} l^i T^0{}_j,$$

$$(\text{resp. } T^i{}_{(j)} = T^i{}_r h^r{}_j = T^i{}_j - \frac{1}{L} l_j T^i{}_0,)$$

is said to be obtained from $T^i{}_j$ by the *indicatorization* with respect to the index i (resp. j). Repeating this process successively on all the indices, we finally get an indicatory tensor $'T^i{}_j = T^{(i)}{}_{(j)}$, which is said to be obtained from $T^i{}_j$ by the *indicatorization*, or called the *indicatorized tensor* of $T^i{}_j$.

It is evident that $T^{(0)}{}_j = T^{(i)}{}_j y_i = 0$, which shows the indicatory property of $T^{(i)}{}_j$ with respect to the index i. If we consider T_{ijk} of (0, 3)-type, for instance, we get the indicatorized tensor

$$'T_{ijk} = T_{pqr}h^p{}_i h^q{}_j h^r{}_k$$

$$(31.10) \qquad = T_{ijk} - \frac{1}{L}(l_i T_{0jk} + l_j T_{i0k} + l_k T_{ij0}) + \frac{1}{L^2}(l_i l_j T_{00k} + l_i l_k T_{0j0} + l_j l_k T_{i00}) - \frac{1}{L^3} l_i l_j l_k T_{000}.$$

<u>PROPOSITION 31.3.</u>

(1) *The projection of a tensor onto the indicatrix I_x coincides with the one of the indicatorized tensor onto I_x.*

(2) *Let T_{ij} be an (r)p-homogeneous indicatory tensor. The indicatorized tensor $'(T_{ij}|_k)$ of v-covariant derivative $T_{ij}|_k$ is given by*

$$'(T_{ij}|_k) = T_{ij}|_k + \frac{1}{L}(l_i T_{kj} + l_j T_{ik} - rT_{ij}l_k).$$

(3) *The indicatorized tensor* $'(T_{ij|k})$ *of h-covariant derivative* $T_{ij|k}$ *of an indicatory tensor* T_{ij} *is given by*

$$'(T_{ij|k}) = T_{ij|k} - \frac{1}{L} T_{ij|0} l_k.$$

Proof.

$$T_{ij}\big|_p h^p{}_k = T_{ij}\big|_p(\delta^p_k - \frac{1}{L} y^p l_k) = T_{ij}\big|_k - \frac{r}{L} T_{ij} l_k$$

$$T_{pj}\big|_k h^p{}_i = T_{pj}\big|_k(\delta^p_i - \frac{1}{L} y^p l_i) = T_{ij}\big|_k - \frac{1}{L} l_i\{(T_{pj}y^p)\big|_k - T_{pj}y^p\big|_k\}$$

$$= T_{ij}\big|_k + \frac{1}{L} l_i T_{kj}.$$

These gives the proof of (2).

$$'(T_{ij|k}) = T_{pq|r}h^p{}_i h^q{}_j h^r{}_k = (T_{pq}h^p{}_i h^q{}_j)_{|r} h^r{}_k = T_{ij|k}h^r{}_k = T_{ij|k} - \frac{1}{L} l_k T_{ij|0},$$

which proves (3).

In the last section we observed that the tensor h_{ij}, not g_{ij}, appears often in special forms of torsion and curvature tensors of hv- and v-type. Further the indicatory tensor T_{hijk} is defined by (28.20) and interesting Theorems 28.3 and 29.5 are obtained. As to these important tensors, we have

PROPOSITION 31.4.

(1) *The indicatorized tensor* $'g_{ij}$ *of fundamental tensor* g_{ij} *is the angular metric tensor* h_{ij}.

(2) *The indicatorized tensor* $'(LC_{hij|k})$ *of* $LC_{hij|k}$ *is the T-tensor* T_{hijk}.

The proof of (2) are instantly obtained from (31.10) and Proposition 31.3.

Now we denote by $T_{\alpha\beta;\gamma}$ the covariant derivative of a tensor field $T_{\alpha\beta}$ on the indicatrix I_x. From (31.6) it is easily seen that if $T_{\alpha\beta}$ is the projection of a Finsler tensor field T_{ij} onto I_x, we have

$$T_{\alpha\beta;\gamma} = T_{ij}\big|_k y^i_\alpha Y^j_\beta Y^k_\gamma - (T_{0j}Y^j_\beta g_{\alpha\gamma} + T_{i0}Y^i_\alpha g_{\beta\gamma}). \tag{31.11}$$

Applying this to the tensor C_{hij}, we get

$$C_{\alpha\beta\gamma;\delta} = C_{hij}\big|_k Y^h_\alpha Y^i_\beta Y^j_\gamma Y^k_\delta.$$

It then follows from (2) of Proposition 31.4 that

$$C_{\alpha\beta\gamma;\delta} = T_{hijk}Y^h_\alpha Y^i_\beta Y^j_\gamma Y^k_\delta, \tag{31.12}$$

which implies that $C_{\alpha\beta\gamma;\delta} = 0$ is equivalent to $T_{hijk} = 0$, because T_{hijk} is indicatory. From (31.7) and (31.8) it is seen that $C_{\alpha\beta\gamma;\delta} = 0$ leads to $R_{\alpha\beta\gamma\delta;\varepsilon} = 0$. Consequently

THEOREM 31.4.

(1) The T-tensor T_{hijk} vanishes, if and only if the projection $C_{\alpha\beta\gamma}$ of C_{ijk} onto I_x is covariant constant.

(2) If the T-tensor T_{hijk} vanishes, the indicatrix I_x is locally symmetric in the sense of Riemannian geometry.

In particular, we consider a three-dimensional Finsler space with $T_{hijk} = 0$. From Theorem 29.5 and Proposition 29.3 we have $S = -1$, so that (29.7') yields $S_{\alpha\beta\gamma\delta} = -(g_{\alpha\gamma}g_{\beta\delta} - g_{\alpha\delta}g_{\beta\gamma})$. Then (31.7) leads to

THEOREM 31.5. *If the T-tensor T_{hijk} of a three-dimensional Finsler space vanishes, the indicatrix I_x is locally flat, provided that C_i has non-zero length.*

REMARK 31.5. As already mentioned in § 28, the importance of T-tensor has been recently noticed. Theorem 31.4 was communicated to the author by H.Kawaguchi in 1973. See H.Kawaguchi [2]. Theorem 31.5 is proved by Matsumoto [20]. Because a locally symmetric Riemannian space reduces to a sphere on some weak condition, Theorem 31.4 may state that a Finsler space with $T_{hijk} = 0$ is Riemannian, if the Finsler metric satisfies some desirable conditions.

The result of Theorem 31.5 is rather strange, because the local flatness of indicatrix lets us image an abnormal figure of the indicatrix. Thus, if the figure of indicatrix is usual, C_i may have zero-length and the space may be Riemannian by Theorem 24.2.

A.Kawaguchi [3] and Kikuchi [2] show various interesting results on indicatrices, and Watanabe [2] gets some characterizations of closed hypersurfaces of M_x which are similar to the indicatrix.

We consider such an n (≥ 4)-dimensional Finsler space that the v-curvature tensor S_{hijk} is written in the simple form

$$L^2 S_{hijk} = S(h_{hj}h_{ik} - h_{hk}h_{ij}), \tag{31.13}$$

which always holds in three-dimensional case (cf. (29.7')). By (31.13) the equation (31.7) is written as

$$R_{\alpha\beta\gamma\delta} = (S + 1)(g_{\alpha\gamma}g_{\beta\delta} - g_{\alpha\delta}g_{\beta\gamma}). \tag{31.14}$$

Accordingly the indicatrix I_x is a Riemannian space of constant curvature $S + 1$, so that S must be constant on I_x. Since S is (0)p-homogeneous, it is constant on M_x.

Conversely, if I_x is of constant curvature S + 1, (31.14) and (31.7) show

$$S_{\alpha\beta\gamma\delta} = S(g_{\alpha\gamma}g_{\beta\delta} - g_{\alpha\delta}g_{\beta\gamma}),$$

and the homogeneous extension leads to (31.13). Therefore

DEFINITION 31.4. If the v-curvature tensor S_{hijk} of an n (≥ 4)-dimensional Finsler space is written in the form (31.13), the space is called *S3-like*.

THEOREM 31.6.

(1) The scalar S in (31.13) of an S3-like Finsler space is a function of position alone.

(2) An n (≥ 4)-dimensional Finsler space is S3-like, if and only if each indicatrix is a Riemannian space of constant curvature.

We are finally concerned with a similar consideration on such an n (≥ 5)-dimensional Finsler space that S_{hijk} is written in the form

$$(31.15) \qquad S_{hijk} = \mathfrak{A}_{(jk)}\{h_{hj}M_{ik} + h_{ik}M_{hj}\},$$

which always holds in four-dimensional case (cf. Theorem 31.2) and in C-reducible Finsler spaces (cf. (30.20)). Then (31.7) is written as

$$R_{\alpha\beta\gamma\delta} = \mathfrak{A}_{(\gamma\delta)}\{g_{\alpha\gamma}(M_{\beta\delta} + \frac{1}{2}g_{\beta\delta}) + g_{\beta\delta}(M_{\alpha\gamma} + \frac{1}{2}g_{\alpha\gamma})\}.$$

From this we have

$$R_{\beta\delta} = (n - 3)(M_{\beta\delta} + \frac{1}{2}g_{\beta\delta}) + (M + \frac{n-1}{2})g_{\beta\delta},$$

$$R = 2(n - 2)(M + \frac{n-1}{2}), \qquad M = M_{\alpha\beta}g^{\alpha\beta}.$$

It follows easily that the Weyl conformal curvature tensor of the (n - 1) (≥ 4)-dimensional Riemannian space I_x vanishes. Consequently

THEOREM 31.7. *If the v-curvature tensor S_{hijk} of an n (≥ 5)-dimensional Finsler space has the form (31.15), the indicatrix is conformally flat.*

REMARK 31.6. (1) of Theorem 31.6 can be also shown by the Bianchi identity (17.19). See Kikuchi [2], Matsumoto [15].

T.Okubo has a conjecture that S3-like Finsler spaces will be Riemannian, provided that the assumption of Theorem 24.3 is satisfied, because I_x is of constant curvature. His papers [2, 3] are interesting in this point of view. This conjecture is regarded as a generalization of Theorem 24.3.

At the end of Chapter V, the author hopes to write a suggestion as follows: The indicatory tensors h_{ij}, C_{ijk}, S_{hijk} and T_{hijk} are all constructed from the fundamental function $L(x, y)$ by partial differentiations with respect to y^i only, so that those tensors can be defined on a single M_x and on a Minkowski space. Any tensor of a Finsler space, however, can be indicatorized by Definition 31.3. For example, we shall deal with the hv-curvature tensor P_{hijk} of $C\Gamma$, from which by indicatorization we obtain an indicatory tensor

$$'P_{hijk} = P_{hijk} - \frac{1}{L}(l_h C_{ijk|0} - l_i C_{hjk|0}).$$

This tensor can not be constructed from $L(x, y)$ without partial differentiation with respect to x^i, so that $'P_{hijk}$ is defined not on a single M_x, but by comparing M_x with neighboring M_{x+dx}. The condition $'P_{hijk} = 0$ must express some special relation between I_x and I_{x+dx}. As a consequence, we have to study the group of motions of Minkowski spaces in detail and introduce a new concept "*a connection of Minkowski spaces*". This new and unknown concept may produce an epoch-making and fundamental progress of Finsler geometry. This might be indeed Rund's dream in his early papers !

CHAPTER VI

THEORY OF TRANSFORMATIONS OF FINSLER SPACES

In this chapter we are concerned with a fundamental theory of generalized transformations of Finsler spaces. After considering usual Lie derivatives in fiber bundles, a generalization of transformations is introduced in such a way that Finsler vector fields, not tangent vector fields in the strict sense, take place in generalized Lie derivatives. As a consequence, an infinitesimal transformation $(x^i, y^i) \to (\bar{x}^i, \bar{y}^i)$ is written

$$\bar{x}^i = x^i + X^i(x, y)dt, \qquad \bar{y}^i = y^i + (\delta_j X^i)y^j dt.$$

Further the concept of rotation arises and an infinitesimal rotation is written in the form

$$\bar{x}^i = x^i, \qquad \bar{y}^i = y^i + X^i_j(x, y)y^j dt.$$

It seems that the theory of transformations is not satisfactorily developed yet in Finsler geometry. We have only few interesting results, for instance, on relations between the existence of transformations of a kind and special properties of the space. It may be expected that the theory of rotations is effective to studying properties of indicatrices, but all are future problems.

§ 32. Lie derivatives

In Riemannian geometry the theory of Lie derivatives has been studied by many authors. To generalize this theory to Finsler geometry, we shall first treat of infinitesimal transformation groups of the tangent bundle $T(M)$ and the linear frame bundle $L(M)$, which are derived from a tangent vector field on the base space M.

Let X be the tangent vector field of a manifold M which is induced by a one-parameter transformation group $\{\mu_t\}$ of M. As it is mentioned in § 2, the relation between X and $\{\mu_t\}$ is given by

(32.1) $$X_x(f) = d_0\{f\circ\mu_t(x)\}$$

for any $f \in \zeta(M)_x$.

DEFINITION 32.1. The *associated mapping* μ^*_t: $(M_{x'})^r_s \to (M_x)^r_s$, $x' = \mu_t(x)$, with a one-parameter transformation group $\{\mu_t\}$ of a manifold M is such that

(1) $r = s = 0$: $\mu^*_t(f) = f\circ\mu_t$,

(2) $r = 1$, $s = 0$: $\mu^*_t(Y) = \mu'_{-t}(Y)$,

(3) $r = 0$, $s = 1$: $\mu^*_t(\alpha) = \alpha\circ\mu'_t$,

(4) $\mu_t^*(T)$, $T \in (M_{x'})^r_s$, is given by

$$\mu_t^*(T)(\alpha_1, \cdots, \alpha_r, Y_1, \cdots, Y_s)$$
$$= T(\mu^*_{-t}(\alpha_1), \cdots, \mu^*_{-t}(\alpha_r), \mu^*_{-t}(Y_1), \cdots, \mu^*_{-t}(Y_s))$$

for any α's $\in (M_x)_*$ and any Y's $\in M_x$.

As to $r = s = 1$, for instance, we get

$$\mu_t^*(T)(\alpha, \ Y) = T(\alpha\circ\mu'_{-t}, \mu'_t Y), \qquad \alpha \in (M_x)_*, \quad Y \in M_x.$$

DEFINITION 32.2. Let X be the tangent vector field of a manifold M which is induced by a one-parameter transformation group $\{\mu_t\}$, and let μ_t^* be the associated mapping with μ_t. The *Lie derivative* L_X with respect to X is an operator on a tensor field T of M such that $L_X T = d_0 t\{\mu_t^*(T)\}$.

Therefore $L_X T$ is a tensor field of the same type with T and the explicit form of $L_X T$ is given by

PROPOSITION 32.1. *The Lie derivative L_X with respect to a tangent vector field X on M is given as follows:*

(1) $L_X f = X(f)$ *for a function f.*

(2) $L_X Y = [X, Y]$ *for a tangent vector field Y.*

(3) $L_X \alpha(Y) = X(\alpha(Y)) - \alpha([X, Y])$ *for a differential one-form* α *, where Y is any tangent vector field.*

(4) For a tensor field T of (r, s)-type

$$L_X T(\alpha_1, \cdots, \alpha_r, Y_1, \cdots, Y_s) = X(T(\alpha_1, \cdots, \alpha_r, Y_1, \cdots, Y_s))$$
$$- \sum_{p=1}^{r} T(\alpha_1, \cdots, L_X\alpha_p, \cdots, \alpha_r, Y_1, \cdots, Y_s)$$
$$- \sum_{q=1}^{s} T(\alpha_1, \cdots, \alpha_r, Y_1, \cdots, L_X Y_q, \cdots, Y_s),$$

where α's are any differential one-forms and Y's are any tangent vector fields.

Proof. (1) From Definitions 32.2 and 32.1, we get

$$L_X f = d_0 t\{\mu_t^*(f)\} = d_0 t\{f\circ\mu_t\},$$

which is equal to X(f) by (32.1).

(2) Similarly we get

$$\mathcal{L}_X Y = -\, d_0 t\{\mu_t' Y\},$$

which is equal to [X, Y] by Proposition 2.2.

(3), (4) For two functions f, g, putting $x' = \mu_t(x)$, we have

$$\{\mu_t^*(f\cdot g)\}_x - \{f\cdot g\}_x = \{f\cdot g\}_{x'} - \{f\cdot g\}_x$$

$$= \{f(x') - f(x)\}g(x') + f(x)\{g(x') - g(x)\}$$

$$= \{(\mu_t^* f)_x - f(x)\}g(x') + f(x)\{(\mu_t^* g)_x - g(x)\},$$

which implies a general formula

$$\mathcal{L}_X(f\cdot g) = (\mathcal{L}_X f)g + f(\mathcal{L}_X g).$$

We are concerned with a tensor field $T = (T^i{}_j)$ of (1, 1)-type, for instance, and $T(\alpha, Y)$, where $\alpha = (\alpha_i)$ is a differential one-form and $Y = (Y^i)$ is a tangent vector field. Then $T(\alpha, Y) = T^i{}_j \alpha_i Y^j$; so we can apply the above general formula to $T(\alpha, Y)$ to obtain

$$\mathcal{L}_X(T(\alpha, Y)) = (\mathcal{L}_X T)(\alpha, Y) + T(\mathcal{L}_X \alpha, Y) + T(\alpha, \mathcal{L}_X Y).$$

The left-hand side is equal to $X(T(\alpha, Y))$ by (1), so that (4) is verified.

<u>EXAMPLE 32.1</u>. We shall recall an infinitesimal conformal transformation of a Riemannian space M with a fundamental tensor $g = (g_{ij}(x))$. A tangent vector field X of M is called an *infinitesimal conformal transformation* of M, if $\mathcal{L}_X g = \tau g$ where τ is a scalar field. From Proposition 32.1, for two tangent vector fields Y, Z, we have

$$\mathcal{L}_X g(Y, Z) = X(g(Y, Z)) - g([X, Y], Z) - g(Y, [X, Z]).$$

If we are concerned with the Riemannian connection γ which is defined by g (Proposition 5.1), for $z \in \pi_L^{-1}(x)$ (5.15) gives

$$[X, Y]_x = \pi_L'[\underline{B}(\tilde{X}), \underline{B}(\tilde{Y})]_z = \pi_L'\{\underline{B}(\underline{\nabla}\tilde{Y}(\tilde{X}) - \underline{\nabla}\tilde{X}(\tilde{Y})) + \underline{Z}(*\underline{R}(\tilde{X}, \tilde{Y}))\},$$

where $\underline{B}(\tilde{X})$ is the basic vector field, corresponding to the tensor function $\tilde{X}$ associated with X. We observe

$$\underline{\nabla}\tilde{Y}(\tilde{X})_z = [(z^{-1})^a_i Y^i{}_{;j} z^j_b\{(z^{-1})^b_k X^k\}]e_a = \{(z^{-1})^a_i Y^i{}_{;k} X^k\}e_a.$$

Thus, if we denote by $\underline{\nabla}_X Y$ the tensor with which $\underline{\nabla}\tilde{Y}(\tilde{X})$ is associated, the components of $\underline{\nabla}_X Y$ are given by $Y^i{}_{;k} X^k$, and we get

$$[X, Y] = \underline{\nabla}_X Y - \underline{\nabla}_Y X.$$

Further $X(f) = df(X) = \underline{\nabla}f(X)$ and $\underline{\nabla}g = 0$ imply

$$X(g(Y, Z)) = \underline{\nabla}(g(Y, Z))X = g(\underline{\nabla}_X Y, Z) + g(Y, \underline{\nabla}_X Z).$$

Therefore the conformality condition is written as

$$g(\underline{\nabla}_Y X, Z) + g(Y, \underline{\nabla}_Z X) = \tau g(Y, Z)$$

for any Y, Z. In terms of components the above is written in the form

$$X_{i;j} + X_{j;i} = \tau g_{ij}, \qquad X_i = g_{ij}X^j,$$

which is a well-known equation.

If τ is a non-zero constant, X is called *homothetic*, and if $\tau = 0$, X is *isometric* or a *Killing vector field*.

Next we are concerned with a single transformation τ of a manifold M. Its differential μ': $M_x \to M_{\mu(x)}$ is regarded as a transformation of the total space T of the tangent bundle T(M).

DEFINITION 32.2. Let μ be a transformation of an n-dimensional manifold M. The *n-ple differential* $\mu^{(n)}$ of μ is a transformation of the total space L of the linear frame bundle L(M) which is given by $\mu^{(n)}(z) = \{\mu'(z_a)\}$ for $z = \{z_a\} \in L$.

It follows instantly from the definitions that the differential μ' and n-ple differential $\mu^{(n)}$ satisfy

(32.2) $$\pi_T \circ \mu' = \mu \circ \pi_T, \qquad \pi_L \circ \mu^{(n)} = \mu \circ \pi_L.$$

If we consider an admissible mapping ${}_z\alpha$: $v \in V^n \mapsto zv \in T$, $z \in L$, we see $\{\mu^{(n)}(z)\}(v) = \mu'(z_a)v^a$, which shows

(32.3) $$\{\mu^{(n)}(z)\}v = \mu'(zv).$$

This is regarded as another definition of $\mu^{(n)}$.

We have the inverse ${}_z\alpha^{-1}$: $X \in T \mapsto z^{-1}X \in V^n$ of ${}_z\alpha$. From (32.3) we get

(32.3') $$\{\mu^{(n)}(z)\}^{-1}X = z^{-1}(\mu^{-1})'X.$$

As for a right translation $\underline{\beta}_g$, we easily get

(32.4) $$\mu^{(n)} \circ \underline{\beta}_g = \underline{\beta}_g \circ \mu^{(n)}.$$

Next we are concerned with $v_* \in V^n_* \mapsto zv_* \in (M_x)_*$ (cf. (3.8')). In terms of inner product we have $(X, zv_*) = (z^{-1}X, v_*)$, $X \in M_x$. Therefore from (32.3') we get

(32.5) $$(X, \{\mu^{(n)}(z)\}v_*) = (z^{-1}(\mu^{-1})'X, v_*), \qquad v_* \in V^n_*, \quad X \in M_x.$$

THEOREM 32.1. *A transformation τ of the total space L of L(M) is the n-ple differential of a transformation of M, if and only if the following two conditions hold:*

(1) *τ is fiber-preserving, i.e., for any $g \in G(n)$ we have $\pi_L \circ \tau \circ \underline{\beta}_g = \pi_L \circ \tau$.*

(2) *The basic form θ on L is τ-invariant, i.e., $\theta \circ \tau' = \theta$.*

Proof. The necessity of (1) is evident from (32.4). Next, for $X_z \in L_z$, (32.3') and (32.2) give

$$\theta \circ (\mu^{(n)})'_z = (\mu^{(n)}(z))^{-1} \pi'_L \circ (\mu^{(n)})' = z^{-1}(\mu^{-1})' \circ \mu' \circ \pi'_L = z^{-1}\pi'_L = \theta.$$

Conversely, a mapping τ satisfying (1) gives rise to the unique transformation μ of M such that $\mu \circ \pi_L = \pi_L \circ \tau$. Then (2) yields

$$z^{-1}\pi'_L = \tau(z)^{-1}\pi'_L \circ \tau' = \tau(z)^{-1}\mu' \circ \pi'_L.$$

If we put $\pi'_L(X_z) = zv$, the above gives $v = \tau(z)^{-1}\mu'(zv)$; so (32.3) shows $\{\mu^{(n)}(z)\}v = (\tau(z))v$, namely, $\mu^{(n)} = \tau$.

DEFINITION 32.4. Let X be a tangent vector field of an n-dimensional manifold M which is induced by a one-parameter transformation group $\{\mu_t\}$.

(1) The one-parameter group of differentials $\{\mu'_t\}$ induces a tangent vector field X' on the total space T of T(M), which is called the *derived vector field* from X.

(2) The one-parameter group of n-ple differentials $\{\mu_t^{(n)}\}$ induces a tangent vector field $X^{(n)}$ on the total space L of L(M), which is called the *n-ple derived vector field* from X.

REMARK 32.1. The derived vector field X' from X is also called the complete lift of X by several authors from the standpoint of differential geometry of tangent bundles.

A main result of the present section states the following relation between Lie derivations and n-ple derived vector fields:

THEOREM 32.2. *Let T be a tensor field of a manifold M and $\tilde{T}$ be the associated tensor function with T. Then we have $\tilde{T}_1 = X^{(n)}(\tilde{T})$, where $T_1 = L_X T$.*

That is, the Lie derivative $L_X T$ is nothing but a tensor field, the associated tensor function with which is the usual derivative of $\tilde{T}$ by $X^{(n)}$. We shall prove this by considering a tensor field T of (1, 1)-type for brevity. It follows from (3.14), Definitions 32.2 and 32.1 that for $v \in V^n$, $v_* \in V^n_*$ and $x = \pi_L(z)$, we have

$$(\tilde{T}_1)_z(v_*, v) = (L_X T)_x(zv_*, zv) = d_0 t\{\mu_t^*(t)\}(zv_*, zv)$$

$$= d_0 t\{T(zv_* \cdot \mu'_{-t}, \ \mu'_t(zv))\}.$$

Then (32.5), (32.3) and (3.14) show

$$= d_0 t\{T((\mu_t^{(n)}(z))v_*, \ (\mu_t^{(n)}(z))v)\} = d_0 t\{\tilde{T}(\mu_t^{(n)}(z))(v_*, \ v)\},$$

which is equal to $X^{(n)}(\tilde{T})(v_*, v)$ from (32.1).

Now we are concerned with the n-ple derived vector field $X^{(n)}$ from $X = X^i\partial/\partial x^i$ in terms of coordinate system. Putting $Y = Y^i\partial/\partial x^i$,

$$Y_1 = L_X Y = [X, Y] = \left(X^j \frac{\partial Y^i}{\partial x^j} - Y^j \frac{\partial X^i}{\partial x^j}\right)\frac{\partial}{\partial x^i}.$$

From (3.14') we obtain

$$(\tilde{Y}_1)_z = (z^{-1})^a_i \{X^j \frac{\partial Y^i}{\partial x^j} - Y^j \frac{\partial X^i}{\partial x^j}\}e_a \in V^n.$$

On the other hand, putting $X^{(n)} = U^i(\partial/\partial x^i) + U^i_a(\partial/\partial z^i_a)$, we have

$$X^{(n)}(\tilde{Y}) = X^{(n)}\{(z^{-1})^a_i Y^i e_a\} = (z^{-1})^a_i\{U^j \frac{\partial Y^i}{\partial x^j} - Y^j(z^{-1})^b_j U^i_b\}e_a.$$

Consequently we have $U^j = X^j$, $U^i_a = z^j_a(\partial X^i/\partial x^j)$ from Theorem 32.2, namely,

(32.6) $$X^{(n)} = X^i \frac{\partial}{\partial x^i} + (\partial_j X^i)z^j_a \frac{\partial}{\partial z^i_a}.$$

Next, from (32.1), (32.3) and (3.17) we have

$$X'_{zv}(f) = d_0 t\{f\circ\mu'_t(zv)\} = d_0 t\{f\circ(\mu_t^{(n)}(z)v)\} = d_0 t\{f\circ\alpha_v(\mu_t^{(n)}(z))\} = X_z^{(n)}(f\circ\alpha_v),$$

where f is a function on T and α_v is an associated mapping $z \in L \mapsto zv \in T$. Consequently

(32.7) $$X'_{zv} = \alpha'_v(X_z^{(n)}).$$

In terms of coordinate system, from (3.17'), (32.6) and (32.7) we obtain an expression of X' at $y = (x^i, y^i)$:

(32.7') $$X' = X^i \frac{\partial}{\partial x^i} + (\partial_j X^i)y^j \frac{\partial}{\partial y^i}.$$

Further (32.2) and (32.4) yield

(32.8) $$\pi'_L(X^{(n)}) = X, \qquad \underline{\beta}'_g(X^{(n)}) = X^{(n)},$$

and (32.4) and Corollary 2.2 give

(32.9) $$[\underline{Z}(A), X^{(n)}] = 0.$$

Suppose that a linear connection $\underline{\Gamma}$ is given in the linear frame bundle L(M). A basic vector field with respect to $\underline{\Gamma}$ is denoted by $\underline{B}(v)$. It is observed from (32.2) and (32.3) that

$$\pi_L'\circ(\mu^{(n)})'(\underline{B}(v)_z) = \mu'\circ\pi_L'(\underline{B}(v)_z) = \mu'(zv) = \mu^{(n)}(z)v = \pi_{\underline{L}}'\underline{B}(v)_{z'},$$

where $z' = \mu^{(n)}(z)$, which shows $(\mu^{(n)})'(\underline{B}(v)) - \underline{B}(v)$ is vertical; so Proposition 2.2 shows that $[X^{(n)}, \underline{B}(v)]$ is vertical. Putting $X^{(n)} = \underline{B}(\tilde{w}) + \underline{Z}(\tilde{A})$ (by identifying G'(n) with $(V^n)^1_1$ (cf. (2.17)), it is obvious from (32.8) that $\tilde{w}$ and $\tilde{A}$ are tensor functions with tensor fields w, A respectively. It then follows from the structure equations (5.14) and (5.15) that

$$[X^{(n)}, \underline{B}(v)] = \underline{B}(\tilde{\underline{T}}(\tilde{w}, v)) + \underline{Z}(\tilde{\underline{R}}(\tilde{w}, v)) - \underline{B}(\tilde{\underline{\nabla}w}(v)) + \underline{B}(\tilde{A}\cdot v) - \underline{Z}(\tilde{\underline{\nabla}A}(v)).$$

Thus we get

$$[X^{(n)}, \underline{B}(v)] = \underline{Z}(\tilde{\underline{R}}(w, v) - \tilde{\underline{\nabla}A}(v)), \qquad \tilde{A}\cdot v = \tilde{\underline{\nabla}w}(v) + \tilde{\underline{T}}(v, \tilde{w}).$$

On the other hand, (32.8) yields $\pi_L'(X_z^{(n)}) = X_x = z\tilde{w}$, namely, X = w. Consequently

PROPOSITION 32.2. *Let $X^{(n)}$ be the n-ple derived vector field from X and $\underline{B}(v)$ be a basic vector field with respect to a linear connection $\underline{\Gamma}$. Then we have*

(1) $\quad X^{(n)} = \underline{B}(\tilde{X}) + \underline{Z}(\tilde{A}), \qquad A = \underline{\nabla}X + \underline{T}(X),$

(2) $\quad [X^{(n)}, \underline{B}(v)] = \underline{Z}(\tilde{\underline{R}}(\tilde{X}, v) - \tilde{\underline{\nabla}A}(v)),$

where $\underline{T}$, $\underline{R}$ are torsion and curvature tensors of $\underline{\Gamma}$.

REMARK 32.2. In a classical representation, a tangent vecor field X of a manifold M is called an infinitesimal affine transformation of a linear connection $\underline{\Gamma}$, if $\mathcal{L}_X\underline{\Gamma}_j{}^i{}_k = 0$ holds for connection coefficients $\underline{\Gamma}_j{}^i{}_k$ of $\underline{\Gamma}$. It is, however, remarked that $\underline{\Gamma}_j{}^i{}_k$ are not components of a tensor field, hence $\mathcal{L}_X\underline{\Gamma}_j{}^i{}_k$ are not given by Definition 32.2. In terms of fiber bundles, such a transformation is formulated as follows:

DEFINITION 32.5. A tangent vector field X of an n-dimensional manifold M is called an *infinitesimal affine transformation* of a linear connection $\underline{\Gamma}$ on M, if the n-ple derived vector field $X^{(n)}$ from X is $\underline{\Gamma}$-preserving, namely, $\mathcal{L}_{\bar{X}}\underline{\Gamma} \subset \underline{\Gamma}$, $\bar{X} = X^{(n)}$.

Because $\mathcal{L}_{\bar{X}}\underline{B}(v) = [X^{(n)}, \underline{B}(v)]$ is vertical in general, the above condition means $\mathcal{L}_{\bar{X}}\underline{B}(v) = 0$, so that (2) of Proposition 32.2 gives

$$\tilde{\underline{R}}(\tilde{X}, v) - \tilde{\underline{\nabla}A}(v) = 0 \tag{32.10}$$

for any $v \in V^n$. In terms of components this is written in the familiar form

$$\mathcal{L}_X\underline{\Gamma}_h{}^i{}_j = \underline{R}_h{}^i{}_{jk}X^k + (X^i{}_{;h} + \underline{T}_h{}^i{}_k X^k)_{;j} = 0. \tag{32.10'}$$

§ 33. V-transformations

A tangent vector field $X = X^i\partial/\partial x^i$ of a manifold M induces an infinitesimal point transformation $(x^i) \mapsto (\bar{x}^i)$ which is given by an old-fashioned equation

$$\bar{x}^i = x^i + X^i(x)dt, \tag{33.1}$$

where dt is an infinitesimal. This is naturally extended to a point transformation $(x^i, y^i) \to (\bar{x}^i, \bar{y}^i)$ of the tangent bundle T(M) such that

$$\bar{x}^i = x^i + X^i(x)dt, \qquad \bar{y}^i = y^i + (\partial_j X^i)y^j dt. \tag{33.2}$$

In fact, from (33.1) we get

$$\frac{d\bar{x}^i}{dt} = \frac{dx^i}{dt} + \partial_j X^i \frac{dx^j}{dt} dt,$$

which yields the second equation of (33.2) by putting $y^j = dx^j/dt$. It is remarked that the so-called *extended point transformation* (33.2) is a transformation induced by the derived vector field X' from X, as it is seen from (32.7').

REMARK 33.1. In Finsler geometry almost all the authors have been concerned with extended point transformations alone. Here we shall cite related papers. Akbar-Zadeh [2, 6, 7, 10, 13, 17], Davies [1], Heil and Laugwitz [1], Hiramatu [1, 2, 3], Knebelman [1, 2, 4], Ku [2], Laptev [1, 2, 3, 5, 7], Matsumoto [2, 3, 9, 12, 16], Maurin [1], Nasu [5, 6], Soós [1, 2, 3, 5], Su [2], K.Takano [2, 3], Tashiro [1], H.-C.Wang [1]. A systematical description is seen in K.Yano's book [1].

Now, Finsler geometry of a manifold M is, of course, a differential geometry of M with a Finsler structure; so it may be natural to study behaviors of Finsler-geometrical quantities under transformations of M.

But, from the viewpoint of our Finsler connection, and also from the viewpoint of Definition 32.5 of affine transformation, it may be more natural for the theory of transformations of Finsler spaces to treat of a tangent vector field $\bar{X}$ of the Finsler bundle F(M) which is derived from a tangent vector field X of the tangent bundle T(M) by some process and to define the affinity of X with respect to a Finsler connection $F\Gamma = (\Gamma^h, \Gamma^v)$ as (Γ^h, Γ^v)-preservingness, because both Γ^h and Γ^v are distributions in the Finsler bundle F(M) and T is its base space. In the following we consider how to derive $\bar{X}$ from X.

It should be remarked that in Finsler geometry tangent vector fields of the tangent bundle T do not appear in essential; we consider Finsler vector fields. Therefore it seems that a point transformation (33.1) should be generalized as

$$\bar{x}^i = x^i + X^i(x, y)dt. \tag{33.3}$$

REMARK 33.2. Recently we have several authors who introduce and study point transformation such as (33.3). See Misra [8], Misra and Mishra [1, 2]. It seems that (33.3) gives a generalization of (33.1) really suitable to Finsler geometry. But, how to derive an equation corresponding to the second of (33.2) ? Although from (33.3) the equation

$$\frac{d\bar{x}^i}{dt} = \frac{dx^i}{dt} + \{\partial_j X^i \frac{dx^j}{dt} + \dot{\partial}_j X^i \frac{dy^j}{dt}\}dt$$

may be derived, we have no information on dy^j/dt.

The purpose of this and next sections is to introduce a reasonable extension of the above generalized point transformation (33.3) from the standpoint of fiber bundles.

DEFINITION 33.1. A *V-transformation* μ_V of the linear frame bundle L(M) over a manifold M is a family $\{\mu_{(v)} \mid v \in V^n\}$ of transformations $\mu_{(v)}$ of L(M), which is parametrized by elements of the standard fiber V^n of the tangent bundle T(M), and satisfies $\underline{\beta}_g \circ \mu_{(v)} = \mu_{(g^{-1}v)} \circ \underline{\beta}_g$.

Here we shall recall Definition 8.5 of V-connection. It is observed that the generalization of n-ple differential $\mu^{(n)}$ to V-transformation μ_V as above is formally analogous to that of linear connection $\underline{\Gamma}$ to V-connection Γ_V: If the generalizations are denoted by arrow, we see

$$\underline{\beta}'_g(\underline{\Gamma}_z) = \underline{\Gamma}_{zg} \longrightarrow \underline{\beta}'_g(\Gamma_{(v)z}) = \Gamma_{(g^{-1}v)zg}.$$

$$\underline{\beta}_g \circ \mu^{(n)} = \mu^{(n)} \circ \underline{\beta}_g \longrightarrow \underline{\beta}_g \circ \mu_{(v)} = \mu_{(g^{-1}v)} \circ \underline{\beta}_g.$$

It has been seen that from a V-connection Γ_V a unique non-linear connection N is obtained by Proposition 8.6. Similar situation occurs about a V-transformation:

DEFINITION 33.2. Let $\mu_V = \{\mu_{(v)} \mid v \in V^n\}$ be a V-transformation of a manifold M. The *associated transformation* μ with μ_V is a transformation of the tangent bundle T(M) given by $\mu(y) = \alpha_v \circ \mu_{(v)}(z)$ for any $y = zv \in T$, where α_v is the associated mapping $z \in L \mapsto zv \in T$ with respect to v.

The associated transformation μ of T is well-defined, independently on the choice of (z, v), because for $(zg, g^{-1}v)$ we see

$$\alpha_{g^{-1}v} \circ \mu_{(g^{-1}v)}(zg) = \alpha_{g^{-1}v} \circ \underline{\beta}_g \circ \mu_{(v)}(z) = \alpha_v \circ \mu_{(v)}(z).$$

Next we shall show a process by means of which a transformation of the Finsler bundle F(M) is derived from a V-transformation. To do so, we are first concerned with the product manifold $L\times V^n$ which is isomorphic with the total space F by the isomorphism $I = (\pi_2, \varepsilon)$, as it is given by Proposition 6.1.

DEFINITION 33.3. The *lift* $\mu^\times$ of a V-transformation μ_V to the product manifold $L\times V^n$ is a transformation of $L\times V^n$ which is given by $\mu^\times(z, v) = (\mu_{(v)}(z), v)$ for $(z, v) \in L\times V^n$.

Therefore, in the notation of (6.17), we have

$$\pi_2^\times\circ\mu^\times = \mu_{(\varepsilon^\times)}\circ\pi_2^\times, \qquad \varepsilon^\times\circ\mu^\times = \varepsilon^\times. \tag{33.4}$$

This is also regarded as another definition of $\mu^\times$.

We transfer the lift $\mu^\times$ to the Finsler bundle F(M) by the isomorphism I:

$$I^{-1}\circ\mu^\times\circ I(y, z) = I^{-1}\circ\mu^\times(z, z^{-1}y) = I^{-1}(\mu_{(z^{-1}y)}(z), z^{-1}y)$$

$$= ((\mu_{(z^{-1}y)}(z))(z^{-1}y), \mu_{(z^{-1}y)}(z)).$$

If we put $v = \varepsilon(y, z) = z^{-1}y$, the above is written as $((\mu_{(v)}(z))v, \mu_{(v)}(z))$, which leads to the following definition:

DEFINITION 33.4. The *lift* $\bar{\mu}$ of a V-transformation μ_V to the Finsler bundle F(M) is a transformation of the total space F which is given by

$$\pi_1\circ\bar{\mu} = \mu\circ\pi_1, \qquad \pi_2\circ\bar{\mu} = \mu_{(\varepsilon)}\circ\pi_2,$$

where μ is the associated transformation of T with μ_V.

To find a condition for a transformation of F to be the lift of a V-transformation, we first consider the lift to the product manifold $L\times V^n$:

THEOREM 33.1. *A necessary and sufficient condition for a transformation τ of the product manifold $L\times V^n$ to be the lift of a V-transformation to $L\times V^n$ is that*

(1) $\beta_g^\times\circ\tau = \tau\circ\beta_g^\times$, *(2)* $\varepsilon^\times\circ\tau = \varepsilon^\times$

are satisfied.

Proof. The necessity of (2) is evident from (33.4). From the definition of right translation $\beta_g^\times$ of $L\times V^n$ it follows that

$$\beta_g^\times\circ\mu^\times(z, v) = \beta_g^\times(\mu_{(v)}(z), v) = (\mu_{(v)}(z)g, g^{-1}v) = (\mu_{(g^{-1}v)}(zg), g^{-1}v)$$

$$= \mu^\times(zg, g^{-1}v) = \mu^\times\circ\beta_g^\times(z, v),$$

which shows (1).

Conversely, assume that a transformation τ of $L\times V^n$ satisfies (1) and (2). If we put

$$\mu_{(v)} = \overset{\times}{\pi}_2\circ\tau\circ\gamma_v, \tag{33.5}$$

where γ_v: $z \in L \mapsto (z, v) \in L\times V^n$, the family $\mu_V = \{\mu_{(v)} \mid v \in V^n\}$ is a V-transformation, because

$$\underline{\beta}_g\circ\mu_{(v)} = \overset{\times}{\pi}_2\circ\overset{\times}{\beta}_g\circ\tau\circ\gamma_v = \overset{\times}{\pi}_2\circ\tau\circ\overset{\times}{\beta}_g\circ\gamma_v = \overset{\times}{\pi}_2\circ\tau\circ\gamma_{g^{-1}v}\circ\underline{\beta}_g = \mu_{(g^{-1}v)}\circ\underline{\beta}_g,$$

which shows the equation of Definition 33.1. The lift $\overset{\times}{\mu}$ of this μ_V to $L\times V^n$ is given by

$$\overset{\times}{\mu}(z, v) = (\mu_{(v)}(z), v) = (\overset{\times}{\pi}_2\circ\tau\circ\gamma_v(z), v)$$

$$= (\overset{\times}{\pi}_2\circ\tau(z, v), \overset{\times}{\varepsilon}\circ\tau(z, v)) = \tau(z, v),$$

which implies $\overset{\times}{\mu} = \tau$.

Now, translating Theorem 33.1 into the languages of the Finsler bundle F(M) by the isomorphism I^{-1}, we immediately obtain a characterization of the lift $\bar{\mu}$ of μ_V to F as follows:

<u>THEOREM 33.2</u>. *A necessary and sufficient condition for a transformation π of the Finsler bundle F(M) to be the lift of a V-transformation to F is that*

$$(1)\quad \beta_g\circ\pi = \pi\circ\beta_g, \qquad (2)\quad \varepsilon\circ\pi = \varepsilon$$

are satisfied.

We deal with the homogeneity of above transformations for the later use.

<u>DEFINITION 33.5</u>. A V-transformation $\mu_V = \{\mu_{(v)} \mid v \in V^n\}$ of L(M) is called *p-homogeneous*, if $\mu_{(av)} = \mu_{(v)}$ holds for any $v \in V^n$ and any $a \in R^+$.

<u>PROPOSITION 33.1</u>. *If a V-transformation μ_V of L(M) is p-homogeneous, the associated transformation μ of T with μ_V, the lifts $\overset{\times}{\mu}$ to $L\times V^n$ and $\bar{\mu}$ to F of μ_V respectively satisfy*

$$(1)\quad \mu\circ{}_a h = {}_a h\circ\mu, \qquad (2)\quad \overset{\times}{\mu}\circ{}_a\overset{\times}{H} = {}_a\overset{\times}{H}\circ\overset{\times}{\mu}, \qquad (3)\quad \bar{\mu}\circ{}_a H = {}_a H\circ\bar{\mu},$$

where ${}_a\overset{\times}{H} = I\circ{}_a H\circ I^{-1}$ is a transformation of $L\times V^n$ such that $(z, v) \mapsto (z, av)$.

Now, starting from the n-ple differential $\underline{\mu}^n$ of a transformation $\underline{\mu}$ of an n-dimensional manifold M, we can also derive respective lifts $(\underline{\mu}^n)^\times$, $\bar{\underline{\mu}}^n$ to $L\times V^n$ and to F of $\underline{\mu}^n$. The lift $\bar{\underline{\mu}}^n$, a transformation of F, should satisfy additional conditions to

(1) and (2) of Theorem 33.2.

THEOREM 33.3. *A transformation π of the Finsler bundle $F(M)$ is the lift to F of n-ple differential of a transformation of the base space M, if and only if the following four conditions hold:*

$$(1)\quad \beta_g\circ\pi = \pi\circ\beta_g, \qquad (2)\quad \varepsilon\circ\pi = \varepsilon, \qquad (3)\quad S_v\circ\pi = \pi\circ S_v,$$

$$(4)\quad \theta^h\circ\pi' = \theta^h,$$

where $S_v : (y, z) \in F \mapsto (y + zv, z) \in F$ is the mapping appeared in Proposition 7.2, and θ^h is the h-basic form in (6.10).

Proof. It follows from Theorem 33.2 that (1) and (2) are necessary and sufficient for π to be the lift $\bar{\mu}$ of a V-transformation μ_V. If we introduce a transformation $S_v^{\times} = I\circ S_v\circ I^{-1}$ of $L\times V^n$ such that $(z, w) \mapsto (z, v + w)$, then, to show the necessity of (3), it is enough to see $S_v^{\times}\circ(\underline{\mu}^{(n)})^{\times} = (\underline{\mu}^{(n)})^{\times}\circ S_v^{\times}$:

$$S_v^{\times}\circ(\underline{\mu}^{(n)})^{\times}(z, w) = S_v^{\times}(\underline{\mu}^{(n)}(z), w) = (\underline{\mu}^{(n)}(z), w + v) = (\underline{\mu}^{(n)})^{\times}(z, w + v),$$

which is equal to $(\underline{\mu}^{(n)})^{\times}\circ S_v^{\times}(z, w)$; so that we have (3). Further

$$\theta^h\circ(\bar{\underline{\mu}}^{(n)})' = \theta\circ\pi_2'\circ(\bar{\underline{\mu}}^{(n)})' = \theta\circ(\underline{\mu}^{(n)})'\circ\pi_2'.$$

Thus Theorem 32.1 shows (4).

Conversely, assume that the lift $\bar{\mu}$ satisfies (3) and (4). Taking $u = (y, z) \in F$ and $w = z^{-1}y \in V^n$, Definition 33.4 and (3) yield

$$\mu_{(w+v)}(z) = \pi_2\circ\bar{\mu}(y + zv, z) = \pi_2\circ\bar{\mu}\circ S_v(y, z) = \pi_2\circ S_v\circ\bar{\mu}(y, z)$$

$$= \pi_2\circ S_v(\mu_{(w)}(z)w, \mu_{(w)}(z)) = \mu_{(w)}(z).$$

Therefore μ_V consists of a single transformation τ. The conditions (1) and (4) show that τ is the n-ple differential of a transformation of the base manifold M from Theorem 32.1. Consequently the proof is completed.

It is noteworthy that two additional conditions (3) and (4) appear in Theorem 33.3. It seems that this suggests the existence of a class of transformations with characteristic properties which are not so restrictive as the case of n-ple differential. Thus we are led to

DEFINITION 33.6. A transformation μ of the linear frame bundle $L(M)$ is called *linear*, if μ commutes with any right translation.

It is obvious that a V-transformation which consists of a single transformation is linear. Thus the condition (1) of Theorem 32.1 remains, while the condition (2)

is removed. Therefore the condition (4) of Theorem 33.3 is removed:

THEOREM 33.4. *A transformation π of the Finsler bundle $F(M)$ is the lift of a linear transformation of the linear frame bundle $L(M)$, if and only if the first three conditions of Theorem 33.3 hold.*

The term "linear" is originated from the following fact: We consider the associated transformation $'\mu$ of a linear transformation μ, namely,

$$'\mu(y) = \mu(z)v, \qquad y = zv. \tag{33.6}$$

It then follows that

$$\pi_T \circ {}'\mu(y) = \pi_T \circ \alpha_v(\mu(z)) = \pi_L(\mu(z)),$$

which implies that $'\mu$ is fiber-preserving. Further, for any a_1, $a_2 \in R$ and any y_1, $y_2 \in \pi_T^{-1}(x)$, we take $v_1 = z^{-1}y_1$, $v_2 = z^{-1}y_2$ and then

$$'\mu(a_1y_1 + a_2y_2) = \mu(z)(a_1v_1 + a_2v_2) = \mu(z)(a_1v_1) + \mu(z)(a_2v_2)$$

$$= a_1{}'\mu(y_1) + a_2{}'\mu(y_2),$$

which shows the linearity of $'\mu$ on each fiber.

PROPOSITION 33.2. *The associated transformation $'\mu$ with a linear transformation μ of $L(M)$ is characterized by (1) $'\mu$ is fiber-preserving and (2) $'\mu$ is linear on each fiber.*

It is enough for the proof to show the sufficiency of (1) and (2). We define a transformation μ of $L(M)$ by $\mu(z) = \{'\mu(z_a)\}$ for $z = \{z_a\}$. Since $'\mu$ is fiber-preserving, we obtain a linear frame $\{'\mu(z_a)\}$. Then it is seen from the linearity that

$$\mu(z)v = {}'\mu(z_a)v^a = {}'\mu(z_av^a), \qquad v = v^ae_a,$$

which implies (33.6). Further

$$\mu(zg) = \{'\mu(z_ag^a_b)\} = \{'\mu(z_a)g^a_b\} = \mu(z)g, \qquad g = (g^a_b).$$

Therefore the proof is completed.

Summarizing up the above all, we obtain desirable four conditions of a transformation π of the Finsler bundle $F(M)$:

(1) $\beta_g \circ \pi = \pi \circ \beta_g$, (2) $\varepsilon \circ \pi = \varepsilon$, (3) $S_v \circ \pi = \pi \circ S_v$,

(4) $\theta^h \circ \pi' = \theta^h$.

From Theorem 33.3 it is seen that the transformation π induced from a transformation of the base manifold M satisfies (1), (2), (3) and (4). Next, from Theorem 33.4 the transformation π induced from a linear transformation of L(M) satisfies (1), (2) and (3). Finally, from Theorem 33.2, the transformation π induced from a V-transformation satisfies (1) and (2) only.

We shall impose a condition on a V-transformation. It has been seen in Proposition 33.2 that a linear transformation is fiber-preserving.

DEFINITION 33.7. If the associated transformation μ with a V-transformation μ_V is fiber-preserving, μ_V is called *fiber-preserving*.

From Definition 33.2 we have

$$\pi_T\circ\mu(y) = \pi_T\circ\alpha_v\circ\mu_{(v)}(z) = \pi_L\circ\mu_{(v)}(z), \qquad y = zv,$$

which instantly implies

PROPOSITION 33.3. *A V-transformation $\mu_V = \{\mu_{(v)} \mid v \in V^n\}$ is fiber-preserving, if and only if $\pi_L\circ\mu_{(v)}$ does not depend on the choice of $v \in V^n$.*

From a fiber-preserving V-transformation μ_V, a transformation $\underline{\mu}$ of the base manifold M is uniquely induced such that $\pi_L\circ\mu_{(v)} = \underline{\mu}\circ\pi_L$ and then $\pi_T\circ\mu = \underline{\mu}\circ\pi_T$ holds. Thus we obtain the differential $\underline{\mu}'$ and the n-ple differential $\underline{\mu}^{(n)}$ of $\underline{\mu}$. It is observed from (32.2) that $\pi_L\circ\mu_{(v)} = \underline{\mu}\circ\pi_L = \pi_L\circ\underline{\mu}^{(n)}$, so that $\mu_{(v)}(z)$ and $\underline{\mu}^{(n)}(z)$ are on the same fiber for any $z \in L$. Therefore there exists $g \in G(n)$ such that $\mu_{(v)}(z) = \underline{\mu}^{(n)}(z)g$, and g is determined by $v \in V^n$ and $z \in L$. This observation leads us to

DEFINITION 33.8. The *deviation* α_V of a fiber-preserving V-transformation $\mu_V = \{\mu_{(v)} \mid v \in V^n\}$ is a family $\{\alpha_{(v)} \mid v \in V^n\}$ of mappings $\alpha_{(v)}: L \to G(n)$, which is parametrized by elements of V^n and given by $\mu_{(v)}(z) = \underline{\mu}^{(n)}(z)\alpha_{(v)}(z)$, where $\underline{\mu}^{(n)}$ is the n-ple differential of transformation $\underline{\mu}$ of the base space M induced from μ_V.

It is noted that the deviation of a linear transformation is a single mapping α: $L \to G(n)$ such that $\mu(z) = \underline{\mu}^{(n)}(z)\alpha(z)$. In general it is observed from Definition 33.1 that

$$\underline{\beta}_g(\underline{\mu}^{(n)}(z)\alpha_{(v)}(z)) = \underline{\beta}_g(\mu_{(v)}(z)) = \mu_{(g^{-1}v)}(zg) = \underline{\mu}^{(n)}(zg)\alpha_{(g^{-1}v)}(zg),$$

which implies from (32.4) that $\underline{\mu}^{(n)}(z)(\alpha_{(v)}(z)g) = \underline{\mu}^{(n)}(z)(g\alpha_{(g^{-1}v)}(zg))$. Thus we have

$$\alpha_{(v)}\circ\underline{\beta}_g = I_{g^{-1}}\circ\alpha_{(gv)}, \tag{33.7}$$

where I_g is an inner automorphism of G(n) by g. Then we are led to

DEFINITION 33.9. Let α_V be a family $\{\alpha_{(v)} \mid v \in V^n\}$ of mappings $\alpha_{(v)}: L \to G(n)$

which is parametrized by elements of V^n and satisfies (33.7). A V-transformation $\rho_V = \{\rho_{(v)} \mid v \in V^n\}$ which is given by $\rho_{(v)}(z) = z(\alpha_{(v)}(z))$ for $z \in L$ is called a *V-rotation*.

The family ρ_V is really a V-transformation of a kind, because from (33.7) we see

$$\underline{\beta}_g \circ \rho_{(v)}(z) = z(\alpha_{(v)}(z)g) = z(g\alpha_{(g^{-1}v)}(zg)) = \rho_{(g^{-1}v)}(zg).$$

From Definition 33.8 and (32.4) it follows that

$$\mu_{(v)}(z) = \underline{\mu}^{(n)}(z)\alpha_{(v)}(z) = \underline{\mu}^{(n)}(z\alpha_{(v)}(z)),$$

which implies

(33.8) $$\mu_{(v)} = \underline{\mu}^{(n)} \circ \rho_{(v)}.$$

Consequently we have an interesting result as follows:

THEOREM 33.5. *A fiber-preserving V-transformation is considered as the composition of a V-rotation and an n-ple differential of a transformation of the base space.*

Therefore such a V-transformation is only the composition of two simple transformations.

By every transformation belonging to a V-rotation, any point of L is moved to a point on a same fiber with the original point, that is, any linear frame at $x \in M$ is moved to a linear frame at the same point x. An n-ple differential is nothing but an extended point transformation, as mentioned at the beginning of the present section.

REMARK 33.3. The theory of linear transformations is developed by Matsumoto [2, 3, 12]. From Theorem 33.5 it is observed that a linear transformation is the composition of a linear rotation and an n-ple differential, so that a generalized transformation of the form (33.3) does not take place. Although Misra and Mishra treat such a generalized transformation given by (33.3), their discussion is not satisfactory. See Matsumoto and Radu [1].

§ 34. Infinitesimal V-transformations

To consider an infinitesimal transformation, we shall start from a *one-parameter V-transformation group* $\{\mu_{Vt}\} = \{\mu_{(v)t} \mid v \in V^n, t \in R\}$. It is assumed that for a fixed $t \in R$ we have a V-transformation μ_{Vt} and for a fixed $v \in V$ we have a one-parameter transformation group $\{\mu_{(v)t}\}$ of the linear frame bundle L(M).

Corresponding to $\{\mu_{Vt}\}$, the lift $\{\bar{\mu}_t\}$ to the Finsler bundle F(M) is obtained. It is a family of lifts $\bar{\mu}_t$ of μ_{Vt} for a fixed t in the sense of Definition 33.4.

DEFINITION 34.1. Let $\{\mu_{Vt}\}$ be a one-parameter V-transformation group and $\{\bar{\mu}_t\}$ be the lift of $\{\mu_{Vt}\}$ to the Finsler bundle F(M). The tangent vector field X_V of F, which is induced by $\{\bar{\mu}_t\}$, is called an *infinitesimal V-transformation.*

From Theorem 33.2, Corollaries 2.1 and 2.2 a characterization of an infinitesimal V-transformation is easily obtained:

THEOREM 34.1. *A tangent vector field X_V of the Finsler bundle F(M) is an infinitesimal V-transformation, if and only if the following two equations hold:*

(1) $\beta'_g(X_V) = X_V$ or $[X_V, Z(A)] = 0$, (2) $X_V(\varepsilon) = 0$.

Let a Finsler connection $F\Gamma$ be given. With respect to it, an infinitesimal V-transformation X_V is written in the form

$$X_V = B^h(v_V) + B^v(w_V) + Z(A_V),$$

where v_V, w_V are V^n-valued functions on F and A_V is a G'(n)-valued function on F. It follows from (6.5), (8.7), (9.8) and (1) of Theorem 34.1 that at a point $u \in F$

$$\beta'_g(X_V)_u = B^h(g^{-1}(v_V)_u)_{ug} + B^v(g^{-1}(w_V)_u)_{ug} + Z(ad(g^{-1})(A_V)_u)_{ug}$$

$$= (X_V)_{ug} = B^h((v_V)_{ug})_{ug} + B^v((w_V)_{ug})_{ug} + Z((A_V)_{ug})_{ug},$$

which easily implies the Finsler tensor field property of v_V, w_V and A_V.

Next, it follows from (6.13), (8.10), (9.16) that (2) of Theorem 34.1 is written as

$$0 = D(v_V) + w_V + C(\varepsilon, w_V) - A_V\cdot\varepsilon.$$

Consequently we first have

PROPOSITION 34.1. *With respect to a Finsler connection $F\Gamma$ an infinitesimal V-transformation X_V is written as*

$$X_V = B^h(v_V) + B^v(w_V) + Z(A_V),$$

where v_V, w_V, A_V are Finsler tensor fields of (1, 0), (1, 0) and (1, 1)-type respectively and satisfy

$$A_V\cdot\varepsilon = D(v_V) + w_V + C(\varepsilon, w_V).$$

REMARK 34.1. It is noted that if $F\Gamma$ satisfies D and C_1-conditions as the Cartan, Berwald and Rund connections, the above equation simply reduces to $A_V\cdot\varepsilon = w_V$, so that X_V is written by only two tensors v_V and A_V.

We consider the homogeneity of X_V. If $\{\bar{\mu}_t\}$ is p-homogeneous in the sense of

Proposition 33.1, it follows from Corollary 2.1 that ${}_aH'(X_V) = X_V$, so that X_V is (0) p-homogeneous. Consider a p-homogeneous Finsler connection $F\Gamma$. Then from Propositions 12.2 and 12.4 it follows that

$$ {}_aH'(X_V)_u = B^h((v_V)_u)_{au} + B^v(a(w_V)_u)_{au} + Z((A_V)_u)_{au} $$

$$ = (X_V)_{au} = B^h((v_V)_{au})_{au} + B^v((w_V)_{au})_{au} + Z((A_V)_{au})_{au}. $$

Consequently we have

PROPOSITION 34.2. *A homogeneous infinitesimal V-transformation X_V is (0)p-homogeneous. With respect to a p-homogeneous Finsler connection the tensor fields v_V, w_V and A_V in Proposition 34.1 are (0), (1) and (0)p-homogeneous respectively.*

Now we deal with special infinitesimal V-transformations. Theorems 33.3 and 33.4 and Proposition 7.2 lead us immediately to

THEOREM 34.2.

(1) If an infinitesimal V-transformation X_V is induced from a one-parameter transformation group of the base space M, then X_V satisfies

(1) $\beta'_g(X_V) = X_V$ *or* $[X_V, Z(A)] = 0$,

(2) $X_V(\varepsilon) = 0$,

(3) $S'_v(X_V) = X_V$ *or* $[X_V, Y(v)] = 0$,

(4) $L_V\theta^h = 0$.

(2) If an infinitesimal V-transformation X_V is induced from a one-parameter linear transformation group, X_V satisfies above (1), (2) and (3).

Throughout the following we shall adopt the notation L_V to denote the Lie derivative with respect to X_V.

By Proposition 32.1, above four conditions are also written respectively as

(34.1)
(1) $L_V Z(A) = 0$, (2) $L_V\varepsilon = 0$,
(3) $L_V Y(v) = 0$, (4) $L_V\theta^h = 0$.

Next, we are concerned with an infinitesimal V-transformation X_V induced from $\{\mu_{Vt}\}$ such that each μ_{Vt} for a fixed t is a V-rotation ρ_{Vt}. In this case X_V is called an *infinitesimal V-rotation* and denoted by X_ρ.

We consider a tangent vector field $X_{(v)}$ of L which is induced by a one-parameter V-rotation group $\{\rho_{(v)t}\}$ for a fixed v. Taking a real-valued function f on L, it follows from Definition 33.9 that

$$X_{(v)}(f)_z = d_0 t\{f \circ \rho_{(v)t}(z)\} = d_0 t\{f \circ z(\alpha_{(v)t}(z))\}.$$

Thus, if we denote by $A_{(v)}(z) \in G'(n)$ the tangent vector of curve $t \in R \to \alpha_{(v)t}(z) \in G(n)$ at the unit e, we have $X_{(v)}(f)_z = \underline{Z}(A_{(v)}(z))_z(f)$ from Proposition 2.5 and (32.1). It then follows from Definition 33.3 that

$$(X_\rho)_{(zv,\ z)} = (I^{-1})'(\underline{Z}(A_{(v)}(z))_z,\ 0).$$

Consequently, from (6.19) and Definition 7.3 we have

<u>PROPOSITION 34.3</u>. *An infinitesimal V-rotation* X_ρ *is written in the form*

$$X_\rho = Y(A_\rho \cdot \varepsilon) + Z(A_\rho),$$

where A_ρ *is a Finsler tensor field of (1, 1)-type.*

It is easy to show the tensor property of A_ρ. An infinitesimal V-rotation X_ρ obviously satisfies (1) and (2) of Theorem 34.2; (1) shows the tensor property of A_ρ and (2) holds instantly from Proposition 7.1 and (6.13). We shall examine the condition (3). It follows from (7.3) and (7.4) that

$$[X_\rho,\ Y(v)] = [Y(A_\rho \cdot \varepsilon),\ Y(v)] + [Z(A_\rho),\ Y(v)]$$

$$= -\ Y(\nabla^0(A_\rho \cdot \varepsilon)v) + Y(A_\rho \cdot v) - Z(\nabla^0 A_\rho(v)).$$

Hence (9.15) leads to

$$(34.2) \qquad [X_\rho,\ Y(v)] = -\ Y(\nabla^0 A_\rho(\varepsilon,\ v)) - Z(\nabla^0 A_\rho(v)).$$

Further we shall introduce a special infinitesimal V-transformation with respect to a given non-linear connection N, which will be important later on.

<u>DEFINITION 34.2</u>. Let a non-linear connection N be given. An infinitesimal V-transformation X_ν is called *N-natural* with respect to N, if the Lie derivative $\mathcal{L}_\nu B^h(v)$ with respect to X_ν of any h-basic vector field $B^h(v)$ of a Finsler connection $F\Gamma = (\Gamma,\ N)$ having the non-linear connection N is quasi-vertical.

It should be verified that the N-naturality property depends only on the non-linear connection N under consideration. From Proposition 15.1 it is seen that two Finsler connections $(\Gamma,\ N)$ and $(*\Gamma,\ N)$ have the common non-linear connection N, if and only if

$$*B^h(v) = B^h(v) + Z(A^h(v))$$

is satisfied for any $v \in V^n$. It follows from the above and Theorem 34.2 that

$$\mathcal{L}_\nu *B^h(v) = [X_\nu,\ *B^h(v)] = \mathcal{L}_\nu B^h(v) + Z(X_\nu(A^h(v))).$$

Hence $L_\nu{*}B^h(v)$ is quasi-vertical, if and only if $L_\nu B^h(v)$ is so.

We consider $L_V B^h(v) = [X_V, B^h(v)]$ in general case. By structure equations of the Finsler connection and Proposition 34.1 we have

$$
\begin{aligned}
L_V B^h(v) &= B^h(T(v_V, v) - \nabla^h v_V(v) - C(v, w_V) + A_V\cdot v) \\
&\quad + B^v(R^1(v_V, v) - P^1(v, w_V) - \nabla^h w_V(v)) \qquad (34.3)\\
&\quad + Z(R^2(v_V, v) - P^2(v, w_V) - \nabla^h A_V(v)).
\end{aligned}
$$

Therefore X_V is N-natural, if and only if

$$T(v_V, v) - \nabla^h v_V(v) - C(v, w_V) + A_V\cdot v = 0.$$

Combining the above with Proposition 34.1, we obtain

PROPOSITION 34.4. *An infinitesimal N-natural transformation X_ν with respect to a non-linear connection N is written in the form*

$$X_\nu = B^h(v_\nu) + B^v(w_\nu) + Z(A_\nu),$$

referring to a Finsler connection with the non-linear connection N, where tensors v_ν, w_ν and A_ν satisfy

(1) $w_\nu = \nabla^h v_\nu(\varepsilon) - T(v_\nu, \varepsilon) - D(v_\nu)$,

(2) $A_\nu\cdot v = \nabla^h v_\nu(v) - T(v_\nu, v) + C(v, w_\nu)$ *for any* $v \in V^n$.

REMARK 34.2. From the above we observe that an infinitesimal N-natural transformation X_ν is uniquely determined by its projection $\pi'_L\circ\pi'_1(X_\nu)$ onto the base space M. In case of the Cartan connection the equations in Proposition 34.4 reduce to

$$w_\nu = \nabla^h v_\nu(\varepsilon), \qquad A_\nu = \nabla^h v_\nu + C(\nabla^h v_\nu(\varepsilon)).$$

The concept of N-naturality is important from the viewpoint of Theorem 34.3. It is noteworthy that a non-linear connection plays an essential role in generalized infinitesimal transformations as well as in the theory of linear connections of Finsler type which has been developed in Chapter IV.

THEOREM 34.3. *Any infinitesimal V-transformation X_V is uniquely written as the sum $X_\nu + X_\rho$ of an infinitesimal N-natural transformation X_ν and an infinitesimal V-rotation X_ρ.*

Proof. We shall refer to the Finsler N-connection FN which is given by Definition 15.1. Its Cartan tensor C vanishes, hence $B^v(v) = Y(v)$, so that from Proposition 34.1 the tangent vector X_V is written

$$X_V = B^h(v_V) + Y(w_V) + Z(A_V),$$

where $A_V \cdot \varepsilon = D(v_V) + w_V$. The above is rewritten in the form

$$X_V = \{B^h(v_V) + Y(\nabla^h v_V(\varepsilon) - T(v_V, \varepsilon) - D(v_V)) + Z(\nabla^h v_V - T(v_V, \delta))\}$$

$$+ \{Y(w_V - \nabla^h v_V(\varepsilon) - T(\varepsilon, v_V) + D(v_V)) + Z(A_V - \nabla^h v_V - T(\delta, v_V))\}.$$

The terms in the first parentheses are written as

$$X_\nu = B^h(v_\nu) + Y(w_\nu) + Z(A_\nu),$$

$$v_\nu = v_V, \qquad w_\nu = \nabla^h v_V(\varepsilon) - T(v_V, \varepsilon) - D(v_V),$$

$$A_\nu = \nabla^h v_V - T(v_V, \delta).$$

These evidently satisfy (1) and (2) of Proposition 34.4, so that X_ν is an infinitesimal N-natural transformation. On the other hand, the terms in the second parentheses are written as

$$X_\rho = Y(w_\rho) + Z(A_\rho),$$

$$A_\rho = A_V - \nabla^h v_V - T(\delta, v_V), \qquad w_\rho = w_V - \nabla^h v_V(\varepsilon) - T(\varepsilon, v_V) + D(v_V).$$

Because $w_\rho = A_\rho \cdot \varepsilon$ is easily shown, X_ρ is really an infinitesimal V-rotation. The uniqueness will be obvious.

Therefore it is concluded that it is enough for studying infinitesimal V-transformations to consider N-natural transformations and V-rotations only.

In terms of coordinate system we consider an infinitesimal N-natural transformation X_ν and a V-rotation X_ρ. As for X_ν, the Finsler N-connection FN is referred. We denote by $X^i(x, y)$ the components of Finsler vector field v_ν in Proposition 34.4. Paying attention to $C = 0$, we have the components of w_ν and A_ν as follows:

$$w_\nu : (\delta_j X^i) y^j + N^i_{\ j} X^j, \qquad A_\nu : \delta_j X^i + F_j{}^i{}_k X^k.$$

Hence, from (9.10), (7.5) and (6.4') we obtain

$$X_\nu = X^i\left(\frac{\partial}{\partial x^i} - N^j_{\ i}\frac{\partial}{\partial y^j} - F_k{}^j{}_i z^k_a \frac{\partial}{\partial z^j_a}\right) + (\delta_j X^i \cdot y^j + N^i_{\ j} X^j)\frac{\partial}{\partial y^j}$$

$$+ (\delta_j X^i + F_j{}^i{}_k) z^j_a \frac{\partial}{\partial z^i_a}.$$

Consequently X_ν is written in the form

$$X_\nu = X^i(x, y)\frac{\partial}{\partial x^i} + (\delta_j X^i) y^j \frac{\partial}{\partial y^i} + (\delta_j X^i) z^j_a \frac{\partial}{\partial z^i_a}. \tag{34.4}$$

It is noted that coefficients N^i_j of the non-linear connection N appear in the δ-derivative.

Next we deal with an infinitesimal V-rotation X_ρ. Denoting A_ρ in Proposition 34.3 by $X^i_j(x, y)$, from Proposition 34.3 we immediately get

$$X_\rho = X^i_j(x, y)y^j \frac{\partial}{\partial y^i} + X^i_j z^j_a \frac{\partial}{\partial z^i_a} . \tag{34.5}$$

REMARK 34.3. It is observed from (34.4) that the projection $\pi'_1(X_\nu)$ of X_ν on the tangent bundle T(M) is given by

$$\pi'_1(X_\nu) = X^i(x, y) \frac{\partial}{\partial x^i} + (\delta_j X^i)y^j \frac{\partial}{\partial y^i} ,$$

which generates the infinitesimal transformation of T associated with an infinitesimal N-natural V-transformation. Therefore, in the old-fashioned notation, this transformation is written

$$\bar{x}^i = x^i + X^i(x, y)dt, \qquad \bar{y}^i = y^i + (\delta_j X^i)y^j dt.$$

This is nothing but what we have wished to obtain as a suitable generalization of extended point transformation.

On the other hand, from (34.5) it is seen that the infinitesimal V-rotation is written in the old-fashioned form

$$\bar{x}^i = x^i, \qquad \bar{y}^i = y^i + X^i_j(x, y)y^j dt.$$

This is completely new transformation; a point is fixed and a line-element y^i rotates, depending on the point and itself.

Laptev [1 - 8] has considered generalized Lie derivatives of spaces of line-elements. See his report [9].

We are concerned with an infinitesimal linear transformation, which satisfies (1), (2) and (3) of Theorem 34.2. From the viewpoint of Theorem 34.3 it is enough to deal with an infinitesimal N-natural linear transformation and an infinitesimal linear V-rotation.

PROPOSITION 34.5.

(1) *An infinitesimal N-natural linear transformation X_ν is a so-called extended point transformation.*

(2) *An infinitesimal V-rotation $X_\rho = Y(A_\rho \cdot \varepsilon) + Z(A_\rho)$ is linear, if and only if $\nabla^0 A_\rho = 0$, namely, X^i_j in (34.5) are functions of position x^i only.*

Proof. (2) immediately follows from (34.2). To show (1), we first treat a general infinitesimal V-transformation X_V. From Proposition 32.1 it follows that

$$(L_V\theta^h)Z(A) = X_V(\theta^h(Z(A))) - \theta^h([X_V, Z(A)]) = -\theta^h(L_VZ(A)).$$

Similarly we get

$$(34.6)\quad \begin{aligned} &(1)\quad (L_V\theta^h)Z(A) = -\theta^h(L_VZ(A)) = 0,\\ &(2)\quad (L_V\theta^h)B^h(v) = -\theta^h(L_VB^h(v)),\\ &(3)\quad (L_V\theta^h)B^v(v) = -\theta^h(L_VB^v(v)),\\ &(4)\quad (L_V\theta^h)Y(v) = -\theta^h(L_VY(v)). \end{aligned}$$

We consider an infinitesimal N-natural linear transformation X_ν. Then quantities of (2) and (4) of (34.6) vanish from the N-naturality and linearity respectively. Therefore $L_V\theta^h = 0$, so that X_ν is an extended one.

To introduce another special V-rotation, we are led to

DEFINITION 34.3. An infinitesimal V-transformation X_σ is called *semi-linear*, if the Lie derivative $L_\sigma Y(v)$ of any induced-fundamental vector field $Y(v)$ with respect to X_σ is vertical.

The semi-linearity condition is surely weaker than the linearity, because $L_VY(v)$ vanishes in the later case. In general, from Proposition 34.1 the Lie derivative $L_VY(v) = [X_V, Y(v)]$ is written

$$(34.7)\quad \begin{aligned} L_VY(v) = &-B^h(\nabla^0 v_V(v)) + Y(P^1(v_V, v) - \nabla^0 w_V(v) + A_V\cdot v)\\ &+ Z(P^2(v_V, v) - \nabla^0 A_V(v)) \end{aligned}$$

in terms of a Finsler connection $F\Gamma$ having the flat vertical connection F^1 $(C = 0)$. Hence the semi-linearity is given by

$$\nabla^0 v_V = 0, \qquad \nabla^0 w_V = A_V + P^1(v_V, \delta).$$

From Proposition 34.1 we then obtain

$$\nabla^0 w_V(\varepsilon) = D(v_V) + w_V + P^1(v_V, \varepsilon).$$

Paying attention to Theorem 12.2, we have

PROPOSITION 34.6. *With respect to a p-homogeneous Finsler connection having the flat vertical connection, an infinitesimal semi-linear transformation X_σ is written in the form*

$$X_\sigma = B^h(v_\sigma) + Y(w_\sigma) + Z(A_\sigma),$$

where

$$\nabla^0 v_\sigma = 0, \qquad A_\sigma = \nabla^0 w_\sigma - P^1(v_\sigma, \delta)$$

hold and w_σ *is (1)p-homogeneous.*

Denoting by X^i and Y^i respective components of v_σ and w_σ of Proposition 34.6, we have $X^i = X^i(x)$ and $Y^i(x, y)$ are (1)p-homogeneous in y^i. The components of A_σ are given by

$$\dot\partial_j Y^i - (\dot\partial_j N^i{}_k - F_j{}^i{}_k)X^k.$$

Then we have

$$X_\sigma = X^i \frac{\partial}{\partial x^i} + (Y^i - N^i{}_j X^j) \frac{\partial}{\partial y^i} + (\dot\partial_j Y^i - X^k \dot\partial_j N^i{}_k) z^j_a \frac{\partial}{\partial z^i_a}.$$

The quantities $Z^i = Y^i - N^i{}_j X^j$ are (1)p-homogeneous in y^i, and we obtain

(34.8)
$$X_\sigma = X^i(x) \frac{\partial}{\partial x^i} + Z^i(x, y) \frac{\partial}{\partial y^i} + (\dot\partial_j Z^i) z^j_a \frac{\partial}{\partial z^i_a}.$$

As a consequence, an infinitesimal semi-linear V-rotation $X_{\sigma\rho}$ is written as

(34.9)
$$X_{\sigma\rho} = Z^i(x, y) \frac{\partial}{\partial y^i} + (\dot\partial_j Z^i) z^j_a \frac{\partial}{\partial z^i_a}.$$

This has a special form of (34.5) because $(\dot\partial_j Z^i) y^j = Z^i$.

EXAMPLE 34.1. There are certain intrinsic tangent vector fields in the Finsler bundle F(M). For instance, we have a special fundamental vector field $Z(\delta) = z^i_a \partial/\partial z^i_a$. This is evidently not an infinitesimal V-transformation because of Proposition 34.1. Next, the intrinsic induced-fundamental vector field $Y(\varepsilon) = y^i \partial/\partial y^i$ is not also. However, the sum $Y(\varepsilon) + Z(\delta)$ is an infinitesimal linear V-rotation because of Proposition 34.5.

Assume that a Finsler connection $F\Gamma$ is given and consider h-and v-basic vector fields. The *intrinsic h-basic vector field* $B^h(\varepsilon)$ is an infinitesimal V-transformation, if and only if $D(\varepsilon) = 0$. This is further N-natural, if and only if $D(v) = T(\varepsilon, v)$ for any $v \in V^n$. Accordingly $B^h(\varepsilon)$ of the Cartan connection $C\Gamma = (\Gamma, N)$ is an infinitesimal N-natural transformation with respect to the N. On the other hand, the *intrinsic v-basic vector field* $B^v(\varepsilon)$ is an infinitesimal V-transformation, if and only if $\varepsilon + C(\varepsilon, \varepsilon) = 0$; in this case $B^v(\varepsilon)$ is a rotation, as it is seen from (8.8).

From the above standpoint of the theory of V-transformations these conditions

$$D(\varepsilon) = 0, \qquad D(v) = T(\varepsilon, v), \qquad \varepsilon + C(\varepsilon, \varepsilon) = 0$$

for a Finsler connection may be important, although we do not pay attention to these conditions in Chapter III.

§ 35. Lie derivatives of torsion and curvature tensors

Let a general infinitesimal V-transformation X_V be given. We consider Lie derivatives with respect to X_V of various quantities arising from a general Finsler connection. We should always pay attention to $L_V Z(A) = 0$ and $L_V \varepsilon = 0$.

We shall put Lie derivatives of basic vector fields in the form

$$(35.1)\qquad \begin{aligned} L_V B^h(v) &= B^h(\alpha(v)) + B^v(\alpha^1(v)) + Z(\alpha^2(v)),\\ L_V B^v(v) &= B^h(\beta(v)) + B^v(\beta^1(v)) + Z(\beta^2(v)). \end{aligned}$$

It follows from (34.3) that the Finsler tensor fields α, α^1 and α^2 of (1, 1)-, (1, 1)- and (1, 2)-type respectively are given by

$$(35.2)\qquad \begin{aligned} \alpha(v) &= T(v_V, v) - \nabla^h v_V(v) - C(v, w_V) + A_V\cdot v,\\ \alpha^1(v) &= R^1(v_V, v) - P^1(v, w_V) - \nabla^h w_V(v),\\ \alpha^2(v) &= R^2(v_V, v) - P^2(v, w_V) - \nabla^h A_V(v). \end{aligned}$$

Similarly the Finsler tensor fields β, β^1 and β^2 are given by

$$(35.3)\qquad \begin{aligned} \beta(v) &= C(v_V, v) - \nabla^v v_V(v) = - \nabla^0 v_V(v),\\ \beta^1(v) &= P^1(v_V, v) - S^1(v, w_V) - \nabla^v w_V(v) + A_V\cdot v,\\ \beta^2(v) &= P^2(v_V, v) - S^2(v, w_V) - \nabla^v A_V(v). \end{aligned}$$

In particular, we are first concerned with the Lie derivative L_λ with respect to an infinitesimal linear transformation X_λ. From Theorem 34.2 and (8.8) we see

$$L_\lambda Y(v) = 0 = B^h(\beta(v)) + B^v(\beta^1(v)) + Z(\beta^2(v) + L_\lambda C(v)),$$

which implies $\beta = \beta^1 = 0$ and $\beta^2 = - L_\lambda C$.

Secondly, if we denote by X_ε an extended point transformation (N-natural and linear), from (1) of Theorem 34.2, (34.6) and (35.1) we get

$$(L_\varepsilon \theta^h) B^h(v) = 0 = - \alpha(v), \qquad (L_\varepsilon \theta^h) B^v(v) = 0 = - \beta(v),$$

where by L_ε we denote the Lie derivative with respect to X_ε. Consequently

PROPOSITION 35.1. *For an infinitesimal V-transformation the tensor fields* α, α^1, α^2, β, β^1 *and* β^2 *are given by (35.2) and (35.3). In particular*

(1) for an infinitesimal linear transformation X_λ *we have* $\beta = \beta^1 = 0$, $\beta^2 = - L_\lambda C$.

(2) *for an extended point transformation* X_ε *we have* $\beta = \beta^1 = \alpha = 0$, $\beta^2 = -L_\varepsilon C$.

Next we consider an infinitesimal N-natural transformation X_ν. From Definition 34.2 it is characterized by the fact that $L_\nu B^h(v)$ is quasi-vertical, so that we have $\alpha = 0$ from (35.1).

Finally we deal with an infinitesimal V-rotation X_ρ. From Proposition 34.3 and (8.8) it follows that X_ρ is written

$$X_\rho = B^v(A_\rho\cdot\varepsilon) + Z(A_\rho + C(A_\rho\cdot\varepsilon)).$$

Therefore, in general equations (35.2) and (35.3) we have $v_V = 0$, $w_V = A_\rho\cdot\varepsilon$ and $A_V = A_\rho + C(A_\rho\cdot\varepsilon)$. Accordingly

$$\alpha(v) = A_\rho\cdot v, \qquad \alpha^1(v) = -P^1(v, A_\rho\cdot\varepsilon) - \nabla^h(A_\rho\cdot\varepsilon)v,$$

(35.4)
$$\alpha^2(v) = -P^2(v, A_\rho\cdot\varepsilon) - \nabla^h(A_\rho + C(A_\rho\cdot\varepsilon))v,$$

$$\beta = 0, \qquad \beta^1(v) = C(A_\rho\cdot\varepsilon, v) - \nabla^v(A_\rho\cdot\varepsilon)v + A_\rho\cdot v,$$

$$\beta^2(v) = S^2(A_\rho\cdot\varepsilon, v) - \nabla^v(A_\rho + C(A_\rho\cdot\varepsilon))v.$$

Consequently we have

PROPOSITION 35.2.

(1) *For an infinitesimal N-natural transformation* X_ν *we have* $\alpha = 0$.

(2) *For an infinitesimal V-rotation* X_ρ *we have* $\beta = 0$ *and* α, α^1, α^2, β^1, β^2 *are given by (35.4).*

The deflection tensor D of a Finsler connection $F\Gamma$ is defined by $D(v) = B^h(v)\varepsilon$. From (35.1) and $L_V\varepsilon = 0$ we have

$$[X_V, B^h(v)]\varepsilon = L_V(B^h(v)\varepsilon) = B^h(\alpha(v))\varepsilon + B^v(\alpha^1(v))\varepsilon + Z(\alpha^2(v))\varepsilon.$$

Paying attention to (9.14), we obtain

$$L_V D(v) = D(\alpha(v)) + \alpha^1(v) + C(\varepsilon, \alpha^1(v)) - \alpha^2(\varepsilon, v). \tag{35.5}$$

Similarly, from the Lie derivative $L_V(B^v(v)\varepsilon)$ we have

$$L_V C(\varepsilon, v) = D(\beta(v)) + \beta^1(v) + C(\varepsilon, \beta^1(v)) - \beta^2(\varepsilon, v). \tag{35.6}$$

As a consequence of (35.5) and (35.6), the following similar to (1) of Theorem 13.3 is concluded:

PROPOSITION 35.3. *If a Finsler connection satisfies the D- and C_1-conditions, in (35.1) we have $\alpha^1(v) = \alpha^2(\varepsilon, v)$ and $\beta^1(v) = \beta^2(\varepsilon, v)$ for any $v \in V^n$.*

REMARK 35.1. The D- and C_1-conditions for a Finsler connection are introduced in § 13 from the viewpoint to give special relations among connections which constitute the Finsler connection, and Proposition 35.3 shows also the importance of these conditions from the viewpoint of the theory of transformations.

The first equation of (35.3) is simple and interesting. From (34.4) it is observed that an infinitesimal N-natural transformation X_ν with $\nabla^0 v_\nu = 0$ is an extended one:

PROPOSITION 35.4. *In the notations of Proposition 34.4, an infinitesimal N-natural transformation X_ν with $\nabla^0 v_\nu = 0$ ($\beta = 0$) is an extended point transformation.*

We shall show another proof of Proposition 35.4 without referring to coordinate system. With respect to the Finsler N-connection FN the equations (11.4) and (13.8) respectively reduce to

$$\nabla^0 T(1, 2, 3) = F(1, 2, 3) - F(2, 1, 3),$$

$$\nabla^0 D(1, 2) = F(\varepsilon, 1, 2).$$

By these equations and (10.22) it is easily shown from Proposition 34.4 that

$$\nabla^0 w_\nu = A , \qquad \nabla^0 A_\nu(1, 2) = F(1, v_\nu, 2).$$

The equation (34.7) is written as

$$L_\nu Y(v) = - B^h(\nabla^0 v_\nu(v)) + Y(- \nabla^0 w_\nu(v) + A_\nu \cdot v) + Z(F(v_\nu, v) - \nabla^0 A_\nu(v)).$$

Thus $L_\nu Y(v)$ vanishes in this case; so X_ν is linear. Consequently (1) of Proposition 34.5 shows the fact that we wish to prove.

As for an infinitesimal V-rotation X_ρ, the following proposition similar to the above is easily shown by (34.2) and (34.5):

PROPOSITION 35.5. *In the notations of Proposition 34.3, an infinitesimal V-rotation X_ρ is linear, if and only if $\nabla^0 A_\rho = 0$, and we have $\beta^1 = 0$.*

Next we consider Jacobi identities satisfied by an infinitesimal V-transformation X_V and basic vector fields of a Finsler connection. First we are concerned with

$$[X_V, [B^h(1), B^h(2)]] + \mathfrak{A}_{(12)}\{[B^h(1), [B^h(2), X_V]]\} = 0.$$

By (35.1) and the structure equation (10.3) this is rewritten in the form which is the sum of h-, v-parts and vertical part. Equating every part to zero, we obtain three equations

$$(35.7)\quad \begin{aligned} L_V T(1, 2) &= - \alpha(T(1, 2)) - \beta(R^1(1, 2)) + \mathfrak{A}_{(12)}\{\nabla^h\alpha(2, 1) \\ &\quad - \alpha^2(1, 2) + T(1, \alpha(2)) + C(1, \alpha^1(2))\}, \end{aligned}$$

$$(35.8)\quad \begin{aligned} L_V R^1(1, 2) &= - \alpha^1(T(1, 2)) - \beta^1(R^1(1, 2)) + \mathfrak{A}_{(12)}\{\nabla^h\alpha^1(2, 1) \\ &\quad + R^1(1, \alpha(2)) + P^1(1, \alpha^1(2))\}, \end{aligned}$$

$$(35.9)\quad \begin{aligned} L_V R^2(1, 2) &= - \alpha^2(T(1, 2)) - \beta^2(R^1(1, 2)) + \mathfrak{A}_{(12)}\{\nabla^h\alpha^2(2, 1) \\ &\quad + R^2(1, \alpha(2)) + P^2(1, \alpha^1(2))\}. \end{aligned}$$

Secondly the Jacobi identity

$$[X_V, [B^h(1), B^v(2)]] + [B^h(1), [B^v(2), X_V]] + [B^v(2), [X_V, B^h(1)]] = 0$$

similarly yields three equations

$$(35.10)\quad \begin{aligned} L_V C(1, 2) &= \nabla^h\beta(2, 1) - \nabla^v\alpha(1, 2) - \beta^2(1, 2) + T(1, \beta(2)) \\ &\quad - \alpha(C(1, 2)) + C(\alpha(1), 2) + C(1, \beta^1(2)) - \beta(P^1(1, 2)), \end{aligned}$$

$$(35.11)\quad \begin{aligned} L_V P^1(1, 2) &= \nabla^h\beta^1(2, 1) - \nabla^v\alpha^1(1, 2) + \alpha^2(2, 1) + R^1(1, \beta(2)) \\ &\quad - \alpha^1(C(1, 2)) - \beta^1(P^1(1, 2)) + P^1(\alpha(1), 2) + P^1(1, \beta^1(2)) \\ &\quad - S^1(2, \alpha^1(1)), \end{aligned}$$

$$(35.12)\quad \begin{aligned} L_V P^2(1, 2) &= \nabla^h\beta^2(2, 1) - \nabla^v\alpha^2(1, 2) - \alpha^2(C(1, 2)) - \beta^2(P^1(1, 2)) \\ &\quad + R^2(1, \beta(2)) + P^2(\alpha(1), 2) + P^2(1, \beta^1(2)) - S^2(2, \alpha^1(1)). \end{aligned}$$

Finally the Jacobi identity

$$[X_V, [B^v(1), B^v(2)]] + \mathfrak{A}_{(12)}\{[B^v(1), [B^v(2), X_V]]\} = 0$$

similarly yeilds two equations

$$(35.13)\quad L_V S^1(1, 2) = - \beta^1(S^1(1, 2)) + \mathfrak{A}_{(12)}\{\nabla^v\beta^1(2, 1) - \beta^2(1, 2)$$

$$- P^1(\beta(2), 1) + S^1(1, \beta^1(2))\},$$

(35.14)
$$L_V S^2(1, 2) = - \beta^2(S^1(1, 2)) + \mathfrak{A}_{(12)}\{\nabla^v\beta^2(2, 1) - P^2(\beta(2), 1) + S^2(1, \beta^1(2))\},$$

and another equation

$$\beta(S^2(1, 2)) + \mathfrak{A}_{(12)}\{C(\beta(2), 1) - \nabla^v\beta(2, 1)\} = 0,$$

which is trivial because of (35.3).

§ 36. Affine transformations of Finsler connections

From the viewpoint of Definition 32.5 we are led to a natural definition of an affine transformation of Finsler connection as follows:

DEFINITION 36.1. An infinitesimal V-transformation X_V is called an *affine transformation* of a Finsler connection $F\Gamma = (\Gamma^h, \Gamma^v)$, if $L_V\Gamma^h \subset \Gamma^h$ and $L_V\Gamma^v \subset \Gamma^v$ are satisfied.

In case of linear connection an infinitesimal transformation X of the base space M is affine, if and only if $L_{\bar{X}}B(v) = 0$, where $\bar{X} = X^{(n)}$ is the n-ple derived vector field from X. In case of Finsler connection and general infinitesimal V-transformation the situation becomes a little complicated as follows: From (35.1) the condition for X_V to be affine is first given by

(36.1) $$\alpha^1 = \beta = 0, \qquad \alpha^2 = \beta^2 = 0.$$

Then (35.1) reduces to

(36.2) $$L_V B^h(v) = B^h(\alpha(v)), \qquad L_V B^v(v) = B^v(\beta^1(v)).$$

Consequently $L_V B^h(v) = L_V B^v(v) = 0$ are generally not a direct result of the affinity property only.

For special V-transformations, from Propositions 35.1 and 35.2 we obtain

PROPOSITION 36.1. *A necessary and sufficient condition for following infinitesimal V-transformations to be affine is that*

(1) linear X_λ: $L_\lambda B^h(v) = B^h(\alpha(v))$, $L_\lambda B^v(v) = 0$.

(2) extended X_ε: $L_\varepsilon B^h(v) = 0$, $L_\varepsilon B^v(v) = 0$.

(3) N-natural X_ν: $L_\nu B^h(v) = 0$, $L_\nu B^v(v) = B^v(\beta^1(v))$.

(4) rotation X_ρ: $L_\rho B^h(v) = B^h(\alpha(v))$, $L_\rho B^v(v) = B^v(\beta^1(v))$.

Throughout the following, it will be assumed that if we consider the affinity property of an infinitesimal N-natural transformation X_ν, the Finsler connection $F\Gamma$ under consideration is restricted such that X_ν is N-natural with respect to the non-linear connection of $F\Gamma$.

REMARK 36.1. The equation $L_V B^h(v) = 0$, for instance, means $[X_V, B^h(v)] = 0$, namely, Lie derivation commutes with h-covariant differentiation. From (2) of Proposition 36.1 it is seen that an affine extended point transformation X_ε which has been treated by almost all the authors is extremely simple. It seems that the equation $L_\varepsilon B^h(v) = 0$ has been first shown by Soós [2].

For an affine transformation, from (35.2) and (35.3) we obtain

$$(36.3)\quad \begin{aligned} &(1)\quad \nabla^0 v_V = 0,\\ &(2)\quad \nabla^h w_V(v) = R^1(v_V, v) - P^1(v, w_V),\\ &(3)\quad \nabla^h A_V(v) = R^2(v_V, v) - P^2(v, w_V),\\ &(4)\quad \nabla^v A_V(v) = P^2(v_V, v) - S^2(v, w_V).\end{aligned}$$

(1) of (36.3) and Proposition 35.4 lead to

THEOREM 36.1. *If an infinitesimal N-natural transformation is affine, it is an extended point transformation.*

From Propositions 34.4 and 35.4 we immediately get

THEOREM 36.2. *An extended point transformation* $X_\varepsilon = B^h(v_\varepsilon) + B^v(w_\varepsilon) + Z(A_\varepsilon)$ *is affine, if and only if*

$$(1)\quad \nabla^h w_\varepsilon(v) = R^1(v_\varepsilon, v) - P^1(v, w_\varepsilon),\qquad (2)\quad \nabla^h A_\varepsilon(v) = R^2(v_\varepsilon, v) - P^2(v, w_\varepsilon);$$

$$(3)\quad \nabla^v A_\varepsilon(v) = P^2(v_\varepsilon, v) - S^2(v, w_\varepsilon)$$

hold for any $v \in V^n$, *where* v_ε *satisfies* $\nabla^0 v_\varepsilon = 0$ *and* w_ε, A_ε *are given by*

$$w_\varepsilon = \nabla^h v_\varepsilon(\varepsilon) - T(v_\varepsilon, \varepsilon) - D(v_\varepsilon),$$

$$A_\varepsilon \cdot v = \nabla^h v_\varepsilon(v) - T(v_\varepsilon, v) + C(v, w_\varepsilon),\qquad \text{for any } v \in V^n.$$

In this case the Lie derivation with respect to X_ε *commutes with h- and v-covariant differentiations.*

On the other hand, from (2) of Proposition 35.2 we get the fundamental theorem on affine V-rotation:

THEOREM 36.3. *An infinitesimal V-rotation* $X_\rho = Y(A_\rho \cdot \varepsilon) + Z(A_\rho)$ *is affine, if and only if*

(1) $\nabla^h(A_\rho \cdot \varepsilon)v = - P^1(v, A_\rho \cdot \varepsilon)$, (2) $\nabla^h(A_\rho + C(A_\rho \cdot \varepsilon))v = - P^2(v, A_\rho \cdot \varepsilon)$,

(3) $\nabla^v(A_\rho + C(A_\rho \cdot \varepsilon))v = - S^2(v, A_\rho \cdot \varepsilon)$

hold for any $v \in V^n$.

In this case we have

$$\text{(36.4)} \qquad \alpha(v) = A_\rho \cdot v, \qquad \beta^1(v) = C(A_\rho \cdot \varepsilon, v) - \nabla^v(A_\rho \cdot \varepsilon)v + A_\rho \cdot v.$$

Thus α never vanish except the identity transformation, so that the Lie derivation with respect to X_ρ does not commute with h-covariant differentiation; it follows from Proposition 36.1 that

$$\text{(36.5)} \qquad L_\rho \nabla^h K(v) - \nabla^h(L_\rho K)v = \nabla^h K(A_\rho \cdot v)$$

for any tensor field K. Similarly the commutation formula of Lie derivation with respect to X_ρ and v-covariant differentiation will be obtained from Proposition 36.1 and (36.4)

We restrict our discussion to a linear rotation. In this case we have $\beta^1 = 0$ from Proposition 35.5. Consequently

THEOREM 36.4. *The Lie derivation with respect to an infinitesimal affine linear rotation* X_ρ *commutes with the v-covariant differentiation.*

In particular, we are concerned with the infinitesimal V-transformation $Y(\varepsilon) + Z(\delta)$ which appear in Example 34.1. This is an infinitesimal linear rotation because of Proposition 35.5. In this case three equations in Theorem 36.3 become

$$D(v) = - P^1(v, \varepsilon), \qquad \nabla^h(C(\varepsilon))v = - P^2(v, \varepsilon),$$

$$\nabla^v C(\varepsilon, v) + C(v + C(\varepsilon, v)) = S^2(\varepsilon, v).$$

The first two equations hold from Theorem 12.2, if the Finsler connection $F\Gamma$ under consideration is p-homogeneous. The last equation holds also, if $F\Gamma$ is p-homogeneous and satisfies the C_2-condition, which is easily shown from the Bianchi identity (11. 7). Therefore $Y(\varepsilon) + Z(\delta)$ is affine, if $F\Gamma$ is strictly p-homogeneous. Since this tangent vector field is linear, it is the one induced from the tangent vector field $\pi_1'(Y(\varepsilon) + Z(\delta)) = 1^V$ of the tangent bundle T(M), which is nothing but the intrinsic vertical vector field given by Definition 3.11. Consequently

THEOREM 36.5. *The intrinsic vertical vector field l^v of the tangent bundle $T(M)$ is an infinitesimal affine linear rotation of any strictly p-homogeneous Finsler connection.*

This may be rather trivial, because this rotation can be written in the form of a homogeneous transformation

$$\bar{x}^i = x^i, \qquad \bar{y}^i = y^i(1 + dt).$$

REMARK 36.2. We shall again recall Example 34.1. The intrinsic h-basic vector field $B^h(\varepsilon)$, which is an infinitesimal V-rotation, provided $D(\varepsilon) = 0$, is never affine, because $\nabla^0\varepsilon = \delta \neq 0$. On the other hand, the intrinsic v-basic vector field $B^v(\varepsilon)$ is an infinitesimal V-rotation, provided $\varepsilon + C(\varepsilon, \varepsilon) = 0$. From Theorem 36.3 it is affine, if and only if

$$D(v) = - P^1(v, \varepsilon), \qquad P^2(v, \varepsilon) = 0, \qquad S^2(v, \varepsilon) = 0$$

hold. The first two hold for a strictly p-homogeneous Finsler connection, but the equation $\varepsilon + C(\varepsilon, \varepsilon) = 0$ is not satisfied in this case.

In case of linear transformation X_λ, from (8.8) and (3) of Theorem 34.2 we have

$$L_\lambda B^v(v) = - Z(L_\lambda C(v)), \tag{36.6}$$

hence $L_\lambda B^v(v) = 0$ means $L_\lambda C(v) = 0$. Consequently

PROPOSITION 36.2. *A necessary and sufficient condition for an infinitesimal linear transformation X_λ to be affine is that the equations $L_\lambda B^h(v) = B^h(\alpha(v))$ and $L_\lambda C = 0$ are satisfied.*

Next we are concerned with a special Finsler space M with the Cartan connection $C\Gamma$ such that $\nabla^h C = 0$ and $\nabla^h K = 0$ hold, where K is the h-curvature tensor of $R\Gamma$ given by (18.2). From $\nabla^h C = 0$ it follows that the space M is a Berwald space, so that the (v) hv-torsion tensor P^1 and the hv-curvature tensor P^2 of $C\Gamma$ vanish. From (18.2) the hv-curvature tensor F of $R\Gamma$ vanishes also, so that covariant differentiations ∇^h and ∇^0 commute with each other. Further it follows from (18.8) that $\nabla^0 K = 0$. After preparing these facts, Soós [3] proves the following theorem:

THEOREM 36.6. *There exists at least one affine extended point transformation in a Finsler space with the Cartan connection $C\Gamma$ such that $\nabla^h C = 0$ and $\nabla^h K = 0$ hold, where K is given by (18.2).*

Proof. Three equations of Theorem 36.2 are written as

$$\nabla^h(\nabla^h v_\varepsilon(\varepsilon))v_1 = R^1(v_\varepsilon, v_1), \tag{36.7}$$

$$(36.8)\qquad \nabla^h(\nabla^h v_\varepsilon(v_1) + C(v_1, \nabla^h v_\varepsilon(\varepsilon)))v_2 = R^2(v_1, v_\varepsilon, v_2),$$

$$(36.9)\qquad \nabla^v(\nabla^h v_\varepsilon(v_1) + C(v_1, \nabla^h v_\varepsilon(\varepsilon)))v_2 = S^2(v_1, \nabla^h v_\varepsilon(\varepsilon), v_2).$$

The equation (36.7) is solely a consequence of (36.8) by putting $v_1 = \varepsilon$ in (36.8). By $\nabla^h C = 0$ the equation (36.8) may be written

$$(36.8')\qquad \nabla^h X(1, 2) = - K(1, 2, v_\varepsilon),$$

where $X = \nabla^h v_\varepsilon$. Here and in the following, we denote $v_1, \cdots, v_5 \in V^n$ by their indices only.

On the other hand, from $\nabla^0\nabla^h = \nabla^h\nabla^0$ and (17.20) the equation (36.9) is written as

$$(36.9')\qquad C(X(1), 2) + C(X(2), 1) + \nabla^0 C(1, 2, X(\varepsilon)) - X(C(1, 2)) = 0.$$

Now, applying the Ricci identity (10.8') to X and paying attention to $\nabla^0 X = 0$, we obtain the integrability condition of (36.8'):

$$(36.10)\qquad K(X(1), 2, 3) - X(K(1, 2, 3)) + K(1, 2, X(3)) - K(1, 3, X(2)) = 0.$$

Next, differentiating (36.9') h-covariantly and substituting from (36.8'), we get

$$(36.11)\qquad \begin{aligned} &C(K(1, 3, v_\varepsilon), 2) + C(1, K(2, 3, v_\varepsilon)) + \nabla^0 C(1, 2, K(\varepsilon, 3, v_\varepsilon)) \\ &- K(C(1, 2), 3, v_\varepsilon) = 0. \end{aligned}$$

Differentiating (36.9') 0-covariantly, we obtain

$$(36.12)\qquad \begin{aligned} &\nabla^0 C(X(1), 2, 3) + \nabla^0 C(1, X(2), 3) + \nabla^0 C(1, 2, X(3)) + \nabla^0\nabla^0 C(1, 2, 3, X(\varepsilon)) \\ &- X(\nabla^0 C(1, 2, 3)) = 0. \end{aligned}$$

Similarly, from (36.10) we get

$$(36.13)\qquad \begin{aligned} &K(K(1, 4, v_\varepsilon), 2, 3) - K(K(1, 2, 3), 4, v_\varepsilon) + K(1, 2, K(3, 4, v_\varepsilon)) \\ &- K(1, 3, K(2, 4, v_\varepsilon)) = 0, \end{aligned}$$

and another equation which is trivial because of $\nabla^0 K = 0$ and $\nabla^0 X = 0$.

We shall discuss whether above five equations (36.9'), (36.10 - 13) are algebraically compatible or not. The Ricci identities

$$\nabla^h\nabla^h K(4, 5) - \nabla^h\nabla^h K(5, 4) = 0, \qquad \nabla^h\nabla^h C(4, 5) - \nabla^h\nabla^h C(5, 4) = 0,$$

$$\nabla^h\nabla^h\nabla^0 C(3, 4, 5) - \nabla^h\nabla^h\nabla^0 C(3, 5, 4) = 0$$

are respectively written as

$$(36.14)\qquad \begin{aligned} &K(K(1, 2, 3), 4, 5) - K(K(1, 4, 5), 2, 3) + K(1, 3, K(2, 4, 5)) \\ &- K(1, 2, K(3, 4, 5)) = 0, \end{aligned}$$

$$(36.15)\quad \begin{aligned} &K(C(1,2),4,5) - C(K(1,4,5),2) - C(1,K(2,4,5)) \\ &- \nabla^0 C(1,2,K(\varepsilon,4,5)) = 0, \end{aligned}$$

$$(36.16)\quad \begin{aligned} &K(\nabla^0 C(1,2,3),4,5) - \nabla^0 C(K(1,4,5),2,3) - \nabla^0 C(1,K(2,4,5),3) \\ &- \nabla^0 C(1,2,K(3,4,5)) - \nabla^0\nabla^0 C(1,2,3,K(\varepsilon,4,5)) = 0. \end{aligned}$$

Now we observe interesting facts as follows: If we put K(v, 4, 5) = X(v) in (36. 14 - 16), they coincide with (36.10), (36.9') and (36.12) respectively. Further (36. 11) and (36.13) are solely identities, as it is seen from (36.15) and (36.14) respectively.

Similarly it will be shown that covariant derivatives of (36.9') and (36.10) of higher order are algebraically compatible. Therefore the proof has been completed.

REMARK 36.3. It is easily verified from (18.2) that conditions $\overset{h}{\nabla}C = 0$ and $\overset{h}{\nabla}K = 0$ yield $\overset{h}{\nabla}R = 0$ and from (18.16) H = K, so that such a Finsler space as in Theorem 36.6 may be called a locally symmetric Berwald space in every sense of RΓ, CΓ and BΓ. Further $\overset{h}{\nabla}C = 0$ (Berwald space) means that the connection is affine. Thus above Sooś' theorem is conjectured by analogy of the geometrical definition of a locally symmetric Riemannian space.

§ 37. Conformal and isometric transformations

Let M be an n-dimensional Finsler space with a fundamental function L(x, y), and let g be the fundamental tensor of M. By (2) of Theorem 34.1, from (16.2) we get

$$\mathcal{L}_V g(\varepsilon, \varepsilon) = 2L(\mathcal{L}_V L).$$

As a consequence, $\mathcal{L}_V g = 0$ implies $\mathcal{L}_V L = 0$, but the inverse is not true in general.

If we are concerned with an infinitesimal linear transformation X_λ, it follows from (3) of Theorem 34.2 that $\mathcal{L}_\lambda$ commutes with ∇^0, hence $\mathcal{L}_\lambda g = \nabla^0\nabla^0(L\mathcal{L}_\lambda L)$, which shows that $\mathcal{L}_\lambda L = 0$ implies $\mathcal{L}_\lambda g = 0$. Consequently $\mathcal{L}_\lambda g = 0$ is equivalent to $\mathcal{L}_\lambda L = 0$, which is a well-known fact in the theory of extended point transformations.

From these points of view, we are led to the following definition giving the concept of isometry:

DEFINITION 37.1. An infinitesimal V-transformation X_V is called an *isometry* of a Finsler space with a metric L(x, y), if $\mathcal{L}_V g = 0$ holds, where g is the fundamental tensor derived from L.

From the general form of X_V given in Proposition 34.1 we obtain

(37.1) $\qquad L_V g(1,\ 2) = \nabla^h g(1,\ 2,\ v_V) + \nabla^v g(1,\ 2,\ w_V) + g(1,\ A_V(2)) + g(2,\ A_V(1)).$

If the Finsler connection under consideration is metrical, the first two terms of right-hand side of (37.1) vanish. Therefore

PROPOSITION 37.1. *Assume that a Finsler connection $F\Gamma$ of a Finsler space with a fundamental tensor g is metrical. Then a necessary and sufficient condition for an infinitesimal V-transformation X_V to be an isometry is that*

$$g(v_1,\ A_V(v_2)) + g(v_2,\ A_V(v_1)) = 0$$

holds for any $v_1,\ v_2 \in V$.

In the following we shall restrict our consideration to Finsler spaces with the Cartan connection $C\Gamma$.

We first deal with an infinitesimal N-natural transformation X_ν. Then from Proposition 34.4 we have $w_\nu = \nabla^h v_\nu(\varepsilon)$, $A_\nu = \nabla^h v_\nu + C(\nabla^h v_\nu(\varepsilon))$, so that the above equation for isometry is written as

(37.2) $$X_{i|j} + X_{j|i} + 2C_i{}^r{}_j X_{r|0} = 0, \qquad X_i = g_{ij}X^j,$$

where $X^i(x,\ y)$ are components of v_ν.

Secondly we consider an infinitesimal V-rotation X_ρ. Then from Proposition 34.3 we get $v_\rho = 0$, $w_\rho = A_\rho\cdot\varepsilon$. Thus the equation for isometry is written as

(37.3) $$X_{ij} + X_{ji} + 2C_i{}^r{}_j X_{r0} = 0, \qquad X_{ij} = g_{ir}X^r_j,$$

where $X^i_j(x,\ y)$ are components of A_ρ.

DEFINITION 37.2.

(1) A Finsler vector field $X^i(x,\ y)$ satisfying (37.2) is called a *Finsler-Killing vector field.*

(2) A Finsler tensor field $X^i_j(x,\ y)$ satisfying (37.3) is called a *Finsler-Killing (1, 1)-tensor field.*

We are concerned with a Finsler-Killing (1, 1)-tensor field $X^i_j(x,\ y)$. Denoting the symmetric and skew-symmetric parts of X_{ij} by Y_{ij} and Z_{ij} respectively, (37.3) may be written in the form

$$Y_{ij} + C_i{}^r{}_j(Y_{r0} + Z_{r0}) = 0,$$

which implies $Y_{i0} = 0$ at once. Thus X_{ij} may be written in the form

(37.4) $$X_{ij} = - C_i{}^r{}_j Z_{r0} + Z_{ij}.$$

It is remarked that the equation (37.3) is a system of algebraic equations for X_{ij}, not differential equations. X_{ij} given by (37.4) obviously satisfy (37.3) for any skew-symmetric tensor field Z_{ij}. Consequently

THEOREM 37.1. *There exist always isometric V-rotations of order $n(n-1)/2$ in an n-dimensional Finsler space with the Cartan connection. An infinitesimal isometric V-rotation is given by (37.4), where Z_{ij} is an arbitrary skew-symmetric Finsler tensor field.*

From (37.2) a similar process yields

$$X_{i|j} = - C_i{}^r{}_j Z_{r0} + Z_{ij}. \tag{37.5}$$

It is noted that (37.2) or (37.5) is a system of differential equations.

REMARK 37.1. The equation (37.2) is given by Knebelman [2] and Soós [1], although they treat extended point transformations only, that is, X^i of (37.2) is an usual tangent vector field of the base space M. The concept of Finsler-Killing (1, 1) -tensor field is introduced by Matsumoto [16].

In Riemannian geometry it is well-known that an isometry of a Riemannian metric g is necessarily affine with respect to the Riemannian connection given by g. A similar fact holds for extended point transformations:

THEOREM 37.2. *If an extended point transformation is an isometry with respect to a Finsler metric L, it is affine with respect to the Cartan connection given by L.*

Proof. Let X_ε be an isometric extended point transformation. From Proposition 35.1 we have $\beta = \beta^1 = \alpha = 0$; (35.7) reduces to

$$\mathfrak{A}_{(12)}\{\alpha^2(1, 2) - C(1, \alpha^1(2))\} = 0.$$

Since we are concerned with CΓ, Proposition 35.3 gives $\alpha^1(v) = \alpha^2(\varepsilon, v)$. Then, denoting the components of α^2 by $\alpha_j{}^i{}_k$ and putting $\alpha_{jik} = g_{ir}\alpha_j{}^r{}_k$, the above equation may be written in the form

$$\alpha_{jik} - \alpha_{kij} = C_i{}^r{}_j \alpha_{0rk} - C_i{}^r{}_k \alpha_{0rj}. \tag{37.6}$$

Next, from the isometry property and (35.1) we have

$$(L_\varepsilon B^h(v))g = - \alpha^2(v)\cdot g = 0,$$

which is written

$$\alpha_{ijk} + \alpha_{jik} = 0. \tag{37.7}$$

We denote the right-hand side of (37.6) by b_{jik}. From the above two equations it is seen that

$$\alpha_{ijk} = -\alpha_{jki} + b_{ijk} = \alpha_{kij} - b_{jki} + b_{ijk}$$

$$= -\alpha_{ijk} + b_{kij} - b_{jki} + b_{ijk},$$

which gives

$$\alpha_{ijk} = -C_j{}^r{}_k\alpha_{0ri} + C_k{}^r{}_i\alpha_{0rj}.$$

Contractions by y^i and then by y^k yield $\alpha_{0jk} = -C_j{}^r{}_k\alpha_{0r0}$, $\alpha_{0j0} = 0$; so $\alpha_{ijk} = 0$. Consequently $\alpha^1 = \alpha^2 = 0$ are obtained. Further $\beta^2 = 0$ holds from Propositions 35.1 and 37.2. Thus we get (36.1).

In general case, however, an isometric V-transformation is not necessarily affine. Even $L_V C = 0$ is not a consequence of $L_V g = 0$ in general (cf. Proposition 36.2). From the equation

$$[X_V, Y(v)]g = 2L_V C_*(v) - \nabla^0(L_V g)v$$

we obtain

PROPOSITION 37.2.

(1) Assume that X_V is an infinitesimal isometric V-transformation. Then $L_V C = 0$ holds, if and only if $(L_V Y(v))g = 0$.

(2) An infinitesimal isometric linear transformation X_λ satisfies $L_\lambda C = 0$.

From the above viewpoint we are naturally led to

DEFINITION 37.3. An infinitesimal V-transformation X_V is called a *strict isometry* of a Finsler space, if $L_V g = 0$ and $L_V C = 0$ hold.

From Proposition 34.1 the equation $L_V C = X_V(C) = 0$ is written as

$$\begin{aligned}(37.8)\qquad &\nabla^h C(1, 2, v_V) + \nabla^v C(1, 2, w_V) - A_V(C(1, 2)) \\ &+ C(A_V(1), 2) + C(1, A_V(2)) = 0.\end{aligned}$$

In case of an infinitesimal N-natural transformation X_ν the above is written

$$\begin{aligned}(37.9)\qquad &C_i{}^h{}_{j|r}X^r + C_i{}^h{}_j|_r X^r{}_{|0} - (X^h{}_{|r} + C_r{}^h{}_s X^s{}_{|0})C_i{}^r{}_j \\ &+ C_i{}^h{}_r(X^r{}_{|j} + C_j{}^r{}_s X^s{}_{|0}) + C_j{}^h{}_r(X^r{}_{|i} + C_i{}^r{}_s X^s{}_{|0}) = 0.\end{aligned}$$

Substituting from (37.5) this may be written

$$(37.9') \qquad C_{hij|r}X^r + C_h{}^r{}_i|_j Z_{r0} - C_i{}^r{}_j Z_{hr} + C_h{}^r{}_i Z_{rj} + C_h{}^r{}_j Z_{ri} = 0,$$

where we use the symmetry property of $C_{hij|r}$.

PROPOSITION 37.3. *An infinitesimal N-natural transformation X_ν is a strict isometry of a Finsler space with the Cartan connection $C\Gamma$, if and only if there exists a Finsler skew-symmetric tensor field Z_{ij} which satisfies (37.5) and (37.9').*

On the other hand, in case of an infinitesimal V-rotation X_ρ the equation (37.8) is written

$$(37.10) \qquad C_i{}^h{}_j|_r X^r_0 - (X^h_r + C_r{}^h{}_s X^s_0)C_i{}^r{}_j + C_i{}^h{}_r(X^r_j + C_j{}^r{}_s X^s_0) + C_j{}^h{}_r(X^r_i + C_i{}^r{}_s X^s_0) = 0.$$

Substituting from (37.4) we have another form of (37.10):

$$(37.10') \qquad C_{hij|r}Z^r_0 + Z_{rh}C_i{}^r{}_j + Z_{ri}C_j{}^r{}_h + Z_{rj}C_h{}^r{}_i = 0.$$

PROPOSITION 37.4. *An infinitesimal V-rotation X_ρ is a strict isometry of a Finsler space with the Cartan connection $C\Gamma$, if and only if there exists a Finsler skew-symmetric tensor field Z_{ij} which satisfies (37.10'). Then X_ρ is given by (37.4).*

We consider a condition for an n-dimensional Finsler space M with $C\Gamma$ to admit the strictly isometric V-rotation group of maximal order $n(n - 1)/2$. Hence it is supposed that (37.10') is satisfied for any skew-symmetric tensor Z_{ij}, that is,

$$(37.11) \qquad \mathfrak{A}_{(rs)}\{C_{hij|r}y_s + g_{hs}C_{irj} + g_{is}C_{jrh} + g_{js}C_{hri}\} = 0.$$

Contractions by g^{hs} and then by g^{ij} yield

$$nC_{irj} = g_{ir}C_j + g_{jr}C_i + C_i|_j y_r, \qquad (n - 2)C_r = C^i|_i y_r.$$

Thus $C^i|_i = 0$, $C_r = 0$ and $C_{irj} = 0$ are successively obtained, provided $n > 2$. Consequently we have

THEOREM 37.3. *If a Finsler space of dimension n (≥ 3) with the Cartan connection $C\Gamma$ admits the strictly isometric V-rotation group of maximal order $n(n - 1)/2$, the space is Riemannian.*

It remains to consider the case of two-dimensional Finsler spaces. Referring to the T-tensor T_{hijk} which is defined by (28.20), the equation (37.11) may be written in the form

$$(37.11') \qquad \mathfrak{A}_{(rs)}\{T_{hijr}l_s + h_{hs}C_{irj} + h_{is}C_{jrh} + h_{js}C_{hri}\} = 0.$$

In two-dimensional case, by (28.3), (28.19) and $h_{ij} = m_i m_j$, this reduces to

$$I_{;2} m_h m_i m_j (l_s m_r - l_r m_s) = 0,$$

which implies $I_{;2} = 0$, namely, $T_{hijk} = 0$ from (28.19). Consequently Theorem 28.3 leads us to the following interesting theorem:

THEOREM 37.4. *A necessary and sufficient condition for a two-dimensional Finsler space with the Cartan connection to admit the strictly isometric V-rotation group of maximal order 1 is that the T-tensor T_{hijk} vanishes or the main scalar I is a function of position only.*

In this case (37.4) is obviously written as

$$X_{ij} = Z(l_i m_j - l_j m_i + I m_i m_j), \tag{37.12}$$

where Z is an arbitrary Finsler scalar field.

REMARK 37.2. Theorem 37.4 is shown by Matsumoto [16]. He first proposed the T-tensor in this paper.

The theory of V-rotations has just begun to study. A V-rotation evidently induces a transformation of every tangent space M_x of Finsler space M, because the point x is fixed by any V-rotation. By comparing Theorem 37.4 with (2) of Theorem 31.4 and from other various viewpoints it may be expected that interesting results will arise from studying V-rotations.

THEOREM 37.5. *Assume that a Finsler space with the Cartan connection admits an affine extended point transformation group A_p of order p. Let $\{X_{r)}\}$, $r = 1,\cdots,p$, be a set of generators of infinitesimal transformations of A_p, and put $K_{r)} = L_r g$, where L_r is the Lie derivation with respect to $X_{r)}$. If the rank q of matrix $(K_{r)ij})$ of row number r and column number (ij) which consists of components of $K_{r)}$ is less than p, then A_p contains at least one isometry group.*

Proof. It follows from Proposition 36.1 that $L_r B^h(v) = L_r B^v(v) = 0$, so that we easily have

$$\nabla^h K_{r)} = \nabla^v K_{r)} = 0. \tag{37.13}$$

From the assumption there are p - q independent solutions $\gamma_{s)r}$, $s = 1,\cdots, p - q$, of the system of linear equations $\gamma_r K_{r)ij} = 0$. Then (37.13) yields

$$\gamma_{s)r|k} K_{r)ij} = 0, \qquad \gamma_{s)r}|_k K_{r)ij} = 0,$$

from which it follows that there exist quantities A_{skt}, B_{skt} such that

$$\gamma_{s)r|k} = A_{skt}\gamma_{t)r}, \qquad \gamma_{s)r}|_k = B_{skt}\gamma_{t)r}.$$

Applying the Ricci identities to the scalars $\gamma_{s)r}$, we obtain

$$\gamma_{s)r|k|h} - \gamma_{s)r|h|k} = - \gamma_{s)r|i}R^i{}_{kh},$$

and so on, which imply

$$\mathfrak{A}_{(kh)}\{A_{skt|h} + A_{sku}A_{uht}\} + B_{sit}R^i{}_{kh} = 0, \tag{37.14}$$

and so on.

Any infinitesimal transformation X belonging to A_p is written as $X = c_r X_{r)}$ with p scalar fields c_r. It is seen that $L_X B^h(v) = - \nabla^h c_r(v)\cdot X_{r)} = 0$, which implies $\nabla^h c_r = 0$. Similarly we have $\nabla^v c_r = 0$, so that c_r should be constant. A necessary and sufficient condition for X to be isometric is that $L_X g = c_r K_{r)} = 0$, so that there exist scalars C_s such that $c_r = C_s \gamma_{s)r}$ and

$$c_{r|k} = 0 = C_{s|k}\gamma_{s)r} + C_s A_{skt}\gamma_{t)r},$$

$$c_r\big|_k = 0 = C_s\big|_k \gamma_{s)r} + C_s B_{skt}\gamma_{t)r}.$$

Therefore $X = c_r X_{r)}$ is isometric, if and only if the system of differential equations

$$C_{s|k} + C_t A_{tks} = 0, \qquad C_s\big|_k + C_t B_{tks} = 0$$

are integrable. As it is easily shown by (37.14), the integrability conditions are all satisfied, hence the proof has been completed.

<u>REMARK 37.3</u>. Theorem 37.5 is proved by Soós [1]. As to the order r of isometric extended point transformation group I_r of a Finsler space with CΓ, the following facts are known:

(1) *If a Finsler space of dimension n (> 2, ≠ 4) admits I_r of $r > n(n - 1)/2 + 1$, the space is a Riemannian space of constant curvature.*

This fundamental theorem of isometry was proved in 1947 by H.-C.Wang [1] by a group-theoretical method. See K.Yano [1, p.182] and Ku [2].

(2) *A necessary and sufficient condition that an n(≠ 4)-dimensional Finsler space admits I_r of $r = n(n - 1)/2 + 1$ is that the fundamental function L(x, y) is given by either*

$$L = f\left((y^1)^2, \frac{w}{\{1 + (K/4)v\}^2}\right)$$

or

$$L = f((y^1)^2, e^{2kx^1}w)$$

in a suitable coordinate system, where K, k(≠ 0) are constant and

$$v = \sum_{a=2}^{n} (x^a)^2, \qquad w = \sum_{a=2}^{n} (y^a)^2.$$

This was proved in 1959 by Tashiro [1] owing to the following interesting principle:

The problem to determine Finsler spaces admitting a transitive group G of isometries is reduced to determining Riemannian spaces admitting the group G as the group of isometries and to finding, in these spaces, Finsler metrics which are left invariant under G.

DEFINITION 37.4. An infinitesimal V-transformation X_V is called a *conformal transformation* of a Finsler space with a fundamental tensor g, if there exists a scalar field τ such that $L_V g = 2\tau g$ holds. The scalar τ is called the *conformal factor*. If τ is constant, X_V is said to be *homothetic*.

We shall first show the following theorem, which is proved by K.Takano [2] and Soós [2] in case of extended point transformation.

THEOREM 37.6. *If an infinitesimal V-transformation X_V is affine and conformal, X_V is homothetic.*

Proof. From (36.2) it follows that

$$[X_V, B^h(v)]g = -2\nabla^h\tau(v)g = 0,$$

which implies $\nabla^h\tau = 0$. Similarly we get $\nabla^v\tau = 0$. Thus τ is constant.

Consider an infinitesimal conformal linear transformation X_λ. It follows from $L_\lambda Y(v) = 0$ and (8.8) that

$$[X_\lambda, B^v(v)]g - L_\lambda C(v)g = 0,$$

that is,

$$\nabla^v(2\tau g)v + L_\lambda C(v)g = 0.$$

If we denote by $\gamma_j{}^i{}_k$ the components of $L_\lambda C$, the above is written as

$$2g_{ij}\dot\partial_k\tau = \gamma_{ijk} + \gamma_{jik}, \qquad \gamma_{ijk} = g_{jr}\gamma_i{}^r{}_k.$$

From the symmetry property of C it follows that $\gamma_j{}^i{}_k = \gamma_k{}^i{}_j$. Thus $\dot\partial_k\tau = 0$ is equivalent to $\gamma_j{}^i{}_k = 0$, as it will be easily seen. Consequently

THEOREM 37.7. *If an infinitesimal linear transformation X_λ is conformal, then $L_\lambda C = 0$ is equivalent to the fact that the conformal factor τ is a function of position only.*

REMARK 37.4. This theorem is interesting in comparison with Knebelman's theorem: The factor π of proportionality of a conformal change of Finsler metric ($\bar g_{ij} = \pi g_{ij}$) must be a point function. See Knebelman [3].

It seems that the concept of *strict conformality* should be introduced by adding a condition $L_V C = 0$.

As to a V-rotation, we have a similar theorem as follows:

THEOREM 37.8. *If an infinitesimal linear rotation is conformal, the conformal factor τ is a function of position only.*

Proof. An infinitesimal rotation X_ρ is conformal, if and only if

$$(Y(A_\rho \cdot \varepsilon) + Z(A_\rho))g = 2\tau g$$

holds in the notations of Proposition 34.3, that is,

$$2\tau g_{ij} = 2C_{ijr}X^r_0 + X_{ij} + X_{ji} \tag{37.15}$$

in the notations of (34.5), where $X_{ij} = g_{ir}X^r_j$. Differentiating (37.15) by y^k and paying attention to Proposition 35.5, we obtain

$$g_{ij}\dot{\partial}_k\tau = -2\tau C_{ijk} + \dot{\partial}_k C_{ijr}X^r_0 + C_{ijr}X^r_k + C_{ikr}X^r_j + C_{jkr}X^r_i.$$

Because the right-hand side is symmetric in j, k, we get $g_{ij}\dot{\partial}_k\tau = g_{ik}\dot{\partial}_j\tau$, which implies $\dot{\partial}_k\tau = 0$.

In a similar way by means of which (37.3) is written in the form (37.4), the equation (37.15) is rewritten as

$$X_{ij} = \tau g_{ij} - C_i{}^r{}_j Z_{r0} + Z_{ij}, \tag{37.15'}$$

where Z_{ij} is a Finsler skew-symmetric tensor field.

THEOREM 37.9.

(1) If X_1 and X_2 are infinitesimal homothetic V-transformations, then $[X_1, X_2]$ is isometric.

(2) If a Finsler space admits a homothety group H_{p+1} of order p + 1, the space admits an isometry group I_p of order p which is a normal subgroup of H_{p+1}.

Proof. Let τ_1, τ_2 be respective conformal factors of X_1, X_2. Then we have

$$[X_1, X_2]g = X_1(2\tau_2 g) - X_2(2\tau_1 g) = 0.$$

Therefore (1) is proved.

Next, let $X_{r)}$, r = 1,⋯,p+1, be generators of H_{p+1} and put $X_{r)}(g) = 2\tau_r g$. The linear combination $X = c_r X_{r)}$ is isometric, if and only if $c_r\tau_r = 0$. Thus H_{p+1} contains an isometry group I_p of order p. From (1) it follows that I_p is a normal subgroup of H_{p+1}.

THEOREM 37.10.

(1) An homothetic extended point transformation is affine.

(2) If an extended point transformation X_ε is a homothetic transformation of Finsler space of non-zero constant curvature, then X_ε is an isometry.

Proof. (1) Similarly to the proof of Theorem 37.2, we obtain the equations (37.6) and (37.7).

(2) From (1) and (35.8) we have $L_\varepsilon R^1 = 0$. Thus (26.7) gives

$$L_\varepsilon R_{ijk} = 2\tau K(y_j g_{ik} - y_k g_{ij}) = 0,$$

which implies $\tau = 0$.

REMARK 37.5. Theorems 37.9 and 37.10 are generalizations of well-known theorems in Riemannian case. See Soós [2, 5] and Hiramatu [1].

Akbar-Zadeh [10] shows

(1) *Let M be an analytic and locally irreducible Finsler space. If an extended point transformation of M leaves invariant the curvature tensors and their covariant derivatives as well as the torsion form, then the transformation is homothetic.*

As to the order of a group of homothetic extended point transformations, the following facts are proved by Hiramatu [1, 2, 3]:

(2) *If an n-dimensional Finsler space admits a group of homothetic transformations of order $n(n + 1)/2 + 1$, the space is locally euclidean.*

This theorem corresponds to a theorem for the isometry group, proved by H.-C. Wang [1].

(3) *There is no essential homothetic group H_r in an n-dimensional Finsler space ($n \neq 4$) of order r such that $n(n + 1)/2 + 1 > r > n(n - 1)/2 + 2$.*

An infinitesimal homothetic transformation is called essential, if the conformal factor does not vanish.

(4) *Let M be a connected Finsler space of dimension n ($\neq 4$). If M admits a group of homothetic transformations of order $n(n - 1)/2 + 2$, then M is locally Minkowski.*

CHAPTER VII

PARALLEL DISPLACEMENTS AND GEODESICS

The first two sections of this final chapter are devoted to a study of parallel displacements and paths, following the procedure in case of linear connections. The conditions satisfied by Finsler connections which were proposed in Chapter III take important effects on parallel displacements.

The theory of geodesics in Finsler spaces is considered in the last four sections. This theory is not satisfactorily developed yet at this stage. Although, from a historical point of view, studying of behavior of geodesics may be most essential to Finsler geometry, the only way of studying has been to make attempts of generalization of results in Riemannian geometry. Therefore the torsion and curvature tensors of hv- and v-type have hardly played a role in this theory. There are two points which need careful treatment. One is that the length of a curve in a Finsler space depends generally on its direction, hence every curve we consider is assumed to be oriented by its parameter. The other is that the square of a fundamental function $L(x, y)$ is not of class C^2 at $y = 0$ in general.

Almost all the proofs of results which mainly need topological treatments are omitted as space is limited.

§ 38. Parallel displacements

In § 6 we introduce the Finsler bundle $F(M) = (F, T, \pi_1, G(n))$ of an n-dimensional manifold M, which is an induced bundle $\pi_T^{-1}(L(M))$ from the linear frame bundle $L(M)$ by the projection $\pi_T\colon T \to M$ of the tangent bundle $T(M)$.

Following the general process to construct an induced bundle, other induced bundle will be introduced from the tensor bundle $T^r_s(M)$ of (r, s)-type (Definition 3.9) by π_T.

DEFINITION 38.1. The induced bundle $\pi_T^{-1}(T^r_s(M))$ from tensor bundle $T^r_s(M)$ of (r, s)-type over an n-dimensional manifold M by the projection $\pi_T\colon T \to M$ of the tangent bundle $T(M)$ over M is called the *Finsler tensor bundle of (r, s)-type* and denoted by $F^r_s(M) = (F^r_s, T, \pi^*_1, (V^n)^r_s, G(n))$.

The bundle structure of $F^r_s(M)$ is as follows: The total space F^r_s is a closed submanifold of the product manifold $T\times T^r_s$ which is defined by

$$F^r_s = \{(y, K) \in T\times T^r_s \mid \pi_T(y) = \pi^r_s(K)\},$$

where $\pi^r_s : T^r_s \to M$ is the projection of the tensor bundle $T^r_s(M)$. Thus a point u* of

F^r_s is a pair (y, K) of a tangent vector $y \in M_x$ and a tensor $K \in (M_x)^r_s$ at a same point $x \in M$.

The projection π^*_1: $F^r_s \to T$ is just a canonical projection $u^* = (y, K) \mapsto y$. Another canonical projection π^*_2: $u^* = (y, K) \in F^r_s \mapsto K \in T^r_s$ is called the *induced mapping*.

The structure group $G(n)$ is a general linear group of order n and the action ξ^r_s of $G(n)$ on $(V^n)^r_s$ is defined by (2.14). The tensor space $(V^n)^r_s$ is the standard fiber of $F^r_s(M)$.

Now a Finsler tensor bundle $F^r_s(M)$ is regarded as an associated bundle with the Finsler bundle $F(M)$ (Definition 3.8). That is, the structure group $G(n)$ of $F(M)$ is also a Lie transformation group of $(V^n)^r_s$ by the action ξ^r_s, so that we obtain a bundle with the standard fiber $(V^n)^r_s$ which is associated with $F(M)$. According to the general process to define an associated bundle in § 3, the total space of the associated bundle is the quotient space $(F\times(V^n)^r_s)/G(n)$; the action γ^* of $G(n)$ on $F\times(V^n)^r_s$ is similar to $\bar{\gamma}$, that is, $((u, w), g) \in (F\times(V^n)^r_s)\times G(n) \mapsto (ug, g^{-1}w) \in F\times(V^n)^r_s$. The canonical projection α^*: $F\times(V^n)^r_s \to (F\times(V^n)^r_s)/G(n)$ is such that $\alpha^*(u, w)$ is an equivalence class $uw = \{(ug, g^{-1}w) \mid g \in G(n)\}$.

The identification of $(F\times(V^n)^r_s)/G(n)$ with F^r_s is defined by $uw \in (F\times(V^n)^r_s)/G(n) \mapsto (y, zw) \in F^r_s$ for $u = (y, z)$. An admissible mapping ${}_u\alpha^*$: $(V^n)^r_s \to F^r_s$ is given by $w \mapsto uw$, and an associated mapping α^*_w: $F \to F^r_s$ by $u \mapsto uw$.

Let Γ be a connection in the Finsler bundle $F(M)$. By Definition 4.9 we get the associated connection Γ^* in a Finsler tensor bundle $F^r_s(M)$, which is given by $\Gamma^*_{u^*} = (\alpha^*_w)'\Gamma_u$ at a point $u^* = uw$.

With respect to a connection Γ in $F(M)$ the concept of lift $\bar{C}$ to F of a curve C of T arises (Definition 4.4). Similarly we obtain a lift C^* to F^r_s of C with respect to the associated connection Γ^* (Definition 4.10). Therefore the concept of parallel displacement is introduced. It is remarked that a lift $\bar{C}$ to F is uniquely determined by its initial point, and a lift C^* to F^r_s is obtained from $\bar{C}$ by $C^* = \alpha^*_w\circ\bar{C}$.

DEFINITION 38.2. Let $\bar{C}$: $t \in [0, 1] \mapsto u_t = (y_t, z_t)$ be a lift to F of a curve C: $t \in [0, 1] \mapsto y_t \in T$. Then a linear frame $\pi_2\circ\bar{C}(t) = z_t$ is said to be obtained from the linear frame z_0 by *parallel displacement along C*. Referring to the projection $\underline{C} = \pi_T\circ C$: $t \in [0, 1] \mapsto x_t$ of C onto M, we say that z_t is obtained from z_0 by *parallel displacement along* $\underline{C}$ *relative to the field of supporting element C*.

We are concerned with a parallel displacement of linear frame in terms of coordinate system. Assume that a given curve C of T is locally written as $y_t = (x^i(t), y^i(t))$ and a lift $\bar{C}$ of C as $u_t = (x^i(t), y^i(t), z^i_a(t))$. A tangent vector $\bar{C}'(d/dt)$ is written in the form

$$\frac{dx^i}{dt}\frac{\partial}{\partial x^i} + \frac{dy^i}{dt}\frac{\partial}{\partial y^i} + \frac{dz^i_a}{dt}\frac{\partial}{\partial z^i_a} .$$

Since $\bar{C}'(d/dt)$ is horizontal, in comparison this with (7.9) we get

$$(38.1) \qquad \frac{dz^i_a}{dt} + z^j_a\left(\Gamma_j{}^i{}_k(x(t), y(t))\frac{dx^k}{dt} + C_j{}^i{}_k(x(t), y(t))\frac{dy^k}{dt}\right) = 0.$$

These are *differential equations satisfied by a field of parallel linear frame* along $C(t) = (x^i(t), y^i(t))$.

DEFINITION 38.3. Let $C^*: t \in [0, 1] \mapsto u^*_t = (y_t, K_t)$ be a lift to F^r_s of a curve $C: t \in [0, 1] \mapsto y_t \in T$. Then a tensor K_t is said to be obtained from the tensor K_0 by *parallel displacement along C*. Referring to the projection $\underline{C} = \pi_T \circ C: t \in [0, 1] \mapsto x_t \in M$ of C, we say that K_t is obtained from K_0 by *parallel displacement along* $\underline{C}$ *relative to the field of supporting element C*.

We are concerned with the parallel displacement of a tangent vector, for instance. In this case the Finsler vector bundle $F^1_0(M)$ is the subject of our consideration. Since a lift C* of C to F^1_0 is given by $C^* = \alpha^*_v \circ \bar{C}$ from a lift $\bar{C}$ to F, we have $C^*(t) = (x^i(t), y^i(t), X^i(t))$, where $\bar{C}(t) = (x^i(t), y^i(t), z^i_a(t))$, $X^i(t) = z^i_a(t)v^a$ and $v = (v^a)$. Therefore, multiplying (38.1) by v^a, we immediately get

$$(38.2) \qquad \frac{dX^i}{dt} + X^j\left(\Gamma_j{}^i{}_k(x(t), y(t))\frac{dx^k}{dt} + C_j{}^i{}_k(x(t), y(t))\frac{dy^k}{dt}\right) = 0.$$

These are *differential equations satisfied by a parallel vector field* along $C(t) = (x^i(t), y^i(t))$.

Next we are concerned with a non-linear connection N in the tangent bundle T(M). Following Definition 4.10, we are led to

DEFINITION 38.4. Let $C: t \in [0, 1] \mapsto y_t$ be a lift to T of a curve $\underline{C}: t \in [0, 1] \mapsto x_t$ of the base space M. Then a tangent vector y_t is said to be obtained from the tangent vector y_0 by *parallel displacement along* $\underline{C}$.

In terms of coordinate system it follows from (8.5) that *differential equations satisfied by a parallel vector field* along $\underline{C}(t) = (x^i(t))$ are given by

$$(38.3) \qquad \frac{dy^i}{dt} + N^i{}_j(x(t), y(t))\frac{dx^j}{dt} = 0.$$

Now we are concerned with a Finsler connection $F\Gamma = (\Gamma, N) = (\Gamma^h, \Gamma^v)$. From (9.6) it is seen that the associated connection Γ^* in $F^r_s(M)$ with Γ is written in the form

$$(38.4)\qquad \Gamma^* = \Gamma^{*h} \oplus \Gamma^{*v}, \qquad (\Gamma^{*h})_{uw} = (\alpha^*_w)'(\Gamma^h_u), \qquad (\Gamma^{*v})_{uw} = (\alpha^*_w)'(\Gamma^v_u).$$

We first deal with a horizontal curve $C\colon t \in [0,1] \mapsto y_t$ of T with respect to N of $F\Gamma$. This is a lift of a curve $\underline{C}\colon t \in [0, 1] \mapsto x_t$ of the base space M with respect to N. It follows from (9.5) that a lift $\bar{C}\colon t \in [0, 1] \mapsto u_t = (y_t, z_t)$ of C to F with respect to Γ of $F\Gamma$ is h-horizontal, namely, the tangent vector $\bar{C}'(d/dt)$ is contained in the h-part Γ^h at any point of $\bar{C}$. Further, from (38.4) it follows that a lift C^*: $t \in [0, 1] \mapsto u^*_t = (y_t, K_t = z_t w)$ to F^r_s is also h-horizontal, namely, the tangent vector $(C^*)'(d/dt)$ is contained in Γ^{*h} at any point of C^*.

In terms of coordinate system we consider the above lift C^*. Since C is horizontal, from (38.3) we get the first equation of the following (38.5). Then from (38.2) it follows that $C^*(t) = (x^i(t), y^i(t), X^i(t))$ satisfies

$$\frac{dX^i}{dt} + X^j\left(\Gamma_j{}^i{}_k \frac{dx^k}{dt} + C_j{}^i{}_k\left(- N^k{}_h \frac{dx^h}{dt}\right)\right) = 0.$$

Therefore (9.3) yields the second of (38.5). Consequently we have

$$(38.5)\qquad \begin{aligned} &\frac{dy^i}{dt} + N^i{}_j(x(t), y(t)) \frac{dx^j}{dt} = 0,\\ &\frac{dX^i}{dt} + X^j F_j{}^i{}_k(x(t), y(t)) \frac{dx^k}{dt} = 0. \end{aligned}$$

These are *differential equations satisfied by a parallel vector field* $X^i(t)$ along a curve $\underline{C}(t) = (x^i(t))$ relative to a field of parallel supporting element $y^i(t)$.

<u>REMARK 38.1</u>. In his monograph [4], Cartan gives the notion of absolute differential of a tangent vector field X^i by

$$DX^i = dX^i + X^j(\Gamma_j{}^i{}_k dx^k + C_j{}^i{}_k dy^k).$$

If $DX^i/dt = 0$ along a curve $C(t) = (x^i(t), y^i(t))$, it is said that X^i undergoes parallel displacement along C. The equation $DX^i/dt = 0$ is nothing but (38.2). His axiom E is such that when the supporting element undergoes parallel displacement, the connection coefficients $\Gamma^*{}_j{}^i{}_k$ are symmetric in j, k. These $\Gamma^*{}_j{}^i{}_k$ are equal to $F_j{}^i{}_k$ in our notation by means of (38.5).

There are various interesting papers concerned with parallel displacement in Finsler spaces. See, for example, Berwald [2], Laugwitz [2, 3], Synge [1], Taylor [1], O.Varga [2, 4].

Secondly we are concerned with a vertical curve $C\colon t \in [0, 1] \mapsto y_t$ of T, that is, C is in a single fiber $\pi_T^{-1}(x_0)$ over a point $x_0 \in M$. The projection $\pi_T \circ C = \underline{C}$ is only a point $x_0 = (x^i_0)$, so that we have $C(t) = (x^i_0, y^i(t))$. In this case a lift $\bar{C}$ of C to F with respect to Γ of $F\Gamma$ is vertical, namely, the tangent vector $\bar{C}'(d/dt)$ is contained in the v-part Γ^v at any point of $\bar{C}$. The similar situation occurs as to a lift C^* of C to F^r_s.

In terms of coordinate system we deal with $C^*(t) = (x^i_0, y^i(t), X^i(t))$. From (38.2) we immediately get

$$\frac{dX^i}{dt} + X^j C_j{}^i{}_k(x_0, y(t)) \frac{dy^k}{dt} = 0. \tag{38.6}$$

These are *differential equations satisfied by a parallel vector field $X^i(t)$* along a vertical curve $C(t) = (x^i_0, y^i(t))$.

In particular we are interested in the Finsler vector bundle $F^1_0(M)$. A point u^* of F^1_0 is a pair (y, X) of two tangent vectors at a same point $x \in M$. From a curve C^*: $t \in [0, 1] \mapsto u^*_t = (y_t, X_t) \in F^1_0$ we obtain two curves

$$C_1 = \pi^*_1 \circ C^*: t \in [0, 1] \mapsto y_t, \qquad C_2 = \pi^*_2 \circ C^*: t \in [0, 1] \mapsto X_t$$

of the tangent bundle T.

DEFINITION 38.5. A curve C_2: $t \in [0, 1] \mapsto X_t \in T$ is said to be *parallel to a curve C_1*: $t \in [0, 1] \mapsto y_t \in T$, if the curve C^*: $t \in [0, 1] \mapsto u^*_t = (y_t, X_t)$ of F^1_0 is a lift of C_1.

REMARK 38.2. This notion of parallelism of two curves of T is not symmetric, that is, if C_2 is parallel to C_1, the latter is not necessarily parallel to the former.

If a Finsler connection $F\Gamma$ satisfies special conditions, the parallel displacements with respect to $F\Gamma$ will behave specially. Further, in a special Finsler space we shall see special circumstances on the parallel displacements. For instance, we are concerned with a Berwald space. From Proposition 25.1, (9.11) and (8.17') it follows that the V-connection Γ_V of the Cartan connection $C\Gamma$ is a linear connection $\underline{\Gamma}$. The quantities $F_j{}^i{}_k(x)$ are nothing but connection coefficients of $\underline{\Gamma}$. Therefore, from the parallel displacement we can immediately introduce and study the so-called holonomy group of $\underline{\Gamma}$ as the one of $C\Gamma$.

REMARK 38.3. To study the *holonomy group of a Finsler space* is still an open problem at this stage. Although we have few papers concerned with holonomy groups of Finsler spaces, it seems to the author that even the concept of holonomy group is not completely generalized to Finsler geometry yet. This is a strange situation.

Akbar-Zadeh [1, 7] discusses only the holonomy group of connection Γ of $F\Gamma = (\Gamma, N)$. On the other hand, Barthel [11] gives a coherent theory of holonomy groups of non-linear connections. Okada [3] defines the holonomy group of a V-connection Γ_V

and develops an interesting theory in Finsler geometry. Since Barthel considers p-homogeneous non-linear connections, Okada's theory is essentially the same with Barthel's in the viewpoint of Theorem 15.3. See Diaz and Grangier [1], Eguchi and Ichijyo [1], Grangier [1], Laugwitz [8], Misra and Pande [1], Nasu [1, 2], Ōtsuki [2], Rund [18, p.223], Szabó [4], Wagner [6].

The vertical connection Γ^v of $C\Gamma$ is solely the Riemannian connection with Christoffel symbols $C_j{}^i{}_k$, so that to study the holonomy group of this Γ^v may be expected to be fruitful from the standpoint of Riemannian geometry.

THEOREM 38.1. *With respect to a p-homogeneous Finsler connection,*

(1) if a curve C_2 of T is parallel to a curve C_1, then C_2 is also parallel to a curve ${}_a h \circ C_1$ for any fixed $a \in R^+$,

(2) if a tangent vector field $\{y_t\}$ is parallel along a curve $\underline{C}$ of M, then a tangent vector field $\{ay_t\}$ is also parallel along $\underline{C}$ for any fixed $a \in R^+$.

Proof. (1) From a homogeneous transformation ${}_aH$ of F a homogeneous transformation ${}_aH^*$ of F^r_s is induced such that ${}_aH^* \circ \alpha^*_w = \alpha^*_w \circ {}_{a_1}H$. From (2) of Proposition 12.4 and (38.4) it follows that if a lift C^* of C_1 to F^1_0 is given by $C^* = \alpha^*_v \circ \bar{C}$ from a lift $\bar{C}$ of C_1 to F, we have ${}_aH^* \circ C^* = \alpha^*_v \circ {}_aH \circ \bar{C}$ and ${}_aH \circ \bar{C}$ is a lift of ${}_a h \circ C_1$. Therefore ${}_aH^* \circ C^*$ is also a lift of ${}_a h \circ C_1$ to F^1_0. From $\pi^*_2 \circ C^* = C_2$ we get $\pi^*_2 \circ ({}_aH^* \circ C^*) = C_2$, so that C_2 is parallel to ${}_a h \circ C_1$.

(2) is evident from (1) of Proposition 12.4.

THEOREM 38.2. *With respect to a strictly p-homogeneous Finsler connection, if a curve C_2 of T is parallel to a curve C_1: $t \in [0, 1] \mapsto y_t \in T$, then C_2 is also parallel to a curve C: $t \in [0, 1] \mapsto a_t y_t \in T$, where $t \mapsto a_t$ is any curve γ of R^+.*

Proof. Let $\bar{C}$: $t \in [0, 1] \mapsto u_t = (y_t, z_t)$ be a lift of C_1 to F, from which a lift $C^* = \alpha^*_v \circ \bar{C}$: $t \in [0, 1] \mapsto u^*_t = (y_t, z_t v)$ of C_1 to F^1_0 and a curve $C_2 = \pi^*_2 \circ C^*$: $t \in [0, 1] \mapsto X_t = z_t v$ are obtained. A tangent vector $\mu'(d/dt)$ of a curve μ: $t \in [0, 1] \mapsto a_t u_t = (a_t y_t, z_t)$ of F is written as

$$\mu'(d/dt) = H'_{u_t}(\gamma'(d/dt)) + {}_{a_t}H'(\bar{C}'(d/dt)).$$

The first term of right-hand side is horizontal by Proposition 13.1, and the second term is also horizontal by (1) of Proposition 12.4. Consequently the curve μ is a lift of C to F^1_0. From $\alpha^*_v \circ \mu$ we obtain C and C_2 by the mappings π^*_1 and π^*_2 respectively, so that the proof is completed.

REMARK 38.4. In (1) of Theorem 38.1 it is supposed that $a \in R^+$ is fixed, while in Theorem 38.2 this supposition is removed on the C_2-condition. In this viewpoint Cartan [5] imposes the C_2-condition from the beginning.

THEOREM 38.3. *If a Finsler connection $F\Gamma$ satisfies the D-condition, any horizontal curve of T is parallel to itself with respect to $F\Gamma$.*

Proof. Let $\bar{C}$: $t \in [0, 1] \mapsto u_t = (y_t, z_t)$ be a lift of a given curve C: $t \in [0, 1] \mapsto y_t$ of T to F. It is sufficient for the proof to show the existence of $v \in V^n$ such that $y_t = z_t v$. To do so, we shall show that $v_t = z_t^{-1} y_t$ is constant. Put $C_L = \pi_2 \circ \bar{C}$: $t \in [0, 1] \mapsto z_t \in L$ and C_V: $t \in [0, 1] \mapsto v_t \in V^n$. Then a tangent vector $C'(d/dt)$ of C is written as

$$C'(d/dt) = \alpha'_{v_t}(C'_L(d/dt)) + {}_{z_t}\alpha'(C'_V(d/dt)).$$

The first term of right-hand side is horizontal from Proposition 13.3 and the second term is obviously vertical. Since $C'(d/dt)$ is horizontal from the assumption, we have $C'_V(d/dt) = 0$, so that C_V is a single point and v_t is constant.

THEOREM 38.4. *With respect to a Finsler connection satisfying the C_1-condition, any vertical curve C of T is not parallel to itself, provided C is not a single point.*

Proof. In the notations of the proof of Theorem 38.3, if a curve C of T is parallel to C, there exists $v \in V^n$ such that $y_t = z_t v$, hence we have

$$C'(d/dt) = \alpha'_v(C'_L(d/dt)).$$

In this case the right-hand side vanishes by Proposition 13.4. Thus C must reduce to a single point.

§ 39. Various paths

It has been seen in Definition 4.7 that a path of a manifold with a linear connection $\underline{\Gamma}$ is the projection of an integral curve of a basic vector field $\underline{B}(v)$, corresponding to a fixed $v \in V^n$.

With respect to a Finsler connection $F\Gamma = (\Gamma^h, \Gamma^v)$ we have two kinds of such basic vector fields; one is an h-basic vector field $B^h(v)$ given by Definition 9.3 and the other a v-basic vector field $B^v(v)$ defined by (8.6). Consequently, following the case of linear connection, two kinds of paths are naturally introduced as follows:

DEFINITION 39.1. Let M be a manifold with a Finsler connection $F\Gamma = (\Gamma^h, \Gamma^v)$. A curve C of the tangent bundle $T(M)$ over M is called an *h* (resp. *v*)*-path*, if C is the projection of an integral curve of an h- (resp. v)-basic vector field $B^h(v)$ (resp. $B^v(v)$), corresponding to a fixed $v \in V^n$.

We first consider an h-path C. It follows from (9.7) that C is horizontal with respect to the non-linear connection N of $F\Gamma$, so that C is a lift to T of a curve

$\underline{C}$ of the base space M. Let $\bar{C}$: $t \mapsto u_t = (y_t, z_t)$ be an integral curve of $B^h(v)$ and $C = \pi_1 \circ \bar{C}$. Then we have

$$C : t \mapsto y_t, \qquad \underline{C} : t \mapsto x_t = \pi_T(y_t).$$

By Definition 39.1 every tangent vector u'_t of $\bar{C}$ is equal to $B^h(v)$, so that Definition 9.3 implies that the respective tangent vectors y'_t, x'_t of C and $\underline{C}$ are written as

$$(39.1) \qquad y'_t = 1_{y_t}(z_t v), \qquad x'_t = z_t v.$$

<u>PROPOSITION 39.1</u>. *An h-path C: $t \mapsto y_t$ is uniquely determined by the initial point y_0 and the projection of the initial direction $\pi'_T(y'_0) = x'_0$.*

Proof. Assume that y_0 and x'_0 are given. Take an arbitrary point $z_0 \in \pi_L^{-1}(x_0)$. Then we obtain an integral curve $\bar{C}$ of the h-basic vector field $B^h(v)$, corresponding to $v = z_0^{-1}x'_0$ and issuing from $u_0 = (y_0, z_0)$. The projection of $\bar{C}$ onto T is, of course, the h-path which we wish to get. If we take another point $z_0 g \in \pi_L^{-1}(x_0)$, $g \in G(n)$, we have $v_1 = g^{-1}v$ and $B^h(v_1)$. By (9.8) it is observed that $B^h(v_1) = \beta'_g B^h(v)$. Thus an integral curve of $B^h(v_1)$ issuing from $u_0 g = (y_0, z_0 g)$ is nothing but the curve $\beta_g \circ \bar{C}$ and its projection onto T coincides with the one of the above $\bar{C}$.

<u>REMARK 39.1</u>. From a standpoint of the theory of differential equations, Proposition 39.1 asserts that an h-path is uniquely determined by the initial point x_0, initial direction x'_0 and initial supporting element y_0. See (39.2).

The following theorem shows an original geometrical meaning of a path, that is, an autoparallel curve:

<u>THEOREM 39.1</u>. *Let C be a horizontal curve of T and put $\underline{C} = \pi_T \circ C$. A necessary and sufficient condition for C to be an h-path is that a curve $\underline{C}'$ of T which is the vector field tangent to $\underline{C}$ is parallel to C.*

Proof. Assume that a horizontal curve C is an h-path. Thus C is the projection of an integral curve $\bar{C}$ of an h-basic vector field $B^h(v)$. Then the curve $C^* = \alpha^*_v \circ \bar{C}$ of F_0^1 is a lift of C to F_0^1 and $C_2 = \pi_2^* \circ C^*$ is parallel to C. We put

$$C : t \mapsto y_t, \qquad \bar{C} : t \mapsto u_t = (y_t, z_t),$$

$$C^* : t \mapsto u^*_t = (y_t, z_t v), \qquad C_2 : t \mapsto X_t = z_t v.$$

It then follows from (39.1) that C_2 is nothing but the curve $\underline{C}'$, and the condition is necessary.

Conversely, if $\underline{C}'$ is parallel to C, the curve C^*: $t \mapsto u^*_t = (y_t, x'_t)$ is a lift of

C to F_0^1, so that there exists $v \in V$ such that $x'_t = z_t v$, where $\bar{C}: t \mapsto u_t = (y_t, z_t)$ is a lift of C to F. Because C is assumed to be horizontal, we have $y'_t = l_{y_t}(x'_t)$, so that

$$u'_t = l_{u_t} \circ l_{y_t}(z_t v) = B^h(v)_{u_t}.$$

Consequently $\bar{C}$ is an integral curve of $B^h(v)$.

In terms of coordinate system we deal with an h-path C. It follows from (9.10) that an integral curve $\bar{C}: t \mapsto u_t = (x^i(t), y^i(t), z_a^i(t))$ of $B^h(v)$ satisfies

$$\frac{dx^i}{dt} = z_a^i v^a, \qquad \frac{dy^i}{dt} = - z_a^j v^a N_j^i, \qquad \frac{dz_a^i}{dt} = - z_b^j v^b F_k{}^i{}_j z_a^k.$$

Therefore the *differential equations of an h-path* are written in the form

$$\text{(39.2)} \qquad \begin{aligned} &\frac{dy^i}{dt} + N^i_j(x(t), y(t)) \frac{dx^j}{dt} = 0, \\ &\frac{d^2x^i}{dt^2} + F_j{}^i{}_k(x(t), y(t)) \frac{dx^j}{dt}\frac{dx^k}{dt} = 0. \end{aligned}$$

REMARK 39.2. The formulation of paths in the present section is due to Matsumoto [4]. It is well-known that Berwald [2, 3] introduces the Finsler connection BΓ from the standpoint of the theory of general paths which is developed by Douglas [1].

Relating to the theory of paths, we recall the concept of normal coordinate system. O.Varga [13] generalizes this concept to Finsler spaces and obtains the complete set of differential invariants in a Finsler space. See also Busemann [10], Rapcsák [2].

In Example 34.1 the intrinsic h-basic vector field $B^h(\varepsilon)$ is considered. The value $B^h(\varepsilon)_u$ at a point $u = (y, z)$ is equal to $B^h(z^{-1}y)_u$.

DEFINITION 39.2. Let $\bar{C}$ be an integral curve of the intrinsic h-basic vector field $B^h(\varepsilon)$. The projection $\underline{C} = \pi_T \circ \pi_1 \circ \bar{C}$ of $\bar{C}$ onto the base space M is called a *path*.

If we put an integral curve $\bar{C}$ of $B^h(\varepsilon)$ as $t \mapsto u_t = (y_t, z_t)$, it then follows from Definition 9.3 that

$$u'_t = B^h(\varepsilon)_{u_t} = l_{u_t} \circ l_{y_t}(z_t(z_t^{-1} y_t)) = l_{u_t} \circ l_{y_t}(y_t),$$

which implies

$$\text{(39.3)} \qquad y'_t = l_{y_t}(y_t), \qquad x'_t = y_t.$$

where x'_t is the tangent vector of the path $\underline{C}: t \mapsto x_t$.

In a similar way to the proof of Proposition 39.1 we have

PROPOSITION 39.2. *A path is uniquely determined by its initial point* x_0 *and initial direction* x'_0.

In Definition 39.2 the concept of path is defined with respect to a Finsler connection $F\Gamma = (\Gamma, N)$. Really speaking, this concept depends only on the non-linear connection N, as it is seen from

THEOREM 39.2. *A necessary and sufficient condition for a curve* $\underline{C}$ *of the base space M to be a path is that a curve* $\underline{C}'$ *which is the tangent vector field of* $\underline{C}$ *is horizontal with respect to the non-linear connection N.*

Proof. For a path $\underline{C}$ equations (39.3) hold, so that $\underline{C}'$: $t \mapsto x'_t = y_t$ is clearly horizontal. Conversely, if $\underline{C}'$ is horizontal, a lift $\bar{C}$: $t \mapsto u_t = (y_t = x'_t, z_t)$ of $\underline{C}'$ to F is such that

$$u'_t = 1_{u_t}(y'_t) = 1_{u_t} \circ 1_{y_t}(x'_t) = 1_{u_t} \circ 1_{y_t}(y_t),$$

which is nothing but $B^h(\varepsilon)$ at u_t.

Theorem 39.2 and (38.3) yields the *differential equations of a path*:

$$(39.4) \qquad \frac{d^2x^i}{dt^2} + N^i_j(x(t), \frac{dx}{dt}) \frac{dx^j}{dt} = 0.$$

We shall turn to consideration of v-paths. Let $\bar{C}$: $t \mapsto u_t = (y_t, z_t)$ be an integral curve of a v-basic vector field $B^v(v)$. Then the projection $C = \pi_1 \circ \bar{C}$: $t \mapsto y_t$ is a v-path. Since C is contained in a fiber of T, the projection $\pi_T \circ C$ onto the base space M is a single point. From (8.6) and Definition 3.10 we obtain

$$(39.5) \qquad u'_t = 1_{u_t} \circ 1^v_{y_t}(z_t v), \qquad y'_t = 1^v_{y_t}(z_t v).$$

PROPOSITION 39.3. *A v-path is uniquely determined by the initial point* y_0 *and initial direction* y'_0.

The proof is similar to the one of Proposition 39.1.

To obtain a theorem as to v-paths corresponding to Theorem 39.1, the following concept is effective:

DEFINITION 39.3. The *point-expression* X^p of a vertical vector $X \in T^v_y$ is given by a point $(1^v_y)^{-1}(X)$ of the fiber through y.

From (3.18) it follows that if we put $X = X^i(\partial/\partial y^i)_y$ and $\pi_T(y) = (x^i)$, the induced coordinates of X^p is given by $(x^i, y^i = X^i)$.

THEOREM 39.3. *A vertical curve C of T is a v-path, if and only if the locus of the point-expression $(C'(d/dt))^p$ of $C'(d/dt)$ tangent to C is parallel to C.*

Proof. If C is a v-path, the equations (39.5) hold, so that the locus $(C')^p$ of the point expression $(C'(d/dt))^p$ is written as $t \mapsto z_t v$. It is obvious that the curve $C^*: t \mapsto u_t^* = (y_t, z_t v)$ is a lift of C to F_0^1, so that the curve $C_2 = \pi_2^* \circ C^*: t \mapsto z_t v = (C')^p$ is parallel to C.

Conversely, if $(C')^p$ is parallel to a vertical curve C, the curve $C^*: t \mapsto u_t^* = (y_t, (1^v_{y_t})^{-1}(y_t'))$ is a lift of C to F_0^1. Thus there exists $v \in V^n$ such that $(1^v_{y_t})^{-1}(y_t') = z_t v$, where $\bar{C}: t \mapsto u_t = (y_t, z_t)$ is a lift of C to F, from which $y_t' = 1^v_{y_t}(z_t v)$ is derived, and $1_{u_t}(y_t') = 1_{u_t} \circ 1^v_{y_t}(z_t v) = B^v(v)_{u_t}$. Therefore C is an integral curve of $B^v(v)$.

We treat a v-path $C: t \mapsto y_t = (x_0^i, y^i(t))$ in terms of coordinate system, where x_0^i are constant. The locus of point-expression of a vector tangent to C is given by $(x_0^i, dy^i/dt)$, so that (38.6) yields the *differential equations of a v-path:*

$$\frac{d^2 y^i}{dt^2} + C_j{}^i{}_k(x_0, y(t)) \frac{dy^j}{dt} \frac{dy^k}{dt} = 0. \tag{39.6}$$

REMARK 39.3. If we consider a Finsler space M with the Cartan connection $C\Gamma$, a v-path is nothing but a geodesic of a tangent Riemannian space M_x, as it is seen from (39.6) and (17.1). Therefore to study v-paths as geodesics may be expected to be fruitful. See Ichijyo [3], Laugwitz and Lorch [1], Matsumoto [28].

With reference to the intrinsic v-basic vector field $B^v(\varepsilon)$ we may introduce a concept analogous to Definition 39.2:

DEFINITION 39.4. The projection $C = \pi_1 \circ \bar{C}$ onto T of an integral curve $\bar{C}$ of the intrinsic v-basic vector field $B^v(\varepsilon)$ is called a *0-path.*

Consider an integral curve $\bar{C}: t \mapsto u_t = (y_t, z_t)$ of $B^v(\varepsilon)$. Then $C = \pi_1 \circ \bar{C}: t \mapsto y_t$ is a 0-path and

$$u_t' = B^v(z_t^{-1} y_t)_{u_t} = 1_{u_t} \circ 1^v_{y_t}(y_t), \qquad y_t' = 1^v_{y_t}(y_t).$$

Accordingly the locus of point-expression $(C'(d/dt))^p$ is nothing but $t \mapsto y_t = C$. Conversely, if the locus of point-expression of vector tangent to a vertical curve C coincides with C itself, we get $y'_t = 1^v_{y_t}(y_t)$, so that $1_{u_t}(y'_t) = B^v(\varepsilon)_{u_t}$. Consequently

THEOREM 39.4. *A vertical curve C of T is a 0-path, if and only if the locus of point-expression of vector tangent to C coincides with C itself.*

As a consequence of Theorem 39.4 we are little interested in the concept of 0-path, because this concept does not depend on any connection. In fact, from $y'_t = 1^v_{y_t}(y_t)$ we instantly obtain the *differential equations of a 0-path* as follows:

$$x^i = \text{const.}, \qquad \frac{dy^i}{dt} = y^i. \tag{39.7}$$

Finally we are concerned with special behavior of paths with respect to Finsler connections satisfying some conditions.

THEOREM 39.5. *With respect to a p-homogeneous Finsler connection, if a curve C is an h-path (resp. v-path), then ${}_a h \circ C$ is also an h-path (resp. v-path) for any fixed $a \in R^+$.*

This will be evident from (2) of Proposition 12.4.

THEOREM 39.6. *If a Finsler connection satisfies the D-condition, the vector field $\underline{C}'$ tangent to a path $\underline{C}$ is an h-path.*

Proof. In terms of coordinate system this is obvious from (39.2), (39.4) and assumption $y^j F_j{}^i{}_k = N^i{}_k$. Another proof is as follows: The curve $\underline{C}'$ is horizontal, so that Theorem 38.3 shows that $\underline{C}'$ is parallel to itself. Since $\pi_T \circ \underline{C}' = \underline{C}$, Theorem 39.1 completes the proof.

§ 40. Critical points of energy functions

Throughout the remainder of the present chapter, we shall restrict our consideration to Finsler spaces with such a fundamental function L(x, y) that satisfies the additional conditions (F 1, 3, 4) of § 16.

REMARK 40.1. Kashiwabara [1] proves a theorem: In a connected differentiable manifold, if the second countability axiom holds, it always admits a positive-definite non-Riemannian Finsler metric, and the converse is also true.

The symmetry condition (F 2) is not assumed. Therefore the orientation of curves is essential to the theory of curves. When a curve C: $t \in [0, 1] \mapsto x_t \in M$ is treated, it is always assumed that C is *oriented by t*; the initial point (resp. terminal point) is $x_0 = C(0)$ (resp. $x_1 = C(1)$). The length s of C means, of course, the one of the oriented curve C, so that s is an increasing function of the parameter t. If the parameter t is positively proportional to s, namely, $t = as$ $(a \in R^+)$, then t is called *normal*.

Further we shall put $L(x, 0) = 0$ for any x. Hence the function $F(x, y) = L^2(x, y)/2$ is defined on T(M) and of class C^∞ on T(M) - M, where M is identified with the closed subset of T(M) consisting of all the zero tangent vectors of M. (Cf. Remarks 12,1 and 16.1). The following proposition shows that F(x, y) is at least of class C^1 at a point of M.

PROPOSITION 40.1. *If a function f(x, y) on the tangent bundle T(M) is of class C^∞ on T(M) - M, positively homogeneous of degree r (> 0) in y^i and has the value f(x, 0) = 0, then f(x, y) is continuous at a point of M.*

Proof. From the homogeneity assumption we have $f(x, u) = a^r f(x, y)$, where $a \in R^+$, $u = (ay^i)$ and $y = (y^i) \neq 0$. Thus $\lim_{a\to 0} f(x, u) = 0$ because of $r > 0$, so that f(x, y) is continuous at y = 0.

EXAMPLE 40.1. In case of a Riemannian metric L(x, y) the function F(x, y) is a quadratic form in y^i.

A Finsler metric

$$L(x, y) = \{(y^1)^4 + \cdots + (y^n)^4\}^{1/4}$$

was already proposed by Riemann. It is easily verified that, as to this L, the function F is not of class C^2 at y = 0.

If a parametrized curve $C: t \to x_t \in M$ is such that $dx/dt \neq 0$ for any point of C, then C is called *regular*. Along a regular curve C we can consider $\dot\partial_i\dot\partial_j F(x(t), dx/dt)$ which is of class C^∞ and equal to g(dx/dt, dx/dt) from (16.2).

DEFINITION 40.1. Let M be a Finsler space with the Cartan connection CΓ. A path of M with a normal parameter is called a *geodesic* of M.

From Theorem 39.2 and Definition 38.4 it follows that a curve C of M is a geodesic, if and only if the vector field C' tangent to C is parallel along C. A geodesic is given by the differential equations (39.4):

$$\frac{d^2x^i}{dt^2} + N^i_j\left(x(t), \frac{dx}{dt}\right)\frac{dx^j}{dt} = 0, \tag{40.1}$$

where and throughout the following the parameter t of a geodesic is assumed to be normal.

REMARK 40.2. Although the Cartan connection is referred in Definition 40.1 and treated in the following, three Finsler connections $B\Gamma$, $C\Gamma$ and $R\Gamma$ have a common non-linear connection N, so that $C\Gamma$ in Definition 40.1 may be changed for $B\Gamma$ or $R\Gamma$. Really speaking, the concept of geodesic is determined by a fundamental function $L(x, y)$ only, independently on the choice of a Finsler connection. In fact, from (17.6) ($D^i{}_0 = 0$) the equations (40.1) are written in the form

$$\frac{d^2x^i}{dt^2} + \gamma_j{}^i{}_k(x(t), \frac{dx}{dt}) \frac{dx^j}{dt}\frac{dx^k}{dt} = 0.$$

The existence, uniqueness and differentiability of geodesic are well-known from the theory of ordinary differential equations.

From the standpoint of equation (38.2), we introduce an operator D_t along a curve C of T(M), which is called the *absolute differentiation* along C. Let $X_t = (X^i(t))$ be a vector field defined along $\pi_T \circ C$. Then

$$D_t X^i = \frac{dX^i}{dt} + X^j\{\Gamma_j{}^i{}_k(x(t), y(t)) \frac{dx^k}{dt} + C_j{}^i{}_k(x(t), y(t)) \frac{dy^k}{dt}\}. \tag{40.2}$$

From (38.2) it is seen that X_t is parallel along C, if and only if $D_t(X_t) = 0$ along C.

We differentiate the scalar product $g(X, Y)$ of vector fields X, Y along a curve $C: t \to y_t$ of T. From (9.3) the condition (17.2) for $C\Gamma$ is written as

$$\partial_k g_{ij} = \Gamma_{ijk} + \Gamma_{jik}, \qquad \Gamma_{ijk} = g_{jr}\Gamma_i{}^r{}_k.$$

Then we immediately get the formula

$$\frac{dg(X, Y)}{dt} = g(D_tX, Y) + g(X, D_tY). \tag{40.3}$$

PROPOSITION 40.2. *The scalar product $g(X, Y)$ of vector fields X, Y and the relative length $\sqrt{g(X, Y)}$ are preserved invariant along a curve C of T, if X, Y are parallel along C with respect to $C\Gamma$.*

DEFINITION 40.2. Let $\Gamma(p, q)$ be a set of all the oriented curves from p to q of a Finsler space M which are of class D^∞. The *tangent set* $\Gamma(p, q)_C$ of $\Gamma(p, q)$ at a point $C \in \Gamma(p, q)$ is a set of all the vector fields along C which are of class D^∞ and vanish at the endpoints p, q.

In the above the endpoints p, q are not necessarily distinct. In general a curve $C: t \in [0, 1] \mapsto x_t \in M$ is said to be *of class D^∞*, if there exists a subdivision $0 = t_0 < t_1 < \cdots < t_p = 1$ of the interval [0, 1] such that every segment $C|[t_{r-1}, t_r]$, r

$= 1, \cdots, p$, is of class C^∞. The concept of vector field of class D^∞ along a curve is similarly defined.

DEFINITION 40.3. For a curve $C: t \in [0, 1] \mapsto x_t$ belonging to $\Gamma(p, q)$ of a Finsler space M with a fundamental function L the integral

$$E_a^b(C) = \int_a^b F(x(t), dx/dt)dt, \qquad 0 \le a \le b \le 1,$$

is called the *energy of C from a to b*, where $F = L^2/2$. The *total energy* $E_0^1(C)$ of C is denoted by E(C).

We recall the well-known Schwarz inequality

$$\left(\int_a^b (fg)dt\right)^2 \le \left(\int_a^b (f^2)dt\right)\left(\int_a^b (g^2)dt\right)$$

for two functions f(t), g(t); the equality holds, if and only if the ratio f(t) : g(t) is constant. If we put f = 1, g = L, we get

$$(40.4) \qquad \{L_a^b(C)\}^2 \le 2(b - a)E_a^b(C),$$

where $L_a^b(C)$ is the length of an oriented curve C from t = a to t = b. The equality holds only when the parameter t is normal, because

$$L(x, dx/dt) = L(x, (dx/ds)(ds/dt)) = ds/dt.$$

DEFINITION 40.4. An *m-parameter variation* α of a curve $C: t \in [0, 1] \mapsto x_t \in M$ belonging to $\Gamma(p, q)$ is a function $U \to \Gamma(p, q)$, where U is a neighborhood of 0 in R^m, satisfying the conditions:

(1) $\alpha(0) = C$.

(2) There exists a subdivision $0 = t_0 < t_1 < \cdots < t_p = 1$ of [0, 1] such that the mapping $\bar{\alpha}: U\times[0, 1] \to M$ defined by $\bar{\alpha}(\xi^1, \cdots, \xi^m, t) = \alpha(\xi^1, \cdots, \xi^m)(t)$ is of class C^∞ on every strip $U\times[t_{r-1}, t_r]$, $r = 1, \cdots, p$.

In the present section we deal with a one-parameter variation $\alpha(\xi)$ of $C \in \Gamma(p, q)$. For a fixed $\xi \in U$ ($\subset R$) we have a curve $\alpha(\xi) \in \Gamma(p, q)$ and the vector field

$$(40.5) \qquad V(t) = \left(\frac{d\alpha}{d\xi}\right)_0 = \left(\frac{\partial\bar{\alpha}}{\partial\xi}\right)_{(0,t)}$$

along C. The vector field V belongs to the tangent set $\Gamma(p, q)_C$ of $\Gamma(p, q)$ at the point C and is called a *variation vector field* induced from the variation $\alpha(\xi)$.

Conversely, it will be obvious that there exists for any $V \in \Gamma(p, q)_C$ a one-parameter variation $\alpha(\xi)$ which induces V. This $\alpha(\xi)$ is called an *associated variation* with V.

Now the total energy E(C) of a curve $C \in \Gamma(p, q)$ is regarded as a non-negative function on the set $\Gamma(p, q)$. It gives a mapping $E^*: \Gamma(p, q)_C \to R_{E(C)}$ (tangent space of R at E(C)) as follows: For a given $V \in \Gamma(p, q)_C$ we consider an associated variation $\alpha(\xi)$ with V. Then we put

$$E^*(V) = \left(\frac{dE(\alpha(\xi))}{d\xi}\right)_0 \left(\frac{d}{dt}\right)_{E(C)}. \tag{40.6}$$

That is, $E^*(V)$ is a tangent vector of R at E(C).

DEFINITION 40.5. A curve $C \in \Gamma(p, q)$ is called a *critical point* of $\Gamma(p, q)$, if $E^*(V) = 0$ for the variation vector field V induced from any one-parameter variation of C.

Therefore it follows from (40.6) that a critical point of $\Gamma(p, q)$ is a critical point of the energy function on $\Gamma(p, q)$ in a usual sense.

The quantity $(dE(\alpha(\xi))/d\xi)_0$ is called the *first variation* of energy function E(C) and denoted by $J^1(C, V)$.

Let S be a surface $(x^i(\xi, t), y^i(\xi, t))$ with two parameters ξ, t of the tangent bundle T(M). Then the *absolute partial derivatives* $D_\xi X^i$ and $D_t X^i$ of a vector field X^i defined on $\underline{S} = \pi_T(S)$ are introduced analogous to (40.2). The formula (40.3) will be generalized with respect to D_ξ and D_t without modification.

In particular, we treat a surface $\underline{S} = (x^i(\xi, t))$ of the base space M with two parameters ξ, t. From $\underline{S}$ a surface $\underline{S}'$ of T is defined by $(x^i(\xi, t), \partial_t x^i)$. Here and in the following we shall use notations ∂_ξ and ∂_t to denote $\partial/\partial\xi$ and $\partial/\partial t$ respectively. Denote D_ξ, D_t with respect to this surface $\underline{S}'$ by D'_ξ, D'_t respectively. Then we easily obtain

$$\begin{aligned} D'_t(\partial_t x^i) &= \partial_t\partial_t x^i + N^i{}_j(x, \partial_t x)\partial_t x^j, \\ D'_\xi(\partial_t x^i) &= \partial_\xi\partial_t x^i + N^i{}_j(x, \partial_t x)\partial_\xi x^j, \end{aligned} \tag{40.7}$$

and further an interesting equation

$$D'_t(\partial_\xi x^i) - D'_\xi(\partial_t x^i) = C_j{}^i{}_k(x, \partial_t x)\partial_\xi x^j D'_t(\partial_t x^k). \tag{40.8}$$

It is noted that $D'_t(\partial_t x^i)$ vanishes identically, if and only if the curve $x^i(\xi, t)$ for a fixed ξ is a geodesic with the normal parameter t.

We now find the first variation along a regular curve C. Putting $\bar{\alpha}(\xi, t) = (x^i(\xi, t))$ and paying attention to $F(x, y) = g(y, y)/2$ for $y \neq 0$, from (40.3) and (40.8) we have

$$\partial_\xi F(x(\xi, t), \partial_t x) = g(D'_t(\partial_\xi x), \partial_t x) - C(\partial_\xi x, D'_t(\partial_t x), \partial_t x).$$

Since we compute the above on the surface $\underline{S}'$, the second term of right-hand side vanishes from $C_{ijk}(x, y)y^k = 0$. Thus

$$\partial_\xi F(x(\xi, t), \partial_t x) = g(D'_t(\partial_\xi x), \partial_t x).$$

Further from (40.3) we have

$$\partial_t g(\partial_\xi x, \partial_t x) = g(D'_t(\partial_\xi x), \partial_t x) + g(\partial_\xi x, D'_t(\partial_t x)),$$

which implies

$$\partial_\xi F(x(\xi, t), \partial_t x) = \partial_t g(\partial_\xi x, \partial_t x) - g(\partial_\xi x, D'_t(\partial_t x)).$$

Consequently we obtain

$$\frac{dE(\alpha(\xi))}{d\xi} = -\sum_{r=1}^{p-1} g(\partial_\xi x, \partial_t x)\Big|_{t_r^-}^{t_r^+} - \int_0^1 g(\partial_\xi x, D'_t(\partial_t x))dt, \tag{40.9}$$

where $\Big|_{t_r^-}^{t_r^+}$ indicates the jump at $t = t_r$.

Evaluating at $\xi = 0$ and using the notation of (40.5), we arrive at an expression of the first variation as follows:

PROPOSITION 40.3. *The first variation $J^1(C, V)$ of the energy function $E(C)$ along a regular curve $C \in \Gamma(p, q)$ with respect to a variation vector V is given by*

$$J^1(C, V) = -\sum_{r=1}^{p-1} g(V, dx/dt)\Big|_{t_r^-}^{t_r^+} - \int_0^1 g(V, D_t(dx/dt))dt,$$

where t_r, $r = 1, \cdots, p-1$, are values of the parameter t of C where dx/dt is discontinuous.

THEOREM 40.1. *A necessary and sufficient condition for a regular curve $C \in \Gamma(p, q)$ to be a critical point of $\Gamma(p, q)$ is that C is a geodesic from p to q.*

Proof. If a regular curve C is a geodesic, it follows from (40.1) and (40.2) that C is of class C^∞ throughout the interval [0, 1] and $D_t(dx/dt) = 0$ identically. Thus for any variation vector field V we clearly get $J^1(C, V) = 0$ from Proposition 40.3.

Conversely, let a regular curve C be a critial point. First, take a variation vector field

$$V(t) = f(t)\cdot D_t(dx/dt),$$

where $f(t)$ is a function assumed to be positive at any $t \neq t_r$, $r = 1, \cdots, p-1$, and equal to zero at every t_r. Then Proposition 40.3 yields

$$J^1(C, V) = -\int_0^1 f(t)\cdot g(D_t x', D_t x')dt = 0,$$

where $x' = dx/dt$. Because the Finsler metric is assumed to be positive-definite, the above leads to $D_t x' = 0$ at any $t \neq t_r$. Thus every segment $C|[t_{r-1}, t_r]$ is a geodesic, and C is a broken geodesic. Therefore on C we have

$$J^1(C, V) = -\Sigma\{g_{ij}(x(t_r), x'(t_r^+))x'^j(t_r^+) - g_{ij}(x(t_r), x'(t_r^-))x'^j(t_r^-)\}V^i(t_r).$$

Every term of the above is equal to

$$\{\dot\partial_i F(x(t_r), x'(t_r^+)) - \dot\partial_i F(x(t_r), x'(t_r^-))\}V^i(t_r).$$

Applying the mean value theorem,it may be written as

$$g_{ij}(x(t_r), x'(t_r^- + \theta\Delta_r))V^i(t_r)\Delta_r^j,$$

where $0 < \theta < 1$ and $\Delta_r = x'(t_r^+) - x'(t_r^-)$. Therefore, if we secondly take a variation vector field V such that $V(t_r) = \Delta_r$, then all the jumps Δ_r must vanish, so that C is of class C^1 at every t_r. Consequently C is a geodesic.

§ 41. Conjugate points

We are concerned with a two-parameter variation α: $U \to \Gamma(p, q)$ of a geodesic $C \in \Gamma(p, q)$, where U is a neighborhood of 0 in R^2. As in Definition 40.4, we put $\bar\alpha$: $U\times [0, 1] \to M$ such that $\bar\alpha(\xi, \eta, t) = \alpha(\xi, \eta)(t)$ and find the *second variation*

$$J^2(C, V_1, V_2) = \left(\frac{\partial^2 E(\alpha)}{\partial\xi\partial\eta}\right)_{(0,0)}, \qquad V_1 = \left(\frac{\partial\alpha}{\partial\xi}\right)_{(0,0)}, \qquad V_2 = \left(\frac{\partial\alpha}{\partial\eta}\right)_{(0,0)}.$$

From (40.9) we first have

$$-\frac{\partial E(\alpha)}{\partial\eta} = \sum_{r=1}^{p-1} g(\partial_\eta x, \partial_t x)\Big|_{t_r^-}^{t_r^+} + \int_0^1 g(\partial_\eta x, D_t'(\partial_t x))dt.$$

By (40.3) we get

$$-\frac{\partial^2 E(\alpha)}{\partial\xi\partial\eta} = \sum_{r=1}^{p-1} \{g(D_\xi'(\partial_\eta x), \partial_t x) + g(\partial_\eta x, D_\xi'(\partial_t x))\}\Big|_{t_r^-}^{t_r^+}$$

$$+ \int_0^1\{g(D_\xi'(\partial_\eta x), D_t'(\partial_t x)) + g(\partial_\eta x, D_\xi'(D_t'(\partial_t x)))\}dt.$$

We evaluate the above at $(\xi, \eta) = (0, 0)$. Because the curve C is assumed to be a geodesic, we have $D_t(x') = 0$ and $x'(t_r^-) = x'(t_r^+)$. Thus both of the first two terms outside and inside the integral vanish. Therefore we have

$$-\frac{\partial^2 E(\alpha)}{\partial\xi\partial\eta}_{(0,0)} = \sum_{r=1}^{p-1} g(V_2, D_\xi' x')\Big|_{t_r^-}^{t_r^+} + \int_0^1 g(V_2, D_\xi'(D_t' x'))dt.$$

Further it is observed from (40.7) that

$$D_\xi' x' = D_t V_1 - C(V_1, D_t x').$$

Thus, along the geodesic C we get $g(V_2, D_\xi' x') = g(V_2, D_t V_1)$.

Next, we refer to the commutation formula for D_ξ' and D_t' on the surface $\underline{S}'$ in the last section. From (40.2), (40.8) and (9.3) we can directly derive a formula

$$D'_\xi(D'_t(\partial_t x^i)) - D'_t(D'_\xi(\partial_t x^i)) = R^i{}_{jk}\partial_t x^j \partial_\xi x^k + P^i{}_{jk}\partial_t x^j D'_\xi(\partial_t x^k) \tag{41.1}$$
$$- P^i{}_{jk}\partial_\xi x^j D'_t(\partial_t x^k).$$

The second term of right-hand side of (41.1) vanishes from (17.22), and the third term does also along the geodesic C. Thus from (40.8) we have

$$D'_\xi(D'_t x') = D_t D_t V_1 + R^2(x', x', V_1)$$

along C. Consequently we conclude

PROPOSITION 41.1. *Let $C \in \Gamma(p, q)$ be a geodesic and $\alpha: U \to \Gamma(p, q)$ be an associated variation with variation vector fields $V_1, V_2 \in \Gamma(p, q)_C$. Then the second variation $J^2(C, V_1, V_2) = \{\partial^2 E(\alpha)/\partial\xi\partial\eta\}_{(0,0)}$ of energy function $E(C)$ along C is given by*

$$J^2(C, V_1, V_2) = -\sum_r g(V_2, \Delta_r(D_t V_1)) - \int_0^1 g(V_2, D_t D_t V_1 + R^2(x', x', V_1))dt,$$

where $x' = dx/dt$ and $\Delta_r(D_t V_1) = (D_t V_1)_{t_r^+} - (D_t V_1)_{t_r^-}$ is the jump of $D_t V_1$ at finite discontinuous points $t = t_r \in (0, 1)$ of $D_t V_1$.

REMARK 41.1. Although we refer to CΓ in the above, it is evident that the term $R^2(x', x', V_1)$ may be changed for $K(x', x', V_1)$ or $H(x', x', V_1)$ referring to RΓ or BΓ.

We have various papers concerned with a geometrical theory of variation calculus. See Rund [18, p.111-124]. More comprehensive bibliography is found in Rund's book "The Hamilton-Jacobi theory in the calculus of variations", D.Van Nostrand, London, 1966.

The second variation $J^2(C, V_1, V_2)$ is defined as the so-called *Hessian* $H(V_1, V_2)$ of energy function E along a geodesic C and well-defined from Proposition 41.1. In fact, it is obviously a symmetric and bilinear function of V_1, V_2. The quadratic form $H(V, V)$ is written as $\{d^2E(\alpha(\xi))/d\xi^2\}_0$, where α is a one-parameter variation associated with V.

DEFINITION 41.1. Let $C \in \Gamma(p, q)$ be a geodesic. A tangent vector field J along C which satisfies the *Jacobi differential equations*

$$D_t D_t J + R^2(x', x', J) = 0$$

is called a *Jacobi field* along C, where x' is the tangent vector field dx/dt of C and components $R_h{}^i{}_{jk}$ of R^2 take the values at (x, x').

It is remarked that the integral appearing in Proposition 41.1 vanishes, if V_1 is a Jacobi field.

Let $z = (z_a)$, $a = 1,\cdots, n$, be an orthonormal frame field which is parallel along C relative to x'. If we put

$$z_{ab} = g(z_a, R^2(x', x', z_b)), \qquad J(t) = J_a(t)z_a(t),$$

Jacobi differential equations are written as

$$\frac{d^2 J_a}{dt^2} + z_{ab}J_b = 0.$$

Because these are linear and of second order, we have 2n linearly independent solutions which are defined and of class C^∞ on [0, 1]. Therefore

PROPOSITION 41.2. *In an n-dimensional Finsler space with $C\Gamma$ all the Jacobi fields along a geodesic $C \in \Gamma(p, q)$ constitute a 2n-dimensional vector space $J(C)$. If the initial values $J(0)$ and $(D_t J)_0$ are given, we obtain a unique Jacobi field J.*

DEFINITION 41.2. The *null space* of Hessian $H(V_1, V_2)$ of energy function E along a geodesic $C \in \Gamma(p, q)$ is a subspace of $\Gamma(p, q)_C$ given by $\{V_1 \in \Gamma(p, q)_C \mid H(V_1, V) = 0$ for any $V \in \Gamma(p, q)_C\}$. The dimension ν of null space is called the *nullity* of $H(V_1, V_2)$. If ν is positive, then $H(V_1, V_2)$ is called *degenerate*.

REMARK 41.2. Let $A(x) = \sum_{i,j=1}^{n} a_{ij}x_i x_j$ be a real symmetric quadratic form of variables $x_1,\cdots, x_n$. If the rank of A(x) is equal to r, there exists a regular linear transformation $(x_i) \to (\bar{x}_i)$ and with respect to $(\bar{x}^i)$ A(x) is written in the canonical form

$$(\bar{x}_1)^2 + \cdots + (\bar{x}_p)^2 - (\bar{x}_{p+1})^2 - \cdots - (\bar{x}_{p+q})^2, \qquad p + q = r.$$

The number $p - q$ or the pair (p, q) is called the *signature* of A(x), and $p + q$ or q is the *index* of A(x) (cf. Definition 43.1). The number $n - r$ is called the *nullity* of A(x)(cf. Definition 41.2).

THEOREM 41.1. *The null space of Hessian $H(V_1, V_2)$ along a geodesic $C \in \Gamma(p, q)$ is a vector space consisting of all the Jacobi fields along C which vanish at the endpoints p, q.*

Proof. Let J be a Jacobi field such $J(0) = J(1) = 0$, hence $J \in \Gamma(p, q)_C$. From Proposition 41.1 it follows that $H(J, V) = 0$ for any $V \in \Gamma(p, q)_C$, because J has no non-differentiable point. Therefore J belongs to the null space.

Conversely, let J be an element of the null space. We take a subdivision $0 = t_0$ $t_1 < \cdots < t_p = 1$ such that every restriction $J|[t_{r-1}, t_r]$, $r = 1,\cdots, p$, is differentiable. Corresponding to this subdivision, we take a variation vector field

$$V(t) = f(t)(D_t D_t J + R^2(x', x', J)),$$

where the function f(t) is assumed to be positive at any $t \neq t_r$, and equal to zero at every t_r. Then we have

$$H(J, V) = - \int_0^1 f(t)\|D_t D_t J + R^2(x', x', J)\| dt,$$

where $\|X\|^2$ denotes g(X, X) relative to x'. Thus equation H(J, V) = 0 implies $D_t D_t J + R^2(x', x', J) = 0$ for $t \neq t_r$, so that every $J|[t_{r-1}, t_r]$ is a Jacobi field. Next, if we take $V \in \Gamma(p, q)_C$ such that

$$V(t_r) = \Delta_r(D_t J), \qquad V(t) = 0 \quad \text{for } t \neq t_r,$$

then it is seen that

$$H(J, V) = - \sum_r \|\Delta_r(D_t J)\|^2 = 0,$$

which implies $\Delta_r(D_t J) = 0$; so $D_t J$ has no jump at t_r. Consequently p Jacobi fields $J|[t_{r-1}, t_r]$ constitute a Jacobi field throughout the unit interval.

REMARK 41.3. From Theorem 41.1 and Proposition 41.2 it follows that the nullity ν is finite and equal to or less than 2n.

DEFINITION 41.3. A point C(b), $0 \leq a < b < 1$, of a geodesic $C \in \Gamma(p, q)$ is called a *conjugate point* of a point C(a) along C, if there exists a non-zero Jacobi field which vanishes at t = a, b. The *multiplicity* μ of conjugate point C(b) is the dimension of a vector space which consists of such Jacobi fields and zero Jacobi field.

Thus the nullity ν of Hessian $H(V_1, V_2)$ is equal to the multiplicity μ of conjugate point C(1) to C(0) along C, if C(1) is conjugate to C(0) along C. As a consequence we obtain

COROLLARY 41.1. *A necessary and sufficient condition for a geodesic $C \in \Gamma(p, q)$ such that the Hessian $H(V_1, V_2)$ is degenerate is that the terminal point q is conjugate to the initial point p along C.*

COROLLARY 41.2. *If a geodesic $C \in \Gamma(p, q)$ has a conjugate point to the initial point p, then C is not a minimal geodesic.*

Because the Hessian $H(V_1, V_2)$ is degenerate along $C|[0, \tau]$, where $C(\tau)$ is conjugate to p along C.

To show how to construct a Jacobi field and to give an intuitive image of conjugate point, we introduce

DEFINITION 41.4. A one-parameter variation $\alpha: U\ (\subset R^1) \to M$ of a geodesic $C \in \Gamma(p, q)$ such that the endpoints p, q are not necessarily fixed is called a *variation of C through geodesics*, if the following three conditions are satisfied:

(1) α is of class C^∞. (2) $\alpha(0) = C$.

(3) $\bar{\alpha}(\xi, t) = \alpha(\xi)(t)$ is a geodesic for every fixed ξ.

THEOREM 41.2. *Let α be a variation of a geodesic C through geodesics. Then the variation vector field induced from α is a Jacobi field along C.*

Proof. Because $\alpha(\xi, t)$ (bar is omitted) is a geodesic for every fixed ξ, we have $D'_t(\partial_t\alpha) = 0$. Then it is seen from (41.1) and (40.8) that

$$0 = D'_\xi(D'_t(\partial_t\alpha)) = D'_t(D'_\xi(\partial_t\alpha)) + R^1(\partial_t\alpha, \partial_\xi\alpha) = D'_t(D'_t(\partial_\xi\alpha)) + R^1(\partial_t\alpha, \partial_\xi\alpha).$$

Evaluating this at $\xi = 0$ and referring to (17.8), we have

$$D_t(D_t(\partial_\xi\alpha)) + R^2(x', x', \partial_\xi\alpha) = 0.$$

Consequently $(\partial_\xi\alpha)_0$ is a Jacobi field.

From Theorem 41.2 we get an intuitive image of conjugate point. That is, roughly speaking, a conjugate point C(b) to C(a) along a geodesic C is such that there exists a field of geodesics with one-parameter which go from C(a) to C(b).

THEOREM 41.3. *Let M be a Finsler space with $C\Gamma$ such that the h-curvature tensor R^2 satisfies an inequality*

$$g(Y, R^2(X, X, Y)) = R_{hijk}(x, y)X^hY^iX^jY^k \le 0$$

for any vectors X, Y at any (x, y), $y \ne 0$. Then one of any two points of M is not conjugate to the other along any geodesic through these points.

Proof. Let $C \in \Gamma(p, q)$ be a geodesic. A Jacobi field J along C satisfies

$$g(J, D_tD_tJ) = -\, g(J, R^2(x', x', J)) \ge 0.$$

From (40.3) it follows that

$$\frac{d}{dt}\, g(D_tJ, J) = g(D_tD_tJ, J) + g(D_tJ, D_tJ).$$

These imply $dg(D_tJ, J)/dt \ge 0$, so that $g(D_tJ, J)$ is a monotone increasing function of t, and it is precisely increasing if $D_tJ \ne 0$.

We now assume that q is conjugate to p along C. Then there exists a non-zero Jacobi field J such that $J(0) = J(1) = 0$. Hence $g(D_tJ, J) = 0$ at the endpoints. From the above it follows that $g(D_tJ, J)$ must vanish identically, so that $D_tJ = 0$ on [0, 1] from $g(D_tJ, D_tJ) = 0$. Since J is a Jacobi field and $J(0) = J(1) = 0$, we have J =

0 on [0, 1], which is a contradiction.

THEOREM 41.4. *Let $C \in \Gamma(p, q)$ be a geodesic such that q is not conjugate to p along C. Then there exists a unique Jacobi field which has the prescribed values at the endpoints.*

Proof. Consider the 2n-dimensional vector space J(C) consisting of all the Jacobi fields along C (cf. Proposition 41.2). We define a linear mapping $J(C) \to M_p \times M_q$ given by $J \mapsto (J(0), J(1))$. Because q is not conjugate to p along C, the kernel of this linear mapping is only zero vector field. Further $M_p \times M_q$ is of the same dimension with J(C). Thus the linear mapping is an isomorphism.

REMARK 41.4. We have few old papers concerned with the theory of geodesics in Finsler spaces, but Myers [1, 2] and Whitehead [1, 2, 3] give very fundamental properties of geodesics. Recently some authors show various interesting results on geodesics. See Dazord [4, 5, 6, 7, 9], Hassan [1, 2], Kern [1], Lehmann [1], Moalla [1, 2, 3, 4].

The following two sections are written, indebted to a systematic description of Hassan's thesis [1], although almost all the proofs must be omitted.

§ 42. Completeness of Finsler spaces

DEFINITION 42.1. A region U of a Finsler space M is called

(1) *simple*, if not more than one geodesic joins any ordered pair (p, q) of points of U without leaving U,

(2) *convex*, if any ordered pair of points of U is joined by at least one geodesic which does not leave U.

It is well-known that in 1932-33 Whitehead [1] proved the existence of a simple and convex neighborhood of any point of a Finsler space by detailed analysis of differential equations of geodesic. The so-called exponential mapping is a very good tool to show this.

DEFINITION 42.2. Let p be a point of a Finsler space M. The *exponential mapping* $\exp_p: M_p \to M$ is such that $\exp_p X$ for $X \in M_p$ is the terminal point C(1) of a geodesic $C: [0, 1] \to M$ with the initial point $C(0) = p$ and the initial direction $C'(0) = X$.

The curve $C: t \in [0, 1] \mapsto \exp_p(tX)$ for $X \in M_p$ is a geodesic from p to $\exp_p X$. Its length is equal to the absoulte length L(X) of X, because from (40.3) we get

$$\frac{d}{dt} g(x', x') = 0 = \frac{d}{dt} L^2(x, x')$$

along C, which implies $L(x, x') = L(p, X)$, so that

$$\int_0^1 L(x, x')dt = L(p, X).$$

From the theory of differential equations it is known that $\exp_p X$ is uniquely determined for $X \in M_p$ such that the absolute length $L(X)$ is enough small. Further $\exp_p$ is of class C^∞ for non-zero X. For X = 0 we have the following fact:

PROPOSITION 42.1. *For any point p of a Finsler space M there exists a neighborhood U'_p of the origin 0 in the tangent space M_p and a neighborhood U_p in M such that the exponential mapping $\exp_p$ is a diffeomorphism of U'_p onto U_p.* (The proof is omitted.)

Such U'_p as well as U_p are called *normal*. The above proposition asserts that $\exp_p$ is of class C^1 even at X = 0.

THEOREM 42.1. *Any point p of a Finsler space M has neighborhoods U_p and V_p such that*

(1) *the closure $\bar{V}_p$ is contained in U_p,*
(2) *V_p is simple and convex,*
(3) *U_p is a normal neighborhood of any point of V_p.*

This is Whitehead's theorem in essential. The proof must be omitted.

The following proposition gives a method to compare the length of a geodesic with the one of any curve.

PROPOSITION 42.2. *Let U_p be a normal neighborhood of a point p of a Finsler space M. Any curve C: $t \in [0, 1] \mapsto U_p - \{p\}$ is written as $C(t) = \exp_p(r(t)X(t))$ by a function r(t) and $X(t) \in M_p$ with the unit absolute length. Then length L(C) of C satisfies an inequality $L(C) \geq r(b) - r(a)$ for any a, $b \in [0, 1]$, provided $r(b) > r(a)$. The equaltity holds, if and only if r(t) is monotone increasing and X(t) is constant.* (The proof is omitted.)

The following shows that an enough small segment of a geodesic is minimal:

THEOREM 42.2.

(1) *For any point p of a Finsler space M there exists a neighborhood U_p and a positive number δ such that any ordered pair of two points of U_p is joined in U_p by a unique geodesic of the length less than δ.*
(2) *Let U_p and δ be as above. Suppose that $C_1, C_2 \in \Gamma(x_0, x_1)$ be in U_p and C_1*

be a geodesic. Then the lengths $L(C_1)$ *and* $L(C_2)$ *satisfy an inequality* $L(C_2) \geq L(C_1)$*; the equality holds only when* C_2 *coincides with* C_1 *as point sets and the parameter of* C_2 *is a monotone increasing function of the arc-length of* C_2. (The proof is omitted.)

Finsler spaces we consider do not necessarily satisfy the symmetry condition (F 2) of § 16, so that the greatest care must be taken in introducing distance function.

DEFINITION 42.3.

(1) The *distance* $d(p, q)$ of an ordered pair $(p, q) \in M \times M$ of a connected Finsler space M is the infinimum of lengths of all the oriented curves from p to q. A topology of M defined by the set $B(p, \delta) = \{q \in M \mid d(p, q) < \delta\}$ is called the *metric topology* of M.

(2) In the above M the *symmetric distance* $d_s(p, q)$ is given by

$$d_s(p, q) = \mathrm{Max}\{d(p, q), d(q, p)\}.$$

As to the distance $d(p, q)$ we have

$$(42.1)\qquad \begin{aligned} &(1) \quad d(p, q) \geq 0,\\ &(2) \quad d(p, q) = 0 \text{ if and only if } p = q,\\ &(3) \quad d(p, q) \leq d(p, r) + d(r, q). \end{aligned}$$

The proofs of (1) and (3) are trivial. To verify (2) we shall first show

LEMMA 42.1. *If an open ball* $B'(p, \delta) = \{X \in M_p \mid L(X) < \delta\}$ *in the tangent space* M_p *of a connected Finsler space M is normal, the normal neighborhood* $U_p = exp_p B'(p, \delta)$ *coincides with* $B(p, \delta)$ *in Definition 42.3.*

Proof. Any point $q \in U_p$ is written as $q = \exp_p X$, $X \in B'(p, \delta)$. Since the length $L(C)$ of a geodesic $C: t \in [0, 1] \mapsto \exp_p(tX)$ from p to q is equal to $L(X)$, we obtain $d(p, q) \leq L(X) < \delta$, so that $q \in B(p, \delta)$.

Conversely, consider any point $q \in B(p, \delta)$, i.e., $d(p, q) < \delta$. There exists a curve $C \in \Gamma(p, q)$ with the length $L(C) < \delta$. Denote by $\underline{u}$ the least upper bound of u such that a part $C_u = C|[0, u]$ of C is contained in a normal neighborhood U_p, and suppose $\underline{u} \in [0, 1]$. Then C_u is uniquely written as $C_u(t) = \exp_p(r(t)X(t))$, where $X(t) \in M_p$ has the unit length. Proposition 42.2 shows $L(C_{\underline{u}}) \geq r(\underline{u}) = \delta$, which yields a contradiction $L(C) \geq \delta$. Therefore C lies in U_p.

Proof of (42.1)(2). Suppose $p \neq q$. It is obvious that there exists a normal neighborhood $U_p = \exp_p B'(p, \delta)$ such that q is not contained in U_p. On any curve C from p to q, there is a point r which is in U_p, but not in $N_p = \exp_p B'(p, \delta/2)$. Since

N_p is normal, we have $N_p = B(p, \delta/2)$ from Lemma 42.1, so that $d(p, r) \geq \delta/2$ and the length of a part of C from p to r is equal to or more than $\delta/2$. Thus $L(C) \geq \delta/2$ and $d(p, q) \geq \delta/2$.

PROPOSITION 42.3. *The metric topology of a connected Finsler space M coincides with the manifold topology of M.*

Proof. A set of all the open balls $B(p, \delta)$ for all $p \in M$ and $0 < \delta < \infty$ forms a basis for the metric topology of M. On the other hand, a set of $\exp_p B'(p, \delta)$, where $B'(p, \delta) = \{X \in M_p \mid L(X) < \delta\}$, forms a fundamental system of neighborhoods of p by Proposition 42.1. Therefore Lemma 42.1 completes the proof.

THEOREM 42.3. *If any two points p, q of a connected Finsler space M are enough near, there exists a unique geodesic from p to q, the length of which is equal to the distance d(p, q).*

Proof. We take a positive number δ such that $B'(p, \delta)$ is normal. It follows from Lemma 42.1 that $\exp_p B'(p, \delta) = B(p, \delta)$ and we have a unique geodesic in $B(p, \delta)$ from p to $q = \exp_p X$ where $L(X) < \delta$. Proposition 42.2 shows that C is of the shortest length in $B(p, \delta)$. For any curve C_1 from p to q which does not lie in $B(p, \delta)$, we have $L(C_1) \geq \delta$ clearly.

The distance function $d: M \times M \to R$ is continuous; this will be verified by means of the following fact:

LEMMA 42.2. *For any* $\varepsilon > 0$ *there exists* $\delta > 0$ *such that* $d(p, q) < \delta$ *implies* $d(q, p) < \varepsilon$. (Proof is omitted.)

We turn to consideration of the symmetric distance $d_s(p, q)$. This satisfies

$$(42.2)\qquad \begin{aligned} &(1)\quad d_s(p, q) \geq 0,\\ &(2)\quad d_s(p, q) = 0 \text{ if and only if } p = q,\\ &(3)\quad d_s(p, q) = d_s(q, p),\\ &(4)\quad d_s(p, q) \leq d_s(p, r) + d_s(q, r). \end{aligned}$$

PROPOSITION 42.4. *The topology of a connected Finsler space M induced by the symmetric distance* d_s *coincides with the manifold topology of M.*

Proof. If we put $B_s(p, \delta) = \{q \in M \mid d_s(p, q) < \delta\}$, then $d(p, q) \leq d_s(p, q)$ implies $B_s(p, \delta) \subset B(p, \delta)$. Conversely, for a given $B_s(p, \varepsilon)$ we have $B(p, \delta)$, where δ

is given in Lemma 42.2. Then $B(p, \delta) \subset B_s(p, \delta')$ for $\delta' = \mathrm{Min}(\varepsilon, \delta)$.

By means of Proposition 42.4 the following will be proved:

THEOREM 42.4. *On a connected Finsler space M there exists a continuous function* $\gamma: M \to R^+$ *such that any two points of* $B(p, \gamma(p))$ *are joined by the minimal geodesic.* (Proof is omitted.)

We are now in a position to introduce the concept of completeness of a Finsler space:

DEFINITION 42.4. A Finsler space M is called *geodesically complete*, if the exponential mapping $\exp_p$ is defined on the whole of M_p for any point p of M.

Then interesting facts are proved:

PROPOSITION 42.5. *Let* $E(p, \delta)$ *be a subset of the closure* $\bar{B}(p, \delta)$ *of* $B(p, \delta)$ *such that there exists a minimal geodesic from p to any point of* $E(p, \delta)$. *If the Finsler space M is geodesically complete, then*

(1) $E(p, \delta)$ *is compact,*
(2) $E(p, \delta) = \bar{B}(p, \delta)$ *for all* δ,
(3) any ordered two points of M are joined by a minimal geodesic.

DEFINITION 42.5. A point sequence $\{p_m\}$ of a Finsler space M is called a *Cauchy sequence*, if there exists for any $\varepsilon > 0$ an integer N such that $d(p_i, p_j) < \varepsilon (i, j > N)$. A Finsler space M is called *metrically complete* with respect to the distance d, if any Cauchy sequence of M converges.

The following main theorem of the present section, a generalization of a well-known theorem in Riemannian geometry, is shown, although the proof must be omitted.

THEOREM 42.5. *The following three conditions for a connected Finsler space M are equivalent:*

(1) M is metrically complete with respect to the distance d.
(2) M is geodesically complete.
(3) Any bounded closed subset of M is compact.

COROLLARY 42.1. *If a connected Finsler space M is compact, M is metrically and geodesically complete.*

§ 43. Index theorem

The Hessian H of energy function E along a geodesic $C \in \Gamma(p, q)$ is the second variation $J^2(C, V_1, V_2)$ and is regarded as a function on $\Gamma(p, q)_C \times \Gamma(p, q)_C$. If H as a symmetric quadratic form is positive-definite on $\Gamma(p, q)_C$, then C is certainly a minimal geodesic relative to curves of $\Gamma(p, q)$. which are enough near to C.

Throughout the present section, the parameter t of a geodesic C is assumed to be *normal and* $t = s/L(C)$, where s is the arc-length of C.

DEFINITION 43.1. Let C be a geodesic from p to q of a Finsler space M. The *index I of the Hessian H along C* is the maximal dimension of subspaces of $\Gamma(p, q)_C$ on which H is negative-definite.

It is already known that there exists an open neighborhood U of every point C(t) of a geodesic $C \in \Gamma(p, q)$ such that any two points of U are joined by a unique minimal geodesic which is differentiably dependent on the endpoints. We choose a subdivision $0 = t_0 < t_1 < \cdots < t_p = 1$ ($C(0) = p$ and $C(1) = q$) of [0, 1] such that

(1) every segments $C|[t_{r-1}, t_r]$, $r = 1,\cdots, p$, is contained in such a neighborhood U,

(2) every segment $C|[t_{r-1}, t_r]$ is a minimal geodesic.

We denote by $J(t_0, t_1,\cdots, t_p)_C$ a subspace of $\Gamma(p, q)_C$ which consists of all the broken Jacobi fields J such that $J(0) = J(1) = 0$ and $J|[t_{r-1}, t_r]$, $r = 1,\cdots, p$, is a Jacobi field along $C|[t_{r-1}, t_r]$. It is obvious from Proposition 41.2 that $J(t_0, t_1,\cdots, t_p)_C$ is of finite dimension.

Further, let $\Gamma^0(p, q)_C$ be a subspace of $\Gamma(p, q)_C$ consisting of all the vector fields which vanish at every t_r.

DEFINITION 43.2. Two vector fields V_1 and V_2 belonging to $\Gamma(p, q)_C$ are called *orthogonal* with respect to the Hessian H, if $H(V_1, V_2) = 0$ along C.

LEMMA 43.1.

(1) $\Gamma(p, q)_C = J(t_0, t_1,\cdots, t_p)_C \oplus \Gamma^0(p, q)_C$.

(2) Any $V_1 \in J(t_0, t_1,\cdots, t_p)_C$ *is orthogonal to any* $V_2 \in \Gamma^0(p, q)_C$ *with respect to the Hessian H.*

(3) The Hessian along C is positive-definite on $\Gamma^0(p, q)_C$.

Proof. (1) For a vector field $V \in \Gamma(p, q)_C$ we take a vector field $V_1 \in J(t_0, t_1,\cdots, t_p)_C$ which is uniquely determined by the values $V_1(t_r) = V(t_r)$, as confirmed by Theorem 41.4. It follows that $V - V_1 \in \Gamma^0(p, q)_C$. Further $J(t_0, t_1,\cdots, t_p)_C \cap \Gamma^0(p, q)_C = 0$ is obvious.

(2) Since V_1 is a broken Jacobi field, Proposition 41.1 yields

$$H(V_1, V_2) = - \sum_r g(V_2, \Delta_r(D_t V_1)).$$

Because $V_2(t_r) = 0$, we have $H(V_1, V_2) = 0$.

(3) Take a vector field $V \in \Gamma^0(p, q)_C$ and the associated variation α with V. Since $V(t_r) = 0$, we can choose α to satisfy $\bar{\alpha}(\xi, t_r) = C(t_r)$. From (40.4) we have

$$2(t_r - t_{r-1})E_{t_{r-1}}^{t_r}(\bar{\alpha}(\xi, t)) \geq \{L_{t_{r-1}}^{t_r}(\bar{\alpha}(\xi, t))\}^2 \geq \{L_{t_{r-1}}^{t_r}(C(t))\}^2.$$

The equality of (40.4) holds on C, so that $E(\bar{\alpha}(\xi, t)) \geq E(C) = E(\bar{\alpha}(0, t))$. Thus we get $(d^2E/d\xi^2)_0 \geq 0$, namely, $H(V, V) \geq 0$.

Assume $H(V, V) = 0$ for a $V \in \Gamma^0(p, q)_C$. (2) shows $H(V_1, V) = 0$ for any $V_1 \in J(t_0, t_1, \cdots, t_p)_C$. Next, for any $V_2 \in \Gamma^0(p, q)_C$ we have

$$0 \leq H(V + cV_2, V + cV_2) = 2cH(V, V_2) + c^2H(V_2, V_2)$$

for any number c, so that $H(V, V_2) = 0$. Consequently V belongs to the null space of H. Then Theorem 41.1 shows that $V \in \Gamma^0(p, q)_C$ is a Jacobi field along C, so that V = 0 from (1).

From Lemma 43.1 the following facts are obvious:

PROPOSITION 43.1.

(1) *The index I as well as the nullity* ν *of Hessian H are equal to those of H restricted to the space* $J(t_0, t_1, \cdots, t_p)_C$.

(2) *The index I of H is finite.*

THEOREM 43.1. *A point* $q = \exp_p X$ *is a conjugate point to p along the geodesic* $C(t) = \exp_p(tX)$, *if and only if the mapping* $\exp_p$ *is critical (not one-to-one) at X* $\in M_p$.

Proof. Let $\exp_p$ be critical at X. Then there exists a non-zero tangent vector $Y \in (M_p)_X$ (tangent space of M_p at X) such that $\exp_p' Y = 0$. We make a curve $\gamma: \xi \mapsto X(\xi)$ in M_p such that $X(0) = X$ and $(dX(\xi)/d\xi)_0 = Y$. Then the one-parameter variation

$$\bar{\alpha} : R^2 \to M, \qquad \bar{\alpha}(\xi, t) = \exp_p(tX(\xi))$$

is a variation of C. The induced variation vector field V(t) is given by

$$V(t) = \{\partial_\xi(\exp_p(tX(\xi)))\}_0,$$

which is a Jacobi field along C, except t = 0, by Theorem 41.2. V(t) is, however, continuous from Proposition 42.1 and a Jacobi field is continuous and defined for all $t \in [0, 1]$, so that the above V(t) is a Jacobi field for all t. Next, in terms of a coordinate system (x^i) we put

$$\exp_p : v^i\left(\frac{\partial}{\partial x^i}\right)_p \mapsto x^i = E^i(v^1, \cdots, v^n),$$

and then

$$\exp_p' : u^i\left(\frac{\partial}{\partial v^i}\right)_v \mapsto u^i \frac{\partial E^j}{\partial v^i}\left(\frac{\partial}{\partial x^j}\right)_x ,$$

where $v = v^i\partial/\partial x^i$. Thus we have

$$Y^j\left(\frac{\partial E^i}{\partial v^j}\right)_{X(0)} = 0, \qquad V^i(t) = \left(\frac{\partial E^i}{\partial v^j}\right)_{tX(0)} tY^j.$$

Therefore $V(0) = V(1) = 0$, but V is not identically equal to zero, because

$$\left(\frac{dV^i}{dt}\right)_0 = \left(\frac{\partial E^i}{\partial v^j}\right)_0 Y^j,$$

and $(\partial E^i/\partial v^j)_0$ is not singular from Proposition 42.1. Consequently the point q is conjugate to p along C.

Conversely, assume that $\exp_p'$ is non-singular at X. Let Y_a, $a = 1,\cdots, n$, be a basis of $(M_p)_X$. Then $\exp_p'Y_a$ are linearly independent. Following the above construction of the curve γ, we have curves $\gamma_a: \xi \mapsto Y_a(\xi)$ in M_p and n Jacobi fields $V_a(t)$. In this case these satisfy $V_a(1) = \exp_p'Y_a$, which are linearly independent. Thus there is no non-trivial Jacobi field along $\exp_p(tX)$ which vanishes at the endpoints. Consequently the point q is not conjugate to p along $\exp_p(tX)$.

As a consequence of Theorem 43.1 together with Proposition 42.1 it is seen that there exists a point non-conjugate to a point p along any geodesic issuing from p.

If we have a conjugate point, its multiplicity must be examined.

<u>THEOREM 43.2</u>. *Let $C: t \in [0, 1] \mapsto C(t)$ be a geodesic from p to q. The index I of the Hessian $H: \Gamma(p, q)_C \times \Gamma(p, q)_C \to R$ is equal to the number of points $C(t)$, $t \in (0, 1)$, such that $C(t)$ is conjugate to p along C, each point being counted with its multiplicity* μ.

To prove this theorem, called the *index theorem*, we shall first show following facts:

<u>LEMMA 43.2</u>. *Let $I(\tau)$ be the index of the Hessian H_0^τ along a part $C_\tau = C|[0, \tau]$ of a geodesic C. As a function of τ the index $I(\tau)$ is such that*

(1) monotone increasing,

(2) $I(\tau) = 0$ for small τ,

(3) $I(\tau - \varepsilon) = I(\tau)$ for small positive ε,

(4) $I(\tau + \varepsilon) = I(\tau) + \nu$ for small positive ε, where ν is the nullity of H_0^τ.

Proof. (1) From Definition 43.1 it follows that $I(\tau)$ is equal to the dimension of a certain vector space V_τ consisting of all the vector fields along C_τ such that they vanish at $t = 0$, τ and H_0^τ is negative-definite on V_τ. Take $X \in V_\tau$ and further $X' \in V_{\tau'}$, $(\tau < \tau')$, defined by

$$X'(t) = X(t) \text{ for } 0 \le t < \tau; \qquad X'(t) = 0 \text{ for } \tau \le t \le \tau'.$$

A vector space V_τ', consisting of such vector fields as X' is of dimension $I(\tau)$ and

$H_0^{\tau'}$ is also negative-definite on it, so that $V'_{\tau'} \subset V_{\tau'}$ and $I(\tau) \leq I(\tau')$.

(2) For a small τ the geodesic segment C_τ is minimal, so that $E(\alpha(\xi)) \geq E(C_\tau) = E(\alpha(0))$ for any one-parameter variation α of C_τ. Thus $(d^2E/d\xi^2)_0 \geq 0$ and $H_0^\tau \geq 0$.

(3) Take a subdivision $0 = t_0 < t_1 < \cdots < t_p = 1$ such that $t_q < \tau < t_{q+1}$ for some q, $q = 0,\cdots, p-1$. From (1) of Proposition 43.1 it follows that $I(1)$ is the index (number of negative sign of the canonical form) of a certain quadratic form Q on the finite-dimensional space $J(t_0, t_1,\cdots, t_p)_C$. Therefore $I(\tau)$ is similarly regarded as the index of a quadratic form Q_τ on such a vector space J_τ. This J_τ is constructed as follows: We choose vectors $X_r \in M_{C(t_r)}$, $r = 1,\cdots, q$, and obtain a unique broken Jacobi field X such that

(i) $X(t_r) = X_r$, $\quad X(0) = X(\tau) = 0$,

(ii) $X|[t_{r-1}, t_r]$ is a Jacobi field.

Let J_τ be a vector space consisting of all the vector fields as above X. If we put

$$M^q = M_{C(t_1)} \times \cdots \times M_{C(t_q)},$$

then the above construction $(X_1,\cdots, X_q) \mapsto X$ is an isomorphism $M^q \to J_\tau$, hence Q_τ and $Q_{\tau'}$ (τ' is near to τ) are regarded as quadratic forms on the same space M^q, and Q_τ varies, of course, continuously in τ. Since $I(\tau)$ is the dimension of a subspace N_τ of M^q on which Q_τ is negative-definite, the quadratic form $Q_{\tau'}$ is also negative-definite on N_τ. Therefore $I(\tau') \geq I(\tau)$. Then (1) implies (3).

(4) The dimension of M^q is equal to nq and Q_τ is positive-definite on a subspace P_τ of M^q of the dimension $I'(\tau) = nq - I(\tau) - \nu$ by Definition 41.2, (1) of Proposition 43.1 and (3) of Lemma 43.1. For τ' near τ the quadratic form $Q_{\tau'}$ is also positive-definite on P_τ, so that

$$I(\tau') \leq \dim M^q - I'(\tau) = I(\tau) + \nu.$$

In the following we shall show the inverse inequality. We construct s ($= I(\tau)$) vector fields Y_a, $a = 1,\cdots, s$, such that

(i) $Y_a(0) = Y_a(\tau) = 0$,

(ii) the matrix $(H_0^\tau(Y_a, Y_b))$, $a, b = 1,\cdots, s$, is negative-definite.

From (ii) it follows that $Y_1,\cdots, Y_s$ are linearly independent.

Further we have ν linearly independent Jacobi fields J_c, $c = 1,\cdots, \nu$, along C which vanish at $t = 0$, τ as it is seen from Theorem 41.1. Since $(D_tJ_c)_\tau$ are linearly independent, we have ν vector fields X_c along $C_{\tau+\varepsilon}$ such that

(i) $X_c(0) = 0$,

(ii) $g((D_tJ_c)_\tau, X_d(\tau)) = \delta_{cd}$ relative to $C'(\tau)$, $c, d = 1,\cdots,\nu$.

Extending Y_a and J_c as vector fields along $C_{\tau+\varepsilon}$ with putting 0 on $\tau \leq t \leq \tau + \varepsilon$, we obtain $s + 2\nu$ vector fields Y_a, J_c and X_c along $C_{\tau+\varepsilon}$. From Proposition 41.1 it follows that

$$H_0^{\tau+\varepsilon}(J_c, Y_a) = 0, \qquad H_0^{\tau+\varepsilon}(J_c, X_d) = -\delta_{cd}.$$

Now we shall show that $s + \nu$ vector fields Y_a and $J_c/k - kX_c$ are linearly independent, where $k \neq 0$ is a small number. In fact, if there exist numbers u_a, v_c such that

$$Z = \sum_a u_a Y_a + \sum_c v_c(J_c/k - kX_c) = 0,$$

then $H_0^{\tau+\varepsilon}(Z, J_d) = 0 = kv_d$, so that $v_d = 0$ and so $u_a = 0$ because of linear independence of Y_a. Thus these vector fields span an $(s + \nu)$-dimensional vector space S. $H_0^{\tau+\varepsilon}$ is negative-definite on S, because the values of $H_0^{\tau+\varepsilon}$ on S are given by the matrix

$$\begin{pmatrix} H_0^{\tau+\varepsilon}(Y_a, Y_b) & -kH_0^{\tau+\varepsilon}(Y_a, X_d) \\ -kH_0^{\tau+\varepsilon}(X_c, Y_b) & -2\delta_{cd} + k^2 H_0^{\tau+\varepsilon}(X_c, X_d) \end{pmatrix},$$

which is negative-definite for enough small k. Therefore $I(\tau + \varepsilon) \geq s + \nu$.

Proof of Theorem 43.2. Because the index function $I(\tau)$ is monotone increasing from (1) of Lemma 43.2 and its value is an integer, it is constant except jump. A jump occurs only at a conjugate point from Corollary 41.2, and the magnitude of the jump is equal to the multiplicity μ of such a point, as it will be easily seen from (3) and (4) of Lemma 43.2. Thus the number of conjugate points to C(0) along C is finite.

Let $C(c_i)$, $i = 1,\cdots, r$, be all the conjugate points to C(0) along C with the nullity ν_i of H. ν_i is the multiplicity μ_i of the conjugate point $C(c_i)$. Since $I(\tau)$ is constant on $0 < \tau < c_1$, (2) of Lemma 43.2 shows that $I(\tau) = 0$ on $0 < \tau \leq c_1$. If τ is near to c_1 and exceeds c_1, (4) of Lemma 43.2 shows that $I(\tau) = \nu_1$, so that $I(\tau) = \nu_1$ on $c_1 < \tau \leq c_2$. By repeating such a process the proof is easily completed.

We now return to the quantity

$$R^2(X, Y, X, Y) = R_{hijk}(x, y)X^hY^iX^jY^k,$$

which was treated in Theorem 41.3. Introduce a quantity

$$\mathrm{Ric}(X, Y) = R_h{}^i{}_{ji}(x, y)X^hY^j.$$

Then Ric(X, X) for a unit vector X is called the *Ricci curvature* of X. If we are concerned with a mapping $Z = (Z^i) \mapsto R^2(X, Y, Z) = (R_h{}^i{}_{jk}X^hY^jZ^k)$, its trace is equal to Ric(X, Y).

Let (X_a), $a = 1,\cdots, n$, be an orthonormal frame at a point of an n-dimensional Finsler space M. Then

$$\sum_{a=1}^{n-1} R^2(X_n, X_a, X_n, X_a) = R_{hijk}X_n^hX_n^j(g^{ik} - X_n^iX_n^k),$$

which implies

$$\text{(43.1)} \qquad \mathrm{Ric}(X_n, X_n) = \sum_{a=1}^{n-1} R^2(X_n, X_a, X_n, X_a).$$

As an application of the index theorem (Theorem 43.2) we shall show the following interesting theorem, a generalization of well-known Myers' theorem in Riemannian geometry:

THEOREM 43.3. *If there exists a positive number r in an n-dimensional Finsler space M such that the inequality*

$$Ric(X, X) \geq (n - 1)/r^2$$

holds for any unit vector X at any point (x, y) of $T(M)$, any geodesic of M with the length more than πr contains a conjugate point to the initial point and is not minimal.

Proof. Let (X_a) be a frame at the initial point $C(0)$ of a geodesic $C: t \in [0, 1] \mapsto C(t)$ which is orthonormal relative to $C'(0)$. By parallel displacement of (X_a) along C relative to C' we obtain a field of orthonormal frame $(X_a(t))$ along C. Further assume that X_n is tangent to C. Then

$$D_t X_a = 0, \qquad C'(t) = L(C(t), C'(t))X_n.$$

Taking n variation vector fields $V_a(t) = \sin(\pi t)X_a(t)$, it is seen from Proposition 41.1 that

$$H(V_a, V_a) = \int_0^1 \{\sin(\pi t)\}^2\{\pi^2 - L^2R^2(X_n, X_a, X_n, X_a)\}dt.$$

Thus (43.1) leads to

$$\sum_{a=1}^{n-1} H(V_a, V_a) = \int_0^1 \{\sin(\pi t)\}^2\{(n - 1)\pi^2 - L^2\mathrm{Ric}(X_n, X_n)\}dt.$$

From the assumption $L > \pi r$ the above is negative, which implies $H(V_a, V_a) < 0$ for some number a $(= 1,\cdots, n - 1)$. Therefore the index I of C is positive and Theorem 43.2 shows the existence of conjugate point to $C(0)$ along C.

COROLLARY 43.1. *If an n-dimensional Finsler space M is geodesically complete and satisfies an inequality $Ric(X, X) \geq (n - 1)/r^2$ for any unit vector X at any point of $T(M)$, then M has the diameter $\leq \pi r$ and is compact.*

REMARK 43.1. Following Lichnerowitz [3] and Akbar-Zadeh [7], various interesting papers concerned with global theory of Finsler spaces have been published. Systematical description of this theory is given by Dazord [7], Hassan [1], Moalla [4].

Theorem 42.5 is proved by Hassan [1] and Moalla [1], and Theorem 43.2 are shown by Lehmann [1]. Some interesting results of the so-called pinching problem such as

Theorem 43.3 are shown by Dazord [5, 6] , Grove, Karcher and Ruh [1], Hassan [1], Moalla [2, 3]. Further Kern [1, 2] introduces an interesting concept called an almost Riemannian Finsler space from the viewpoint of pinching problem by means of osculating Riemannian metric.

The problem to generalize the Gauss-Bonnet formula to Finsler spaces has attracted every geometrician. See Nazim [2], In relation to this problem, Lichnerowicz [3] introduces the concept of Berwald space ($S_{hijk} = 0$) (cf. Theorem 24.3). Dazord's papers [8, 9] make some contributions to this problem. Rund's recent work [29] increases interest in this subject, although he asserts a negative conjecture in general Finsler spaces.

BIBLIOGRAPHY

AGRAWAL, P.
[1] On the concircular geometry in Finsler spaces. (Tensor, N.S.23 (1972), 333-336).
AKBAR-ZADEH, H.
[1] Sur la réductibilité d'une variété finslérienne. (C.R.Acad. Sci. Paris 239 (1954), 945-947).
[2] Sur les isométries infinitésimales d'une variété finslérienne. (Ibid. 242 (1956), 608-610).
[3] Sur une connexion euclidienne d'espace d'éléments linéaires. (Ibid. 245 (1957), 26-28).
[4] Sur une connexion coaffine d'espace d'éléments linéaires. (Ibid. 247 (1958), 1707-1710).
[5] Sur les espaces de Finsler isotropes. (Ibid. 252 (1961), 2061-2063).
[6] Transformations infinitésimales conformes des variétés finslériennes compactes. (Ibid. 252 (1961), 2807-2809).
[7] Les espaces de Finsler et certaines de leurs généralisations. (Ann. Sci. École Norm. Sup. (3) 80 (1963), 1-79).
[8] Une généralisation de la géométrie finslérienne. (Topol. Géom. Diff. Sém. Ehresmann 6-6 (1964), 9p.).
[9] Sur les automorphismes de certaines structures presque cosymplectiques. (Canad. Math. Bull. 8 (1965), 39-57).
[10] Sur les homothéties infinitésimales des variétés finslériennes. (C.R.Acad. Sci. Paris 262 (1966), 1058-1060).
[11] Sur quelques théorèmes issus du calcul des variations. (Ibid. 264 (1967), 517-519).
[12] Sur les sous-variétés des variétés finslériennes. (Ibid. 266 (1968), 146-148).
[13] Sur les invariants conformes des variétés finslériennes. (Ibid. 268 (1969), 402-404).
[14] Sur le noyau de l'opérateur de courbure d'un variété finslérienne. (Ibid. 272 (1971), 807-810).
[15] Espaces de nullité de certains opérateurs en géométrie des sous-variétés. (Ibid. 274 (1972), 490-493).
[16] Espaces de nullité en géométrie finslérienne. (Tensor, N.S. 26 (1972), 89-101).
[17] Sur les isométries infinitésimales d'une variété finslérienne compacte. (C. R. Acad. Sci. Paris 278 (1974), 871-874).
[18] Sur les transformations infinitésimales projectives des variétés finslériennes compactes. (Ibid. 280 (1975), 591-593).
[19] Transformations infinitésimales projectives des variétés finslériennes compactes. (Ibid. 280 (1975), 661-663).
[20] Transformations infinitésimales conformes des variétés finslériennes compactes. (Ibid. 281 (1975), 655-657).
[21] Remarques sur les isométries infinitésimales d'une variété finslérienne compacte. (Ibid. 284 (1977), 451-453).
[22] Sur les transformations infinitésimales conformes de certaines variétés finslériennes compactes. (Ibid. 286 (1978), 177-179).
[23] Transformations infinitésimales conformes des variétés finslériennes compactes. (Ann. Polon. Math. 36 (1979), 213-229).
AKBAR-ZADEH, H.; BONAN, E.
[1] Structure presque kählérienne naturelle sur le fibré tangent à une variété finslérienne. (C.R.Acad. Sci. Paris 258 (1964), 5581-5582).
AKBAR-ZADEH, H.; WEGRZYNOWSKA, A.
[1] Sur la géométrie du fibré tangent à une variété finslérienne. (Ibid. 282 (1976), 325-328).
[2] Sur le géométrie du fibré tangent à une variété finslérienne compactes. (Ann.

Polon. Math. 36 (1979), 231-244).
ANOSOV, D.V.
[1] Geodesic in Finsler geometry. (Russian) (Proc. Intern. Congr. Math., Vancouver, 1974, II 293-297).
[2] Geodesics in Finsler geometry. (Amer. Math. Soc. Translat. II, 109 (1977), 81-85).
ATKIN, C.J.
[1] Bounded complete Finsler structures. I. (Studia Math. 62 (1978), 219-228).
AUSLANDER, L.
[1] On curvature in Finsler geometry. (Trans. Amer. Math. Soc. 79 (1955), 378-388).
AWASTHI, G.D.
[1] A study of certain special Finsler spaces. (Univ. Nac. Tucumán, Rev. A 24 (1974), 163-166).
BARBILIAN, D.
[1] Les J-géométries naturelles finslériennes. (Romanian) (Acad. R.P.Romine. Stud. Cerc. Mat. 11 (1960), 7-47).
BARBILIAN, D.; RADU, N.
[1] Les J-métriques finslériennes naturelles et la fonction de représentation de Riemann. (Romanian) (Ibid. 13 (1962), 21-36).
BARTHEL, W.
[1] Zum Inhaltsbegriff in der Minkowskischen Geometrie. (Math. Z. 58 (1953), 358-375).
[2] Über eine Parallelverschiebung mit Längeninvarianz in lokal-Minkowskischen Räumen I, II. (Arch. Math. 4 (1953), 346-365).
[3] Über Minkowskische und Finslersche Geometrie. (Conveg. Geom. Diff. Roma 1953, 71-76).
[4] Zur Flächentheorie in Finslerschen Räumen. (Proc. Intern. Math. Congr. Amsterdam 1954, 194-195).
[5] Über die Minimalflächen in gefaserten Finslerräumen. (Ann. Mat. Pura Appl. (4) 36 (1954), 159-190)
[6] Variationsprobleme der Oberflächenfunktion in der Finslerschen Geometrie. (Math. Z. 62 (1955), 23-36).
[7] Extremalprobleme in der Finslerschen Inhaltsgeometrie. (Ann. Univ. Sarav. Naturwiss. Sci. 3-IV (1955), 171-183).
[8] Zur isodiametrischen und isoperimetrischen Ungleichung in der Relativgeometrie. (Comm. Math. Helv. 33 (1959), 241-257).
[9] Zur Minkowski-Geometrie, begründet auf dem Flächeninhaltbegriff. (Monatsh. Math. 63 (1959), 317-343).
[10] Natürliche Gleichungen einer Kurve in der metrischen Differentialgeometrie. (Arch. Math. 10 (1959), 392-400).
[11] Nichtlineare Zusammenhänge und deren Holonomiegruppen. (J. Reine Angew. Math. 212 (1963), 120-149).
BAŠILOVA, V.I.
[1] An example of a Minkowski geometry with a non-symmetric indicatrix. (Russian) (Collect. Art. Diff. Geom., 1974, 93-99).
BEEM, J.K
[1] Indefinite Minkowski spaces. (Pacific J. Math. 33 (1970), 29-41).
[5] Symmetric perpendicularity for indefinite Finsler metrics on Hilbert manifold. (Geom. Dedicata 4 (1975), 45-49).
[6] Characterizing Finsler spaces which are pseudo-Riemannian of constant curvature. (Pacific J. Math. 64 (1976), 67-77).
BEEM, J.K.; KISHTA, M.A.
[1] On generalized indefinite Finsler spaces. (Indiana Univ. Math. J. 23 (1974), 845-853).
BERWALD, L.
[1] Ueber die erste Krümmung der Kurven bei allgemeiner Massbestimmung. (Lotos Prag. 67/68 (1919/20), 52-56).
[2] Über Parallelübertragung in Räumen mit allgemeiner Massbestimmung. (Jber. Deutsch. Math.-Verein. 34 (1926), 213-220).

[3] Untersuchung der Krümmung allgemeiner metrischer Räume auf Grund des in ihnen herrschenden Parallelismus. (Math. Z. 25 (1926), 40-73; 26 (1927), 176).
[4] Zur Geometrie ebener Variationsprobleme. (Lotos Prag 74 (1926), 43-52).
[5] Über zweidimensionale allgemeine metrische Räume. I, II. (J. Reine Angew. Math. 156 (1927), 191-222).
[6] Sui differenziali secondi covarianti. (Lincei Rend. (6) 5 (1927), 763-768).
[7] Parallelübertragung in allgemeinen Räumen. (Atti Congr. Intern. Mat. Bologna (1928), 4, 263-270).
[8] Una forma normale invariante della seconda variazione. (Lincei Rend. (6) 7 (1928), 301-306).
[9] Über eine charakterische Eigenschaft der allgemeinen Räume konstanter Krümmung mit geradlinigen Extremalen. (Monatsh. Math. Phys. 36 (1929), 315-330).
[10] Über die n-dimensionalen Geometrien konstanter Krümmung, in denen die Geraden die kürzesten sind. (Math. Z. 30 (1929), 449-469).
[11] Über Finslersche und verwandte Räume. (Cas. Mat. Pys. 64 (1935), 1-16).
[12] Über die Hauptkrümmungen einer Fläche im dreidimensionalen Finslerschen Raum. (Monatsh. Math. Phys. 43 (1936), 1-14)
[13] Ueber Finslersche und Cartansche Geometrie I. Geometrische Erklärungen der Krümmung und des Hauptskalars eines zweidimensionalen Finslerschen Raumen. (Math. Timişoara 17 (1941), 34-58).
[14] On Finsler and Cartan geometries. III. Two-dimensional Finsler spaces with rectilinear extremals. (Ann. of Math. (2) 42 (1941), 84-112).
[15] Ueber die Beziehungen zwischen den Theorien der Parallelübertragung in Finslerschen Räumen. (Nederl. Akad. Wetensch. Proc. 49 (1946), 642-647 = Indag. Math. 8 (1946), 401-406).
[16] Ueber Systeme von gewöhnlichen Differentialgleichungen zweiter Ordnung deren Integralkurven mit dem System der geraden Linien topologisch aequivalent sind. (Ann. of Math. (2) 48 (1947), 193-215).
[17] Ueber Finslersche und Cartansche Geometrie IV. Projektivkrümmung allgemeiner affiner Räume und Finslersche Räume skalarer Krümmung. (Ann. of Math. (2) 48 (1947), 755-781).

BLIZNIKAS, V.I.
[1] Finsler spaces and their generalizations. (Russian) (Algebra, Topology, Geometry 9 (1971), 75-136). (English Transl.: Progr. Math. 9 (1971), 75-136).

BRICKELL, F.
[1] A new proof of Deicke's theorem on homogeneous functions. (Proc. Amer. Math. Soc. 16 (1965), 190-191).
[2] On the differentiability of affine and projective transformations. (Ibid. 16 (1965), 567-574).
[3] A theorem on homogeneous functions. (J. London Math. Soc. 42 (1967), 325-329).
[4] Differentiable manifolds with an area measure. (Canad. J. Math. 19 (1967), 540-549).
[5] A relation between Finsler and Cartan structures. (Tensor, N.S. 25 (1972), 360-364).
[6] Area measures on a real vector space. (Global Analy. & its Appl. II 23-32. Intern. Atomic Energy Agency, Vienna, 1974).

BRICKELL, F.; CLARK, R.S.; AL-BORNEY, M.S.
[1] (G, E) Structures. (Topics in Diff. Geom., 1976, 29-43).

BRICKELL, F.; YANO, K.
[1] Concurrent vector fields and Minkowski structures. (Kōdai Math. Sem. Rep. 26 (1974), 22-28).

BROWDER, F.E.
[1] Infinite dimensional manifolds and non-linear elliptic eigenvalue problems. (Ann. of Math. (2) 82 (1965), 459-477).
[2] Lusternik-Schnirelman category and nonlinear elliptic eigenvalue problems. (Bull. Amer. Math. Soc. 71 (1965), 644-648).
[3] Existence theorems for nonlinear partial differential equations. (Proc. Symp. Pure Math. 16 (1968), 1-60).

BROWN, G.M. (THORNLY)
[1] A study of tensors which characterize a hypersurface of a Finsler space. (Canad. J. Math. 20 (1968), 1025-1036).
[2] Gaussian curvature of a subspace in a Finsler space. (Tensor, N.S. 19 (1968), 195-202).

BUCUR, I.
[1] Sur une propriété globale des ligues géodésiques d'un espace de Finsler. (Romanian) (Com. Acad. R.P. Romine 5 (1955), 965-968).

BUSEMANN, H.
[1] Metric conditions for symmetric Finsler spaces. (Proc. Nat. Acad. Sci. U.S.A. 27 (1941), 533-535).
[2] Metric methods in Finsler spaces and in the foundations of geometry. (Ann. of Math., Studies 8, Princeton, 1942, 243p.)
[3] Intrinsic area. (Ann. of Math. (2) 48 (1947), 234-267).
[4] The geometry of Finsler spaces. (Bull. Amer. Math. Soc. 56 (1950), 5-16).
[5] The isoperimetric problem for Minkowski area. (Amer. J. Math. 71 (1949), 743-762).
[6] The foundations of Minkowskian geometry. (Comment. Math. Helv. 24 (1950), 156-187).
[7] On geodesic curvature in two-dimensional Finsler spaces. (Ann. Mat. Pura Appl. (4) 31 (1950), 281-295).
[8] Metrics on the torus without conjugate points. (Spanish) (Bol. Soc. Mat. Mexicana 10, 1-2 (1953), 12-29). (English: Ibid. 10, 3-4 (1953), 1-18).
[9] The geometry of geodesics. (Pure and Appl. Math. 6, Acad. Press Inc., New York, 1955, 422p.).
[10] On normal coordinates in Finsler spaces. (Math. Ann. 129 (1955), 417-423).
[11] The synthetic approach to Finsler space in the large. (C.I.M.E. Geom. Calcolo Variaz. 1-2 (1961), 70p.).
[12] On Hilbert's fourth problem. (Russian) (Uspehi Mat. Nauk. 21 (1966), 1 (127), 155-164).
[13] Problem IV: Desarguesian spaces. (Math. Dev. Hilbert Probl., Proc. Symp. Pure Math. 28 (1974), 131-141 (1976)).

CARTAN, É.
[1] Sur un problème d'équivalence et la théorie des espaces métriques généralisés. (Math. Cluj 4 (1930), 114-136. Oeuvres Compl. III 2, 1131-1153).
[2] Observations de M.Élie Cartan sur la Communication précédente. (C.R. Acad. Sci. Paris 196 (1933), 27-28).
[3] Sur les espaces de Finsler. (Ibid. 196 (1933), 582-586. Oeuvres Compl. III 2, 1245-1248).
[4] Les espaces de Finsler. (Actualités 79, Paris, 1934, 41p. 2nd edit. 1971).
[5] Les espaces de Finsler. (French and Russian) (Abh. Sem. Vektor u. Tensor Anal. Moskau 4 (1937), 70-81, 82-94. Oeuvres Compl. III 2, 1385-1396).

ČEGODAEV, Ju.M.
[1] A geodesic field of q-dimensional directions in a Finsler space, and an included system of paths. (Russian) (Izv. Vysš. Učebn Zaved. Mat. (1975), 4 (155), 74-86).

ČETYRKINA, Z.N.
[1] Homotheties and motions in two-dimensional Finsler spaces. (Russian) (Volz. Mat. Sb. Vyp. 5 (1966), 366-373).
[2] Conformal transformations in Finsler spaces. (Russian) (Ukrain. Geom. Sb. Vyp. 11 (1971), 95-98).

CHAWLA, M.; BEHARI, R.
[1] Almost complex space with Finsler metric. (Indian J. Pure Appl. Math. 2 (1971), 401-408).

CHERN, S.
[1] On the Euclidean connections in a Finsler space. (Proc. Nat. Acad. Sci. U.S.A. 29 (1943), 33-37).
[2] Local equivalence and Euclidean connections in Finsler spaces. (Sci. Rep. Nat. Tsing Hua Univ. A 5 (1948), 95-121).

CHU, Y.-H. (JYOO)
[1] Studies on Finsler spaces. (Korean) (Hak-sul Chi 11 (1970), 645-655).

ČOMIĆ, I.
[1] Hyperplanes of Finsler spaces of the constant intrinsic curvature. (Serbo-Croatian) (Mat. Vesnik 7 (22) (1970), 257-267).
[2] Relations between induced curvature tensors of Finsler hypersurface F_{n-1} and curvature tensors of imbedding space F_n. (Publ. Inst. Math. Beograd, N.S. 11 (25) (1971), 43-52).
[3] The induced curvature tensors of a subspace in a Finsler space. (Tensor, N.S. 23 (1972), 21-34).
[4] The intrinsic curvature tensors of a subspace in a Finsler space. (Ibid. 24 (1972), 19-28).
[5] Relations between induced and intrinsic curvature tensors of a subspace in the Finsler space. (Math. Vesnik 1 (14) (29) (1977), 65-72).
DAVIES, E.T.
[1] Lie derivation in generalized metric spaces. (Ann. Mat. Pura Appl. (4) 18 (1939), 261-274).
[2] Subspaces of a Finsler space. (Proc. London Math. Soc. (2) 49 (1947), 19-39).
DAZORD, P.
[1] Tenseur de structure d'un G-structure dérivée. (C.R. Acad. Sci. Paris 258 (1964), 2730-2733).
[2] Sur une généralisation de la notion de "spray". (Ibid. 263 (1966), 543-546).
[3] Connexion de direction symétrique associée à un "spray" généralisé. (Ibid. 263 (1966), 576-578).
[4] Variétés finslériennes à géodésiques fermées. (Ibid. 266 (1968), 348-350).
[5] Variétés finslériennes de dimension paire δ-pincées. (Ibid. 266 (1968), 496-498).
[6] Variétés finslériennes en forme de sphères. (Ibid. 267 (1968), 353-355).
[7] Propriétés globales des géodésiques des espaces de Finsler. (Thèse, Lyon, 1969, 193p.).
[8] Sur la formule de Gauss-Bonnet en géométrie finslérienne. (C.R. Acad. Sci. Paris 270 (1970), 1241-1243).
[9] Tores finslériens sans points conjugués. (Bull. Soc. Math. France 99 (1971), 171-192, 397).
DEICKE, A.
[1] Über die Finsler-Räume mit $A_i = 0$. (Arch. Math. 4 (1953), 45-51).
[2] Über die Darstellung von Finsler-Räumen durch nichtholonome Mannigfaltigkeiten in Riemannschen Räumen. (Ibid. 4 (1953), 234-238).
[3] Finsler spaces as non-holonomic subspaces of Riemannian spaces. (J. London Math. Soc. 30 (1955), 53-58).
DHAWAN, M.; PRAKASH, N.
[1] Adjoint complex Finsler manifolds. (Indian J. Pure Appl. Math. 1 (1970), 8-16).
DIAZ, J.-G.
[1] Quelques propriétés des tenseurs de courbure en géométrie finslérienne. (C.R. Acad. Sci. Paris 274 (1972), 569-572).
DIAZ, J.-G.; GRANGIER, G.
[1] Courbure et holonomie des variétés finslériennes. (Tensor, N.S. 30 (1976), 95-109).
DOUGLAS, J.
[1] The general geometry of paths. (Ann. of Math. (2) 29 (1927/28), 143-168).
DOWLING, J.
[1] Finsler geometry on Sobolev manifolds. (Proc. Symp. Pure Math. 15 (1968), 1-10).
DRAGUNOV, V.K.
[1] A certain way of representing the geometry of the hyperbolic plane locally as a Minkowski geometry. (Russian) (Collect. Art Diff. Geom. 100-106. Kalinin. Gos. Univ., Kalinin, 1974).
EARLE, C.J.; EELLS, J.
[1] Foliations and fibrations. (J. Diff. Geom. 1 (1967), 33-41).
EARLE, C.J.; HAMILTON, R.S.
[1] A fixed point theorem for holomorphic mappings. (Proc. Symp. Pure Math. 16 (1968), 61-65).

EELLS, J.
[1] On the geometry of function spaces. (Symp. Intern. Topol. Alg. Mexico City, 1958, 303-308).
[2] A setting for global analysis. (Bull. Amer. Math. Soc. 72 (1966), 751-807).
EGUCHI, K.; ICHIJYŌ, Y.
[1] General projective connections and Finsler metric. (J. Math. Tokushima Univ. 3 (1969), 1-20).
ELIASSON, H.I.
[1] Variation integrals in fibre bundles. (Proc. Symp. Pure Math. 16 (1968), 67-89).
[2] Introduction to global calculus of variations. (Global Anal. Appl. Intern Sem. Course Trieste, 1972, II, 113-135 (1974)).
ELIOPOULOS, H.A.
[2] Sur la définition de la courbure totale d'une hypersurface plongée dans un espace de Finsler localement minkowskien. (Acad. Roy. Belg. Bull. Cl. Sci. (5) 45 (1959), 205-214).
[3] Subspaces of a generalized metric space. (Canad. J. Math 11 (1959), 235-255).
ERMAKOV, Ju.I.
[1] Three-dimensional space with a cubic semimetric. (Russian) (Dokl. Akad. Nauk SSSR, N.S. 118 (1958), 1070-1073).
[2] Spaces X_n with an algebraic metric and semimetric. (Russian) (Ibid. 128 (1959), 460-463).
[3] Hypersurfaces of a space with a cubic metric. (Russian) (Diff. Geom. 1, 20-35, 135. Izdat. Saratov. Univ., Saratov, 1974).
FAVA, F.; MISRA, R.B.
[1] Eulerian curvature tensors and the conformal mappings. (Univ. Politec. Torino. Rend. Sem. Mat. 35 (1976-77), 311-326).
FERNANDES, A.M.
[1] Derivazione tensoriale composta negli spazii non puntuali. (Lincei Rend. (6) 21 (1935), 555-562).
[2] Axiomatics of spaces of linear elements. (Portuguese) (Portugaliae Math. 2 (1941), 7-12).
FINSLER, P.
[1] Über Kurven und Flächen in allgemeinen Räumen. (Dissertation, Göttingen, 1918, 121p. Verlag Birkhäuser, Basel, 1951).
[2] Über die Krümmungen der Kurven und Flächen. (Reale Accad. Italia, Fond. A. Volta, Atti dei Conveg. 9 (1939), 463-478).
[3] Über eine Verallgemeinerung des Satzes von Meusnier. (Vierteljschr. Naturforsch. Ges. Zürich 85, Beibl. 32 (1940), 155-164).
FREEMAN, J.G.
[1] First and second variations of the length integral in a generalized metric space. (Quart. J. Math. Oxford 15 (1944), 70-83).
[2] Theory of a ruled two-space in a generalized metric space. (Ibid. 17 (1946), 119-128).
[3] A generalization of minimal varieties. (Edinburgh Math. Proc. (2) 8 (1948), 66-72).
[4] Finsler-Riemann systems. (Quart. J. Math. Oxford (2) 7 (1956), 100-109).
[5] Complete Finsler-Riemann systems. (Ibid. (2) 8 (1957), 161-171).
[6] Parallel transport in a Finsler space defined by a field of oriented vectors. (Math. Nachr. 45 (1970), 161-166).
FUKUI, M.; YAMADA, T.
[1] On the indicatrized tensor of $S_{ijkh|1}$ in Finsler spaces. (Tensor, N.S. 33 (1979), 373-379).
FUNK, P.
[1] Über den Begriff „extremale Krümmung" und eine kennzeichnende Eigenschaft der Ellipse. (Math. Z. 3 (1919), 87-92).
[2] Über Geometrien, bei denen die Geraden die Kürzesten sind. (Math. Ann. 101 (1929), 226-237).
[3] Über Geometrien, bei denen die Geraden die kürzesten Linien sind und Äquidistanten zu einer Geraden wieder Gerade sind. (Monatsh. Math. Phys. 37 (1930), 153-158).

[4] Über zweidimensionale Finslersche Räume, insbesondere über solche mit geradlinigen Extremalen und positiver konstanter Krümmung. (Math. Z. 40, (1935), 86-93).
[5] Beiträge zur zweidimensionalen Finsler'schen Geometrie. (Monatsh. Math. Phys. 52 (1948), 194-216).
[6] Eine Kennzeichnung der zweidimensionalen elliptischen Geometrie. (Österreich. Akad. Wiss. Math. Natur. Kl. S.-B. II 172 (1963), 251-269).

GALVANI, O.
[1] Sur la réalisation des espaces de Finsler. (C.R. Acad. Sci. Paris 222 (1946), 1067-1069).
[2] Les connexions finslériennes de congruences de droites. (Ibid. 222 (1946), 1200-1202).
[3] Sur l'immersion du plan de Finsler dans certains espaces de Riemann à trois dimensions. (Ibid. 223 (1946), 1088-1090).
[4] La réalisation des connexions euclidiennes d'éléments linéaires et des espaces de Finsler. (Ann. Inst. Fourier 2 (1951), 123-146).
[5] La réalisation des espaces de Finsler. (C.R. Congr. Soc. Savantes Paris et Départ. Grenoble, 1952, Sect. Sci. 57-60).
[6] Rèalisations euclidiennes des planes de Finsler. (Ann. Inst. Fourier 5 (1955), 421-454).

GHEORGHIU, Gh.Th.
[1] Sur les espaces de Finsler. (Math.-Rev. Anal. Numér. Théor. Approx. Math. 19 (42) (1977), 45-53).

GHINEA, I.
[1] Connexions Finslériennes conformes métriques. (Anal. Univ. Timişoara, Ser. Şti. Mat. 14 (1976), 101-115).
[2] Connexions Finslériennes presque symplectiques. (Ibid. 15 (1977), 93-102).
[3] Connexions Finslériennes presque complexes. (Math.-Rev. Anal. Numér. Théor. Approx. Math. 19 (42) (1977), 55-61).
[4] Connexions Finslériennes presque Hermite. (Ibid. 19 (42) (1977), 153-161).

GOŁĄB, S.
[1] Quelques problèmes metriques de la géométrie de Minkowski. (Polish) (Trav. Acad. Mines Cracovie Fasc. 6 (1932), 1-79).
[2] Einige Bemerkungen über Winkelmetrik in Finslerschen Räumen. (Verh. Intern. Math. Kongr. Zurich, 1932, II 178-179).
[3] Sur la représentation conforme de l'espace de Finsler sur l'espace euclidien. (C.R. Acad. Sci. Paris 196 (1933), 25-27).
[4] Sur un invariant intégral relatif aux espaces métriques généralises. (Lincei Rend. (6) 17 (1933), 515-518).
[5] Sur la représentation conforme de deux espaces de Finsler. (C.R. Acad. Sci. Paris 196 (1933), 986-988).
[6] Contribution à un théorème de M.N.S.Knebelman. (Prace Mat. Fiz. 41 (1933), 97-100).
[7] Sur une condition nécessaire et suffisante afin qu'un espace de Finsler soit un espace riemannien. (Lincei Rend. (6) 21 (1935), 133-137).
[8] Sur la mesure des aires dans les espace de Finsler. (C.R. Acad. Sci. Paris 200 (1935), 197-199).
[9] Les transformations par polaires réciproques dans la géométrie de Finsler. (Ibid. 200 (1935), 1462-1464).
[10] Sur le rapport entre les notions des mesures des angles et des aires dans les espaces de Finsler. (Ibid. 201 (1935), 250-251).
[11] On Finsler's measurement of an angle. (Ann. Soc. Polon. Math. 24 (1951), 78-84).
[12] Zur Theorie des Übertragungen. (Schr. Forschungsinst. Math. 1 (1957), 162-177).
[13] Sur la longueur de l'indicatrice dans la géométrie plane de Minkowski. (Colloq. Math. 15 (1966), 141-144).
[14] On a certain problem of Minkowski geometry. (Polish) (Prace Nauk. Inst. Mat. Fiz. Teor. Politech. Wrocław. 2 Ser. Stud. Materiały 2 (1970), 3-5).

[15] Sur un problème de la métrique angulaire dans la géométrie de Minkowski. (Aequations Math. 6 (1971), 121-129).
[16] Sur quelques conditions suffisantes pour que l'espace de Minkowski à deux dimensions soit euclidien. I. Le rôle de l'axiome de L.Dubikajtis. (Tensor, N.S. 24 (1972), 383-388).

GOŁĄB, S.; TAMÁSSY, L.
[1] Eine Kennzeichnung der euklidischen Ebene unter den Minkowskischen Ebenen. (Publ. Math. Debrecen 7 (1960), 187-193).

GORŠKOVA, L.S.
[1] Finsler spaces $F_4(x, \dot{x})$ that admit groups of motions G_r, $r \geqq 3$. (Russian) (Penz. Ped. Inst. Učen. Zap. 124 (1971), 31-35).
[2] Four-dimensional Finsler spaces with groups of motion G_4. (Russian) (Ibid. 124 (1971), 36-41).
[3] Finsler spaces F_4 that admit groups of motions with a nonabelian three-parameter subgroup. (Russian) (Ibid. 124 (1971), 42-46).
[4] Finsler spaces F_4 that admit five-parameter groups of motions with an abelian subgroup G_3. (Russian) (Volz. Mat. Sb. Vyp. 23 (1973), 6-25).
[5] Motions in Finsler spaces. (Russian) (Ibid. 23 (1973), 3-6).

GRANGIER, G.
[1] Sur l'holonomie des variétés finslériennes. (C.R. Acad. Sci. Paris 276 (1973), 289-292).

GRIFONE, J.
[1] Connexions non linéaires conservatives. (Ibid. 268 (1969), 43-45).
[2] Prolongement linéaire d'une connexion de directions. (Ibid. 269 (1969), 90-93).
[3] Structure presque-tangente et connexions, I. (Ann. Inst. Fourier 22, 1 (1972), 287-334).
[4] Sur les connexions d'une variétés finslérienne et d'un système mécanique. (C.R. Acad. Sci. Paris 272 (1971), 1510-1513).
[5] Structure presque tangente et connexions - II. (Ann. Inst. Fourier 22, 3 (1972), 291-338).
[6] Transformations infinitésimales conforme d'une variétés finslérienne. (C.R. Acad. Sci. Paris 280 (1975), 519-522).
[7] Sur les transformations infinitésimales conformes d'une variétés finslérienne. (Ibid. 280 (1975), 583-585).
[8] Sur les connexions induite et intrinsèque d'une sous-variétés d'une variété finslérienne. (Ibid. 282 (1976), 599-602).

GROSSMAN, N.
[1] On real projective spaces as Finsler manifolds. (Proc. Amer. Math. Soc. 18 (1967), 325-326).

GROVE, K.; KARCHER, H.; RUH, E.A.
[1] Jacobi fields and Finsler metrics on compact Lie groups with an application to differentiable pinching problems. (Math. Ann. 211 (1974), 7-21).

GUGGENHEIMER, H.M.
[1] Pseudo-Minkowski differential geometry. (Ann. Mat. Pura Appl. (4) 70 (1965), 305-370).
[2] Approximation of curves. (Pacific J. Math. 40 (1972), 301-303).

HAIMOVICI, M.
[1] Formules fondamentales dans la théorie des hypersurfaces d'un espace de Finsler. (C.R. Acad. Sci. Paris 198 (1934), 426-427).
[2] Sur les espaces généraux qui se correspondent point par point avec conservation du parallélisme de M.Cartan. (Ibid. 198 (1934), 1105-1108).
[3] Sur quelques types de métriques de Finsler. (Ibid. 199 (1934), 1091-1093).
[4] Les formules fondamentales des la théorie des hypersurfaces d'un espace général. (Ann. Sci. Univ. Jassy 20 (1935), 39-58).
[5] Sur les espaces de Finsler à connexion affine. (C.R. Acad. Sci. Paris 204 (1937), 837-839).
[6] Le parallélisme dans les espaces de Finsler et la différentiation invariante de M.Levi-Civita. (Ann. Sci. Univ. Jassy I 24 (1938), 214-218).
[7] Sulle superficie totalmente geodetiche negli spazi di Finsler. (Lincei Rend.

(6) 27 (1938), 633-641).
[8] Variètès totalement extrèmales et variétés totalement géodésiques dans les espaces de Finsler. (Ann. Sci. Univ. Jassy I 25 (1939), 559-644).
HASHIGUCHI, M.
[1] On parallel displacement in Finsler spaces. (J. Math. Soc. Japan 10 (1958), 365-379).
[2] On determinations of Finsler connections by deflection tensor fields. (Rep. Fac. Sci. Kagoshima Univ. (Math., Phys., Chem.) 2 (1969), 29-39).
[3] On the hv-curvature tensors of Finsler spaces. (Ibid. 4 (1971), 1-5).
[4] On Wagner's generalized Berwald space. (J. Korean Math. Soc. 12 (1975), 51-61).
[5] On conformal transformations of Finsler metrics. (J. Math. Kyoto Univ. 16 (1976), 25-50).
HASHIGUCHI, M.; HŌJŌ, S.; MATSUMOTO, M.
[1] On Landsberg spaces of two dimensions with (α, β)-metric. (J. Korean Math. Soc. 10 (1973), 17-26).
HASHIGUCHI, M.; ICHIJYŌ, Y.
[1] On some special (α, β)-metrics. (Rep. Fac. Sci. Kagoshima Univ. (Math., Phys., Chem.) 8 (1975), 39-46).
[2] On conformal transformations of Wagner spaces. (Ibid. 10 (1977), 19-25).
HASHIGUCHI, M.; VARGA, T. (KÁNTOR, S.)
[1] On Wagner spaces of W-scalar curvature. (Studia Sci. Math. Hungar. 14 (1979), 11-14).
HASSAN, B.T.M.
[1] The theory of geodesics in Finsler spaces. (Thesis, Southampton, 1967, 108p.).
[2] The cut locus of a Finsler manifold. (Lincei Rend. (8) 54 (1973), 739-744).
[3] A theorem on compact Finsler surfaces. (Proc. Math. Phys. Soc. Egypt 40 (1975), 1-6).
[4] Isometric immersions of a compact Finsler space into a Finsler space. (Ibid. 40 (1975), 15-18).
[5] Connections associated with linear maps on the induced bundle of a Finsler space. I, II. (Ibid. 47 (1979), 1-6).
HEIL, E.
[1] A relation between Finslerian and Hermitian metrics. (Tensor, N.S. 16 (1965), 1-3).
[2] Eine Charakterisierung lokal-Minkowskischer Räume. (Math. Ann. 167 (1966), 64-70).
[3] Scheitelsätze in der euklidischen, affinen und Minkowskischen Geometrie. (Darmstadt, 1967, 73p.).
[4] Der Vierscheitelsatz in Relativ- und Minkowski-Geometrie. (Monatsh. Math. 74 (1970), 97-107).
HEIL, E.; LAUGWITZ, D.
[1] Finsler spaces with similarity are Minkowski spaces. (Tensor, N.S. 28 (1974), 59-62).
HIRAMATU, H.
[1] Groups of homothetic transformations in a Finsler space. (Ibid. 3 (1954), 131-143).
[2] On some properties of groups of homothetic transformations in Riemannian and Finslerian spaces. (Ibid. 4 (1954), 28-39).
[3] On n-dimensional Finslerian manifolds admitting homothetic transformation groups of dimension > n(n-1)/2 + 1. (Kumamoto J. Sci. A 4 (1959), 4-10).
HŌJŌ, S. (HOZYO)
[1] Pair connections on homogeneous spaces which are invariant under tangential transformations. (Mem. Fac. Eng. Yamaguchi Univ. 14 (1964), 135-139).
[2] On the connection mapping of the induced (C1)-pair connections on homogeneous spaces. (Ibid. 16 (1965), 33-37).
[3] On geodesics of certain Finsler metrics. (Tensor, N.S. 34 (1980), 211-217).
[4] Structure of fundamental functions of S3-like Finsler spaces. (J. Math. Kyoto Univ. 21 (1981), 787-807).

HOSOKAWA, T.
[1] On the various linear displacements in the Berwald-Finsler's manifold. (Sci. Rep. Tokyo 19 (1930), 37-51).
[2] Conformal property of a manifold B_n. (Jap. J. Math. 9 (1932), 59-62).
HU, H.-S.
[1] A Finslerian product of two Riemannian spaces. (Sci. Record, N.S. 3 (1959), 446-448).
ICHIJYŌ, Y.
[1] Almost complex structures of tangent bundles and Finsler metrics. (J. Math. Kyoto Univ. 6 (1967), 419-452).
[2] Finsler manifolds modeled on a Minkowski space. (Ibid. 16 (1976), 639-652).
[3] Finsler manifolds with a linear connection. (J. Math. Tokushima Univ. 10 (1976), 1-11).
[4] On special Finsler connections with the vanishing hv-curvature tensor. (Tensor, N.S. 32 (1978), 146-155).
[5] On the Finsler connection associated with a linear connection satisfying $P^h_{ikj} = 0$. (J. Math. Tokushima Univ. 12 (1978), 1-7).
[6] On the condition for a {V,H}-manifold to be locally Minkowskian or conformally flat. (Ibid. 13 (1979), 13-21).
IKEDA, F.
[1] On two-dimensional Landsberg spaces. (Tensor, N.S. 33 (1979), 43-48).
[2] On the tensor T_{ijkl} of Finsler spaces. (Ibid. 33 (1979), 203-209).
[3] On the tensor T_{ijkl} of Finsler spaces, II. (Ibid. 34 (1980), 85-93).
INGARDEN, R.S.
[1] Problème p 58. (Colloq. Math. 1 (1948), 334).
[2] Über die Einbettung eines Finslerschen Raumes in einem Minkowskischen Raum. (Bull. Acad. Polon. Sci. III 2 (1954), 305-308).
IONUŠAUSKAS, A. (JONUSAUSKAS)
[1] Über die existenz von invarianten Finslerschen Metriken in homogenen Räumen. (Russian) (Liet Mat. Rinkinys 5 (1965), 45-55).
[2] Existenz von invarianten Finslerschen Metriken in homogenen Räumen mit linearer Isotropiegruppe tensorischen Typus. (Russian) (Ibid. 6 (1966), 51-57).
[3] The existence of invariant Finsler metrics in certain uniform spaces. (Russian) (Ibid. 6 (1966), 621-622).
[4] Existenz von invarianten Finslerschen Metriken in homogenen Räumen mit linearer Isotropiegruppe tensorischen Typus. II. (Russian) (Ibid. 7 (1967), 619-631).
ISPAS, C.I.
[1] Identités de type Ricci dans l'espace de Finsler. (Romanian) (Com. Acad. R.P.R. 2 (1952), 13-18).
[2] Les identités de Veblen dans les espaces généralises. (Romanian) (Bul. Şti. Secţ. Şti. Mat. Fiz. 4 (1952), 533-539).
[3] Les déformations itérées. Sur les déformations des géodésiques dans les espaces généralises. (Romanian) (Acad. R.P.R. Baza Timişoara, Lucr. Consf. Geom. Dif., 1955, 253-261).
[4] Die Gleichwertigkeit der Bianchischen und Veblenschen Identitäten in den Finslerschen Räumen. (Rev. Roumaine Math. Pures Appl. 12 (1967), 1467-1478).
[5] On the equivalence of Bianchi and Veblen identities in Finsler-Cartan-Kawaguchi spaces (I). (Tensor, N.S. 26 (1972), 468-476).
[6] On the equivalence of Bianchi and Veblen identities in Finsler-Cartan-Kawaguchi spaces (II). (Proc. Inst. Math. Iaşi, 1974, 51-55).
ISPAS, C.I.; ISPAS, M.I.
[1] Les espaces Finsler-Berwald. (Romanian). (Bul. Şti. Inst. Const. Bucureşti 16 (1973), 23-28).
IZUMI, H.
[1] Conformal transformations of Finsler spaces. I. Concircular transformations of a curve with Finsler metric. (Tensor, N.S. 31 (1977), 33-41).
[2] On *P-Finsler spaces, I, II. (Memo. Defense Acad. 16 (1976), 133-138; 17

(1977), 1-9).
IZUMI, H.; SRIVASTAVA, T.N.
[1] On R3-like Finsler spaces. (Tensor, N.S. 32 (1978), 339-349).
IZUMI, H.; YOSHIDA, M.
[1] On Finsler spaces of perpendicular scalar curvature. (Ibid. 32 (1978), 219-224).
KAVANOV, N.I.
[1] A singular Finsler space defined to within a Carathéodory mapping. (Russian) (Sibirsk. Mat. Z. 2 (1961), 655-671).
[2] Differential-geometric methods in the calculus of variations. (Russian) (Algebra, Topology, Geometry, 1968, 193-224).
KAGAN, F.I.
[1] On two-dimensional Finsler spaces admitting a singular embedding in a three-dimensional affine space with a vector metric. (Russian) (Izv. Vysš. Učebn. Zaved. Mat. (1964), 1 (38), 46-55).
[2] On groups of pseudo-motions in spaces of Finsler and Riemann. (Russian) (Volž. Mat. Sb. Vyp. 3 (1965), 190-196).
[3] On the operation of an infinitesimal S-extension with respect to a one-dimensional S-distribution on X_n. (Russian) (Ibid. 4 (1966), 103-108).
[4] The operation of an infinitesimal S-extension. (Russian) (Izv. Vysš. Učebn. Zaved. Mat. (1966), 1 (50), 66-78).
KALJUŽNYĬ, V.N.
[1] Commutative groups of isometries of Minkowski spaces. (Russian) (Sibirsk. Mat. Z. 15 (1974), 1138-1142, 1182). (English: Siber. Math. J. 15 (1974), 801-803 (1975)).
KASHIWABARA, S.
[1] On Euclidean connections in a Finsler manifold. (Tôhoku Math. J. (2) 10 (1958), 69-80).
KATOK, A.B.
[1] Ergodic perturbations of degenerate integrale Hamiltonian systems. (Russian) (Izv. Akad. Nauk SSSR Mat. 37 (1973), 539-576). (English: Math. USSR, Izv. 7 (1973), 535-571).
KAUL, R.N.
[1] Curvatures in Finsler space. (Bull. Calcutta Math. Soc. 50 (1958), 189-192).
KAWAGUCHI, A.
[1] Beziehung zwischen einer metrischen linearen Übertragung und einer nicht-metrischen in einem allgemeinen metrischen Raume. (Akad. Wetensch. Amsterdam Proc. 40 (1937), 596-601).
[2] On the theory of non-linear connections I. Introduction to the theory of general non-linear connections. (Tensor, N.S. 2 (1952), 123-142).
[3] On the theory of non-linear connections II. Theory of Minkowski spaces and of non-linear connections in a Finsler space. (Ibid. 6 (1956), 165-199).
KAWAGUCHI, A.; LAUGWITZ, D.
[1] Remarks on the theory of Minkowski spaces. (Ibid. 7 (1957), 190-199).
KAWAGUCHI, H.
[1] On the Finsler spaces with the vanishing second curvature tensor. (Ibid. 26 (1972), 250-254).
[2] A Minkowski space closely related to a certain special Finsler space. (Ibid. 32 (1978), 114-118).
[3] Some characterization of the vector field A_i proper to Finsler spaces. (Ibid. 34 (1980), 367-372).
KAWAGUCHI-KANAI, T.
[1] On Finsler spaces with special line elements. (Gakuen Rev. 15 (1969), 39-44).
[2] On the covariant differentiation of rheonomic Finsler spaces. (Ibid. 15 (1969), 45-51).
[3] On the connection parameters and the stretch tensor of rheonomic Finsler spaces. (Ibid. 16 (1970), 21-28).
[4] On the curvature strong tensors and the identities of Ricci and Bianchi in rheonomic Finsler spaces. (Ibid. 18 (1971), 63-75).
KERN, J.
[1] Fastriemannsche Finslersche Metriken. (Manuscripta Math. 4 (1971), 285-303).

[2] Das Pinchingproblem in Fastriemannschen Finslerschen Mannigfaltigkeiten. (Ibid. 4 (1971), 341-350).

KIKUCHI, S.

[1] On the theory of subspace in a Finsler space. (Tensor, N.S. 2 (1952), 67-79).

[2] Theory of Minkowski space and of non-linear connections in a Finsler space. (Ibid. 12 (1962), 47-60).

[3] On some special Finsler spaces. (Ibid. 19 (1968), 238-240).

[4] On the condition that a space with (α, β)-metric be locally Minkowski. (Ibid. 33 (1979), 242-246).

KISHTA, M.A.

[1] Finsler spaces of constant singularities. (Arabian J. Sci. Eng. 3 (1978), 93-97).

KITAMURA, S.

[1] On Finsler spaces with the fundamental function $L = \sqrt{g_{ij}(x)\dot{x}^i\dot{x}^j} + A_i(x)\dot{x}^i$. (Japanese) (Thesis, Kyoto, 1960, 36p.).

KNEBELMAN, M.S.

[1] Groups of collineations in a space of paths. (Proc. Nat. Acad. Sci. U.S.A. 13 (1927), 396-400).

[2] Motion and collineations in general space. (Ibid. 13 (1927), 607-611).

[3] Conformal geometry of generalized metric space. (Ibid. 15 (1929), 376-379).

[4] Collineations and motions in generalized spaces. (Amer. J. Math. 51 (1929), 527-564).

KROPINA, V.K.

[1] On projective Finsler spaces with a metric of some special form. (Russian) (Naučn. Dokl. Vysš. Skoly. Fiz.-Mat. Nauki, 1959, no.2, 38-42).

[2] Projective two-dimensional Finsler spaces with special metric. (Russian) (Trudy Sem. Vektor. Tenzor. Anal. 11 (1961), 277-292).

[3] Projective Finsler spaces. (Russian) (Uch. Zap. Arkhang. Gos. Ped. Inst. 4 (1959), 111-118).

[4] The introduction of absolute differentiation in Finsler space. (Russian) (Uch. Zap. Yaroslavsk. Gos. Ped. Inst. 34 (1960), 113-123).

KU, C.-H. (GU)

[1] Imbedding of a Finsler space in a Minkowski space. (Chinese) (Acta Math. Sinica 6 (1956), 215-232).

[2] On Finsler spaces admitting a group of motions of the greatest order. (Sci. Record, N.S. 1 (1957), 215-218).

[3] Imbedding of Finsler manifolds in a Minkowski space. (Chinese) (Acta Math. Sinica 8 (1958), 272-275). (English: Amer. Math. Soc. Transl. II 37 (1964), 253-257; Chinese Math. 8 (1966), 878-882).

KURITA, M.

[1] On the dilatation in Finsler spaces. (Osaka Math. J. 15 (1963), 87-98).

[2] Theory of Finsler spaces based on the contact structure. (J. Math. Soc. Japan 18 (1966), 119-134).

LAGET, B.

[1] Sur une méthode de déformation en géométrie Banachique. (Publ. Dépt. Math. Lyon 8 (1971), 71-85).

LANDSBERG, G.

[1] Über die Totalkrümmung. (Jber. Deutsch. Math. Verein. 16 (1907), 36-46).

[2] Krümmungstheorie und Variationsrechnung. (Ibid. 16 (1907), 547-551).

[3] Über die Krümmung in der Variationsrechnung. (Math. Ann. 65 (1908), 313-349).

LAPTEV, B.L.

[1] Covariant integration in a Finsler space of two and three dimensions. (Izv. Fiz.-Mat. Obshch., Kazan 3 (9) (1937), 61-76).

[2] La dérivée de S. Lie des objets géométriques qui dépendent de point et direction. (Russian) (Bull. Soc. Phys.-Math. Kazan (3) 10 (1938), 37-38).

[3] Une forme invariante de la variation et la dérivée de S. Lie. (Russian) (Ibid. 12 (1940), 3-8).

[4] Lie derivative of geometric objects in a support element space. (Proc. 3rd All-Union Math. Congr. 1 (1956), 157).

[5] The Lie derivation in the space of supporting elements. (Russian) (Trudy Sem. Vektor. Tenzor. Anal. 10 (1956), 227-248).

[6] Lie derivative of geometric objects in a support element space. (Uch. Zap. Kazansk. Univ. 112 (2) (1957), 16-18).
[7] The covariant differential and the theory of differential invariants in the space of tensor support elements. (Russian) (Kazan Gos. Univ. Učen. Zap. 118 (1958), no.4, 75-147).
[8] Application of Lie differentiation to investigation of geodesic displacement in a space of linear elements. (Russian) (Izv. Vysš. Učebn. Zaved. Mat. (3) 2 (1958), 173-181).
[9] Lie differentiation. (Russian) (Algebra, Topology, Geometry, 1965, 429-465). (English: Progr. Math. 6 (1970), 229-269).

LAUGWITZ, D.
[1] Konvexe Mittelpunktsbereiche und normierte Räume. (Math. Z. 61 (1954), 235-244).
[2] Zur geometrischen Begründung der Parallelverschiebung in Finslerschen Räumen. (Arch. Math. 6 (1955), 448-453).
[3] Die Vektorübertragung in der Finslerschen Geometrie und der Wegegeometrie. (Nederl. Akad. Wetensch. Proc. A 59 = Indag. Math. 18 (1956), 21-28).
[4] Grundlagen für die Geometrie der unendlichdimensionalen Finslerräume. (Ann. Mat. Pura Appl. (4) 41 (1956), 21-41).
[5] Zur projektiven und konformen Geometrie der Finsler-Räume. (Arch. Math. 7 (1956), 74-77).
[6] Zur Differentialgeometrie der Hyperflächen in Vektorräumen und zur affingeometrischen Deutung der Theorie der Finsler-Raume. (Math. Z. 67 (1957), 63-74).
[7] Eine Beziehung zwischen affiner und Minkowskischer Differentialgeometrie. (Publ. Math. Debrecen 5 (1957), 72-76).
[8] Geometrische Behandlung eines inversen Problems der Variationsrechnung. (Ann. Univ. Sarav. 5 (1956), 235-244).
[9] Die Geometrie von H.Minkowski. (Mathematikunterricht Heft 4 (1958), 27-42).
[10] Beiträge zur affinen Flächentheorie mit Anwendung auf die allgemeinmetrische Differentialgeometrie. (Bayer. Akad. Wiss. Math.-Natur. Kl. Abh. 93 (1959), 59p.).
[11] Differentialgeometrie. (Math. Leitfäden, B.G.Teubner, Stuttgart, 1960, 183p.).
[12] Geometrical methods in the differential geometry of Finsler spaces. (C.I.M.E. Geom. Calcolo Varia. 1-3 (1961), 49p.).
[13] Über die Erweiterung der Tensoranalysis auf Mannigfaltigkeiten unendlicher Dimension. (Tensor, N.S. 13 (1963), 295-304).
[14] Differentialgeometrie in Vektorräumen, unter besonderer Berücksichtigung der unendlichdimensionalen Räumen. (Fredr. Vieweg & Sohn, Braunschweig; VEB Deutsch. Verlag der Wiss., Berlin, 1965, 89p.).
[15] A characterization of Minkowski spaces. (Boll. Un. Mat. Ital. (4) 12 (1975), Suppl. 3, 267-270).

LAUGWITZ, D.; LORCH, E.R.
[1] Riemann metrics associated with convex bodies and normed spaces. (Amer. J. Math. 78 (1956), 889-894).

LEHMANN, D.
[1] Théorie de Morse en géométrie finslérienne.(Topol. Géom. Diff. Sém. Ehresmann 6 (1964), 9p.).

LIBER, A.E.
[1] On two-dimensional spaces with an algebraic metric. (Russian) (Trudy Sem. Vektor. Tenzor. Anal. 9 (1952), 319-350).

LICHNEROWICZ, A.
[1] Sur une généralisation des espaces de Finsler. (C.R. Acad. Sci. Paris 214 (1942), 599-601).
[2] Sur une extension de la formule d'Allendoerfer-Weil à certains variétés finslériennes. (Ibid. 223 (1946), 12-14).
[3] Quelques théorèmes de géométrie différentielle globale. (Comment. Math. Helv. 22 (1949), 271-301).

LIPPMANN, H.
[1] Zur Winkeltheorie in zweidimensionalen Minkowski- und Finsler Räumen. (Nederl. Akad. Wetensch. Proc. A 60 = Indag. Math. 19 (1957), 162-170).
[2] Metrische Eigenschaften verschiedener Winkelmasze im Minkowski- und Finsler-

raum. I, II. (Ibid. 61 = ibid. 20 (1958), 223-238).
LUMISTE, Ju.G.
[1] The theory of connections in fibered spaces. (Russian) (Algebra, Topology, Geometry, 1969, 123-168).
MAEBASHI, T.
[1] A weakly osculating Riemann space of the Finsler space and its application to a theory of subspaces in the Finsler space. (Tensor, N.S. 9 (1959), 62-72).
[2] Vector fields and space forms. (J. Fac. Sci. Hokkaido Univ. I 15 (1960-61), 62-92).
[3] A certain type of vector field. I, II, III. (Proc. Japan Acad. (1961), 23-26, 137-143).
[4] On line geometry. (Kumamoto J. Sci. Math. 12 (1977), 56-61).
MANGIONE, V.
[1] Proprietã geometriche negli spazi di Finsler quasi hermitiani. (Rend. Mat. (6) 4 (1971), 867-876).
MAREI, L.
[1] Sous-variété finslériennes et principe de moindre courbure. (Thèse, Grenoble, 1980, 45p.).
MATSUMOTO, M.
[1] A global foundation of Finsler geometry. (Memo. Coll. Sci. Univ. Kyoto A 33 (1960/61), 171-208).
[2] Affine transformations of Finsler spaces. (J. Math. Kyoto Univ. 3 (1963), 1-35).
[3] Linear transformations of Finsler connections. (Ibid. 3 (1963/64), 145-167).
[4] Paths in a Finsler space. (Ibid. 3 (1963/64), 305-318).
[5] On R.Sulanke's method deriving H.Rund's connection in a Finsler space. (Ibid. 4 (1965), 355-368).
[6] A Finsler connection with many torsions. (Tensor, N.S. 17 (1966), 217-226).
[7] Connections, metrics and almost complex structures of tangent bundles. (J. Math. Kyoto Univ. 5 (1966), 251-278).
[8] Theory of Finsler spaces and differential geometry of tangent bundles. (Ibid. 7 (1967), 169-204).
[9] Intrinsic transformations of Finsler metrics and connections. (Tensor, N.S. 19 (1968), 303-313).
[10] On F-connections and associated nonlinear connections. (J. Math. Kyoto Univ. 9 (1969), 25-40).
[11] A geometric meanings of a concept of isotropic Finsler spaces. (Ibid. 9 (1969), 405-411).
[12] The theory of Finsler connections. (Publ. of the Study Group of Geometry, 5. Deptt. Math. Okayama Univ., 1970, 120p.).
[13] On h-isotropic and C^h-recurrent Finsler spaces. (J. Math. Kyoto Univ. 11 (1971), 1-9).
[14] On some transformations of locally Minkowskian spaces. (Tensor, N.S. 22 (1971), 103-111).
[15] On Finsler spaces with curvature tensors of some special forms. (Ibid. 22 (1971), 201-204).
[16] V-transformations of Finsler spaces. I. Definition, infinitesimal transformations and isometries. (J. Math. Kyoto Univ. 12 (1972), 479-512).
[17] On C-reducible Finsler spaces. (Tensor, N.S. 24 (1972), 29-37).
[18] A theory of three-dimensional Finsler spaces in terms of scalars. (Demonst. Math. 6 (1973), 223-251).
[19] Theory of non-linear connections from the standpoint of Finsler connections. (Tensor, N.S. 28 (1974), 69-77).
[20] On three-dimensional Finsler spaces satisfying the T- and B^p-conditions. (Ibid. 29 (1975), 13-20).
[21] On Finsler spaces with Randers' metric and special forms of important tensors. (J. Math. Kyoto Univ. 14 (1974), 477-498).
[23] Metrical differential geometry. (Japanese) (Kiso Sūgaku Sensho 14, Shokabō, Tokyo, 1975, 229p.).
[24] On the indicatrices of a Finsler space. (Period. Math. Hungar. 8 (1977), 185-191).

[25] What is the Finsler geometry ? (Bull. Korean Math. Soc. 13 (1976), 121-126).
[26] Strongly non-Riemannian Finsler spaces. (Anal. Şti. Univ. Iaşi 23 (1977), 141-149).
[27] Finsler spaces with the hv-curvature tensor P_{hijk} of a special form. (Rep. on Math. Phys. 14 (1978), 1-13).
[28] On geodesics in the tangent space of a Finsler space. (J. Korean Math. Soc. 14 (1978), 167-183).
[29] The length and the relative length of tangent vectors of Finsler spaces. (Rep. on Math. Phys. 15 (1979), 375-386).
[30] Differential-geometric properties of indicatrix bundle over Finsler space. (Publ. Math. Debrecen 28 (1981), 281-293).
[31] Fundamental functions of S3-like Finsler spaces. (Tensor, N.S. 34 (1980), 141-146).
[32] Projective changes of Finsler metrics and projectively flat Finsler spaces. (Ibid. 34 (1980), 303-315).

MATSUMOTO, M.; EGUCHI, K.
[1] Finsler spaces admitting a concurrent vector field. (Ibid. 28 (1974), 239-249).

MATSUMOTO, M.; HŌJŌ, S.
[1] A conclusive theorem on C-reducible Finsler spaces. (Ibid. 32 (1978), 225-230).

MATSUMOTO, M.; MIRON, R.
[1] On an invariant theory of the Finsler spaces. (Period. Math. Hungar. 8 (1977), 73-82).

MATSUMOTO, M.; NUMATA, S.
[1] On Finsler spaces with cubic metric. (Tensor, N.S. 33 (1979), 153-162).
[2] On semi-C-reducible Finsler spaces with constant coefficients and C2-like Finsler spaces. (Ibid. 34 (1980), 218-222).

MATSUMOTO, M.; OKADA, T.
[1] Connections in Finsler spaces. (Japanese) (Sem. Diff. Geom. 4, Kyoto Univ. 1965, 146p.).

MATSUMOTO, M.; SHIBATA, C.
[1] On the curvature tensor R_{ijkh} of C-reducible Finsler spaces. (J. Korean Math. Soc. 13 (1976), 141-144, 189).
[2] On semi-C-reducibility, T-tensor = 0 and S4-likeness of Finsler spaces. (J. Math. Kyoto Univ. 19 (1979), 301-314).

MATSUMOTO, M.; SHIMADA, H.
[1] On Finsler spaces with the curvature tensors P_{hijk} and S_{hijk} satisfying special conditions. (Rep. on Math. Phys. 12 (1977), 77-87).
[2] On Finsler spaces with 1-form metric. (Tensor, N.S. 32 (1978), 161-169).
[3] On Finsler spaces with 1-form metric II. Berwald-Moór's metric $L = (y^1 y^2 \dots y^n)^{1/n}$. (Ibid. 32 (1978), 275-278).

MATSUMOTO, M.; TAMÁSSY, L.
[1] Scalar and gradient vector fields of Finsler spaces and holonomy groups of nonlinear connections. (Demonst. Math. 13 (1980), 551-564; 14 (1981), 529).

MAURIN, K.
[1] Eingliedrige Gruppen der homogenen kanonischen Transformationen und Finslersche Räume.

MEHER, F.M.
[1] Some commutation formulae arising from the generalised Lie differentiation in Finsler space. (Rev. Univ. Istanbul 37 (1972), 121-125).
[2] Projective motion in a symmetric Finsler space. (Tensor, N.S. 23 (1972), 275-278).

MERCURI, F.
[1] Closed geodesics on Finsler manifolds. (Lincei Rend. (8) 60 (1976), 111-118).
[2] The critical points theory for the closed geodesic problem. (Math. Z. 156 (1977), 231-245).

MERZA, J.
[1] On a special Finsler geometry. (Demonst. Math. 6 (1973), 761-769).

MIRON, R.
[1] Sur les connexions pseudo-euclidiennes des espaces de Finsler à métrique indéfinie. (Romanian) (Acad. R.P. Romine Fil. Iaşi, Şti. Mat. 12 (1961), 125-

134).
[2] Espaces de Finsler ayant un groupe de Lie comme groupe d'invariance. (Rev. Roum. Math. Pures Appl. 13 (1968), 1409-1412).
[3] On a class of Finsler connections. (Bul. Inst. Politeh. Iaşi 25 (1979), 53-55).

MIRON, R.; HASHIGUCHI, M.
[1] Metrical Finsler connections. (Rep. Fac. Sci. Kagoshima Univ. (Math.,Phys., Chem.) 12 (1979), 21-35).

MISHRA, R.S.; PANDE, H.D.
[1] Conformal identities. (Rev. Univ. Istanbul 31 (1966), 39-48).
[2] The Ricci identity. (Ann. Mat. Pura Appl. (4) 75 (1967), 355-361).
[3] Recurrent Finsler space. (J. Indian Math. Soc., N.S. 32 (1968), 17-22).
[4] Certain projective changes in Finsler space. (Rev. Mat. Hisp.-Amer. (4) 28 (1968), 49-55).

MISHRA, R.S.; SINHA, R.S.
[1] Bianchi identities satisfied by the first and second of Cartan's curvature tensors. (Bull. Calcutta Math. Soc. 57 (1965), 99-101).
[2] Some identities satisfied by Cartan's curvature tensors. (Proc. Nat. Acad. Sci. India A 36 (1966), 534-538).

MISRA, R.B.
[1] The projective transformation in a Finsler space. (Ann. Soc. Sci. Bruxelles I 80 (1966), 227-239).
[2] Projective tensors in a conformal Finsler space. (Acad. Roy. Belg. Bull. Cl. Sci. (5) 52 (1966), 1275-1279).
[3] The commutation formulae in a Finsler space. I, II. (Ann. Mat. Pura Appl. (4) 75 (1967), 363-383).
[4] The Bianchi identities satisfied by curvature tensors in a conformal Finsler space. (Tensor, N.S. 18 (1967), 187-190).
[5] Some problems in Finsler spaces. (Thesis, Allahabad, 1967).
[6] On the deformed Finsler space. (Tensor, N.S. 19 (1968), 241-250).
[7] Hyper-asymptotic curves of a subspace of a Finsler space. (Bull. Polon. Math. Astro. Phys. 17 (1969), 65-69).
[8] The generalised Killing equation in Finsler space. (Rend. Circ. Mat. Palermo (2) 18 (1969), 99-102).
[9] Projective invariants in a conformal Finsler space. (Tensor, N.S. 21 (1970), 186-188).
[10] On a recurrent Finsler space. (Rev. Roumaine Math. Pures Appl. 18 (1973), 701-712).
[11] On the generalised Lie differentiation arising from Su's infinitesimal transformation. (Rev. Univ. Istanbul 35 (1970), 5-15).
[12] A symmetric Finsler space. (Tensor, N.S. 24 (1972), 346-350).
[13] A projectively symmetric Finsler space. (Math. Z. 126 (1972), 143-153).

MISRA, R.B.; MEHER, F.M.
[1] Projective motion in an RNP-Finsler space. (Tensor, N.S. 22 (1971), 117-120).
[2] On the existence of affine motion in a HR-F_n. (Indian J. Pure Appl. Math. 3 (1972), 219-225).
[3] A recurrent Finsler space of second order. (Rev. Roumaine Math. Pures Appl. 18 (1973), 563-569).
[4] Lie differentiation and projective motion in the projective Finsler space. (Tensor, N.S. 23 (1972), 57-65).

MISRA, R.B.; MISHRA, R.S.
[1] Lie-derivatives of various geometric entities in Finsler space. (Rev. Univ. Istanbul 30 (1965), 77-82).
[2] The Killing vector and generalised Killing equation in Finsler space. (Rend. Circ. Mat. Palermo (2) 15 (1966), 216-222).
[3] Curvature tensor arising from non-linear connections in a Finsler space. (Bull. Polon. Math. Astro. Phys. 17 (1969), 755-760).

MISRA, R.B.; PANDE, K.S.
[1] On the Finsler space admitting a holonomy group. (Ann. Mat. Pura Appl. (4) 85 (1970), 327-346).
[2] On Misra's covariant differentiation in a Finsler space. (Lincei Rend. (8) 48 (1970), 199-204).

MIZOGUCHI, N.; KATO, S.
[1] Umbilical points on subspaces of Finslerian and Minkowskian spaces. (Memo. Kitami Inst. Tech. 2 (1968), 319-325).
MOALLA, M.F.
[1] Espaces de Finsler complets. (C.R. Acad. Sci. Paris 258 (1964), 2251-2254).
[2] Espaces de Finsler complets à courbure de Ricci positive. (Ibid. 258 (1964), 2734-2737).
[3] Espaces de Finsler sans points conjugués. (Ibid. 260 (1965), 6510-6512).
[4] Sur quelques théorèmes globaux en géométrie Finslérienne. (Ann. Mat. Pura Appl. (4) 73 (1966), 319-365).
MOÓR, A.
[1] Espaces métriques dont le scalaire de courbure est constant. (Bull. Sci. Math. (2) 74 (1950), 13-32).
[2] Finslersche Räume mit der Grundfunktion L = f/g. (Comment. Math. Helv. 24 (1950), 188-195).
[3] Généralisation du scalaire de courbure et du scalaire principal d'un espace finslérien à n dimensions. (Canad. J. Math. 2 (1950), 307-313)
[4] Einführung des invarianten Differentials und Integrals in allgemeinen metrische Räumen. (Acta Math. 86 (1951), 71-83).
[5] Finslersche Räume mit algebraischen Grundfunktionen. (Publ. Math. Debrecen 2 (1952), 178-190).
[6] Quelques remarques sur la généralisation du scalaire de courbure et du scalaire principal. (Canad. J. Math. 4 (1952), 189-197).
[7] Über die Dualität von Finslerschen und Cartanschen Räumen. (Acta Math. 88 (1952), 347-370).
[8] Über oskulierende Punkträume von affinzusammenhängenden Linienelementmannigfaltigkeiten. (Ann. of Math. (2) 56 (1952), 397-403).
[9] Ergänzung zu meiner Arbeit: „Über die Dualität von Finslerschen und Cartanschen Räumen". (Acta Math. 91 (1954), 187-188).
[10] Entwicklung einer Geometrie der allgemeinen metrischen Linienelementräume. (Acta Sci. Math. Szeged 17 (1956), 85-120).
[11] Über die Torsions- und Krümmungsinvarianten der dreidimensionalen Finslerschen Räume. (Math. Nachr. 16 (1957), 85-99).
[12] Über affine Finslerräume von skalarer Krümmung. (Acta Sci. Math. Szeged 22 (1961), 157-189).
[13] Untersuchungen über Finslerräume von rekurrenter Krümmung. (Tensor, N.S. 13 (1963), 1-18).
[14] Über Finslerräume von zweifach rekurrenter Krümmung. (Acta Math. Acad. Sci. Hungar. 22 (1971), 453-465).
[15] Unterräume von rekurrenter Krümmung in Finslerräumen. (Tensor, N.S. 24 (1972), 261-265).
[16] Untersuchungen in vierdimensionalen verallgemeinerten Finslerräumen mit Hilfe eines natürlichen Vierbeins. (Publ. Math. Debrecen 20 (1973), 241-258).
[17] Finslerräume von identischer Torsion. (Acta Sci. Math. Szeged 34 (1973), 279-288).
[18] Finslerräume von rekurrenter Torsion. (Publ. Math. Debrecen 21 (1974), 255-265).
[19] Übertragungstheorien in Finslerschen und verwandten Räumen. (Berichte Math.-Stat. Sekt. Forschungszentrum Graz 42 (1975), 14p.).
[20] Über die Charakterizierung von speziellen 3- und 4-dimensionalen Finslerräumen durch Invarianten. (Boll. Un. Mat. Ital.(4) 12 (1975), Suppl.3, 189-199).
[21] Über spezielle Type von Hyperflächen in verallgemeinerten Finslerräumen. (Publ. Math. Debrecen 23 (1976), 27-39).
[22] Über die durch Kurven bestimmten Sektionalkrümmungen der Finslerschen und Weylschen Raume. (Publ. Math. Debrecen 26 (1979), 205-214).
MYERS, S.B.
[1] Arc length in metric and Finsler manifolds. (Ann. of Math. (2) 39 (1938), 463-471).
[2] Arcs and geodesics in metric spaces. (Trans. Amer. Math. Soc. 57 (1945), 217-227).

NAGAI, T.
[1] On groups of rotations in Minkowski space I, II. (J. Fac. Sci. Hokkaido Univ. I 15 (1960), 29-61).
NAGATA, Y.
[1] Normal curvature of a vector field in a hypersurface in a Finsler space. (Tensor, N.S. 5 (1955), 17-22)
[2] Remarks on the normal curvature of a vector field. (Ibid. 8 (1958), 177-183).
[3] Formulas of Frenet for a vector field in a Finsler space. (Memo. Muroran Inst. Tech. 3 (1960), 157-160).
[4] On some geometrical properties in three dimensional Finsler space. (Memo. Ehime Univ. Nat. Sci. A 8-3 (1978), 1-2).
NAGATA, Y.; ANH, K.Q.
[1] On the curvatures of a vector field on a subspace of a Finsler space. (Tensor, N.S. 32 (1978), 307-310).
NAGY, P.T. (NAD')
[1] Über die Äquivalenztheorie der verallgemeinerten Linienelementräume. (Publ. Math. Debrecen 16 (1969), 79-89).
[2] On the theory of Finsler connections especially their equivalence. (Acta Sci. Math. Szeged 37 (1975), 331-338).
NASU, Y.
[1] On the structure of a Finsler space whose holonomy group preserves invariant more than one point. (Japanese) (Holonomy-gun Kenkyu 12 (1949), 40-47).
[2] Non Euclidean geometry in Finsler spaces. (Kumamoto J. Sci. A 1 (1952), 8-12).
[3] On the torse-forming directions in Finsler spaces. (Tôhoku Math. J. (2) 4 (1952), 99-102).
[4] On the normality in Minkowski spaces. (Kumamoto J. Sci. A 2 (1954), 11-17).
[5] On similarities and transitive abelian groups of motions in Finsler spaces. (Ibid. 4 (1959), 103-110).
[6] On similarities in a Finsler space. (Tensor, N.S. 9 (1959), 175-189).
[7] On local properties in a metric space. (Ibid. 10 (1960), 81-89).
[8] On characterization of spaces of constant curvature. (Ibid. 11 (1961), 6-15)
[9] On Desarguesian spaces. (Math. J. Okayama Univ. 11 (1962), 19-26).
NAZIM TERZIOĞLU, A.
[1] Über Finslersche Räume. (Dissertation, München, 1936).
[2] Über den Satz von Gauss-Bonnet im Finslerschen Raum. (Univ. Istanbul, Fac. Sci. (1948), 26-32).
NUMATA, S.
[1] On the curvature tensor S_{hijk} and the tensor T_{hijk} of generalized Randers spaces. (Tensor, N.S. 29 (1975), 35-39).
[2] On Landsberg spaces of scalar curvature. (J. Korean Math. Soc. 12 (1975), 97-100).
[3] On the torsion tensors R^{i}_{hjk} and P^{i}_{hjk} of Finsler spaces with a metric ds = $(g_{ij}(dx)dx^{i}dx^{j})^{1/2} + b_{i}(x)dx^{i}$. (Tensor, N.S. 32 (1978), 27-31).
[4] On C3-like Finsler spaces. (Rep. on Math. Phys. 18 (1980), 1-10).
O'BYRNE, B.
[1] On Finsler geometry and applications to Teichmüller spaces. (Proc. Conf. Stony Brook, N.Y., 1969, 317-328).
OHKUBO, T.
[1] Geometry in a space with a generalized metric. (Japanese) (Tensor, O.S. 3 (1940), 48-55).
[2] On a symmetric displacement in a Finsler space. (Japanese) (Ibid. 4 (1941), 53-55).
[3] Geometry in a space with generalized metrics II. (J. Fac. Sci. Hokkaido Imp. Univ. I 10 (1941), 157-178).
[4] On relations among various connections in Finslerian space. (Kumamoto J. Sci. A 1 (1954), no.3, 1-6).
OKADA, T.
[1] Connection in frame bundle on tangent vector bundle and Finsler connection. (Japanese) (Thesis, Kyoto, 1962, 21p.)

[2] Theory of pair-connections. (Japanese) (Sci. Eng. Rev. Doshisha Univ. 5 (1964), 133-152).
[3] The holonomy group of Finsler connection I. The holonomy group of the V-connection. (Tensor, N.S. 27 (1973), 229-239; 29 (1975), 314).

OKUBO, K.
[1] On Finsler metrics. (Japanese) (Memo. Fac. Education, Shiga Univ. 17 (1967), 1-5).
[2] On Finsler spaces with the curvature S_{ijkl} of a special form. (Ibid. 25 (1975), 1-3).
[3] Some theorems of $S_3(K)$ metric spaces. (Rep. on Math. Phys. 16 (1979), 401-408).

OKUBO, T.
[1] On the connections of the tangent bundle of a Finsler space. (Yokohama Math. J. 12 (1964), 23-37).

OKUBO, T.; HOUH, C.
[1] Some cross-section theorems on the tangent bundle over a Finsler manifold. (Ann. Mat. Pura Appl. (4) 92 (1972), 129-138).

ŌTSUKI, T.
[1] On geodesic coordinates in Finsler spaces. (Math. J. Okayama Univ. 6 (1957), 135-145).
[2] Theory of affine connections of the space of tangent directions of a differentiable manifold I, II, III. (Ibid. 7 (1957), 1-74, 95-122).
[3] Note on homotopies of some curves in tangent bundles. (Ibid. 7 (1957), 191-194).

PALAIS, R.S.
[1] Lusternik-Schnirelman theory on Banach manifolds. (Topol. 5 (1966), 115-132).
[2] Critical point theory and the minimax principle. (Proc. Symp. Pure Math. 15 (1970), 185-212).

PENOT, J.-P.
[1] Topologie faible sur des variétés de Banach. (C.R. Acad. Sci. Paris 274 (1972), 405-408).
[2] Topologie faible sur des variétés de Banach. Applications aux géodésiques des variétés de Sobolev. (J. Diff. Geom. 9 (1974), 141-168).

PINL, M.
[1] Quasimetrik auf totalisotropen Flächen I, II, III. (Akad. Wetensch. Proc. 35 (1932), 1181-1188; 36 (1933), 550-557; 38 (1935), 171-180).
[2] In memory of Ludwig Berwald. (Scripta Math. 27 (1964), 193-203; Czechoslovakian: Casopis Pesto. Mat. 92 (1967), 229-238).

PRAKASH, N.
[1] Generalised normal curvature of a curve and generalised principal directions in Finsler space. (Tensor, N.S. 11 (1961), 51-56).
[2] Kaehlerian Finsler manifolds. (Math. Student 30 (1962), 1-12).

PRAKASH, N.; BEHARI, R.
[1] Deviations from parallelism and equidistance in Finsler space. (Proc. Indian Acad. Sci. A 52 (1960), 209-227).

PRAKASH, N.; DHAWAN, M.
[1] A study of contravariant and covariant almost analytic vectors in a Finsler space. (J. Math. Sci. 4 (1969), 69-76).

RADU, C.F.
[1] On extensions of the Lie derivative to tensors that depend on point and direction. (Romanian) (An. Univ. Bucureşti, Mat.-Mec. 22-1 (1973), 95-101).

RANI, N.
[1] Deviations from parallelism and equidistance of the congruence of curves. (Progr. Math. Allahabad 2 (1968), no.2, 15-20).
[2] A note on a complex Finsler manifold. (Tensor, N.S. 20 (1969), 91-94).
[3] Theory of Lie derivatives in a Minkowskian space. (Ibid. 20 (1969), 100-102).

RAPCSÁK, A.
[1] Kurven auf Hyperflächen im Finslerschen Raume. (Hungar. Acta Math. 1 (1949), no.4, 21-27).
[2] Eine neue Definition der Normalkoordinaten im Finslerschen Raum. (Hungarian)

(Acta Univ. Debrecen 1 (1954), 109-116. ad.1 (1955), 17).
[3] Invariante Taylorsche Reihe in einem Finslerschen Raum. (Publ. Math. Debrecen 4 (1955), 49-60).
[4] Eine neue Charakterisierung Finslerscher Räume skalarer und konstanter Krümmung und projektiv-ebene Räume. (Acta Math. Acad. Sci. Hungar. 8 (1957), 1-18).
[5] Metrische Charakterisierung der Finslerschen Räume mit verschwindender projektiver Krümmung. (Acta Sci. Math. Szeged 18 (1957), 192-204).
[6] Hyperebenen in Finslerschen Räumen. (Hungarian) (Acta Univ. Debrecen 4 (1959), 85-87).
[7] Die Bestimmung der Grundfunktionen projektiv-ebener metrischer Räume. (Publ. Math. Debrecen 9 (1962), 164-167).
[8] Über die Metrisierbarkeit affinzusammenhängender Bahnräume. (Ann. Mat. Pura Appl. (4) 57 (1962), 233-238).

RASTOGI, S.C.
[1] On the Bianchi and Veblen identities in a Finsler space. (An. Univ. Timişoara Şti. Mat. 8 (1970), 95-100).
[2] Submanifolds of a Finsler manifold. (Tensor, N.S. 30 (1976), 140-144).

RIEDLER, K.
[1] Globale Finslersche Geometrie. (Ber. Math.-Stat. Sekt. Forschungszentrum Graz 43 (1975), 3-18).

RIZZA, G.B.
[1] Strutture di Finsler sulle varietà quasi complesse. (Lincei Rend. (8) 33 (1962), 271-275).
[2] Strutture di Finsler di tipo quasi Hermitiano. (Riv. Mat. Univ. Parma (2) 4 (1963), 83-106).
[3] F-forme quadratiche ed hermitiane. (Rend. Mat. Appl. (5) 23 (1964), 221-249)
[4] Monogenic functions on the real algebras and conformal mappings. (Boll. U.M.I. (4) 12 (1975), Suppl. 3, 437-450).

RUND, H.
[1] Finsler spaces considered as generalized Minkowskian spaces. (Thesis, Cape Town, 1950).
[2] Zur Begründung der Differentialgeometrie der Minkowskischen Räume. (Arch. Math. 3 (1952), 60-69).
[3] Über die Parallelverschiebung in Finslerschen Räumen. (Math. Z. 54 (1951), 115-128).
[4] Eine Krümmungstheorie der Finslerschen Räume. (Math. Ann. 125 (1952), 1-18).
[5] A theory of curvature in Finsler spaces. (Colloq. Topol. Strasbourg, 1951, no.4, 12p.).
[6] The theory of subspaces of a Finsler space. I. (Math. Z. 56 (1952), 363-375).
[7] The theory of subspaces in a Finsler space. II. (Ibid. 57 (1953), 193-210).
[9] The scalar form of Jacobi's equations in the calculus of variations. (Ann. Mat. Pura Appl. (4) 35 (1953), 183-202).
[11] On the analytical properties of curvature tensors in Finsler spaces. (Math. Ann. 127 (1954), 82-104).
[12] Über nicht-holonome allgemeine metrische Geometrie. (Math. Nachr. 11 (1954), 61-80).
[13] On the geometry of generalised metric spaces. (Convegno Intern. Geom. Diff. Italia, 1953, 114-121).
[14] Hypersurfaces of a Finsler space. (Canad. J. Math. 8 (1956), 487-503).
[15] Über allgemeine nicht-holonome und dissipative dynamische Systeme. (Schr. Forschungsinst. Math. 1 (1957), 269-279).
[16] Some remarks concerning the theory of non-linear connections. (Nederl. Akad. Wetensch. Proc. A 61 = Indag Math. 20 (1958), 341-347).
[17] Some remarks concerning Carathéodory's method of "equivalent" integrals in the calculus of variations. (Ibid. 62 = ibid. 21 (1959), 135-141).
[18] The differential geometry of Finsler spaces. (Grundlehr. Math. Wiss. 101, Springer Verlag, Berlin-Göttingen-Heidelberg, 1959, 284p.).
[19] The theory of problems in the calculus of variations whose Lagrangian function involves second order derivatives: A new approach. (Ann. Mat. Pura Appl. (4) 55 (1961), 77-104).

[21] Über Finslersche Raume mit speziellen Krümmungseigenschaften. (Monatsh. Math. 66 (1962), 241-251).

[23] Curvature properties of hypersurfaces of Finsler and Minkowskian spaces. (Tensor, N.S. 14 (1963), 226-244).

[24] The intrinsic and induced curvature theories of subspaces of a Finsler space. (Ibid. 16 (1965), 294-312).

[25] Finsler spaces of scalar and constant curvature in geodesic correspondence. (Tydskr. Natuurwet. 6 (1966), 243-254).

[26] Generalized metrics on complex manifolds. (Math. Nachr. 34 (1967), 55-77).

[27] Invariant theory of variational problems for geometric objects. (Tensor, N.S. 18 (1967), 239-258).

[28] The curvature theory of direction-dependent connections on complex manifolds. (Ibid. 24 (1972), 189-205).

[29] A divergence theorem for Finsler metrics. (Monatsh. Math. 79 (1975), 233-252).

SANINI, A.

[1] Derivazioni su distribuzioni e connessioni di Finsler. (Univ. Politec. Torino, Rend. Sem. Mat. 31 (1971-72/1972-73), 157-184).

[2] Su un tipo di struttura quasi Hermitiana del fubrato tangente ad uno spazio di Finsler. (Ibid. 32 (1973-74), 303-316).

[3] Alcuni tipi di connessioni su varietã quasi prodotto. (Atti. Accad. Sci. Torino, Cl. Sci. Fis. 106 (1971-72), 317-332).

[4] Connessioni lineari del tipo di Finsler e strutture quasi hermitiane. (Riv. Mat. Univ. Parma 30 (1974), 239-252).

ŠAPUKOV, N.N.

[1] Extremal displacement of a minimal hypersurface in Riemannian and Finsler spaces. (Russian) (Izv. Vysš. Učebn. Zaved. Mat. 5(24) (1961), 112-116).

SCHNEIDER, R.

[1] Über die Finslerräume mit $S_{ijkl} = 0$. (Arch. Math. 19 (1968), 656-658).

SCHÖNE, W.

[1] Über eine Verallgemeinerung der Minkowskischen Distanzfunktion und ihre Anwendung in der Finslerschen Geometrie. (Math. Nachr. 40 (1969), 305-325).

[2] Ein Eigenwertproblem des metrischen Tensors und seine geometrische Deutung. (Demonst. Math. 6 (1973), 343-365).

SEN, R.N.

[1] Application of an algebraic system in Finsler geometry. (Tensor, N.S. 18 (1967), 191-195).

[2] On curvature tensors in Finsler geometry. (Ibid. 18 (1967), 217-226).

[3] Some generalised formulae for curvature tensors in Finsler geometry. (Indian J. Math. 9 (1967), 211-221).

[4] Finsler spaces of recurrent curvature. (Tensor, N.S. 19 (1968), 291-299).

SHIBATA, C.

[1] On Finsler spaces with Kropina metric. (Rep. on Math. Phys. 13 (1978), 117-128).

[2] On the curvature tensor R_{hijk} of Finsler spaces of scalar curvature. (Tensor, N.S. 32 (1978), 311-317).

SHIBATA, C.; SHIMADA, H.

[1] The g-hypercone of a Minkowski space. (Tensor, N.S. 35 (1981), 73-85).

SHIBATA, C.; SHIMADA, H.; AZUMA, M.; YASUDA, H.

[1] On Finsler spaces with Randers' metric. (Ibid. 31 (1977), 219-226).

SHIMADA, H.

[1] On the Ricci tensors of particular Finsler spaces. (J. Korean Math. Soc. 14 (1977), 41-63).

[2] Finsler spaces of recurrent torsion and constant curvature. (Research Rep. Kushiro Tech. College 13 (1979), 173-180).

[3] On Finsler spaces with the metric $L = \sqrt[m]{a_{i_1 i_2 \dots i_m}(x) y^{i_1} y^{i_2} \cdots y^{i_m}}$. (Tensor, N.S. 33 (1979), 365-372).

SHING, D.

[1] On the symmetric properties in some Finsler spaces. (Chinese) (Acta Math. Sinica 9 (1959), 191-198). (English: Chinese Math. 9 (1967), 498-506).

SINGH, U.P.
[1] On principal directions in Finsler subspaces. (Proc. Nat. Inst. Sci. India A 35 (1969), Suppl. 2, 130-137).
[2] On intrinsic and induced theories of a Finsler subspace. (Univ. Nac. Tucumán, Rev. A 21 (1971), 119-121).
[3] On Finsler spaces admitting parallel vectorfields. (Tensor, N.S. 27 (1973), 170-172).
SINGH, U.P.; PRASAD, B.N.
[1] On hyper Darboux lines in Finsler subspace. (Rev. Univ. Istanbul 34 (1969), 83-90).
[2] The Bianchi and Veblen identities in Finsler space. (Acad. Roy. Belg. Bull. Cl. Sci. (5) 56 (1970), 1100-1109).
SINHA, B.B.
[1] Projective invariants. (Math. Student 33 (1965), 121-127).
[2] A generalization of Gauss and Codazzi's equations for intrinsic curvature tensor. (Indian J. Pure Appl. Math. 2 (1971), 270-274).
[3] On curvature mapping in a Finsler space. (Tensor, N.S. 22 (1971), 326-328).
SINHA, B.B.; SINGH, S.P.
[1] On recurrent Finsler spaces. (Rev. Roum. Math. Pures Appl. 16 (1971), 977-986).
[2] Recurrent Finsler space of second order. I, II. (Yokohama Math. J. 19 (1971), 79-85; Indian J. Pure Appl. Math. 4 (1973), 45-50).
SINHA, R.S.
[1] Projective curvature tensors of a Finsler space. (Acad. Roy. Belg. Bull. Cl. Sci. (5) 54 (1968), 272-279).
[2] Infinitesimal conformal transformation in Finsler spaces. (Univ. Nac. Tucumán Rev. A 21 (1971), 283-287).
[3] Infinitesimal projective transformation in Finsler spaces. (Progr. Math. Allahabad 5 (1971), 30-34).
[4] Some formulae in the theory of Lie derivatives in Finsler spaces. (Tensor, N.S. 25 (1972), 332-336).
ŠINKŪNAS, J.
[1] The space of support linears of a Finsler structure. (Russian) (Liet. Mat. Rinkinys. 12 (1972), 221-227, 243).
SLOBODJAN, Ju.S.
[1] Certain properties of the locally geodesic hypersurfaces of a Finsler space. (Russian) (Ukrain. Geom. Sb. 10 (1971), 65-71).
[2] The Gauss-Codazzi conditions for totally geodesic surfaces in a Finsler space. (Russian) (Ibid. 12 (1972), 124-131, 170).
SOÓS, G.
[1] Über Gruppen von Affinitäten und Bewegungen in Finslerschen Räumen. (Acta Math. Acad. Sci. Hungar. 5 (1954), 73-84).
[2] Über Gruppen von Automorphismen in affinzusammenhängenden Räumen von Linienelementen. (Publ. Math. Debrecen 4 (1956), 294-302).
[3] Über eine spezielle Klasse von Finslerschen Räumen. (Ibid. 5 (1957), 150-153).
[4] Über einfache Finslersche Räume. (Ibid. 7 (1960), 364-373).
[5] Über die homothetische Gruppen von Finslerschen Räuemn. (Acta Math. Acad. Sci. Hungar. 10 (1959), 391-394).
[6] On the theory of fiber spaces of Finsler type. (Hungarian) (Magyer Tud. Akad. Mat. Fiz. Oszt. Közl. 13 (1963), 17-64).
SOROKIN, V.A.
[1] Certain questions of a Minkowski geometry with a non-symmetric indicatrix. (Russian) (Orehovo-Zuev. Ped. Inst. Učen. Zap. Kaf. Mat. 22 (1964), no.3, 138-147).
[2] Jung's and Blaschke's theorems for a Minkowski plane. (Russian) (Moskov. Gos. Ped. Inst. Učen. Zap. 271 (1967), 145-153).
SRIVASTAVA, T.N.
[1] On a Finsler space I. (Tensor, N.S. 28 (1974), 13-18).
SRIVASTAVA, T.N.; WATANABE, S.
[1] Some properties of indicatrices in a Finsler space. (Canad. Math. Bull. 18 (1975), 715-721).

SU, B.
[1] Geodesic deviation in generalized metric space. (Acad. Sinica Sci. Record 2 (1949), 220-226).
[2] On the isomorphic transformations of minimal hypersurfaces in a Finsler space. (Chinese) (Acta Math. Sinica 5 (1955), 471-488).
[3] Differential geometry of generalized spaces. (Chinese) (Science Publ., Peking, 1958, 162p.)

SULANKE, R.
[1] Die eindeutige Bestimmtheit des von Hanno Rund eingeführten Zusammenhangs in Finsler-Räumen. (Wiss. Z. Hunboldt-Univ. Math.-Nat. R. 4 (1954/55), 229-233; 5 (1955/56), 269).
[2] Eine Ableitung des Cartanschen Zusammenhangs eines Finslerschen Raumes. (Publ. Math. Debrecen 5 (1957), 197-203).

SUMITOMO, T.
[1] On the reduction of Minkowski space. (J. Fac. Sci. Hokkaido Univ. I 14 (1958), 50-58).

SYNGE, J.L.
[1] A generalisation of the Riemannian line-element. (Trans. Amer. Math. Soc. 27 (1925), 61-67).

SZABÓ, Z.I.
[1] Ein Finslerschen Raum ist gerade dann von skalarer Krümmung, wenn seine Weylsche Projektivkrümmung verschwindet. (Acta Sci. Math. Szeged 39 (1977), 163-168).
[2] Finslersche Projektivgeometrie I. Eine globale Begründung der Finslerschen Projektivconnexionen. (Acta Math. Acad. Sci. Hungar. 35 (1980), 79-96).
[3] Finslersche Projektivgeometrie II. Über Finslersche Projektivbundel mit Weylscher Projektivkrümmung Null. (Ibid. 35 (1980), 97-108).
[4] Positive definite Finsler spaces satisfying the T-condition are Riemannian. (Tensor, N.S. 35 (1981), 247-248).
[5] Positive definite Berwald spaces. (Structure theorems on Berwald spaces.) (Ibid. 35 (1981), 25-39).

SZIGETI, F.
[1] Differentiable approximation on Banach manifolds. (Ricerche Mat. 19 (1970), 171-178).

SZOLCSÁNYI, E.
[1] Hyperflächen mit Riemannscher Massbestimmung im Finslerschen Raum. I, II. (Publ. Math. Debrecen 22 (1975), 133-150; 25 (1978), 203-210).

TACHIBANA, S.
[1] On Finsler spaces which admit a concurrent vector field. (Tensor, N.S. 1 (1950), 1-5).

TAKANO, K.
[1] Homogeneous contact transformations and a metric space. (Japanese) (Rep. Univ. Electro-Commun. 2 (1950), 269-278).
[2] Homothetic transformations in Finsler spaces. (Japanese) (Holonomy-gun Kenkyu 25 (1952), 10-13).
[3] Homothetic transformations in Finsler spaces. (Rep. Univ. Electro-Commun. 4 (1952), 61-70).

TAMÁSSY, L.
[1] Ein Problem der zweidimensionalen Minkowskischen Geometrie. (Ann. Polon. Math. 9 (1960/61), 38-48, 175).
[2] Zur Metrisierbarkeit der affinzusammenhängenden Räume. (Demonst. Math. 6 (1973), 851-859).

TAMÁSSY, L.; MATSUMOTO, M.
[1] Direct method to characterize conformally Minkowski Finsler spaces. (Tensor, N.S. 33 (1979), 380-384).

TAMIM, A.A,E.
[1] Study of special type of Finsler spaces. (Thesis, Cairo Univ., 1978, 45p.)

TASHIRO, Y.
[1] A theory of transformation groups on generalized spaces and its applications to Finsler and Cartan spaces. (J. Math. Soc. Japan 11 (1959), 42-71).

TAYLOR, J.H.
[1] A generalization of Levi-Civita's parallelism and the Frenet formulas. (Trans. Amer. Math. Soc. 27 (1925), 246-264).
[2] Reduction of Euler's equations to a canonical form. (Bull. Amer. Math. Soc. 31 (1925), 257-262).
[3] Parallelism and transversality in a sub-space of a general (Finsler) space. (Ann. of Math. (2) 28 (1927), 620-628).

TEODORESCU, N.
[1] Sur les géodésiques de longueur nulle de certains éléments linéaires finslériens. (Bull. École Polytech. Bucurest 12 (1941), 9-16).
[2] Géométrie finslérienne et propagation des ondes. (Acad. Roum. Bull. Sci. 23 (1942), 138-144).
[3] A propos d'espace de Finsler. (Romanian) (Acad. R.P. Romine Stud. Cerc. Mat. 13 (1962), 499-510).

TOMONAGA, Y.
[1] Jacobi fields in a Finsler space. (TRU Math. 5 (1969), 37-42).

TONOOKA, K.
[1] On a geometry of three-dimensional space with an algebraic metric. (Tensor, N.S. 6 (1956), 60-68).
[2] On three and four dimensional Finsler spaces with the fundamental forms $\sqrt[3]{a_{\alpha\beta\gamma}x'^{\alpha}x'^{\beta}x'^{\gamma}}$. (Ibid. 9 (1959), 209-216).
[3] Theories of surface of three-dimensional space with an algebraic metric. (Ibid. 14 (1963), 219-225).
[4] Subspace theory of an n-dimensional space with an algebraic metric. (J. Fac. Sci. Hokkaido Univ. I 18 (1964), 34-40).
[5] On a geometry of three-dimensional space based on the differential form of the fourth order. (Tensor, N.S. 25 (1972), 148-154).

TROMBA, A.J.
[1] Almost-Riemannian structures on Banach manifold: The Morse lemma and the Darboux theorem. (Canad. J. Math. 28 (1976), 640-652).

UHLENBECK, K.
[1] Harmonic maps; a direct method in the calculus of variations. (Bull. Amer. Math. Soc. 76 (1970), 1082-1087).
[2] Morse theory on Banach manifolds. (J. Funct. Anal. 10 (1972), 430-445).
[3] Bounded set and Finsler structures for manifolds of maps. (J. Diff. Geom. 7 (1972), 585-595).

VANSTONE, J.R.
[1] A generalization of Finsler geometry. (Canad. J. Math. 14 (1962), 87-112).

VARGA, O.
[1] Beiträge zur Theorie der Finslerschen Räume und der affinzusammenhängenden Räume von Linienelementen. (Lotos Prag 84 (1936), 1-4).
[2] Zur Herleitung des invarianten Differentials in Finslerschen Räumen. (Monatsh. Math. Phys. 50 (1942-43), 165-175).
[3] Zur Differentialgeometrie der Hyperflächen in Finslerschen Räumen. (Deutsch. Math. 6 (1941), 1920212).
[4] Bestimmung des invarianten Differentials in Finsler'schen Räumen. (Hungarian) (Mat. Fiz. Lapok 48 (1941), 423-435).
[5] Zur Begründung der Minkowskischen Geometrie. (Acta Univ. Szeged Sci. Math. 10 (1941-43), 149-163).
[6] Aufbau der Finslerschen Geometrie mit Hilfe einer oskulierenden Minkowskischen Massbestimmung. (Hungarian) (Math. Naturwiss. Anz. Ungar. Akad. Wiss. 61 (1942), 14-22).
[7] Über eine Klasse von Finslerschen Räumen, die die nichteuklidischen verallgemeinern. (Comment. Math. Helv. 19 (1947), 367-380).
[8] Linienelementräume deren Zusammenhang durch eine beliebige Transformationsgruppe bestimmt ist. (Acta Univ. Szeged Sci. Math. 11 (1946), 55-62).
[9] Über affinzusammenhängende Mannigfaltigkeit von Linienelementen insbesondere deren Äquivalenz. (Publ. Math. Debrecen 1 (1949), 7-17).
[10] Affinzusammenhängende Mannigfaltigkeiten von Linienelementen, die ein

Inhaltsmass besitzen. (Nederl. Akad. Wetensch. Proc. A 52 (1949), 868-874 = Indag. Math. 11 (1949), 316-322).
[11] Über den Zusammenhang der Krümmungsaffinoren in zwei eineindeutig aufeinander abgebildeten Finslerschen Räumen. (Acta Sci. Math. Szeged 12 (1950), 132-135).
[12] Über das Krümmungsmass in Finslerschen Räumen. (Publ. Math. Debrecen 1 (1949), 116-122).
[13] Normalkoordinaten in allgemeinen differentialgeometrischen Räumen und ihre Verwendung zur Bestimmung sämtlicher Differentialinvarianten. (Hungarian and German) (C.R. Premier Congr. Math. Hongrois, 1950, 131-162).
[14] Eine geometrische Charakterisierung der Finslerschen Räume skalarer und konstanter Krümmung. (Acta Math. Acad. Sci. Hungar. 2 (1951), 143-156).
[15] Eine Charakterisierung der Finslerschen Räume mit absolutem Parallelismus der Linienelemente. (Arch. Math. 5 (1954), 128-131).
[16] Bedingungen für die Metrizierbarkeit von affinzusammenhängenden Linienelementmannigfaltigkeit. (Acta Math. Acad. Sci. Hungar. 5 (1954), 7-16).
[17] Eine Charakterisierung der Finslerschen Räume mit absolutem Parallelismus der Linienelemente. (Hungarian) (Acta Univ. Debrecen 1 (1954), 105-108. Add., ibid. 1 (1955), 16).
[18] Die Krümmung der Eichfläche des Minkowskischen Raumes und die geometrische Deutung des einen Krümmungstensors des Finslerschen Raumes. (Abh. Math. Sem Univ. Hamburg 20 (1956), 41-51).
[19] Hilbertsche Verallgemeinerte nicht-euklidische Geometrie und Zusammenhang derselben mit der Minkowskischen Geometrie. (Intern. Congr. Math. Edinburgh, 1958, 111).
[20] Über die Zerlegbarkeit von Finslerschen Räumen. (Acta Math. Acad. Sci. Hungar. 11 (1960), 197-203).
[21] Bemerkung zur Winkelmetrik in Finslerschen Räumen. (Ann. Univ. Sci. Budapest. Sect. Math. 3-4 (1961), 379-382).
[22] Über den innern und induzierten Zusammenhang für Hyperflächen in Finslerschen Räumen. (Publ. Math. Debrecen 8 (1961), 208-217).
[23] Über eine Charakterisierung der Finslerschen Räume konstanter Krümmung. (Monatsh. Math. 65 (1961), 277-286).
[24] Zur Begründung der Hilbertschen Verallgemeinerung der nichteuklidischen Geometrie. (Monatsh. Math. 66 (1962), 265-275).
[25] Herleitung des Cartanschen Zusammenhangs in Finslerräumen mit Hilfe der Riemannschen Geometrie. (Acta Phys. Chim. Debrecina 8 (1962), 121-124).
[26] Eine einfache Herleitung der Cartanschen Übertragung der Finslergeometrie. (Math. Notae 18 (1962), 185-196).
[27] Über Hyperflächen konstanter Normalkrümmung in Minkowskischen Räumen. (Tensor, N.S. 13 (1963), 246-250).
[28] Hyperflächen mit Minkowskischer Massbestimmung in Finslerräumen. (Publ. Math. Debrecen 11 (1964), 301-309).
[29] Die Methode des beweglichen n-Beines in der Finsler-Geometrie. (Acta Math. Acad. Sci. Hungar. 18 (1967), 207-215).
[30] Hyperflächen konstanter Normalkrümmung in Finslerschen Räumen. (Math. Nachr. 38 (1968), 47-52).
[31] Zur Invarianz des Krümmungsmasses der Winkelmetrik in Finsler-Raümen bei Einbettungen. (Math. Nachr. 43 (1970), 11-18).
[32] Beziehung der ebenen verallgemeinerten nichteuklidischen Geometrie zu gewissen Flächen im pseudominkowskischen Raum. (Aequations Math. 3 (1969), 112-117).

VARGA, T. (KÁNTOR, S.)
[1] Über Berwaldsche Räume I, II. (Publ. Math. Debrecen 25 (1978), 213-223; 26 (1979), 41-50).

VASIL'EV, A.M.
[1] Invariant affine connections in a space of linear elements. (Russian) (Mat. Sb., N.S. 60 (102) (1963), 411-424).

VASIL'EVA, M.V.
[1] Finslerian geometry in invariant presentation. (Russian) (Moskov. Gos. Ped. Inst. Učen. Zap. 243 (1965), 38-54).
[2] An invariant description of certain Finslerian geometries. (Russian) (Ibid.

243 (1965), 55-68).
[3] The connections between the various Finsler geometries, and the bundle of Finsler geometries. (Russian) (Ibid. 271 (1967), 49-54).

VERMA, M.
[1] Hyper Darboux lines of subspaces of a Finsler space. (Math. Nachr. 47 (1970), 293-298).
[2] Generalised Laguerre functions of a hypersurface in a Finsler space. (Tensor, N.S. 24 (1972), 169-172).

VILMS, J.
[1] Connections on tangent bundles. (J. Diff. Geom. 1 (1967), 235-243).
[2] Curvature of nonlinear connections. (Proc. Amer. Math. Soc. 19 (1968), 1125-1129).
[3] Nonlinear and direction connections. (Ibid. 28 (1971), 567-572).
[4] On the existence of Berwald connections. (Tensor, N.S. 26 (1972), 118-120).

VRǍNCEANU, G.
[1] Espaces à connexion affine généralisée. (Rev. Roum. Math. Pure Appl. 22 (1977), 1337-1342).

WAGNER, V.V. (VAGNER)
[1] A generalization of non-holonomic manifolds in Finslerian space. (Abh. Tscherny. Staatsuniv. Saratow I (14) 2 (1938), 67-96).
[2] Two-dimensional space with the metric defined by a cubic differential form. (Russian and English) (Ibid. I (14) 35 (1938), 29-40).
[3] Über Berwaldsche Räume. (Rec. Math. (Mat. Sb.) (2) 3 (1938), 655-662).
[4] On generalized Berwald spaces. (C.R. Dokl. Acad. Sci. URSS, N.S. 39 (1943), 3-5).
[5] The inner geometry of non-holonomic manifolds. (Rec. Math. (Mat. Sb.), N.S. 13 (55) (1943), 135-167).
[6] Les espaces de Finsler à deux dimensions à groupes d'holonomie finis et continus. (C.R. Dokl. Acad. Sci. URSS, N.S. 39 (1943), 210-212). (Japanese: Holonomy-gun Kenkyu 22 (1951), 24-27).
[7] Geometry of field of local curves in X_3 and the simplest case of Lagrange's problem in the calculus of variations. (Ibid. 48 (1945), 229-232).
[8] Homological transformations of Finslerian metric. (Ibid. 46 (1945), 263-265).
[9] Geometry of field of local central plane curves in X_3. (Ibid. 48 (1945), 382-384).
[10] The geometrical theory of the simplest n-dimensional singular problem of the calculus of variations. (Russian) (Mat. Sb., N.S. 21 (63) (1947), 321-364).
[11] Theory of field of local (n-2)-dimensional surfaces in X_n and its applications to the problem of Lagrange in the calculus of variations. (Ann. of Math. (2) 49 (1948), 141-188).
[12] The geometry of Finsler as a theory of the field of local hypersurfaces in X_n. (Russian) (Trudy Sem. Vektor. Tenzor. Anal. 7 (1949), 65-166).
[13] On the embedding of a field of local surfaces in X_n in a constant field of surfaces in affine space. (Russian) (Dokl. Akad. Nauk SSSR, N.S. 66 (1949), 785-788).
[15] The theory of composite manifolds. (Russian) (Trudy Sem. Vektor. Tenzor. Anal. 8 (1950), 11-72).
[16] The theory of a field of local hyperstrips. (Russian) (Ibid. 8 (1950), 197-272).

WANG, H.-C.
[1] On Finsler spaces with completely integrable equations of Killing. (J. London Math. Soc. 22 (1947), 5-9).

WANG, Y.
[1] Finsler spaces of constant curvature and totally extremal hypersurfaces. (Sci. Record, N.S. 2 (1958), 211-214).

WATANABE, S.
[1] On special Kawaguchi spaces III. Generalizations of affine spaces and Finsler spaces. (Tensor, N.S. 11 (1961), 144-153).
[2] On indicatrices of a Finsler space. (Ibid. 27 (1973), 135-137).

WATANABE, S.; IKEDA, F.

[1] On some properties of Finsler spaces based on the indicatrices. (Publ. Math. Debrecen 28 (1981), 129-136).

[2] On Finsler spaces satisfying the C-reducibility condition and the T-condition. (Tensor, N.S. 34 (1980), 103-108).

WEGENER, J.M.

[1] Hyperflächen in Finslerschen Räumen als Transversalflächen einer Schar von Extremalen. (Monatsh. Math. Phys. 44 (1936), 115-130).

[2] Untersuchungen der zwei- und dreidimensionalen Finslerschen Räume mit der Grundform $L = \sqrt[3]{a_{ikl}x'^i x'^k x'^l}$. (Akad. Wetensch. Proc. 38 (1935), 949-955).

[3] Untersuchung über Finslersche Räume. (Lotos Prag 84 (1936), 4-7).

WEI, H.(X.)

[1] Veblen identities in Finsler spaces and in generalized Finsler spaces. (Acta Sci. Nat. Univ. Amoiensis 12 (2) (1965), 23-31).

WHITEHEAD, J.H.C.

[1] Convex regions in the geometry of paths. (Quart. J. Math. Oxford 3 (1932), 33-42. Add., ibid. 4 (1933), 226-227).

[2] The Weierstrass E-function in differential metric geometry. (Ibid. 4 (1933), 291-296).

[3] On the covering of a complete space by the geodesics through a point. (Ann. of Math. (2) 36 (1935), 679-704).

WRONA, W.

[1] Neues Beispiel einer Finslerschen Geometrie. (Prace Mat. Fiz. 46 (1938), 281-290).

[2] On multi-isotropic Finsler spaces. (Bull. Polon. Math. Astro. Phys. 11 (1963), 285-288).

[3] A necessary and sufficient condition for the Finsler space with distant parallelism of line elements to be multi-isotropic. (Ibid. 11 (1963), 289-292).

[4] Generalized F.Schur theorem in Finsler spaces. (Ibid. 11 (1963), 293-295).

[5] On geodesics of a certain singular n-dimensional Finsler space. (Zeszyty Nauk. Politech. Warszawsk. Mat. 11 (1968), 271-276).

YAMADA, T.

[1] On the indicatrized tensor of $S_{ijkh}|_1$ in semi-C-reducible Finsler spaces. (Tensor, N.S. 34 (1980), 151-156).

YANO, K.

[1] The theory of Lie derivatives and its applications. (North-Holland Publ. Amsterdam, 1957, 293p.).

YANO, K.; DAVIES, E.T.

[1] On the connection in Finsler space as an induced connection. (Rend. Circ. Mat. Palermo (2) 3 (1954), 409-417).

[2] On the tangent bundles of Finsler and Riemannian manifolds. (Ibid. 12 (1963), 211-228).

[3] Metrics and connections in the tangent bundle. (Kōdai Math. Sem. Rep. 23 (1971), 493-504).

[4] Differential geometry on almost tangent manifolds. (Ann. Mat. Pura Appl. (4) 103 (1975), 131-160).

YANO, K.; MUTŌ, Y.

[1] Homogeneous contact manifolds and almost Finsler manifolds. (Kōdai Math. Sem. Rep. 21 (1969), 16-45).

YANO, K.; OKUBO, T.

[1] On tangent bundles with Sasakian metrics of Finslerian and Riemannian manifolds. (Ann. Mat. Pura Appl. (4) 87 (1970), 137-162).

YASUDA, H.

[1] On extended Lie systems, III (Finsler spaces). (Tensor, N.S. 23 (1972), 115-130).

[2] Finsler spaces as distributions on Riemannian manifolds. (Hokkaido Math. J. 1 (1972), 280-297).

[3] On Finsler spaces with absolute parallelism of line-elements. (J. Korean Math. Soc. 13 (1976), 179-188).

[4] On the indicatrix bundle endowed with the K-connection over a Finsler space. (Ann. Rep. Asahikawa Med. College 1 (1979), 117-124).
[5] On the indicatrices of a Finsler space. (Tensor, N.S. 33 (1979), 213-221).
[6] On Landsberg spaces. (Ibid. 34 (1980), 77-84).

YASUDA, H.; SHIMADA, H.
[1] On Randers spaces of scalar curvature. (Rep. on Math. Phys. 11 (1977), 347-360).

YASUDA, H.; YAMANOI, T.
[1] On curves in the indicatrix bundle over a Finsler space. (Tensor, N.S. 33 (1979), 357-364).

YOSHIDA, M.
[1] On an R3-like Finsler space and its special cases. (Ibid. 34 (1980), 157-166).

YOUSSEF, N.L.
[1] Distribution de nullité du tenseur de courbure d'une connexion. (Thèse, Univ. Grenoble, 1978, 60p.).

ZAGUSKIN, V.I. (ZAGUSKII)
[1] On a certain Finsler space and the motion in a Minkowski space. (Russian) (Naučn. Dokl. Vysš. Školy. Fiz.-Mat. Nauki, 1958, no.3, 50-52).
[2] Certain aspects of Finsler geometry. (Russian) (Uch. Zap. Yaroslavsk. Gos. Ped. Inst. 34 (1960), 83-110).

ŽOTIKOV, G.I.
[1] The differential singular Finsler metric definable in X_n by a field of local singular hypersurfaces of singularity n - m - 1. (Russian) (Baskir. Gos. Univ. Učen. Zap. Vyp. 20 (1965), Ser. Mat. no.2, 32-45).

ADDITIONS TO BIBLIOGRAPHY

AIKOU, T.; HASHIGUCHI, M.
[1] On generalized Berwald connections. (Rep. Fac. Sci. Kagoshima Univ. (Math., Phys., Chem.) 17 (1984), 9-13).

DRAGOMIR, S.
[1] The theorem of E.Cartan on Finsler manifolds. (Coloc. Nat. Geom. Topol. Busteni, 1981, 103-112).
[2] The theorem of K.Nomizu on Finsler manifolds. (An. Univ. Timişoara, Ser. Mat. 19 (1981), 117-127).

FUKUI, M.; YAMADA, T.
[2] On Finsler spaces of constant curvature. (Tensor, N.S. 38 (1982), 129-134).

HASHIGUCHI, M.
[6] Wagner connections and Miron connections of Finsler spaces. (Rev. Roum. Math. Pures Appl. 25 (1980), 1387-1390).

HASHIGUCHI, M.; ICHIJYŌ, Y.
[3] Randers spaces with rectilinear geodesics. (Rep. Fac. Sci. Kagoshima Univ. (Math., Phys., Chem.) 13 (1980), 33-40).
[4] On generalized Berwald spaces. (Ibid. 15 (1982), 19-32).

HASSAN, B.T.M.
[6] Hypersurfaces of a Minkowski space. (Tensor, N.S. 41 (1984), 1-9).

HŌJŌ, S.
[5] On generalizations of Akbar-Zadeh's theorem in Finsler geometry. (Ibid. 37 (1982), 285-290).

ICHIJYŌ, Y.
[7] On the g-connections and motions on a {V,G}-manifold. (J. Math. Tokushima Univ. 14 (1980), 11-23).
[8] Almost Hermitian Finsler manifolds. (Tensor, N.S. 37 (1982), 279-284).
[9] On holonomy mappings associated with a non-linear connection. (J. Math. Tokushima Univ. 17 (1983), 1-9).

IKEDA, F.
[4] On S3- and S4-like Finsler spaces with the T-tensor of a special form. (Tensor, N.S. 35 (1981), 345-351).

IZUMI, H.
[3] Conformal transformations of Finsler spaces. II. An h-conformally flat

Finsler spaces. (Ibid. 34 (1980), 337-359).

IZUMI, H.; YOSHIDA, M.

[2] Remarks on Finsler spaces of perpendicular scalar curvature and the property H. (Ibid. 40 (1983), 215-220).

KOBAYASHI, S.

[1] Negative vector bundles and complex Finsler structures. (Nagoya Math. J. 57 (1975), 153-166).

LIN, Y.

[1] On the reducibility of Finsler spaces. (Chinese) (J. Xiamen Univ., Nat. Sci. 22 (1983), 133-143).

MATSUMOTO, M.

[33] Projectively flat Finsler spaces of constant curvature. (J. Nat. Acad. Math. India 1 (1983), 142-164).

[34] The induced and intrinsic Finsler connections of a hypersurface and Finslerian projective geometry. (J. Math. Kyoto Univ. 25 (1985), 107-144).

[35] V-transformations of Finsler spaces, II. Finsler spaces whose isometries are all affine. (To appear in Anal. Şti. Univ. Iaşi).

[36] Theory of Y-extremal and minimal hypersurfaces in Finsler spaces. On Wegener's and Barthel's theories. (To appear in J. Math. Kyoto Univ.).

MATTHIAS, H.-H.

[1] Eine Finslermetrik auf S^2 mit nur zwei geschlossenen Geodätischen. (Bonn. Math. Schrift. 102 (1978), 27-50).

MIRON, R.

[4] On transformation groups of Finsler connections. (Tensor, N.S. 35 (1981), 235-240).

[5] A nonstandard theory of hypersurfaces in Finsler spaces. (Anal. Şti. Univ. Iaşi 30 (1984), 35-53).

MIRON, R.; BEJANCU, A.

[1] A new method in geometry of Finsler subspaces. (Ibid. 30 (1984), 55-59).

OKADA, T.

[4] Minkowskian product of Finsler spaces and Berwald connection. (J. Math. Kyoto Univ. 22 (1982), 323-332).

[5] On models of projectively flat Finsler spaces of constant curvature. (Tensor, N.S. 40 (1983), 117-124).

OKADA, T.; NUMATA, S.

[1] On generalized C-reducible Finsler spaces. (Ibid. 35 (1981), 313-318).

SAKAGUCHI, T.

[1] Remarks on Finsler spaces with 1-form metric. (Ibid. 40 (1983), 173-183).

SHIBATA, C.

[3] On invariant tensors of β-changes of Finsler metrics. (J. Math. Kyoto Univ. 24 (1984), 163-188).

SHIMADA, H.

[4] On Finsler spaces with the m-th root metric and a generalization of Akbar-Zadeh's theorem of Hōjō. (J. Nat. Acad. Math. India 1 (1983), 45-56).

SZABÓ, Z.I.

[6] Generalized spaces with many isometries. (Geom. Dedicata 11 (1981), 369-383).

TAMÁSSY, L.; BÉLTEKY, K.

[1] On the coincidence of two kinds of ellipses in Minkowskian spaces and in Finsler planes. (Publ. Math. Debrecen 31 (1984), 157-161).

TAMÁSSY, L.; KIS, B.

[1] Relations between Finsler and affine connections. (Rend. Circ. Mat. Palermo (2), Suppl. 3 (1984), 329-337).

TAZAWA, Y.

[1] A Finsler space with non-convex indicatrices where geodesics play no significant role. (Tensor, N.S. 41 (1984), 161-165).

VASIL'EVA, M.V.

[4] Classification of Finsler spaces. (Colloq. Math. Soc. J.Bolyai, 31. Diff. Geom. Budapest, 1979, 769-785).

[5] The holonomy groups of an n-dimensional Finsler space. (Russian) (Geom. imbedd manifolds, 14-21. Moskov. Gos. Ped. Inst. 1979).

VERMA, M.
[3] On subspaces of the tangent space of a Finsler space. (Proc. Nat. Acad. Sci. India, Sect. A 53 (1983), 235-238).
WEI, H. (X.)
[2] Geodesic mappings between Finsler spaces. (Chinese) (Acta Sci. Nat. Univ. Amoiensis 20 (1981), 146-154).
YAMADA, T.; FUKUI, M.
[1] On quasi-C-reducible Finsler spaces. (Tensor, N.S. 35 (1981), 177-182).
YASUDA, H.
[7] On TMA-connections of a Finsler space. (Ibid. 40 (1983), 151-158).
YOSHIDA, M.
[2] On Finsler spaces of Hp-scalar curvature. (Ibid. 38 (1982), 205-210).

Proc. Nat. Sem. Finsler Spaces, I, II, III. (Univ. Braşov, Romania, 1980, 1982, 1984).

Anal. Şti. Univ. Iaşi 30, s.Ia, Mat. 1984-4. Romanian-Japanese Colloq. Finsler Geometry, Iaşi, 1984.

Proc. Romanian-Japanese Colloq. Finsler Geometry. (Univ. Braşov, 1984).

BIBLIOGRAPHY OF APPLICATIONS

ALBU, I.D.; OPRIŞ, D.

[1] Géométrie des spineurs de Finsler. (An. Univ. Timişoara, Şti. Mat. 19 (1981), 5-22).

AMARI, S.

[1] A theory of deformations and stresses of ferromagnetic substances by Finsler geometry. (RAAG Mem. 3-D (1962), 193-214).

[2] On the analysis of the plastic structures of ferromagnetic substances by Finsler geometry. (Japanese) (Japan Material Sci. 1 (1964), 27-34).

ASANOV, G.S.

[1] The gravitation field in Finsler space, which is based on the notion of volume. (Russian) (Vestnik Moskov. Univ. (3) 17 (1976), 288-296).

[2] Finsler space with the algebraic metric defined by the field of the ennuples. (Russian) (Problems of Geom. 8 (1977), 67-87).

[3] The Finslerian structure of space-time, defined by its absolute parallelism. (Ann. der Phys. (7) 34 (1977), 169-174).

[4] Motion of the rest frame of the electric charge defined by the Finslerian structure of the electromagnetic field. (Rep. on Math. Phys. 11 (1977), 221-226).

[5] Scalar and electromagnetic fields in Finslerian space-time with the absolute parallelism. (Ibid. 14 (1978), 237-246).

[6] Some consequences of the Finslerian structure of the space-time with the absolute parallelism. (Ibid. 13 (1978), 13-23).

[7] On Finslerian relativity. (Nuovo Cimento 49 (1979), 221-246).

[8] New examples of S3-like Finsler spaces. (Rep. on Math. Phys. 16 (1979), 329-333).

[9] On 1-form Finsler spaces. (Inst. Math. Polish. Acad. Sci., preprint 195, 1979).

[10] C-reducible Finsler spaces. Finsler spaces with Randers metrics and Kropina metrics. (Russian) (Problems of Geom. 11 (1980), 65-88). (English Transl.: J. Soviet Math. 17 (1981), 1610-1624).

[11] Observables in the general theory of relativity. II. Finsler approach. (Russian) (Izv. Vyss. Učebn. Zaved. Fizika 1976, no.7, 84-88).

[12] Clebsch representations and energy-momentums of the classical electromagnetic and gravitational fields. (Found. Phys. 10 (1980), 855-863).

[13] A Finslerian extension of general relativity. (Ibid. 11 (1981), 137-154).

[14] Russian translation of Rund's book [18] with the supplements: §7 of Chap.VI, Chapt. VII, VIII, IX, written by H.Rund; Additions I, II written by G.S. Asanov, 1981).

[15] Variational principle for the Finslerian extension of general relativity. (Aequations Math. 24 (1982), 207-229).

[16] Gravitational field equations based on Finsler geometry. (Found. Phys. 13 (1983), 501-527).

[17] Examples of Finslerian metric tensors of the space-time signature. (Ann. der Phys. (7) 41 (1984)).

BEEM, J.K.

[2] Indefinite Finsler spaces and timelike spaces. (Canad. J. Math. 22 (1970), 1035-1039).

[3] Motions in two dimensional indefinite Finsler spaces. (Indiana Univ. Math. J. 21 (1971), 551-555).

[4] On the indicatrix and isotropy group in Finsler spaces with Lorentz signature. (Lincei Rend. (8) 54 (1973), 385-392).

BEEM, J.K.; WOO, P.Y.

[1] Doubly timelike surfaces. (Mem. Amer. Math. Soc. 92 (1969), 115p.).

BERNŠTEĬN, I.N.; GERVER, M.L.

[1] A problem on integral geometry for a family of geodesics and an inverse kinematic seismics problem. (Russian) (Dokl. Akad. Nauk SSSR 243 (1978), 302-305).

BLASCHKE, W.

[1] Integralgeometrie 12. Über vollkommene optische Instrumente. (Abh. Math. Sem. Hamburg Univ. 11 (1936), 409-412).

BOGOSLOVSKII, G.Ju.

[1] The special relativity theory of anisotropic space-time. (Russian) (Dokl. Akad. Nauk SSSR 213 (1974), 1055-1058). (English Transl.: Soviet Phys. Dokl. 18 (1973), 810-811).

[2] A special-relativistic theory of the locally anisotropic space-time. I: The metric and group of motions of the anisotropic space of events. II: Mechanics and electrodynamics in the anisotropic space. (Nuovo Cimento 40 B (1977), 99-115, 116-134).

[3] On the gauge invariance of the observables in the general relativistic theory of the locally anisotropic space-time. (Hadronic J. 7 (1984), 1078-1117).

CAVALLERI, G.; SPINELLI, G.

[1] Relativistic Lagrangian equations of motion with constraints: Check on the continuum. (Nuovo Cimento 39 (1977), 87-92).

[2] Gravity theory allowing for point particles and zitterbewegung. (Ibid. 39 (1977), 93-104).

ELIOPOULOS, H.A.

[1] Methods of generalized metric geometry with applications to mathematical physics. (Thesis, Toronto, 1956).

[4] A generalized metric space for electromagnetic theory. (Acad. Roy. Belg. Bull. Cl. Sci. (5) 51 (1965), 986-995).

[5] Multi-particle theory derived from the geometry of a locally Minkowskian Finsler space. (Ibid. 52 (1966), 69-75).

FUJINAKA, M.

[1] On Finsler spaces and dynamics with special reference to equations of hunting. (Proc. 3rd Japan Nat. Congr. Appl. Mech., 1953, 433-436).

GOHMAN, A.V.

[1] Differential-geometric foundations of the classical dynamics of systems. (Russian) (Izdat. Saratov. Univ., Saratov, 1969, 93p.).

HERMES, H.

[1] The geometry of time-optimal control. (SIAM J. Control 10 (1972), 221-229).

HESKIA, S.

[1] A new viewpoint on the space-time model of elementary particles. I, II. (Progr. Theor. Phys. 45 (1971), 277-294, 640-648).

HORVÁTH, J.I.

[1] A geometrical model for the unified theory of physical fields. (Phys. Rev. (2) 80 (1950), 901).

[2] On the theory of the electromagnetic field in moving dielectrica. (Bull. Acad. Polon. Sci. III 4 (1956), 447-452).

[3] Contribution to Stephenson-Kilmister's unified theory of gravitation and electromagnetism. (Nuovo Cimento (10) 4 (1956), 571-576).

[4] Contributions to the unified theory of physical fields. (Ibid. 4 (1956), 577-581).

[5] Classical theory of physical fields of second kind in general spaces. (Acta Phys. Chem. Szeged 4 (1958), 3-17).

[6] New geometrical methods of the theory of physical fields. (Nuovo Cimento (10) 9 (1958), Suppl. 444-496).

[7] A possible geometrical interpretation of the isospace and of its transformations. (Acta Phys. Chem. Szeged 7 (1961), 3-16).

[8] Internal structure of physical fields. (Ibid. 9 (1963), 3-24).

[9] Zero-point kinetic energy of relativistic Fermion gases. (Ibid. 13 (1967), 3-19).

[10] On the hyper-geometrization of relativistic phase-space formalism. I, II, III. (Acta Phys. Acad. Sci. Hungar. 24 (1968), 205-223, 347-371; 25 (1968), 1-15).

HORVÁTH, J.I.; GYULAI, J.

[1] Über die Erhaltungssätze des elektromagnetischen Feldes in bewegten Dielektriken. (Acta Phys. Chem. Szeged 2 (1956), 39-48).

HORVÁTH, J.I.; MOÓR, A.
[1] Entwicklung einer einheitlichen Feldtheorie begründet auf die Finslerschen Geometrie. (Z. Phys. 131 (1952), 544-570).
[2] Entwicklung einer Feldtheorie begründet auf einen allgemeinen metrischen Linienelementraum. I, II. (Nederl. Akad. Wetensch. Proc. A 58 = Indag. Math. 17 (1955), 421-429, 581-587).
HOSOKAWA, T.
[3] Finslerian wave geometry and Milne's world-structure. (J. Sci. Hiroshima Univ. A 8 (1938), 249-270).
HURT, N.E.
[1] Topology of quantizable dynamical systems and the algebra of observables. (Ann. Inst. Poincaré, N.S. A 16 (1972), 203-217).
[2] Homogeneous fibered and quantizable dynamical systems. (Ibid. 16 (1972), 219-222).
HYLAND, G.J.
[1] A nonlocal spinor field theory of matter. (General Relativity and Gravitation 10 (1979), 231-252).
IKEDA, S.
[1] Prolegomena to applied geometry. (Mahā Shobō, Koshigaya-city, Japan, 1975).
[2] A structurological consideration on a unified field. (Lett. Nuovo Cimento 13 (1975), 497-500). Also ibid. 15 (1976), 523-626; 18 (1977), 29-32; 21 (1978), 165-168, 567-571; 23 (1978), 449-454; 25 (1979), 21-25, 26-31; 26 (1979), 277-281, 313-316; 28 (1980), 541-544; (11) 61B (1981), 220-228.
[3] Some structurological remarks on a nonlocal field. (Intern. J. Theor. Phys. 15 (1976), 377-387).
[4] A differential geometrical consideration on a "nonlocal" field. (Rep. on Math. Phys. 18 (1980), 103-110).
[5] On the theory of fields in Finsler spaces. (Post-RAAG Rep. 130, 1981, 27p.).
[6] On the theory of gravitational field in Finsler spaces. (Proc. Einstein Centenary Symp., Napur, India, 1980, 155-164).
[7] On the conservation law in the theory of fields in Finsler spaces. (J. Math. Phys. 22 (6) (1981), 1211-1214). Also ibid. 22 (6) (1981), 1215-1218; 22 (12) (1981), 2831-2834; 26 (5) (1985), 958-960.
[8] Some structurological features induced by the internal variable in the theory of fields in Finsler spaces. (Progr. Theor. Phys. 65 (1981), 2075-2078). Also ibid. 66 (1981), 2284-2286.
[9] On the theory of fields in Finsler spaces. (Acta Phys. Polonica B 13 (1982), 321-329). Also ibid. 15 (1984), 757-765.
[10] Some structural features induced by the space-time metrical fluctuation in the theory of gravitational fields. (Found. Phys. 13 (1983), 629-636).
[11] On a Finslerian metrical structure of the gravitational field. (Proc. Einstein Found Intern. 2-1 (1985), 1-7).
[12] The principles of Applied Geometry. (Nippatsu Shuppan, Tokyo, 1985).
INGARDEN, R.S.
[3] On the geometrically absolute optical representation in the electron microscope. (Trav. Soc. Sci. Lettr. Wrocław B 45 (1957), 60p.).
[4] Differential geometry and physics. (Tensor, N.S. 30 (1976), 201-209).
[5] On the Finsler geometry of non-equilibrium thermodynamics. (Lecture Notes in Math. Conf. Clausthal, 1983, Springer).
ISHIKAWA, H.
[1] Indefinite Finsler type metrics and their peculiar properties. (Progr. Theor. Phys. 62 (1979), 295-296).
[2] Einstein tensor in scalar curvature Finsler spaces. (Ann. der Phys. 37 (1980), 151-154).
[3] Einstein equation in lifted Finsler spaces. (Nuovo Cimento 56 B (1980), 252-262).
[4] Note on Finslerian relativity. (J. Math. Phys. 22 (1981), 995-1004).
ISPAS, C.I.; ISPAS, M.I.
[2] About some problems of plasticity theory. (Proc. Nat. Sem. Finsler Spaces, Braşov, 1980, 107-119).

ISPAS, C.I.; ONCESCU, Gh.F.
[1] A new formulation of the problem of continuous deformations in Finsler spaces. (Romanian) (Bul. Inst. Petrol, Gaze Geologie 16 (1967), 257-267).

ISPAS, M.I.
[1] Systeme de coordonees „convective" dans la theorie de la plasticite, dans le cadre des espaces de Finsler. (Romanian) (Bul. Şti. Inst. Const. Bucureşti 12 (1969), 23-25).
[2] Plasticity equations in Finsler's three-dimensional spaces with convective coordinates. (Romanian) (Ibid. 12 (1969), 27-30).

KADOMTSEV, S.B.; POZNYAK, Eh.G.; SOKOLOV, D.D.
[1] Certain questions of Lobachevskij geometry, connected with physics. (Russian) (Itogi Nauki Tekh., ser Probl. Geom. 13 (1982), 157-188). (English Transl.: J. Soviet Math. 25 (1984), 1331-1350, specially p.1343).

KAWAGUCHI, M.; YAMANOI, T.; KAWAGUCHI, T.; KUDO, T.; MIZUTA, M.; KURINO, Y.
[1] Visual illusion having a Finsler space as mathematical model. (Japanese) (Res. Rep. Res. Assoc. Multiv. Anal. 13 (1979), 1-14, Hokkaido Univ.)

KAWAGUCHI, M.; YAMANOI, T.; SATO, Y.; KUDO, T.; MIZUTA, M.; MATSUI, S.
[1] A visual space involving visual illusion as an application of Minkowski space. (Ibid. 12 (1979), 1-3).

KAWAGUCHI, T.
[1] On the application of Finsler geometry to engineering dynamical systems. (Period. Math. Hungar. 8 (1977), 281-289).

KERN, J.
[1] Lagrange geometry. (Arch. Math. 25 (1974), 438-443).

KILMISTER, C.W.; STEPHENSON, G.
[1] An axiomatic criticism of unified field theories - I, II. (Nuovo Cimento 11 (1954), Suppl. 91-105, 118-140).

KIRZHNITS, D.A.; CHECHIN, V.A.
[1] Ultra-high-energy cosmic rays and a possible generalization of relativistic theory. (Soviet J. Nuclear Phys. 15 (1972), 585-589).

KLEIN, J.
[1] Sur les trajectoires d'un système dynamique dans un espace finslérien ou variationnel généralisé. (C.R. Acad. Sci. Paris 238 (1954), 2144-2146).
[2] Espaces variationnels et mécanique. (Ann. Inst. Fourier 12 (1962), 1-124).

KONDO, K.
[1] On the theoretical investigation based on abstract geometry of dynamical systems appearing in engineering. (Proc. 3rd Japan Nat. Congr. Appl. Mech., 1953, 425-432).
[2] A Finslerian approach to space-time and some microscopic as well as macroscopic criteria with reference to quantization, mass spectrum and plasticity. (RAAG Memo. 3-E-VIII (1962), 307-318).
[3] Non-Riemannian and Finslerian approaches to the theory of yielding. (Intern. J. Engin. Sci. 1 (1963), 71-99, 422).
[4] A pseudo-Finslerian picture of energy conservation. (Post RAAG Memo. 4 (1973), 24p.)
[5] On the osculatory character of a geometrical picture of the Voigt transformation and its modification. (Ibid. 14 (1974), 30p.)
[6] Reorganization of the higher order space picture of the world of elementary particles. I: Imbedding the quantum-mechanical formalism in the geometry of higher order spaces. (Ibid. 109 (1980), 37p.)

KONDO, K,: AMARI, S.
[1] A constructive approach of the non-Riemannian feature of dislocation and spin distributions in terms of Finsler's geometry and a possible extension of the space-time formalism. (RAAG Memo. 4-D (1968), 225-238).

KONDO, K.; FUJINAKA, M.
[1] Geometrization of dissipation and general non-linearity in dynamical equations with a preliminary reference to hysteresis. (Ibid. 1-B (1955), 335-355).

KONDO, K.; KAWAGUCHI, T.
[1] On the basic principle of the analysis of magnetic hysteresis by a Finslerian approach. (RAAG Res. Notes 3-111 (1966), 20p.)

[2] On the origin of the hysteresis in the Finslerian magnetic dynamical system. (RAAG Memo. 4-B (1968), 65-72).

KÜNZLE, H.P.

[1] Degenerate Lagrangean systems. (Ann. Inst. Poincaré, N.S. A 11 (1969), 393-414).

KURITA, M.

[3] Formal foundation of analytical dynamics based on the contact structure. (Nagoya Math. J. 37 (1970), 107-119).

LICHNEROWICZ, A.; THIRY, Y.

[1] Problèms de calcul des variations liés à la dynamique classique et à la théorie unitaire du champ. (C.R. Acad. Sci. Paris 224 (1947), 529-531).

LOVELOCK, D,; RUND, H.

[1] Variational principles in the general theory of relativity. (Proc. 13th Biennial Sem. Canad. Math. Congr. I 51-68, 1971).

MATSUMOTO, M.

[22] On Einstein's gravitational field equation in a tangent Riemannian space of a Finsler space. (Rep. on Math. Phys. 8 (1975), 103-108).

MAUGIN, G.

[1] Étude des déformations d'un milieu continu magnétiquement saturé, employant la notion d'espace de Finsler. (C.R. Acad. Sci. Paris 273 (1971), 474-476).

MAURIN, K.

[2] Calculus of variations and classical field theory. Part I. (Lecture Notes 34, Aarhus : Mat. Inst. Aarhus Univ., 1976, 100p.).

MCKIERNAN, M.A.

[1] Fatigue spaces in electromagnetic-gravitational theory. (Canad. Math. Bull. 9 (1966), 489-507).

MERCIER, A.; TREDER, H.-Ju.

[1] Future of GRG (general relativity and gravitation). (Math. Student 40 (1972), 49-64).

MONTESINOS, A.

[1] Geometry of spacetime founded on spacelike metric. (J. Math. Phys. 20 (5) (1979), 953-965).

MOSHARRAFA, A.M.

[1] On the metric of space and the equations of motion of a charged particle. (Proc. Math. Phys. Soc. Egypt 3 (1945), 19-24).

[2] The metric of space and mass deficiency. (Philos. Mag. (7) 39 (1948), 728-738).

NEWHOUSE, S.; PALIS, J.

[1] Bifurcations of Morse-Smale dynamical systems. (Proc. Symp. Univ. Bahia, Salvador, 1971, 303-366).

NUTKU, Y.

[1] Geometry of dynamics in general relativity. (Ann. Inst. Poincaré, N.S. A 21 (1974), 175-183).

PIHL, M.

[1] Classical mechanics in a geometrical description. (Danish) (Danske Vid. Selsk. Mat.-Fys. Medd. 30 (1955), no.12, 26p.).

PIMENCO, R.I.

[1] Finsler kinematics. (Russian) (Sibilsk. Mat. Z. 22 (1981), no.3, 136-146, 237).

RANDERS, G.

[1] On an asymmetric metric in the four-space of general relativity. (Phys. Rev. (2) 59 (1941), 195-199).

ROXBURGH, I.W.; TAVAKOL, R.

[1] The gravitational theories of Poincaré and Milne and the nonRiemannian kinematic models of the universe. (Mon. Not. R. Astro. Soc. 170 (1975), 599-610).

RUIZ, M., O.R.

[1] Existence of brake orbits in Finsler mechanical systems. (Geometry and Topology, 542-567. Lecture Notes in Math. 597 (1977), Springer).

RUND, H.
[8] Die Hamiltonsche Funktion bei allgemeinendynamischen Systemen. (Arch. Math. 3 (1952), 207-215).
[10] Application des méthods de la géométrie métrique généralisée à la dynamique théoretique. (Colloq. Intern. Géom. Diff. Strasbourg, 1953, 41-51).
[20] Dynamics of particles with internal "spin". (Ann. Physik (7) 7 (1961), 17-27).
[22] Note on the Lagrangian formalism in relativistic mechanics. (Nuovo Cimento (10) 23 (1962), 227-232).
RUND, H.; LOVELOCK, D.
[1] Variational principles in the general theory of relativity. (Jber. Deutsch. Math.-Verein. 74 (1972), 1-65).
SAKATA, S.
[1] A constructive approach to non-teleparallelism and non-metric representations of plastic material manifold by generalized diakoptical tearing II. Finslerian tearing. (RAAG res. Notes 3-149 (1970), 27p.)
SCHAER, J.
[1] De la possibilité d'une théorie unitaire finslérienne de l'électromagnétisme et de la gravitation. (Arch. Sci. Geneve 13 (1960), 542-549).
SEVRJUK, V.P.
[1] Covariant derivatives of a spinor in Finsler geometry. (Russian) (Izv. Vysš. Učebn. Zaved. Fizika (1973), no.2, 43-49).
SHARMA, I.D.; TRIVEDI, H.K.N.
[1] On the Einstein's law of gravitation in Finsler space. (Ganita 23 (1972), 13-18).
STEPHENSON, G.
[1] Affine field structure of gravitation and electromagnetism. (Nuovo Cimento (9) 10 (1935), 354-355).
[2] La géométrie de Finsler et les théories du champ unifié. (Ann. Inst. Poincaré 15 (1956), 205-215).
STEPHENSON, G.; KILMISTER, G.W.
[1] A unified field theory of gravitation and electromagnetism. (Nuovo Cimento (9) 10 (1935), 230-235).
SUZUKI, Y.
[1] Finsler geometry in classical physics. (J. Coll. Arts. Sci. Chiba Univ. Natur. Sci. Ser. 2 (1956), 12-16).
TAKANO, Y.
[1] Spinor field in Finsler spaces. (Progr. Theor. Phys. 32 (1964), 365-366).
[2] Theory of fields in Finsler spaces I. Geometrical foundation of non-local field theory. (Japanese) (Soryushiron Kenku 36 (1967), 29-66).
[3] Theory of fields in Finsler spaces. I. (Progr. Theor. Phys. 40 (1968), 1159-1180).
[4] Gravitational field in Finsler spaces. (Lett. Nuovo Cimento (2) 10 (1974), 747-750).
[5] Variation principle in Finsler spaces. (Ibid. (2) 11 (1974), 486-490).
[6] On the theory of fields in Finsler spaces. (Proc. Intern. Symp. Relativity & Unified Field Theory, Calcutta, 1975, 17-26).
[7] Gauge fields in Finsler spaces. (Lett. Nuovo Cimento 35 (1982), 213-217).
[8] The differential geometry of spaces whose metric tensor depends on spinor variables and the theory of spinor gauge fields. (Tensor, N.S. 40 (1983), 249-260).
TORÓ, T.
[1] On spinorial equations in Finsler spaces. (Romanian) (An. Univ. Timişoara, Şti. Fiz.-Chim. 7 (1969), 107-113).
[2] Sur l'équation spinorielle pour le neutrino dans les espaces non euclidiens. (Ibid. 9 (1971), 131-138).
VELTE, W.
[1] Bemerkung zu einer Arbeit von H. Rund. (Arch. Math. 4 (1953), 343-345).
WAGNER, V.V. (VAGNER)
[14] The theory of a field of local hyperstrips in X_n and its application to the mechanics of a system with nonlinear anholonomic connection. (Russian)

(Dokl. Akad. Nauk SSSR, N.S. 66 (1949), 1033-1036).
WALKER, A.G.
[1] The principle of least action in Milne's kinematical relativity. (Proc. Roy. Soc. London A 147 (1934), 478-490).
YAMANOI, T.; KUDO, T.; YAMAZAKI, T.; KAWAGUCHI, M.
[1] On the interpretation of some visual illusions by a geometrical model of Minkowski spaces. (Tensor, N.S. 37 (1982), 257-262).

See also Grifone [4], Rund [15], [19] and Wagner [7], [11].

APPENDIX

English translation.

(p.87)

A treatment of geometry under the most general possible assumptions has a merit to let us know to what extent every theorem depends on special assumption of the metric or the number of dimension.

(p.104)

In the astronomy we measure the distance in a time-mass, in particular, in a light-year. When we take a second as the unit, the unit surface is a sphere with the radius of 300,000 km. To every point of our space is associated such a sphere; this defines the distance (measured in time) and the geometry of our space is the simplest one, namely, the euclidean geometry. Next, when a ray of light is considered as the shortest line in the gravitational field, the geometry of our space is Riemannian geometry. Furthermore, in an anisotropic medium the speed of the light depends on its direction, and the unit surface is not any longer a sphere.

Now, on a slope of the earth surface we sometimes measure the distance in a time-mass, namely, the time required such as seen on a guidepost. Then the unit curve, taken a minute as the unit, will be a general closed curve without center, because we can walk only a shorter distance in an uphill road than in a downhill road. This defines a general geometry, although it is not exact. The shortest line along which we can reach the goal, for instance, the top of a mountain as soon as possible will be a complicated curve.

(p.106)

The basic idea of this paper is as follows: A Minkowski space is an affine space, with each direction of which is associated an euclidean distance. In § 1 we study Minkowski spaces with reference to cartesian coordinates. In § 2 we refer to curvilinear coordinates. Then it is remarkable that all the results are seemingly of Finslerian form. This is analogous to the fact that, referring to curvilinear coordinates in an euclidean space, all the results are seemimgly of Riemannian form.

INDEX